Nuclear Physics in Astrophysics V

AA002759

Journal of Physics: Conference Series Volume 337

Eilat, Israel
3-8 April 2011

ISBN: 978-1-62276-671-0
ISSN: 1742-6588

Printed from e-media with permission by:

Curran Associates, Inc.
57 Morehouse Lane
Red Hook, NY 12571

Some format issues inherent in the e-media version may also appear in this print version.

Printed by Curran Associates, Inc. (2013)

For permission requests, please contact the Institute of Physics
at the address below.

Institute of Physics
Dirac House, Temple Back
Bristol BS1 6BE UK

Phone: 44 1 17 929 7481
Fax: 44 1 17 920 0979

techtracking@iop.org

Additional copies of this publication are available from:

Curran Associates, Inc.
57 Morehouse Lane
Red Hook, NY 12571 USA
Phone: 845-758-0400
Fax: 845-758-2634
Email: curran@proceedings.com
Web: www.proceedings.com

TABLE OF CONTENTS

NUCLEAR STRUCTURE AND ASTROPHYSICS

BIG BANG AND EXPLOSIVE SYNTHESIS

NEUTRINO PHYSICS

SOLAR REACTIONS

RARE EVENTS

GENERAL ASTROPHYSICS

Journal of Physics: Conference Series **337** (2012) 011001 doi:10.1088/1742-6596/337/1/011001

Foreword

The fifth edition of the bi-annual "Nuclear Physics in Astrophysics (NPA)" conference series was held in Eilat, Israel on April 3-8, 2011. This Conference is also designated as the 24th Nuclear Physics Divisional Conference of the EPS. The main purpose of this conference, as that of the four previous ones in this series, is to deal with those aspects of nuclear physics that are directly related to astrophysics. The concept of such meeting was conceived by the Nuclear Physics Board of the European Physical Society in 1998. At that time, the idea of such a conference was quite new and it was decided that this meeting will be sponsored by the EPS. The first meeting, in January 2001, was planned and organized in Eilat, Israel. Due to international circumstances the conference was moved to Debrecen, Hungary. Subsequent conferences were held in Debrecen again, in Dresden Germany, and in Frascati, Italy (moved from Gran Sasso due to the tragic earthquake that hit the L'Aquila region). After 10 years the conference now returned to Eilat, the originally envisioned site.

Eilat is a resort town located on the shore of the Gulf of Eilat which connects Israel to the Red Sea and further south to the Indian Ocean. It commands spectacular views of desert and mountains, offering unique touristic attractions. The local scientific backdrop of the conference is the fact that the Israeli scientific scene exhibits a wide variety of research activities in many areas of nuclear physics and astrophysics. A new accelerator SARAF at Soreq Nuclear Research Center is presently undergoing final acceptance tests. SARAF will serve as a platform for production of radioactive ion beams and nuclear-astrophysics research in Israel. The meeting in Eilat was organized by four Israeli scientific institutions, Hebrew University, Soreq Nuclear Research Center, Tel Aviv University and the Weizmann Institute of Science.
The welcome reception and lectures were held at the King Solomon hotel and the conference dinner banquet at the Dan hotel. An excursion to the "Red Canyon" in the Eilat Mountains on Wednesday afternoon was one of the social highlights of the Conference.

A total number of 140 scientists attended NPA5 and about 30 accompanying persons; about 25% of these were young participants (less than 36 years old). 23 participants were from Israel, 27 from outside of Europe (including two from Africa).

The subjects covered at the conference in Eilat concentrated mainly on the spirit of the original idea - to probe experimental and theoretical activity in nuclear structure and reactions that is directly related to the physics of the Universe. There were also sessions of general interest in astrophysics, as well as a poster session on Tuesday evening featuring 40 posters.

The topics included:
Nuclear Structure -Theory and Experiment
Big-Bang Nucleosynthesis and Formation of First Stars
Stellar Reactions and Solar Neutrinos
Explosive Nucleosynthesis, Radioactive Beams and Exotic Nuclei-New Facilities and Future Possibilities for Astrophysics
Neutrino Physics- the Low and High-Energy Frontiers

Rare events, Dark Matter, Double beta-decay, Symmetries.

The conference started with an excellent expose of the progress made in the discovery of super-heavy elements and the study of their properties. The progress in this field is enormous, and this subject should be communicated to more general audiences.
The role of the nuclear equation of state and of the precise determination of nuclear masses in nucleosynthesis was emphasized in several talks.

The role of neutrinos in astrophysics was discussed extensively in several sessions. One of the highlights of this was the presentation about the IceCube and DeepCore detectors operating deep in the Antarctic ice. These facilities are able to detect cosmogenic neutrinos in a wide energy range from, 10 GeV to 10^{10} GeV.

The subject of solar neutrinos was discussed in number of talks. Topics related to properties of neutrinos, such as double-beta decay, neutrino mixing were well represented at the conference.
One of the central problems in modern cosmology and astrophysics is the search for dark matter. Several talks dealt with this subject and with methods to detect dark matter.
Another intriguing and rather novel subject that was discussed at the meeting was time variation of fundamental physical constants. Two speakers have examined the sensitivity of Big Bang Nucleosynthesis to the variation of the values of the fundamental constants.
The role of some specific nuclei (such as Ni 56) in cosmology was pointed out.

Many of the presentations at the conference described experimental studies of reactions relevant to nucleosynthesis at various stages of cosmic evolution. As reflected in the conference, these activities are widespread, encompassing many laboratories. Rare Isotope Beam (RIB) facilities are in the forefront of these studies.
To understand the various processes of nucleosynthesis one has to have a good theory of nuclei far from the stability line. A number of presentations dealt with the description of such exotic nuclei. It is clear from the presentations that the future of experimental nuclear astrophysics looks promising as existing experimental facilities are being upgraded and new facilities are being built.
X-Ray and Gamma-Ray Bursts and cosmic explosions were the subject of several talks.
A discussion of various experiments attempting to measure time reversal violation has been the subject of one lecture. The solution of the puzzle why the universe is asymmetric with respect to matter-antimatter requires the knowledge of the limit of time reversal conservation.
Recent measurements of neutron widths in the compound nucleus that reveal deviations from the Porter-Thomas distribution have been addressed in another lecture. Beta –decay studies in traps, stellar electron capture, electron screening effects were discussed in several presentations. It is clear that the breadth and vitality of the field of nuclear physics in astrophysics is enormous and the future looks very bright.

The late John Bahcall was a great astrophysicist and a supporter of the series conferences "Nuclear physics in Astrophysics". On the last day of the conference, following a talk by Neta Bahcall from Princeton University on dark matter in the Universe, a short commemoration for John was held.

Journal of Physics: Conference Series **337** (2012) 011001 doi:10.1088/1742-6596/337/1/011001

Detailed information about the NPA5 conference and its scientific program can be found at:
www.weizmann.ac.il/conferences/NPA5

Naftali Auerbach
Michael Hass
Michael Paul
Editors

International Advisory Committee:

- A. Bonaccorso (University of Pisa, Italy)
- J. Dobaczewski (Warsaw University, Poland)
- Z. Fulop (Atomki, Debrecen, Hungary)
- P.-H. Heenen (Free University, Brussels, Belgium)
- J. Jose (Universitat Politecnica de Catalunya, Barcelona, Spain)
- M. Lewitowicz (GANIL, France)
- M. Lindroos (Lund University, Sweden)
- A. Richter (ECT Institute, Trento, Italy)
- C. Spitaleri (INFN, Catania, Italy)
- J. Truran (University of Chicago, USA)

Program Advisory Committee:

- P. Descouvemont (NPB, Free University, Brussels, Belgium)
- R. Diehl (Max Planck Institut für extraterrestrische Physik, Garching, Germany)
- A. Formicola (Gran Sasso Laboratory, Italy)
- M.C. Gonzalez-Garcia (SUNY, Stony Brook, USA)
- B. Fulton (U. of York, UK)
- H. Goutte (NPB, CEA Bruyères-le-Châtel, France)
- B. Jennings (TRIUMF, Canada)
- A. Leviatan (Hebrew University of Jerusalem, Israel)
- H. Schatz (MSU, USA)
- M. Wiescher (U. of Notre Dame, USA)

Local Organizing Committee:

- Naftali Auerbach – Tel Aviv University, Co-Chairperson
- Dan Berkovits – Nuclear Research Center, Soreq
- Michael Hass – The Weizmann Institute, Chairperson
- Michael Paul – Hebrew University, Jerusalem
- Eli Waxman – The Weizmann Institute

The Conference was sponsored by:
European Physical Society

The Conference was supported by:
The Weizmann Institute of Science
The Nuclear Research Center, Soreq
The Hebrew University of Jerusalem
Tel Aviv University

Journal of Physics: Conference Series **337** (2012) 011001 doi:10.1088/1742-6596/337/1/011001

List of Participants

Name	Email	Institution
ARCONES, Almudena	a.arcones@unibas.ch	University of Basel
ASIDA, Shimon	sasida@phys.huji.ac.il	Racah Ins. of Physics Hebrew University of Jerusalem
AUERBACH, Naftali	auerbach@post.tau.ac.il	Tel Aviv University
AUMANN, Thomas	t.aumann@gsi.de	Technische Universität Darmstadt
AVIV, Ofer	oaviv@weizmann.ac.il	The Weizmann Institute of Science
AYRANOV, Marin	marin.ayranov@psi.ch	Paul Scherrer Institut, Villigen
BAHCALL, Neta	neta@astro.princeton.edu	Princeton University
BALANTEKIN, Baha	baha@physics.wisc.edu	Univ. Wisconsin- Madison
BAN, Gilles	ban@lpccaen.in2p3.fr	CNRS-IN2P3, CAEN
BANU, Adriana	banula@jmu.edu	James Madison University, Harrisonburg
BEDNARCZYK, Piotr	piotr.bednarczyk@ifj.edu.pl	IFJ PAN, Krakow
BELLINI, Alessandro	alessandro.bellini@ge.infn.it	INFN, Savona
BEMMERER, Daniel	d.bemmerer@fzd.de	Forschungszentrum Dresden-Rossendorf
BENLLIURE, Jose	j.benlliure@usc.es	University of Santiago de Compostela
BERKOVITS, Dan	berkova@soreq.gov.il	Soreq NRC
BERTULANI, Carlos	carlos_bertulani@tamu-commerce.edu	Texas A and M University
BETAK, Emil Betak	betak@savba.sk	Institute of Physics SAS
BLOCK, Michael	m.block@gsi.de	GSI, Darmstadt
BONDILI, Sreenivasa Nara Singh	snsb500@york.ac.uk	University of York
BOZIK, Daniel	ksi3cd@gmail.com	Charles University, Prague
BROWN, Alex	brown@nscl.msu.edu	Michigan State University
BUDACA, Andreea-Ioana	abudaca@theory.nipne.ro	IFIN-HH, Bucharest
CACIOLLI, Antonio	caciolli@pd.infn.it	INFN Padua, CGT - University of Siena
CARMONA GALLARDO, Mariano	mcarmona@iem.cfmac.csic.es	Instituto de Estructura de la Materia (CSIC)
CLARK, Jason	jclark@phy.anl.gov	Argonne National Laboratory
COC, Alain	alain.coc@csnsm.in2p3.fr	CSNSM, Orsay
CRUZ, Joao	jdc@fct.unl.pt	FCT-UNL / CFNUL, Lisbon
CZERSKI, Konrad	czerski@physik.tu-berlin.de	University of Szczecin
DE LA RIPELLE, Fabre	ana.weksler@gmail.com	Institute of Nuclear Physics, Orsay
DEMBINSKI, Hans	hans.dembinski@kit.edu	IEKP, Karlsruhe Institute of Technology
DEPPMAN, Airton	deppman@if.usp.br	S□o Paulo University
DESCOUVEMONT, Pierre	pdesc@ulb.ac.be	Universite Libre de Bruxelles
DHAL, Anukul	anukuldhal@gmail.com	The Weizmann Institute of Science
DILLMANN, Iris	i.dillmann@gsi.de	GSI, Darmstadt
DMITRIEV, Vladimir	dmitriev@inp.nsk.su	Budker Institute of Nuclear Physics, Novosibirsk
DUGUET, Thomas	thomas.duguet@cea.fr	CEA/Saclay
EISEN, Yossi	yosef@soreq.gov.il	Soreq NRC
FAESSLER, Amand	amand.faessler@uni-tuebingen.de	University of Tuebingen
FEINBERG, Gitai	gitai.feinberg@gmail.com	Soreq NRC
FIASE, Joseph Orbugh	fiasejo@yahoo.com	Benue State University
FORMICOLA, Alba	formicola@lngs.infn.it	INFN Laboratori Nazionali del Gran Sasso
FORNAL, Bogdan	bogdan.fornal@ifj.edu.pl	Institute of Nuclear Physics PAN, Krakow
FRIEDMAN, Moshe	moshe.friedman@mail.huji.ac.il	The Hebrew University Jerusalem
FULOP, Zsolt	fulop@atomki.hu	ATOMKI, Debrecen
GAI, Moshe	moshe.gai@yale.edu	Yale and UConn
GAL-YAM, Avishay	avishay.gal-yam@weizmann.ac.il	The Weizmann Institute of Science
GALAVIZ REDONDO, Daniel	galaviz@cii.fc.ul.pt	Centro de F□sica Nuclear, University of Lisbon
GAVIN, Lotay	g.lotay@ed.ac.uk	University of Edinburgh
GAZIZOV, Askhat	gazizov@lngs.infn.it	LNGS, INFN, Assergi
GEORGIEV, Georgi	georgi.georgiev@csnsm.in2p3.fr	CSNSM, Orsay
GIACHERO, Andrea	andrea.giachero@mib.infn.it	INFN Sezione di Milano Bicocca
GLASNER, Ami	glasner@phys.huji.ac.il	The Hebrew University Jerusalem
GOLDBERG, Vladilen	goldberg@comp.tamu.edu	Texas A and M University, Cyclotron Institute
GOLDRING, Gvirol	gvirol.goldring@weizmann.ac.il	The Weizmann Institute of Science
GORIELY, Stephane	sgoriely@astro.ulb.ac.be	Universit' Libre de Bruxelles
GREINER, Walter	greiner@fias.uni-frankfurt.de	(FIAS), J.W. Goethe-University Frankfurt am Main
GUSTAVINO, Carlo	carlo.gustavino@lngs.infn.it	INFN, Rome
GYURKY, Gyorgy	gyurky@atomki.hu	ATOMKI, Debrecen
HALZEN, Francis	kim.kreiger@icecube.wisc.edu	University of Wisconsin-Madison
HALASZ, Zoltan	fiser@atomki.hu	HAS-ATOMKI, Debrecen
HASS, Michael	michael.hass@weizmann.ac.il	The Weizmann Institute of Science
HEENEN, Paul-Henri	phheenen@ulb.ac.be	Universite Libre de Bruxelles
HEUSLER, Andreas	a.heusler@mpi-hd.mpg.de	MPI-Kernphysik, Heidelberg

HIRSH, Tsviki	tsviki.hirsh@weizmann.ac.il	The Weizmann Institute of Science
HJORTH-JENSEN, Morten	morten.hjorth-jensen@fys.uio.no	University of Oslo
ISERN, Jordi	isern@ieec.cat	Institut for Space Sciences, Bellaterra
JOKINEN, Ari	ari.s.a.jokinen@jyu.fi	University of Jyvaskyla
JOSE, Jordi	jordi.jose@upc.edu	Univ. Politecnica de Catalunya
JUNKER, Matthias	matthias.junker@lngs.infn.it	Laboratori Nazionali del Gran Sasso
KAMINSKI, Grzegorz	grzegorz.kaminski@ifj.edu.pl	Joint Institute for Nuclear Research, Dubna
KIRSON, Michael W.	michael.kirson@weizmann.ac.il	The Weizmann Institute of Science
KISS, Miklos	kiss-m@chello.hu	Berze High School, Hungary
KNAPP, Frantisek	frantisek.knapp@mff.cuni.cz	Charles University, Prague
KORSCH, Wolfgang	korsch@pa.uky.edu	University of Kentucky
KOWALSKA, Magdalena	kowalska@cern.ch	CERN
KRTICKA, Milan	krticka@ipnp.troja.mff.cuni.cz	Charles University, Prague
KVASIL, Jan	kvasil@ipnp.troja.mff.cuni.cz	Charles University, Prague
LEDERER, Claudia	claudia.lederer@univie.ac.at	Faculty of Physics, University of Vienna
LONGLAND, Richard	richard.longland@upc.edu	Universitat Polytecnica de Catalunya
LUIS, Helio	heliofluis@itn.pt	ITN, Lisbon
MADEIRA SZANTO, Eloisa	eloisa@dfn.if.usp.br	Universidade de Sao Paulo
MAJ, Adam	adam.maj@ifj.edu.pl	IFJ PAN Krakow
MARGANIEC, Justyna	j.marganiec@gsi.de	EMMI, GSI Darmstadt
MELNIK, David	davmelnik@013.net.il	Soreq, NRC
MORSELLI, Aldo	aldo.morselli@roma2.infn.it	INFN, Roma
NEFF, Thomas	t.neff@gsi.de	GSI, Darmstadt
NEUHAEUSER, Ralph	rne@astro.uni-jena.de	Astrophysics Institute, University Jena
OGANESSIAN, Yuri	oganessian@jinr.ru	Joint Institute for Nuclear Research, Dubna
OLIVEIRA, Francois	oliveira@ganil.fr	GANIL, CAEN
OPPER, Allena	akopper@gwu.edu	The George Washington University
ORIO, Filippo	filippo.orio@roma1.infn.it	INFN, Roma
OSER, Scott	oser@phas.ubc.ca	University of British Columbia
OSNES, Eivind	eivind.osnes@fys.uio.no	University of Oslo
PAAR, Nils	npaar@phy.hr	University of Zagreb
PAPISH, Oded	papish@tx.technion.ac.il	Technion, Haifa
PAUL, Michael	paul@vms.huji.ac.il	The Hebrew University, Jerusalem
PENIONZHKEVICH, Yuri	pyuer@lnr.jinr.ru	JINR, Dubna
PEoA GARAY, Carlos	penya@ific.uv.es	IFIC (CSIC-UVEG), Valencia
PIRAN, Tsvi	tsvi@phys.huji.ac.il	The Hebrew University, Jerusalem
RAUT, Rajarshi	raut@tunl.duke.edu	Triangle Universities Nuclear Laboratory, Duke University
REHM, K. Ernst	rehm@anl.gov	Argonne National Laboratory
SCOTT, David	d.scott-3@sms.ed.ac.uk	The University of Edinburgh
SHAVIV, Giora	gioras@physics.technion.ac.il	Technion, Haifa
SHERRILL, Bradley	sherrill@frib.msu.edu	FRIB, East Lansing
SHLOMO, Shalom	shlomo@comp.tamu.edu	Cyclotron Institute, Texas A and M University
SHOR, Asher	shor@soreq.gov.il	Soreq NRC
SORLIN, Olivier	sorlin@ganil.fr	GANIL, CAEN
SPITALERI, Claudio	spitaleri@lns.infn.it	LNS-INFN and Universitא di Catania
SZANTO DE TOLEDO, Alejandro	szanto@dfn.if.usp.br	Universidade de Sao Paulo
TALMI, Igal	igal.talmi@weizmann.ac.il	The Weizmann Institute of Science
TANG, Xiaodong	x.tang@nd.edu	University of Notre Dame
TARGOSZ-SLECZKA, Natalia	natalia.targosz@wmf.univ.szczecin.pl	University of Szczecin
TRACHE, Livius	livius_trache@tamu.edu	Texas A and M University
TRURAN, James	truran@nova.uchicago.edu	University of Chicago and Argonne National Laboratory
TUMINO, Aurora	tumino@lns.infn.it	LNS-INFN and Universitא di Enna
UJIC, Predrag	ujic@ganil.fr	GANIL, CAEN
USMAN, Iyabo	usman@tlabs.ac.za	iTHEMBA LABS, Cape Town
VAINTRAUB, Sergey	sergeyva@weizmann.ac.il	The Weizmann Institute of Science
VANGIONI, Elisabeth	vangioni@iap.fr	Institut d'Astrophysique de Paris-CNRS
VON NEUMANN-COSEL, Peter	vnc@ikp.tu-darmstadt.de	Institut fuer Kernphysik, Technische Universitaet Darmstadt
VON OERTZEN, Wolfram	oertzen@helmholtz-berlin.de	Helmholtz Zentrum, Berlin
VOTANO, Lucia	lucia.votano@lngs.infn.it	INFN - Laboratori Nazionali del Gran Sasso
WEISSMAN, Leo	leo.weissman@gmail.com	Soreq NRC
WOODS, Philip	pjw@ph.ed.ac.uk	University of Edinburgh
YARON, Ofer	ofer.yaron@weizmann.ac.il	The Weizmann Institute of Science
ZAMICK, Larry	zamick@physics.rutgers.edu	Rutgers University, Piscataway
ZARUBIN, Pavel	zarubin@lhe.jinr.ru	JINR, Dubna
ZARUBINA, Irina	zarubina@lhe.jinr.ru	JINR, Dubna

Self-consistent Gorkov Green's function calculations of one-nucleon spectral properties

V. Somà
CEA-Saclay, IRFU/Service de Physique Nucléaire, F-91191 Gif-sur-Yvette, France

T. Duguet
CEA-Saclay, IRFU/Service de Physique Nucléaire, F-91191 Gif-sur-Yvette, France
National Superconducting Cyclotron Laboratory and Department of Physics and Astronomy, Michigan State University, East Lansing, MI 48824, USA

C. Barbieri
Department of Physics, University of Surrey, Guildford GU2 7XH, UK

Abstract. Results from the newly developed Gorkov self-consistent Green's function approach are presented. Ab-initio spectral strength distributions for one-nucleon addition or removal calculated in doubly-closed shell ^{40}Ca and in semi-magic ^{44}Ca are briefly discussed. The object of the present communication is to illustrate the potential spectroscopic reach of the method.

1. Introduction

In the last decade the reach of ab-initio nuclear structure calculations has extended up to the region of medium-mass systems. Despite the significant progress both from the theoretical and computational point of view, methods as coupled-cluster (CC) [1] or Dyson self-consistent Green's function [2] (Dyson-SCGF) are however currently limited to a few doubly-magic nuclei. A few tens of neighbor isotopes with ±1 or ±2 nucleons can also be reached with particle attachment or removal formalisms [3, 4].

One way of tackling truly open-shell nuclei may involve the development of a multi-reference scheme, such as e.g. multi-reference CC [5]. Alternatively, one can build the correlated many-body state starting from a reference state that already incorporates zeroth-order pairing correlations, as in the case of e.g. Bogoliubov theory. We are currently following this second route and developing [6] a many-body method based on the self-consistent Green's function theory in the Gorkov formalism. Such method explicitly accounts for pairing correlations through the introduction of anomalous propagators, treated on the same footings as (and coupled to) the normal propagators of Dyson theory. This generalization of the standard Green's function approach is then applicable to several hundreds of (semi-magic +/-1) nuclei, including systems up to, e.g., the tin isotopic chain.

In the current implementation the method starts from realistic interactions and consists in the self-consistent solution of Gorkov's equations on the basis of first- and second-order self-energies.

2. Gorkov Green's function theory

Let us introduce a basis $\{a_a^\dagger\}$ of the one-body Hilbert space $\mathcal{H}_1$ that can be divided into two blocks according to the value (or more precisely to the sign) of an appropriate quantum number. In the present calculations such a basis will be identified with the harmonic oscillator basis; notice however that the formalism is more generally valid for any basis that possesses this property.

To any state a belonging to the first block, a single-particle state $\bar{a}$ belonging to the second block and having the same quantum numbers as a, except for the one differentiating the two blocks, can be associated through a transformation, i.e. time-reversal in the present case. With that in mind one can define a basis $\{\bar{a}_a^\dagger\}$ dual to the basis $\{a_a^\dagger\}$ through

$$\bar{a}_a^\dagger(t) \equiv \eta_a a_{\bar{a}}^\dagger(t) , \qquad \bar{a}_a(t) \equiv \eta_a a_{\bar{a}}(t) , \tag{1}$$

which correspond to exchanging the state a by its partner $\bar{a}$ up to the phase η_a. By convention $\bar{\bar{a}} = a$ with $\eta_a \, \eta_{\bar{a}} = -1$.

Using the second quantized operators $\{a_a^\dagger, a_a\}$ and their dual counterparts $\{\bar{a}_a^\dagger, \bar{a}_a\}$ one defines an "annihilation" column vector

$$\mathbf{A}_a(t) \equiv \begin{pmatrix} a_a(t) \\ \bar{a}_a^\dagger(t) \end{pmatrix} , \tag{2a}$$

and a "creation" row vector

$$\mathbf{A}_a^\dagger(t) \equiv \begin{pmatrix} a_a^\dagger(t) & \bar{a}_a(t) \end{pmatrix} . \tag{2b}$$

The four Gorkov propagators [7] are then constructed by means of these generalized operators through

$$i\,\mathbf{G}_{ab}(t,t') \equiv \langle\Psi_0|T\left\{\mathbf{A}_a(t)\mathbf{A}_b^\dagger(t')\right\}|\Psi_0\rangle = i \begin{pmatrix} G_{ab}^{11}(t,t') & G_{ab}^{12}(t,t') \\ G_{ab}^{21}(t,t') & G_{ab}^{22}(t,t') \end{pmatrix} . \tag{3}$$

Here $|\Psi_0\rangle$ represents the ground state of the grand-canonical-like potential $\Omega = H - \mu N$, where μ is the chemical potential and N the particle-number operator, having the number $N = \langle\Psi_0|N|\Psi_0\rangle$ of particles in average. Notice that the state $|\Psi_0\rangle$ is a priori not an eigenstate of N, i.e. it is likely to break particle number symmetry.

Similarly to the Dyson case, self-consistent, i.e. *dressed*, Gorkov propagators are solution of an equation of motion taking, in the energy representation, the typical form

$$\mathbf{G}_{ab}(\omega) = \mathbf{G}_{ab}^{(0)}(\omega) + \sum_{cd} \mathbf{G}_{ac}^{(0)}(\omega)\, \tilde{\mathbf{\Sigma}}_{cd}(\omega)\, \mathbf{G}_{db}(\omega) . \tag{4}$$

Here $\mathbf{G}_{ab}^{(0)}(\omega)$ indicates the unperturbed Gorkov Green's function associated with a reference state of the Bogoliubov type, whereas $\tilde{\mathbf{\Sigma}}_{cd}(\omega)$ defines normal and anomalous irreducible self-energies through

$$\tilde{\mathbf{\Sigma}}_{ab}(\omega) \equiv \begin{pmatrix} \tilde{\Sigma}_{ab}^{11}(\omega) & \tilde{\Sigma}_{ab}^{12}(\omega) \\ \tilde{\Sigma}_{ab}^{21}(\omega) & \tilde{\Sigma}_{ab}^{22}(\omega) \end{pmatrix} . \tag{5}$$

Any actual solution of Gorkov equations involves a truncation of the diagrammatic expansion of the irreducible self-energies. Such an expansion is performed in terms of dressed propagators, which implies that at every order an infinite subset of self-energy insertions is implicitly resummed. It is easy to see that such a prescription goes beyond perturbation theory. In more practical terms, the solution of Gorkov equations is typically achieved via an iterative procedure that ends when a suitably chosen quantity (e.g. the chemical potential) has reached the desired degree of convergence.

Figure 1 (2) shows the corresponding diagrammatic contribution to the normal (anomalous) self-energy. In such diagrams, double lines denote self-consistent normal (two arrows in the same direction) and anomalous (two arrows in opposite directions) propagators solutions of Eq. 4. The first- and second-order contributions to the self-energy are computed and inserted into Eq. 4, which is then solved iteratively to convergence. Extensive details regarding both the formalism and the computational scheme will be reported in a forthcoming publication [6].

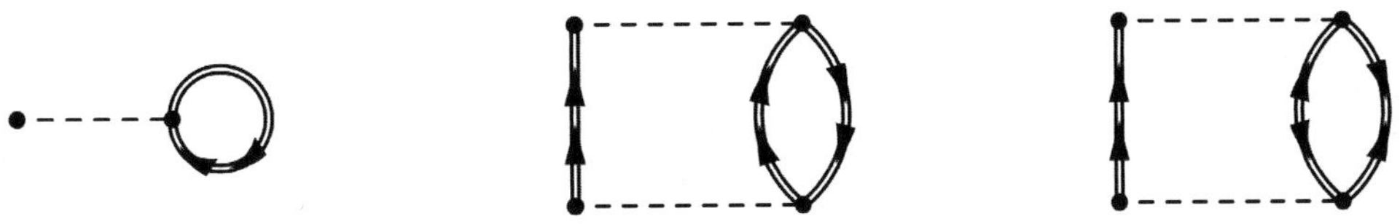

Figure 1. First- and second-order contributions to the normal self-energy Σ^{11}.

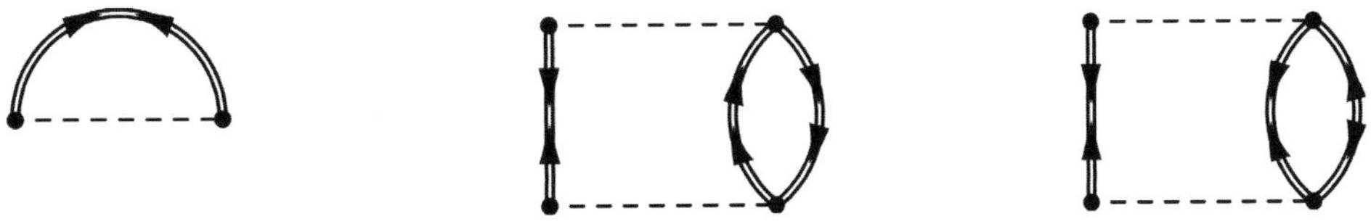

Figure 2. First- and second-order contributions to the anomalous self-energy Σ^{21}.

3. Results

The knowledge of the single-particle propagator gives access not only to the total energy and to all one-body observable of the targeted A-body ground state but also to one-nucleon separation energies and spectroscopic amplitudes to eigenstates of $A \pm 1$ systems. In order to retrieve the spectroscopic content of the one-body Gorkov Green's function it is convenient to express it in its so-called spectral or Lehmann representation [8, 6]

$$\mathbf{G}_{ab}(\omega) = \sum_k \left\{ \frac{\mathbf{X}_a^k \, \mathbf{X}_b^{k\dagger}}{\omega - E_k^+ - \mu + i\eta} + \frac{\mathbf{Y}_a^{k\dagger} \, \mathbf{Y}_b^k}{\omega - E_k^- - \mu - i\eta} \right\} , \tag{6}$$

with $\mathbf{X}_a^{k\dagger} \equiv \langle \Psi_k | \mathbf{A}_a^\dagger | \Psi_0 \rangle$, $\mathbf{Y}_a^k \equiv \langle \Psi_k | \mathbf{A}_a | \Psi_0 \rangle$ and where $E_k^\pm$ represent one-nucleon addition and removal separation energies. A pole's position and residue of the Gorkov propagator therefore carry information respectively on the separation energies to $A \pm 1$ systems excited states and the spectroscopic amplitudes associated with such states.

In the present calculations we restrict ourselves to the case of a $J^\Pi = 0^+$ many-body target state, with J being its total angular momentum and Π its parity. A natural choice for labeling single-particle basis states in this context is $a \equiv (n_a, \pi_a, j_a, m_a, q_a)$, where n_a represents the principal quantum number, π_a is the parity, j_a is the total angular momentum, m_a is the projection of the total angular momentum along the z axis and q_a is the isospin projection. In this case all relevant objects in the theory turn out to be independent of m and block-diagonal in j, π and q. In particular one can define block-diagonal spectroscopic amplitudes according to $\mathbf{X}_a^k \equiv \delta_{\alpha\kappa} \delta_{m_a m_k} \mathbf{X}_{n_a[\kappa]}^{n_k}$ and $\mathbf{Y}_a^k \equiv \delta_{\alpha\kappa} \delta_{m_a m_k} \mathbf{Y}_{n_a[\kappa]}^{n_k}$, where we have introduced $\alpha \equiv \{j_a, \pi_a, q_a\}$.

Suitable quantities to analyze the spectroscopic information contained in the single-particle Green's function are the spectroscopic factors

$$\mathbf{F}^{-}_{n_k[\kappa]} \equiv \sum_{n_a} \left| \mathbf{Y}^{n_k}_{n_a[\kappa]} \right|^2 , \qquad \mathbf{F}^{+}_{n_k[\kappa]} \equiv \sum_{n_a} \left| \mathbf{X}^{n_k}_{n_a[\kappa]} \right|^2 , \tag{7}$$

and the spectral strength distribution (SSD)

$$\mathbf{S}_{[\kappa]}(\omega) \equiv \sum_{n_k} \mathbf{F}^{-}_{n_k[\kappa]} \, \delta(\omega - E^{-}_k) + \sum_{n_k} \mathbf{F}^{+}_{n_k[\kappa]} \, \delta(\omega - E^{+}_k) , \tag{8}$$

that can be interpreted as a generalized probability distribution for adding/removing a nucleon to/from the ground state while leaving the $A \pm 1$ system in the many-body state κ with (relative) energy ω.

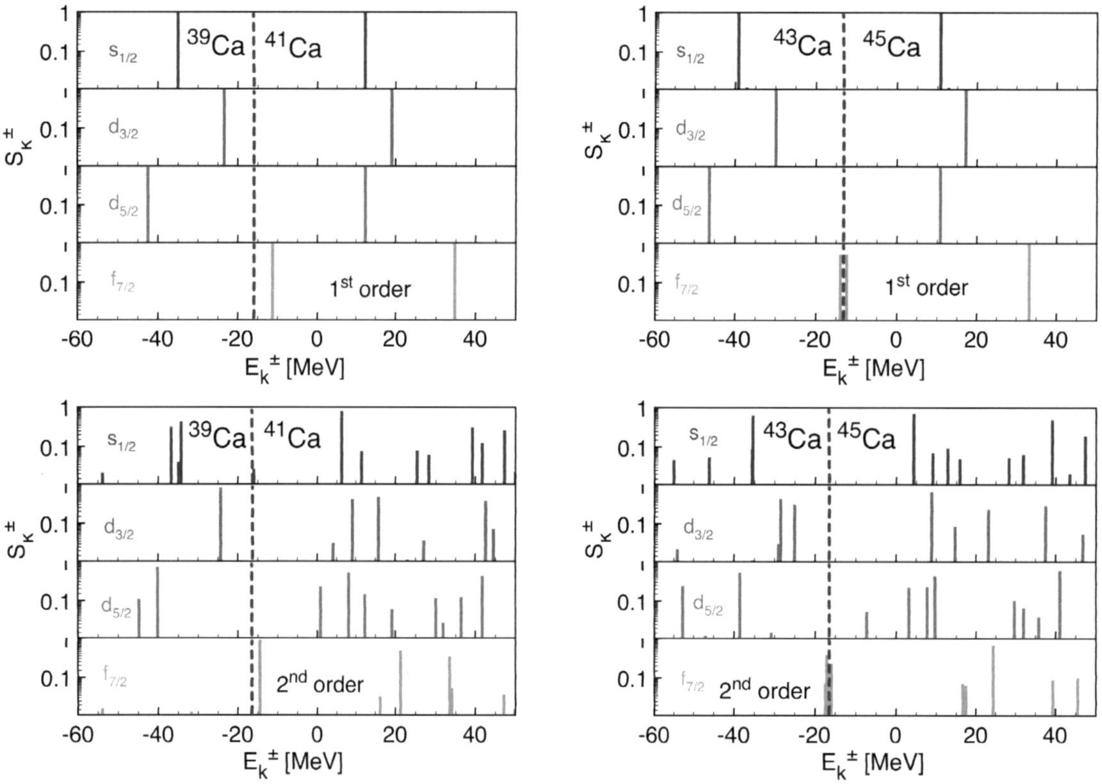

Figure 3. Neutron spectral strength distributions for different values of angular momentum and parity in ^{40}Ca (left) and ^{44}Ca (right). Results are displayed for self-consistent first-order (top) and second-order (bottom) approximations to the self-energy. The dashed vertical line denotes the Fermi energy. Calculations are performed using the $\Lambda = 500$ MeV chiral two-nucleon interaction [9] evolved down to $\Lambda = 2.1$ fm^{-1} through renormalization group technique [10]. Three-nucleon forces are omitted.

In Fig. 3 results for the normal neutron SSDs $\mathrm{S}^{11}_{[\kappa]}(\omega)$ in ^{40}Ca and ^{44}Ca are plotted for different values of angular momentum and parity as a function of ω. SSDs are shown for two different truncations of the self-energy expansion, i.e. at first and second order, respectively.

Calculations are performed using the $\Lambda = 500$ MeV chiral two-nucleon interaction [9] evolved down to $\Lambda = 2.1$ fm^{-1} through renormalization group technique [10]; three-nucleon forces are omitted. The model space consists of 7 major oscillator shells.

Both spectroscopic factors and SSDs represent an analysis tool to investigate correlations arising in the many-body system. Spectral peaks close to unity denote states that possess a single-particle character, i.e. that can be described to a good approximation as one nucleon added to or removed from an $A-$particle core. This is the case of ^{40}Ca when only first-order self-energies are included (top left in Fig. 3). When going from a doubly-closed shell systems as ^{40}Ca to one with a genuinely open-shell structure as ^{44}Ca, however, the arising of (static) pairing correlations can be detected. This is evident when focusing on the $f_{7/2}$ orbital in the top plots of Fig. 3: the lowest quasiparticle peak in ^{40}Ca breaks into two peaks of approximately equal strength right across the Fermi surface in ^{44}Ca, reflecting the partial filling of the neutron shell that can not be described in a single quasi-particle picture. The inclusion of the energy-dependent normal and anomalous second-order self-energies further fragment the single-particle strength into many states distributed over a wide range of energies. In most cases a single quasi-particle peak that carries 80 to 90 % of the strength is still visible. For the $f_{7/2}$ orbital the splitting of such main peak across the Fermi surface do not seem to be much affected by dynamical pairing correlations and remains similar to the first-order case.

4. Conclusions

Spectral strength distributions for one-nucleon addition and removal in ^{40}Ca and ^{44}Ca from Gorkov self-consistent Green's functions calculations have been presented. Although quantitatively not yet realistic due to the limited model space and the absence of three-nucleon forces, such calculations already show many of the features one requires from an ab-initio method that has access to spectroscopic quantities. The onset of static pairing correlations when filling up the neutron $f_{7/2}$ orbital going from ^{40}Ca to ^{48}Ca is reflected in the splitting of the quasi-particle peak across the Fermi surface in ^{44}Ca. The inclusion of energy-dependent self-energies accounts for dynamical correlations that result into a diffused fragmentation of the single-particle strength over a considerable range of energies.

More systematic application of Gorkov self-consistent Green's functions method are in progress [6] and will aim to describe for the first time in an ab-initio fashion the properties of long isotopic chains in the medium-mass region.

Acknowledgments

Calculations have been performed at the Centre de Calcul Recherche et Technologie (CCRT) of the Commissariat à l'Energie Atomique et aux Energies Alternatives (CEA), France. C.B. acknowledges support from the United Kingdom Science and Technology Facilities Council (STFC) through grant No. ST/I003363.

References

[1] G. Hagen, T. Papenbrock, D. J. Dean, M. Hjorth-Jensen, Phys. Rev. C82, 034330 (2010).
[2] C. Barbieri, M. Hjorth-Jensen, Phys. Rev. C79, 064313 (2009).
[3] C. Barbieri, C. Giusti, F. D. Pacati, and W. H. Dickhoff, Phys. Rev. C70, 014606 (2004).
[4] G. R. Jansen, M. Hjorth-Jensen, G. Hagen, and T. Papenbrock, Phys. Rev. C 83, 054306 (2011).
[5] B. Jeziorski, H. J. Monkhorst, Phys. Rev. A24, 1668 (1981).
[6] V. Somà, T. Duguet, C. Barbieri, *unpublished* (2011).
[7] L. P. Gorkov, Sov. Phys. JETP 7, 505 (1958).
[8] H. Lehmann, Nuovo Cim. 11, 342 (1954).
[9] D. R. Entem, R. Machleidt, Phys. Rev. C68, 041001 (2003).
[10] S. K. Bogner, T. T. S. Kuo, A. Schwenk, Phys. Rept. 386, 1 (2003).
[11] T. Duguet, J. Sadoudi, V. Somà, G. Hagen, C. Barbieri, *unpublished* (2011).

Nuclear Physics in Astrophysics V IOP Publishing
Journal of Physics: Conference Series **337** (2012) 012002 doi:10.1088/1742-6596/337/1/012002

SUPERHEAVY NUCLEI AND BEYOND: HYPERMATTER AND ANTIMATTER

Walter Greiner

Frankfurt Institute for Advanced Studies, J. W. Goethe–Universität, D-60438, Frankfurt a.M., Germany

E-mail: greiner@fias.uni-frankfurt.de

Abstract. The extension of the periodic system into various new areas is investigated. Experiments for the synthesis of superheavy elements and the predictions of magic numbers are reviewed. Further on, investigations on hypernuclei and the possible production of antimatter-clusters in heavy-ion collisions are reported. Various versions of the meson field theory serve as effective field theories at the basis of modern nuclear structure and suggest structure in the vacuum which might be important for the production of hyper- and antimatter.

1. Introduction

There are fundamental questions in science, like e. g. "how did life emerge" or "how does our brain work" and others. However, the most fundamental of those questions is "how did the world originate?". The material world has to exist be- fore life and thinking can develop. Of particular importance are the substances themselves, i. e. the particles the elements are made of (baryons, mesons, quarks, gluons), i. e. elementary matter. The vacuum and its structure is closely related to that. I want to report on these questions, beginning with the discussion of modern issues in nuclear physics.

The elements existing in nature are ordered according to their atomic (chemical) properties in the **periodic system**, which was developed by Dmitry Mendeleev and Lothar Meyer. The heaviest element of natural origin is uranium. Its nucleus is composed of Z = 92 protons and a certain number of neutrons (N = 128–150). They are called the different uranium isotopes. The transuranium elements reach from neptunium (Z = 93) via californium (Z = 98) and fermium (Z = 100) up to lawrencium (Z = 103). The heavier the elements are, the larger are their radii and their number of protons. Thus, the Coulomb repulsion in their interior increases, and they undergo fission. In other words: the transuranium elements become more unstable as they get bigger. In the late sixties, the dream of the superheavy elements arose. Theoretical nuclear physicists around S. G. Nilsson (Lund) and from the Frankfurt school [1, 2] predicted that so-called closed proton and neutron shells should counteract the repelling Coulomb forces. Atomic nuclei with these special **"magic" proton and neutron numbers** and their neighbours could again be rather stable. These magic proton (Z) and neutron (N) numbers were thought to be Z = 114 and N = 184 or 196. Typical predictions of their life-times varied between seconds and many thousand years. Figure 1 summarizes the expectations at the time. One can see the islands of superheavy elements around Z = 114, N = 184 and 196, respectively, and the one around Z = 164, N = 318.

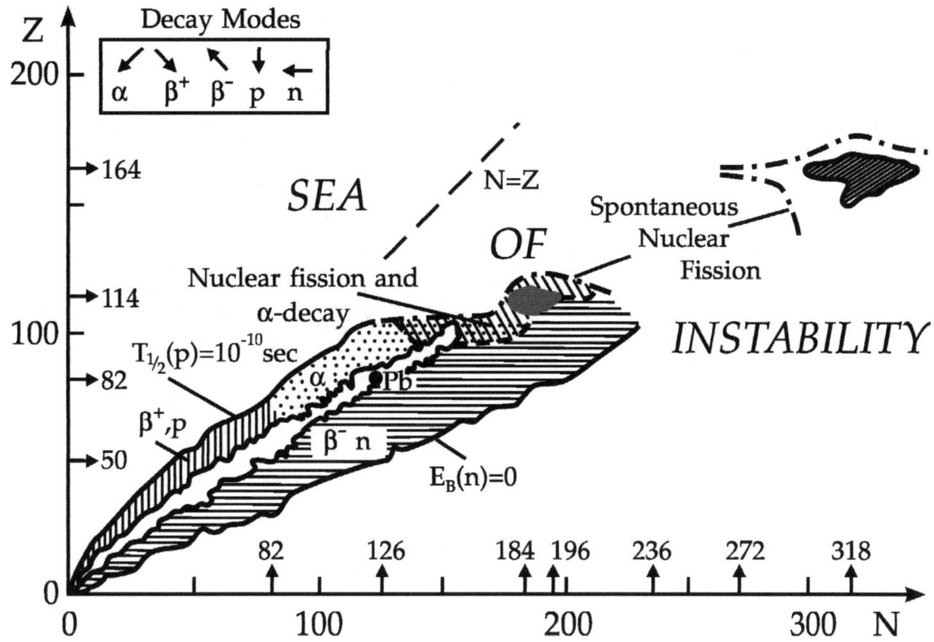

Figure 1. The periodic system of elements as conceived by the Frankfurt school in the late sixties. The islands of superheavy elements (Z = 114, N = 184, 196 and Z = 164, N = 318) are shown as dark hatched areas.

2. Cold Valleys in the Potential

The important question was how to produce these superheavy nuclei. There were many attempts, but only little progress was made. It was not until the middle of the seventies that the Frankfurt school of theoretical physics together with foreign guests (R. K. Gupta (India), A. Sandulescu (Romania)) [3] theoretically understood and substantiated the concept of bombarding of double magic lead nuclei with suitable projectiles, which had been proposed intuitively by the Russian nuclear physicist Y. Oganessian [4]. The two-center shell model, which is essential for the description of fission, fusion and nuclear molecules, was developed in 1969–1972 by W. Greiner and his students U. Mosel [1] and J. Maruhn [5]. It showed that the shell structure of the two final fragments was visible far beyond the barrier into the fusioning nucleus. The collective potential energy surfaces of heavy nuclei were calculated in the framework of the two-center shell model, exhibit pronounced valleys. These valleys provide promising doorways to the fusion of superheavy nuclei for certain projectile-target combinations (Fig. 2). If projectile and target approach each other through those **"cold" valleys** [3, 6], they get only minimally excited and the barrier, which has to be overcome (fusion barrier) is lowest (as compared to the neighbouring projectile-target combinations). In this way, the correct projectile- and target-combinations for fusion were predicted. In- deed, Gottfried Münzenberg and Sigurd Hofmann and their group at GSI [7] have followed this approach. With the help of the SHIP mass-separator and the position sensitive detectors, which were especially developed by them, they produced the pre-superheavy elements Z = 106, 107, . . . 112, each of them with the theoretically predicted projectile-target combinations, and only with these. Everything else failed. This is an impressive success, which crowned the laborious construction work of many years. The last but one example of this success, was the discovery of element 112 and its long α–decay chain. The Dubna group produced the six isotopes of Z = 113–118 by bombarding ^{244}Pu with ^{48}Ca [8]. This is also a

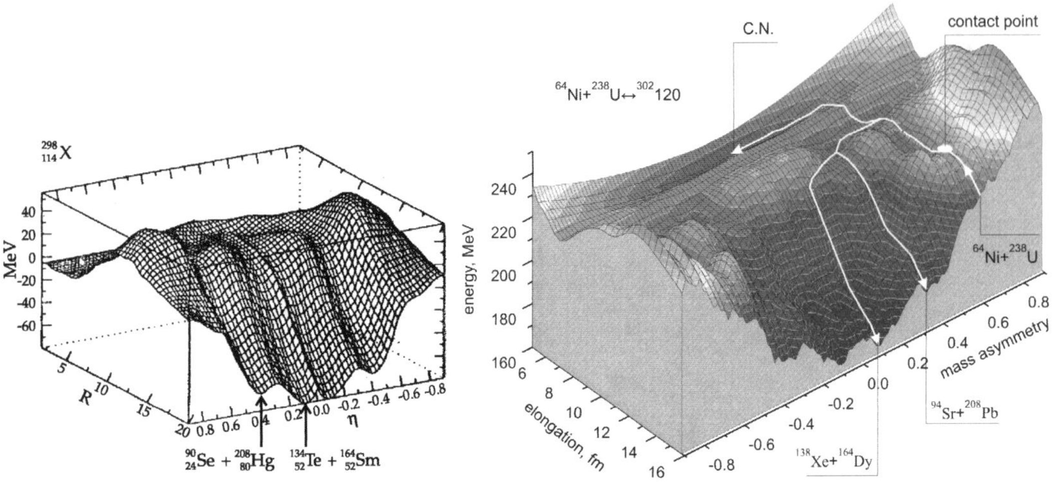

Figure 2. Left: The collective potential energy surface of $^{184}114$ calculated within the two center shell model by J. Maruhn *et al.* shows clearly the cold valleys, which reach up to the barrier and beyond. Right: Collective potential energy surface of the element $^{302}120$ calculated by A. Karpov, Y. M. Palenzuela and L. F. Ruiz.

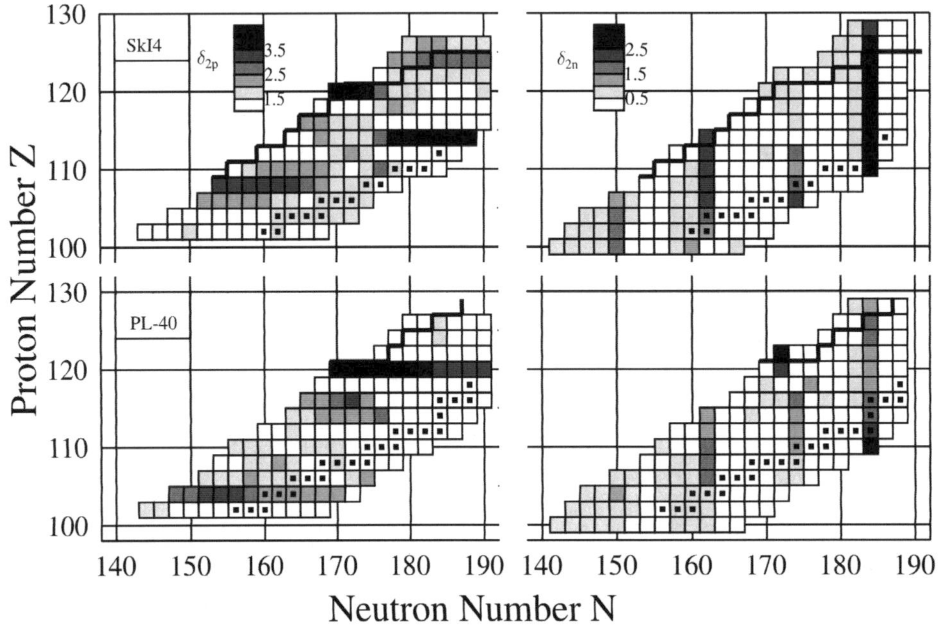

Figure 3. Grey scale plots of proton gaps (left column) and neutron gaps (right column) in the N-Z plane for deformed calculations with the forces SkI4 and PL- 40. Besides the spherical shell closures one can see the deformed shell closures for protons at Z = 104 (PL-40) and Z = 108 (SkI4) and the ones for neutrons at N = 162 for both forces.

coldvalley reaction (in this case due to the combination of a spherical and a deformed nucleus), as predicted by Gupta, Sandulescu and Greiner in 1977 [3]. There exist also cold valleys for which both fragments are deformed [6], or have non-axial orientations [9], but these have not been verified experimentally. The cold valleys also play an important role in nuclear fission giving rise to asymmetric and superasymmetric [10, 11] fission and to cluster radioactivity [12].

3. Shell structure in the superheavy region

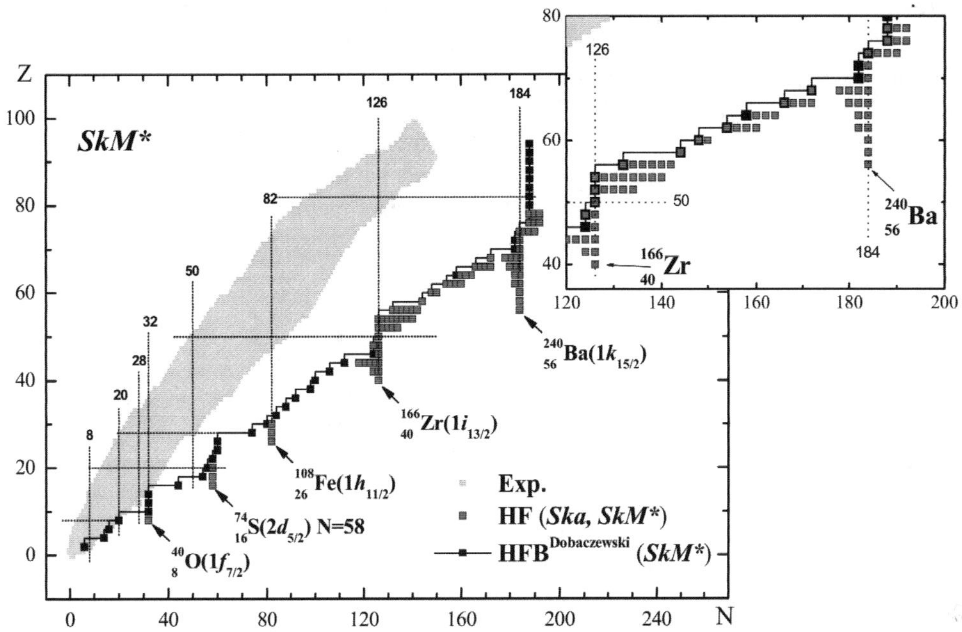

Figure 4. Fragment of the neutron drip line and elements (red squares) that are stable against one neutron emission [17]. One can see the formation of stability peninsulas along neutron magic numbers.

Studies of the shell structure of superheavy elements in the framework of the meson field theory and the Skyrme-Hartree-Fock approach have recently shown that the magic shells in the superheavy region are very isotope dependent [3]. Additionally, there is a strong dependence on the parameter set and the model. Some forces hardly show any shell structure, while other predict the magic numbers Z = 114, 120 and 126. Using the heaviest known even-even nucleus Hassium $^{264}_{156}108$ as a criterium to find the best parameter sets in each model, it turns out that PL-40 and SkI4 produce best its binding energy. However, these two forces make conflicting predictions for the magic number in the superheavy region: SkI4 predicts Z = 114, 120 and PL-40 Z = 120. Most interesting, Z = 120 **as magic proton number seems to be as probable as** Z = 114. Calculations of deformed sys- tems within the two models [13]reveal again different predictions: Though both parametrizations predict N = 162 as the deformed neutron-shell closure, the deformed proton-shell closures are Z = 108 (SkI4) and Z = 104 (PL-40) (see Fig. 3). Calculations of the potential energy surfaces [13] show single humped barriers; their heights and widths strongly depend on the predicted magic number. Furthermore, recent investigations in a chirally symmetric mean-field theory (see also below) result also in the prediction of these two magic numbers [14, 15]. The corresponding magic neutron numbers are predicted to be N = 172 and to a lesser extend N = 184. Thus, this region provides an open field of research. The charge distribution of the Z = 120,N = 184 nucleus indicates a hollow inside. This may suggest that it might be essentially a fullerene consisting of 60 α–particles and one binding neutron per alpha. The cold valleys in the collective potential energy surface are basic for understanding this exciting area of nuclear physics! It is a master example for under- standing the **structure of elementary matter**, which is so important for other fields, especially astrophysics, but even more so for enriching our "Weltbild", i.e. the status of our understanding of the world around us.

The investigation of the neutron drip line by extended Hartree-Fock+BCS calculations led to a surprise: extremely neutron rich nuclei along the magic neutron numbers become stable against one-and tow-neutron separation [17], see Fig. 4. The standard production of superheavy nuclei by fusing two smaller stable nuclei leads automatically to neutron poor isotopes near the proton drip line (therefore the lifetime of the produced superheavies is so small). Only a few suoerheavy atoms are produced this way. This leads us directly to the question of how superheavies with larger neutron numbers (and therefore having larger lifetimes: up to thousands of years) can be produced. One also wants to produce such long–living superheavies in macroscopic quantities (milligrams, grams,...) so that they can eventually be used technically. This can be done either by double (or multiple) underground atomic bomb explosions or by pulsed reactors with very high neutron flux ($\simeq 10^{21}$ neutrons/ sec cm^2) see Figs. 5 and 6.

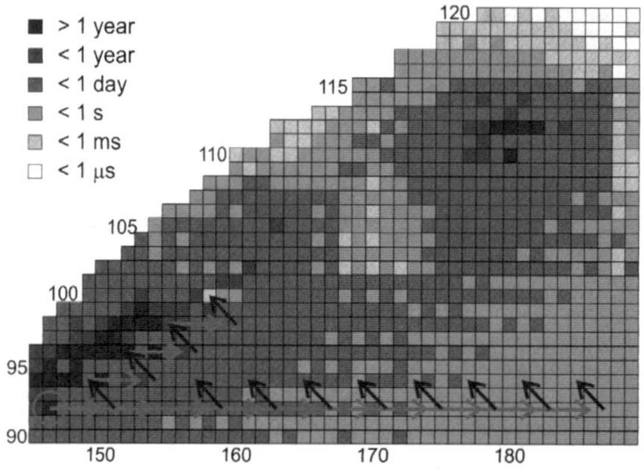

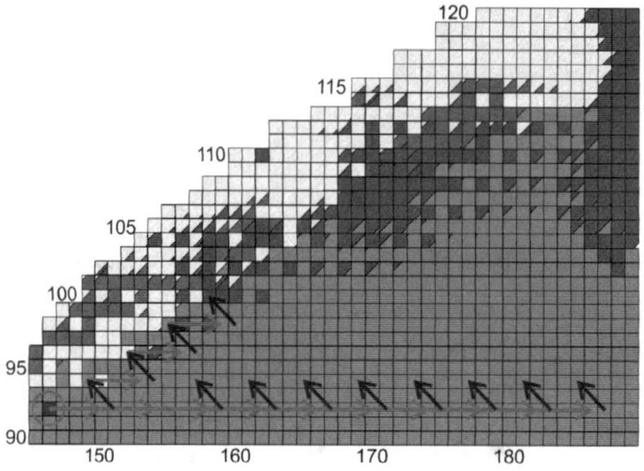

Figure 5. Half-lives (up) and decay modes (bottom) of nuclei in the upper part of the nuclear map. Schematic view of slow (terminated at the short-lived fission Fermium isotopes) and fast neutron capture processes with subsequent beta-minus decays are shown by the arrows.

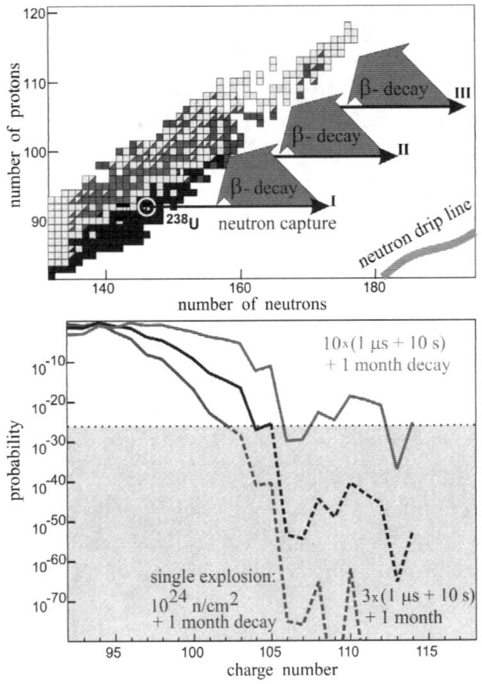

Figure 6. Schematic picture for multiple neutron irradiation of initial ^{238}U material (up) and probability for formation of heavy nuclei (bottom) in such process (one, three and ten subsequent explosions). Dotted line denotes the level of few atoms.

4. Extension of the periodic system into the field of hyper- and antimatter

Nuclei that are found in nature consist of nucleons (protons and neutrons) which themselves are made of u (up) and d (down) quarks. However, there also exist s (strange) quarks and even heavier flavours, called charm, bottom, top. The latter has just recently been discovered. Let us stick to the s quarks. They are found in the "strange" relatives of the nucleons, the so-called hyperons ($\Lambda, \Sigma, \Theta, \Omega$). The Λ–particle, e. g., consists of one u, d and s quark, the $Theta$–particle even of an u and two s quarks, while the Ω (sss) contains strange quarks only.

If such a hyperon is taken up by a nucleus, a **hyper-nucleus** is created. Hypernuclei with one hyperon have been known for 20 years [16]. Several years ago, Carsten Greiner, Jürgen Schaffner and Horst Stöcker [20] theoretically investigated nuclei with many hyperons, **hypermatter**, and found that the binding energy per baryon of strange matter is in many cases even higher than that of ordinary matter (composed only of u and d quarks). This leads to the idea of extending the periodic system of elements in the direction of strangeness.

One can also ask for the possibility of building atomic nuclei out of **antimatter**, that means searching e. g. for anti-helium [21], anti-carbon, anti-oxygen. Figure 7 depicts this idea. Due to charge conjugation symmetry, antinuclei should have the same magic numbers and the same spectra as ordinary nuclei. However, as soon as they get in touch with ordinary matter, they annihilate with it and the system explodes. Now the important question arises, how these strange matter and antimatter clusters can be produced. First, one thinks of collisions of heavy nuclei, e. g. lead on lead, at high energies (energy per nucleon $\geq$200 GeV). Calculations with the URQMD-model of the Frankfurt school show that through **nuclear shock waves** [22] nuclear matter gets compressed to 510 times of its usual value, $\rho_0 \simeq 0.17$ fm^{-3}, and heated up to temperatures of $kT \simeq 200$ MeV. As a consequence, about 10 000 pions, 100 Λ's, 40 $Sigma$'s and

Journal of Physics: Conference Series **337** (2012) 012002 | doi:10.1088/1742-6596/337/1/012002

Θ's and about as many antiprotons and many other particles are created in a single collision. It seems conceivable that it is possible in such a scenario for some Λ's to get captured in a nuclear cluster. This happens indeed rather frequently for one or two Λ–particles; however, more of them get built into nuclei with rapidly decreasing probability only. This is due to the low probability for finding the right conditions for such a capture in the phase space of the particles: the numerous particles travel with all possible momenta (velocities) in all directions. The chances for hyperons and antibaryons to meet get rapidly worse with increasing number. In order to produce multi–Λ–nuclei and antimatter nuclei, one has to look for a different source.

In the framework of the meson field theory, the energy spectrum of baryons has a peculiar structure, depicted in upper part of Fig. 8. It consists of an upper and a lower continuum, as it is known for electrons (see, e. g. Ref.[23]). Of special interest in the case of the baryon spectrum is the potential well, built of the scalar and the vector potential, which rises from the lower continuum. Naftali Auerbach and collaborators noticed this first [24]. It is known since P. A. M. Dirac (1930) that the negative energy states of the lower continuum have to be occupied by particles (electrons or, in our case, baryons). Otherwise our world would be unstable, because the "ordinary" particles are found in the upper states which can decay through the emission of photons into lower lying states. However, if the "underworld" is occupied, the Pauli-principle will prevent this decay. Holes in the occupied "underworld" (Dirac sea) are antiparticles.

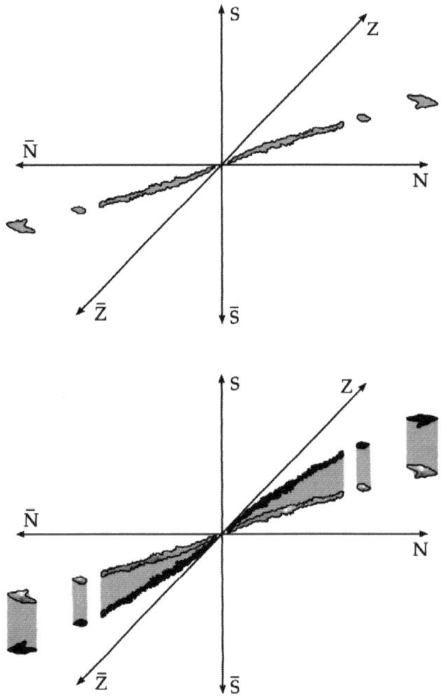

Figure 7. The extension of the periodic system into the sectors of strangeness $S, \overline{S}$ and antimatter $\overline{ZN}$. The stable valley winds out of the known proton (Z) and neutron (N) plane into the S and $\overline{S}$ sector, respectively. The same can be observed for the antimatter sector. In the upper part of the figure only the stable valley in the usual proton (Z) and neutron (N) plane is plotted, however, extended into the sector of antiprotons and antineutrons. In the second part of the figure it has been indicated, how the stable valley winds out of the Z-N-plane into the strangeness sector.

The occupied states of this underworld, including up to 40 000 occupied bound states of the lower potential well, represent the **vacuum**. The peculiarity of this strongly correlated vacuum structure in the region of atomic nuclei is that — depending on the size of the nucleus — more than 20 000 up to 40 000 (occupied) bound nucleon states contribute to this polarization effect. Obviously, we are dealing here with a **highly correlated vacuum**. Pronounced shell structure can be recognized [25]. Holes in these states have to be interpreted as bound antinucleons (antiprotons, antineutrons). If the primary nuclear density rises due to compression, the lower well increases while the upper decreases and soon is converted into a repulsive barrier. This compression of nuclear matter can only be carried out in relativistic nucleus-nucleus collision with the help of shock waves, which have been proposed by the Frankfurt school (see W. Scheid et al., Ref. [26]) and which have since then been confirmed extensively (see, e. g. Ref. [27]). These **nuclear shock waves** are accompanied by heating of the nuclear matter. Indeed, density and temperature are intimately coupled in terms of the hydrodynamic Rankine–Hugoniot equations. Heating as well as the violent dynamics cause the creation of many holes in the very deep (measured from $-M_Bc^2$) vacuum well. These numerous bound holes resemble antimatter clusters which are bound in the medium; their wave functions have large overlap with antimatter clusters. When the primary matter density decreases during the expansion stage of the heavy–ion collision, the potential wells, in particular the lower one, disappear.

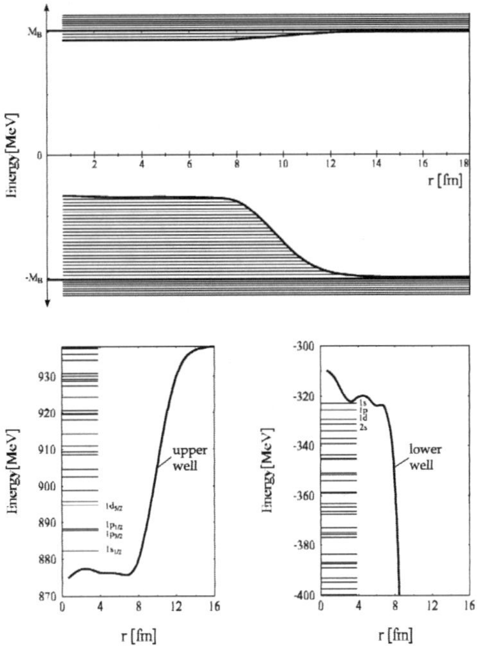

Figure 8. Baryon spectrum in a nucleus. Below the positive energy continuum exists the potential well of real nucleons. It has a depth of 50-60 MeV and shows the correct shell structure. The shell model of nuclei is realized here. However, from the negative continuum another potential well arises, in which about 40 000 bound particles are found, belonging to the vacuum. A part of the shell structure of the upper well and the lower (vacuum) well is depicted in the lower figures.

The bound antinucleons are then pulled down into the (lower) continuum. In this way antimatter clusters may be set free. Of course, a large part of the antimatter will annihilate on ordinary matter present in the course of the expansion. However, it is important that this mechanism for the production of antimatter clusters out of the highly correlated vacuum does not proceed via the phase space. The required coalescence of many particles in phase space suppresses the production of clusters, while it is favoured by the direct production out of the highly correlated vacuum. In a certain sense, the highly correlated vacuum is a kind of cluster vacuum (vacuum with cluster structure). The shell structure of the vacuum levels (see Fig. 8) supports this latter suggestion. Fig. 9 illustrates this idea. Recently the STAR Collaboration at RHIC observed Anti–^{4}He with production rate in excess of coalescent nucleosynthesis production [21].

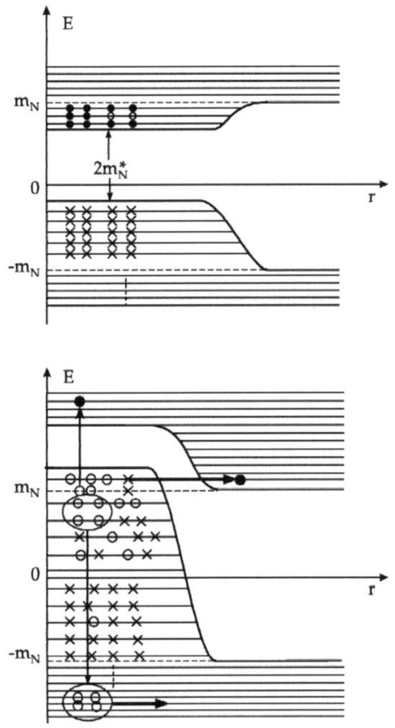

Figure 9. Due to the high temperature and the violent dynamics, many bound holes (antinucleon clusters) are created in the highly correlated vacuum, which can be set free during the expansion stage into the lower continuum. In this way, antimatter clusters can be produced directly from the vacuum. The horizontal arrow in the lower part of the figure denotes the spontaneous creation of baryon-antibaryon pairs, while the antibaryons occupy bound states in the lower potential well. Such a situation, where the lower potential well reaches into the upper continuum, is called supercritical. Four of the bound holes states (bound antinucleons) are encircled to illustrate a "quasi-antihelium" formed. It may be set free (driven into the lower continuum) by the violent nu- clear dynamics.

The mechanism is similar for the production of multi-hyper nuclei $(\Lambda, \Sigma, \Theta, \Omega)$. Meson field theory predicts also for the Λ energy spectrum at finite primary nucleon density the existence of upper and lower wells. The lower well belongs to the vacuum and is fully occupied by Λ's.

Dynamics and temperature then induce transitions ($\Lambda\overline{\Lambda}$ creation) and deposit many Λ's in the upper well. These numerous bound Λ's are sitting close to the primary baryons: in a certain sense a giant multi–Λ hypernucleus has been created. When the system disintegrates (expansion stage) the Λ's distribute over the nucleon clusters (which are most abundant in peripheral collisions). In this way multi–Λ hypernuclei can be formed. Of course this vision has to be worked out and probably refined in many respects. This requires a much more and thorough investigation in the future. It is particularly important to gain more experimental information on the properties of the lower well by (e, e p) or (e, e p p) and also ($\overline{p}_c p_b$, $p_c\overline{p}_b$) reactions at high energy ($\overline{p}_c$ denotes an incident antiproton from the continuum, p_b is a proton in a bound state; for the reaction products the situation is just the opposite). Also the reaction (p, p' d), (p, p' ^{3}He), (p, p' ^{4}He) and others of similar type need to be investigated in this context. The systematic studies of antiproton scattering on nuclei can contribute to clarify these questions. Various effective theories, e. g. of the Walecka-type on the one side and theories with chiral invariance on the other side, have been constructed to describe dense strongly interacting matter [4]. It is important to note that they seem to give different strengths of the potential wells and also different dependence on the baryon density.

According to chirally symmetric meson field theories, the antimatter-cluster- production and multi–hypermatter–cluster production out of the highly correlated vacuum takes place at approximately the same heavy-ion energies as compared to the predictions of the Dürr-Teller-Walecka type meson field theories. This in itself is a most interesting, quasi-fundamental question to be clarified. In the future, the question of the nucleonic substructure (form factors, quarks, gluons) and its influence on the highly correlated vacuum structure has to be studied. The nucleons are possibly strongly modified in the correlated vacuum: the Δ resonance correlations are probably important. Is this highly correlated vacuum state, especially during the compression, a preliminary stage to the quark-gluon cluster plasma? To which extent is it similar or perhaps even identical with it?

5. Concluding remarks – outlook

The extension of the periodic system into the sectors of hypermatter (strangeness) and antimatter is of general and astrophysical importance. Indeed, microseconds after the big bang, the new dimensions of the periodic system we have touched upon, certainly have been populated in the course of the baryo- and nucleo-genesis. In the early history of the universe, even higher dimensional extensions (charm, bottom, top) may have played a role, which we did not pursue here. It is an open question, how the depopulation (the decay) of these sectors influences the structure and composition of our world today. Our conception of the world will certainly gain a lot through the clarification of these questions.

References

[1] U. Mosel and W. Greiner, Z. Phys. **217** (1968) 256; **222** (1969) 261; see also *Memorandum zur Errichtung eines gemeinsamen Ausbildungszentrums für Kernphysik Hessischer Hochschulen*, this was the initiative for building GSI.

[2] S. G. Nilsson et al., Nucl. Phys. **A 115** (1968) 545; S. G. Nilsson, S. G. Thompson and C. F. Tsang, Phys. Lett. **B 28**, 7 (1969) 458.

[3] A. Săndulescu, R. K. Gupta, W. Scheid and W. Greiner, Phys. Lett. **B 60**, 3 (1976) 225; R. K. Gupta, A. Săndulescu and W. Greiner, Phys. Lett. **B 67**, 3 (1976) 257.

[4] Yu. Ts. Oganesian, A. G. Demin et al., Nucl. Phys. **A 239** (1975) 157.

[5] J. A. Maruhn and W. Greiner, Phys. Rev. Lett. **32** (1974) 548; J. A. Maruhn, Ph. D. Thesis, J. W. Goethe Universität, Frankfurt an Main (1973).

[6] R. K. Gupta, G. Münzenberg and W. Greiner, J. Phys. **G 23** (1997) L13.

[7] S. Hofmann and G. Münzenberg, Rev. Mod. Phys. **72** (2000) 733.

[8] Yu. Ts. Oganesian, V. K. Ityonkov and K. J. Moody, Sci. Am. **282** (Jan. 2000) 4.

[9] Ş. Mişicu and W. Greiner, Phys. Rev. **C 66** (2002) 044606.

[10] M. G. Itkis, S. M. Luk'yanov, V. N. Okolovich et al., Yad. Fiz. **52**, 23 (1990) [Sov. J. Nucl. Phys. **52**, 15 (1990)].

[11] F. Gönnenwein *et al.*, AIP Conference Proceedings **1224**, (2010) 338.

[12] D. N. Poenaru and W. Greiner, *Cluster Radioactivity* in Clusters in Nuclei 1, C. Beck, ed., vol. **818** of Lecture Notes in Physics, pp. 1–56, Springer, 2010.

[13] K. Rutz, M. Bender, T. Bürvenich, T. Schilling, P.-G. Reinhard, J. A. Maruhn and W. Greiner, Phys. Rev. **C 56** (1997) 238; T. Bürvenich, K. Rutz, M. Bender, P.- G. Reinhard, J. A. Maruhn and W. Greiner, Eu. Phys. J. **A 3** (1998) 139; M. Bender, K. Rutz, P.-G. Reinhard, J. A. Maruhn and W. Greiner, Phys. Rev. **C 58** (1998) 2126.

[14] B. D. Serot and J. D. Walecka, Advances in Nuclear Physics, 16, Plenum Press, New York (1986); J. Theis, G. Graebner, G. Buchwald, J. Maruhn, W. Greiner, H. Stöcker and J. Polonyi, Phys. Rev. **D 28** (1983) 2286; S. Klimt, M. Lutz and W. Weise, Phys. Lett. **B 249** (1990) 386; I. N. Mishustin, Proc. Int. Conf. Structure of Vacuum and Elementary Matter, South Africa, Wilderness, World Scientific, Singapore (1996) p. 499; P. Papazoglou, D. Zschiesche, S. Schramm, J. SchaffnerBielich, H. Stöcker and W. Greiner, Phys. Rev. **C 59** (1999) 411; I. N. Mishustin, L. M. Satarov, H. Stöcker and W. Greiner, Phys. Rev. **C 62** (2000) 034901.

[15] P. Papazoglou, Ph. D. Thesis, University of Frankfurt (1998); Ch. Beckmann, P. Papa- zoglou, D. Zschiesche, S. Schramm, H. Stöcker and W. Greiner, Phys. Rev. **C 65** (2002) 024301.

[16] B. Povh, Rep. Progr. Phys. **39** (1976) 823.

[17] K.A. Gridnev, V.N. Tarasov, D.V. Tarasov, D.K. Gridnev, V.V. Pilipenko, W. Greiner, Int. Jour. Mod. Phys. **E19** (2010) 449-457; V.N. Tarasov, K.A. Gridnev, D.K. Gridnev, D.V. Tarasov, S. Schramm and W. Greiner, *Stability Peninsulas near Neutron Drip Line*, ssubmitted to Phys. Rev. Lett.

[18] A. Botvina, I. Mishustin, V. Zagrebaev, and W. Greiner, Int. J. Mod. Phys. **E19** (2010) 20632075

[19] V. I. Zagrebaev, A. V. Karpov, I. N. Mishustin, and W. Greiner, *Superheavies: Theoretical incitements and predictions* (2011) arXiv:1105.1659 [nucl-th].

[20] J. Schaffner, C. Greiner and H. Stöcker, Phys. Rev. **C 46** (1992) 322.

[21] The STAR Collaboration, Nature **aop**, (2011) doi:10.1038/nature10079.

[22] H. Stöcker, W. Greiner and W. Scheid, Z. Phys. **A 286** (1978) 121.

[23] W. Greiner, B. Müller and J. Rafelski, QED of Strong Fields, Springer Verlag, Heidel- berg (1985). For a more recent review see W. Greiner and J. Reinhardt, Supercritical Fields in Heavy-Ion Physics, Proc. 15th Advanced ICFA Beam Dynamics Workshop on Quantum Aspects of Beam Physics, World Scientific, Singapore (1998).

[24] N. Auerbach, A. S. Goldhaber, M. B. John- son, L. D. Miller and A. Picklesimer, Phys. Lett. **B 182** (1986) 221.

[25] I. Mishustin, L. M. Satarov, J. Schaffner, H. Stöcker and W. Greiner, J. Phys. **G 19** (1993) 1303; P. K. Panda, S. K. Patra, J. Reinhardt, J. Maruhn, H. Stöcker and W. Greiner, Int. J. Mod. Phys. **E 6** (1997) 307.

[26] W. Scheid, I. Hoffmann and W. Greiner, in Proc. Symp. Physics with Relativistic Heavy Ions, Lawrence Berkeley Lab., Berkeley, Cal. (1974).

[27] H. Stöcker and W. Greiner, Phys. Rep. **137** (1986) 279.

Nuclear Physics in Astrophysics V IOP Publishing
Journal of Physics: Conference Series **337** (2012) 012003 doi:10.1088/1742-6596/337/1/012003

Applications of the USD Hamiltonians to rapid-proton capture rates for sd-shell nuclei

B Alex Brown[1], W. A. Richter[2,3]

[1] Department of Physics and Astronomy, and National Superconducting Cyclotron Laboratory, Michigan State University, East Lansing, Michigan 48824-1321, USA
[2] iThemba LABS, P.O.Box 722, Somerset West 7129, South Africa
[3] Department of Physics, University of the Western Cape, Private Bag X17, Bellville 7535, South Africa

E-mail: brown@nscl.msu.edu

Abstract. We discuss the use of the USD, USDA and USDB Hamiltonians for calculating rapid-proton capture rates for the sd-shell nuclei. The isobaric-mass-multiplet equation is used to make associations between the levels in the proton-rich nuclei of interest and those in the nuclei with $N \geq Z$. The theoretical uncertainties associated with the calculated spectroscopic factors and gamma-decay widths are discussed. Results are shown for the ^{25}Al(p,γ)^{26}Si resonance-capture rate.

1. Introduction

The rapid proton capture (rp-process) is a sequence of proton capture reactions and β^+ decays passing through proton-rich nuclei. The rp-process reaction rates are crucial nuclear nuclear physics input to astrophysical models of nucleosynthesis in novae, supernovae and explosive hydrogen burning conditions such as X-ray bursters. A compilation of nuclear structure input and the resulting rates has recently been made [1].

In this paper we will review some applications of nuclear structure theory for the sd-shell for nuclei in the mass region A=20-38. Improved theory input together with new experimental results has greatly reduced the uncertainty for many of the crucial reaction rates. As an example we will show results for the ^{25}Al(p,γ)^{26}Si resonance-capture rate. In 1995 this rate was uncertain by many orders of magnitude [2]. A combination of theoretical and experimental improvements has reduced this uncertainty to about 30%.

The resonant reaction rate for capture on a nucleus in an initial state i, for an isolated narrow resonances is calculated as a sum over all relevant compound nucleus states f above the proton-decay threshold [3]

$$N_A < \sigma v >_{\mathrm{res}\,i} = 1.540 \times 10^{11} (\mu T_9)^{-3/2} \times \sum_f \omega\gamma_{if} \; \mathrm{e}^{-\mathrm{E_{res}}/(kT)} \;\; \mathrm{cm^3\,s^{-1} mole^{-1}}. \quad (1)$$

Here T_9 is the temperature in GigaK, $E_{res} = E_f - E_i$ is the resonance energy in the center

Published under licence by IOP Publishing Ltd

of mass system, the resonance strengths in MeV for proton capture are

$$\omega\gamma_{if} = \frac{(2J_f+1)}{(2J_p+1)(2J_i+1)} \frac{\Gamma_{\mathrm{p}\,if}\Gamma_{\gamma f}}{\Gamma_{\mathrm{total}\,f}}. \tag{2}$$

$\Gamma_{\mathrm{total}\,f} = \Gamma_{\mathrm{p}\,if} + \Gamma_{\gamma f}$ is the total width and J_i, J_p and J_f are target, the proton projectile ($J_p = 1/2$), and states in the final nucleus, respectively. The proton decay width depends on the resonance energy via the single-particle proton width and can be calculated from the proton spectroscopic factor C^2S_{if} and the single-particle proton width $\Gamma_{\mathrm{sp}\,if}$ as $\Gamma_{\mathrm{p}\,if} = C^2S_{if}\Gamma_{\mathrm{sp}\,if}$.

We will discuss the ingredients in this equation in the following sections. In most cases the input comes from a combination of theory and experiment. When this is done the theory has to be good enough so that one can associate a specific final state observed in experiment with a specific theoretical state. The theoretical energy uncertainty is on the order of 150 keV. However, due its exponential dependence in the cross section, it is necessary to know the energies of the individual final states just above the proton-decay threshold to within an uncertainty of about 5 keV. For the proton-rich nuclei in the rp-process path this is achievable in experiments based on transfer reactions or the observation of the protons and gammas in their decay after being formed in beta decay or knockout reactions [4], [5]. But there are often uncertainties in the J^π values. The experimental data for the neutron-rich mirror nucleus is usually known much better (since they lie closer to stability). Thus one tries to make match experimental mirror levels so that information on the J^π value, spectroscopic factors and Γ_γ from the neutron-rich nucleus can be utilized. The Coulomb displacement energy has state-dependent shifts up to about 300 keV as shown in Fig. (2) for the case of ^{26}Mg and ^{26}Si. Thus when the level density becomes on the order of 300 keV or less the precise matching of mirror states can become ambiguous. We will discuss in Sec. 3 how theory can be used improve the association of mirror levels. In the next section we review the general properties of the sd-shell Hamiltonians and how they have been used to compare with experiment. Finally in Sec. 4 we discuss the cross-section for the ^{25}Al(p,γ)^{26}Si example. These results are discussed in more detail [6].

2. sd-shell Hamiltonians

For nearly 30 years the USD Hamiltonian for the ($0d_{5/2}, 0d_{3/2}, 1s_{1/2}$) *sd*-shell has been used for structure input to rp-process rates. More recently two new Hamiltonians have been obtained from single-valued decomposition (SVD) fits of parameters determined by the 63 two-body matrix and three single-particle energies to a complete set of energy data [7]. The USDA Hamiltonian is based on the SVD fit with 30 linear combinations of parameters and the USDB Hamiltonian is based on 56 linear combinations of parameters.

Comparison of spectroscopic and electromagnetic observables for low-lying states with USD, USDA and USDB has been made [8]. For these low-lying states the comparison of theory vs theory shows that Hamiltonian uncertainty in the spectroscopic factors is on the order of 5%, the B(E2) values on the order of 10% and the B(M1) values on the order of 30%. Comparison of theory vs experiment shows that the theoretical uncertainties are on the order of 20% for spectroscopic factors and 30% for electromagnetic transition strengths.

For ^{26}Mg assignments between theory and experiment for about 50 levels in ^{26}Mg up to 10 MeV in excitation have been made, based on a comparison of the experimental and theoretical level energies, electromagnetic transition strengths and electron scattering data [9]. These comparisons show that specific experimental and theoretical states can be matched one to one up to about 8 MeV in excitation energy. For the level lifetimes (or Γ_γ) theory and experiment agree within about 30% for levels up to 8 MeV, but there are often large experimental errors for the states above 6 MeV.

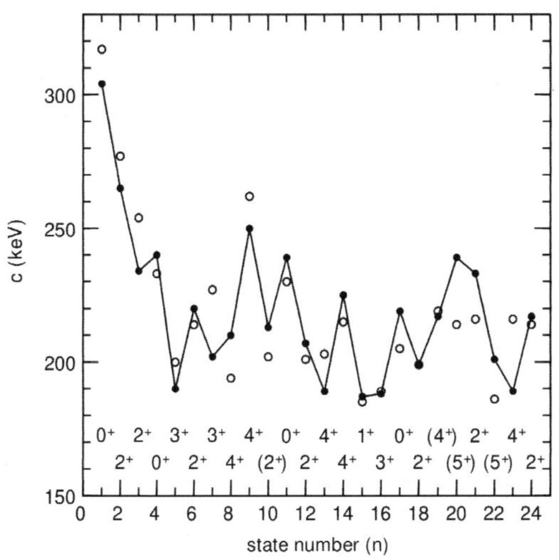

Figure 1. *c*-coefficients from the isobaric mass multiplet equation versus state number (in order of increasing energy) in ^{26}Si based on experimental energies (closed circles) and energies calculated from USDB (open circles).

3. Mirror level associations

The energies of states in an isobaric multiplet are connected by the isobaric-mass-multiplet equation (IMME):

$$B(T_z) = a + bT_z + cT_z^2, \quad (3)$$

where B are the binding energies. Ormand and Brown studied the b- and c-coefficients of the IMME we use the USD Hamiltonian [10]. They add Coulomb, charge-dependent and charge-asymmetric parts to the nuclear Hamiltonian. For the nuclei considered in [10], A=18-22 and A=34-39, the 42 b-coefficients were reproduced with an rms deviation of 27 keV and the 26 c-coefficients were reproduced with an rms deviation of 9 keV. There is considerable state-dependence in the c-coefficients (ranging in values from 130 keV to 350 keV) that is nicely reproduced by the calculations [see Fig. (9) in [10]]. When the only experimental information available is the energy of the neutron-rich nucleus $B(T_z = T)$, one can use the calculated b-coefficient to predict the energy of the proton-rich nucleus; $B(T_z = -T) = B(T_z = T) - 2T_z b_{th}$. This method was used by Herndl et al. [11].

In the example of A=26, the there is a $T = 1$ triplet of states in ^{26}Si, ^{26}Al and ^{26}Mg. In Fig. (1) values of c from experiment and theory are compared for $T = 1$ states $A = 26$ ordered according to increasing experimental energy. The experimental values are obtained for states where all three members of the multiplet are known. In general a good correspondence can be seen, the largest deviations being less than 30 keV. The agreement with experiment and theory [Fig. (1)] for A=26 appears to be better than obtained in [12] for the odd-odd cases with mass 28, 32 and 36. Predicted binding energies for states in ^{26}Si can be then be obtained from Eq. (3), with experimental values of binding energy for corresponding states in ^{26}Al and ^{26}Mg (when they are known in both) together with the calculated c-coefficient:

$$B_{\rm th}(^{26}{\rm Si}) = 2B_{\rm exp}(^{26}{\rm Al}) - B_{\rm exp}(^{26}{\rm Mg}) + 2c_{th}. \quad (4)$$

This method was used previously in [12].

Journal of Physics: Conference Series **337** (2012) 012003 doi:10.1088/1742-6596/337/1/012003

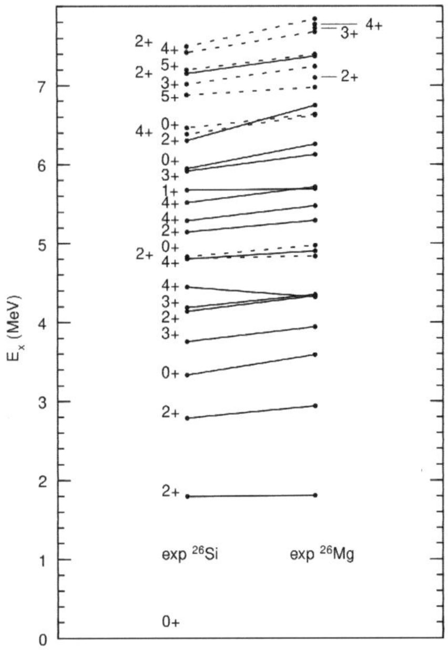

Figure 2. Experimental excitation energies in ^{26}Si and ^{26}Mg [15]. Solid lines are for the states in ^{26}Si with firm J^{π} values. Dashed lines are for states in ^{26}Si with uncertain J^{π} values with the most likely mirror associations in ^{26}Mg with known J^{π} values.

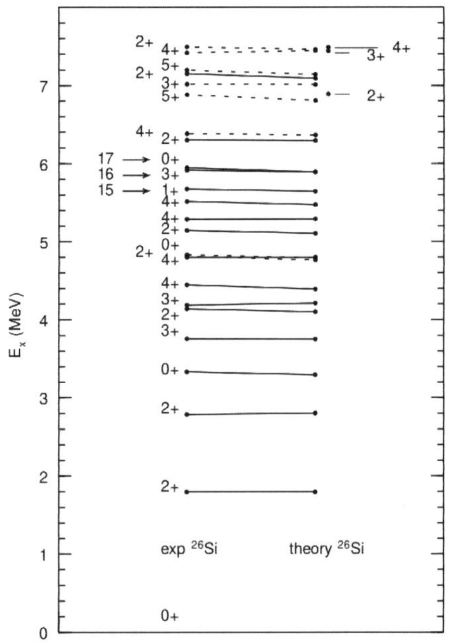

Figure 3. Experimental excitation energies in ^{26}Si [15] versus predicted energies E_{th} based on Eq. (4). Dashed lines are for states in ^{26}Si with uncertain J^{π} values with the most likely mirror associations in ^{26}Mg with known J^{π} values.

Fig. (3) shows the excitation energies for ^{26}Si obtained from Eq. (4) on the right compared to experiment on the left. The calculated values can then be used as a guide to the correct spin/parity assignments for measured levels in ^{26}Si. Where no levels in ^{26}Si are known, levels can be predicted. Two such levels (the 2^+ and 4^+) are indicated on the right-hand side of Fig. (3). The energy of the 3^+ state shown in the right-hand side of Fig. (3) was obtained from the average shift (250 keV) of the five highest states in Fig. (2). Above eight MeV where the property of states in ^{26}Mg or ^{26}Al become uncertain we use the energies obtained from the USDB Hamiltonian. This includes the addition of about 170 states with $J^{\pi} \leq 5^+$ up to 14 MeV in excitation energy.

The three levels that are just above the proton-decay separation energy of 5.51 MeV and of potential importance for the capture reaction at low temperatures are indicated by the arrows in Fig. (3). The J^{π} of levels 16 and 17 are from the recent analysis of Wrede [13] where arguments for the J^{π} are based on all available data for these states. This included the analysis of Bardayan et al. [14] for the ^{28}Si(p,t) data where an assignment $J^{\pi} = 2^+$ or 3^+ was made for state 16. From the associations made in Fig. (3) we can rule out 2^+.

4. Results for the ^{25}Al(p,γ)^{26}Si reaction rate

The production mechanism and production site for the long-lived radioactive isotope ^{26}Al has been of interest since the first indications of ^{26}Al enrichment in meteoritic inclusions was observed [16]. Understanding its origin would serve as a unique signature for nucleosynthesis in novae and supernovae. The main reaction sequence leading to ^{26}Al is ^{24}Mg(p,γ)^{25}Al($\beta^+ + \nu$)^{25}Mg(p,γ)^{26}Al.

At the high-temperature conditions expected for shell carbon burning and explosive neon burning the ^{25}Al(p,γ)^{26}Si reaction becomes faster than the ^{25}Al β decay. Since ^{26}Si β decays to the short-lived 0^+ state of ^{26}Al, the production of the long-lived (5^+) state is by-passed. In a recent paper [15] energies of levels in ^{26}Si were measured and used together with previous data and theoretical input to obtain a cross section for the ^{25}Al(p,γ)^{26}Si reaction. An updated theoretical calculation was presented in [6] and is reviewed here.

The total resonance-capture reaction rates have been calculated for each of the interactions USD, USDA and USDB. Fig. (4) shows the results for the resonance-capture rate obtained using the properties of ^{26}Si given in Table I of [6]. The Γ_p and Γ_γ in this case are all based on the USDB Hamiltonian. In Fig. (5) we show some sensitivity studies. The upper two panels (a) and (b) show the results based on Γ_p and Γ_γ from the USDA and USD Hamiltonians relative to USDB. In panel (c) we compare the rate obtained when the theoretical $\Gamma_\gamma(^{26}\text{Si})$ are replaced by $\Gamma_\gamma(^{26}\text{Mg})$ with the excitation energies the same in both cases. These Γ_γ differ because the electromagnetic matrix elements have a small mirror asymmetry. This comparison shows that at the level of 10% it is adequate to take the Γ_γ information from the mirror nucleus.

For the region of $\text{Log}_{10}(\text{T9}) < -0.8$ the resonance-capture rate comes entirely from the 5.675 MeV 1^+ state (number 15). Since $\Gamma_\gamma >> \Gamma_p$ for this case the rate is determined by Γ_p. The large change shown on the left-hand side of panels (a) and (b) in Fig. (5) is due to the change in the relatively small spectroscopic factors; 0.0048 (USDB), 0.0027 (USDA) and 0.0035 (USD). For this region of T9 we recommend the USD rate with an uncertainty of 40% coming from the spread of the theoretical spectroscopic factors.

For the region of $-0.7 < \text{Log}_{10}(\text{T9}) < 0.5$ the rate is dominated by the properties of the 3^+ state at 5.915 MeV (number 16). Since $\Gamma_\gamma < \Gamma_p$ the rate is determined by Γ_γ. The gamma-decay half-life of the analogue 3^+ level in the mirror nucleus ^{26}Mg has been measured [17]. Experiment and theory are compared for lower energy states of ^{26}Mg in Table I and Fig. (2) of [9]. The experimental half-life of this 3^+ level in ^{26}Mg of 14(6) ps [17] is larger than the USDB result of 4.0 ps. The USDA and USD values for the half-life are 4.5 and 5.0 fs, respectively. Based on the comparisons shown in Fig. (2) of [9] for the lifetimes of other levels in ^{26}Mg above five MeV (levels 10-22), this deviation is larger than expected. However, the experimental uncertainty is relatively large for this lifetime. It has only been measured once by the Doppler shift attenuation method [17]. Thus, we choose to use the USDB value. Since Matic et al., [15] use the experimental value, their rate is about a factor lower than ours in this region of temperature. It would be important to improve the experimental uncertainty in this lifetime. In addition, one could measure the gamma-decay decay branching for the 403 keV resonance in ^{26}Si which is predicted to be 3% relative to proton decay. The theoretical uncertainty for this energy range coming from Γ_γ is about 20%. But one should confirm the experimental result for ^{26}Mg which deviates from the theory outside of this error.

For the region of $0.7 < \text{Log}_{10}(\text{T9})$ the rate comes from the contribution of many states with $\Gamma_\gamma << \Gamma_p$. The $\omega\gamma$ depends on the $(2J_f + 1)$ level density and the associated Γ_γ. The sd-shell provides a fairly realistic model for the positive-parity level density up to 10 MeV, but there will be contributions from pf shell intruder states starting with the possible state at 7.2 MeV in ^{26}Mg [9]. Starting with the known 3^- state in ^{26}Mg at 6.8 MeV there will be contributions from negative-parity states. At $\log_{10}(\text{T9})$=1 we estimate that the effective level density and the effective rate is about a factor of two higher than that given by the sd-shell model. Above $\log_{10}(\text{T9}) = 1$ one should base the rate on a Hauser-Feshbach formulation with level densities adjusted to match the known level density in the region of 6−9 MeV excitation energy.

The astrophysical implications for novae and x-ray bursts in terms of the competition between the ^{25}Al(p,γ)^{26}Si and the ^{25}Al β decay rates is shown in Fig. (9) of [15]. Our factor of three higher rate for ^{25}Al(p,γ)^{26}Si compared to that of [15] in the temperature range of interest, $\log_{10}(\text{T9})$=0.1-0.3, will relatively reduce the population of the long-lived 5^+ state of ^{26}Al

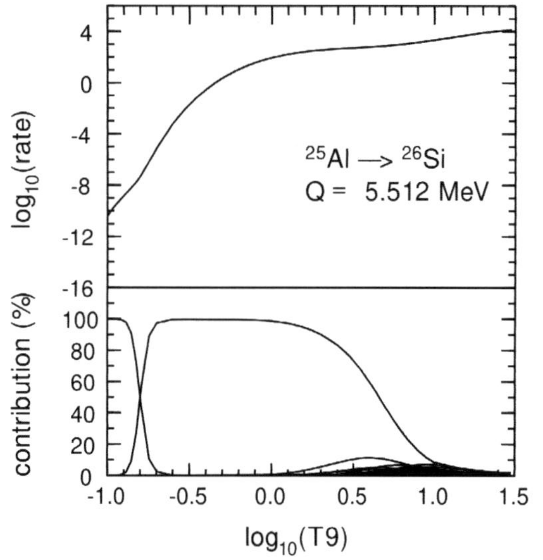

Figure 4. The total resonance-capture reaction rate versus temperature T9 (GigaK) (top panel) and the contribution of each of the final states (lower panel) with USDB.

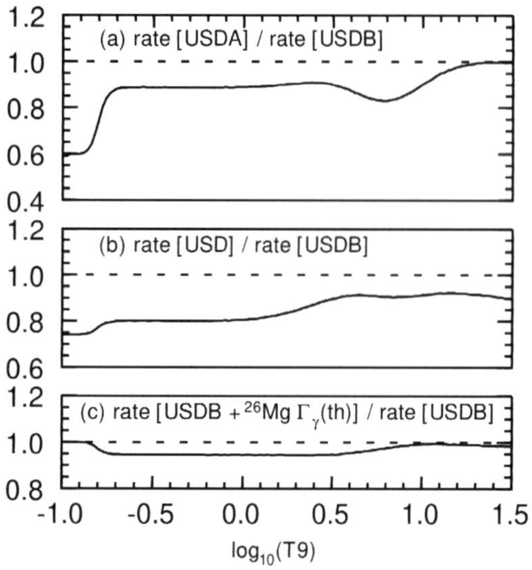

Figure 5. Rates calculated with different assumptions divided by the USDB results given in Fig. (4).

by bypassing its production. It would be interesting to apply these new rates to various astrophysical scenarios to find the quantitative consequences. For the next generation of rp capture cross section calculations it will be important to consider theoretical errors coming from uncertainties within the model-space assumptions as well as those that come from the limitations of the model-space truncations.

Acknowledgments This work is partly supported by NSF Grant PHY-0758099, NSF Grant No. PHY08-22648 (Joint Institute for Nuclear Astrophysics), and the National Research Foundation of South Africa.

References

[1] C. Iliadis, R. Longland, A. E. Champagne and A. Coc, Nucl. Phys. **A841**, 251 (2010).
[2] A. Coc, R. Mochkovitch, Y. Oberto, J. P. Thibaud and E. Vangioni-Flam, Astron. Astrophys. **299**, 479 (1995).
[3] W. A. Fowler and F. Hoyle, Ap. J. Suppl. **9**, 201 (1964).
[4] H. Schatz, C. A. Bertulani, B. A. Brown, R. R. C. Clement and B. M. Sherrill, Phys. Rev. C **72**, 065804 (2005).
[5] R. R. C. Clement *et. al.* Phys. Rev. Lett. **92**, 172502 (2004).
[6] W. A. Richter, B. A. Brown, A. Signoracci and M. Wiescher, Phys. Rev. C **83**, 065803 (2011).
[7] B. A. Brown and W. A. Richter, Phys. Rev. C **74**, 034315 (2006).
[8] W. A. Richter, S. Mkhize and B. A. Brown, Phys. Rev. C **78**, 064302-1 (2008).
[9] W. A. Richter and B. A. Brown, Phys. Rev. C **80**, 034301 (2009).
[10] W. E. Ormand and B. A. Brown, Nucl. Phys. A **491**, 1 (1989).
[11] H. Herndl, J. Goerres, M. Wiescher, B. A. Brown and L. Van Wormer, Phys. Rev. C **52**, 1078 (1995).
[12] C. Iliadis, P. M. Endt, N. Prantzos and W. J. Thompson, The Astrophysical Jour. **524**, 434 (1999).
[13] C. Wrede, Phys. Rev. C **79**, 035803 (2009).
[14] D. W. Bardayan et al., Phys. Rev. C **74**, 045804 (2006).
[15] A. Matic et al., Phys. Rev. C **82**, 025807 (2010).
[16] T. Lee, D. A. Papanantassiou and G. J. Wasserburg, Astrophys. J. Lett. **211**, L107 (1977); **220**, L21 (1978).
[17] F. Glatz, S. Norbert, E. Bitterwolf, A. Burkard, F. Heidinger, Th. Kern, R. Lehmann, H. Ropke, J. Siefert and C. Schneider, Z. Phys. A **324**, 187 (1986).

Journal of Physics: Conference Series **337** (2012) 012004 doi:10.1088/1742-6596/337/1/012004

Complete dipole strength distributions in ^{208}Pb from high-resolution polarized proton scattering at 0°

Peter von Neumann-Cosel

Institut für Kernphysik, Technische Universität Darmstadt, D-64289 Darmstadt, Germany

E-mail: vnc@ikp.tu-darmstadt.de

Abstract. Small-angle polarized proton scattering including 0° at incident energies of a few 100 MeV/nucleon is established as a new spectrospcopic tool for the study of E1 and M1 strength distributions. Experiments of this type have been realized recently at RCNP, Osaka, Japan with high energy resolution of the order 25 - 30 keV (FWHM). Using ^{208}Pb as an example, the physics potential of such data is discussed. It includes information on the properties of the Pygmy Dipole Resonance but also on complete E1 and M1 strength distributions and thus the gamma strength function. The E1 polarizability can be extracted with a precision of about 4% providing important experimental constraints on the neutron skin thickness in ^{208}Pb.

1. Introduction

The electric dipole (E1) response of nuclei is dominated by the giant dipole resonance (GDR), a highly excited collective mode above the particle emission threshold [1]. Its properties are well understood but recent interest focusses on evidence for a soft mode in neutron-rich nuclei below the GDR termed pygmy dipole resonance (PDR). Because of the saturation of nuclear density, excess neutrons might form a skin whose oscillations against an isospin-saturated core should give rise to a low-energy E1 mode [2]. Therefore, the PDR may shed light onto the formation of neutron skins in nuclei [3]. Another quantity related to nuclear E1 modes is the symmetry energy acting as restoring force. The E1 strength distribution carries information on its poorly known magnitude and density dependence [4], indispensable ingredients for the modeling of the equilibrium properties of neutron stars [5].

A case of special interest is the doubly magic nucleus ^{208}Pb. In a measurement of parity-violating elastic electron scattering at JLAB, the PREX collaboration [6] aimed at the first model-independent determination of the neutron skin thickness in ^{208}Pb. However, the recent result $r_{\rm skin} = 0.34^{+0.15}_{-0.17}$ fm suffers still from limited statistics. Studies of energy density functionals (EDFs) [7] using Skyrme forces [8] or a relativistic framework [9] suggest the nuclear dipole polarizability α_D as an alternative observable constraining both neutron skin and symmetry energy. The polarizability is related to the photoabsorption cross section σ_{abs} [10]

$$\alpha_D = \frac{\hbar c}{2\pi^2 e^2} \int \frac{\sigma_{abs}}{\omega^2}\, d\omega, \tag{1}$$

where ω denotes the photon energy. Because of the inverse energy weighting, α_D depends on the E1 strength at low energies.

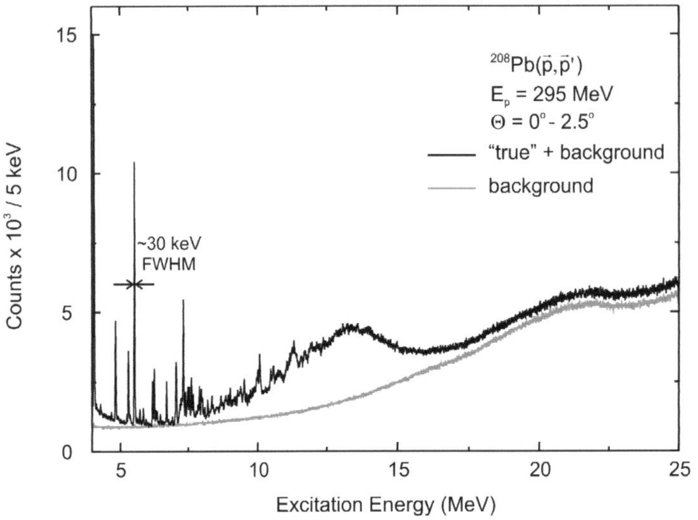

Figure 1. Spectrum of the ^{208}Pb($\vec{p},\vec{p}'$) reaction at $E = 295$ MeV and $\Theta = 0°$. The grey line indicates the experimental background.

The centroid of the PDR lies typically in the vicinity of the neutron emission threshold (S_n). Data on the PDR in very neutron-rich nuclei are still scarce [11, 12, 13]. Stable nuclei at different shell closures have been explored with the (γ, γ') reaction (Ref. [14] and refs. therein). While this technique provides important information on the fine structure of the PDR, it is essentially limited to excitation energies up to S_n, and unobserved branching ratios of the γ decay to excited states may require corrections for the total strength [15]. Measurements of decay neutrons are constrained to energies $E_x > S_n$ and uncertainties in the vicinity of S_n are large. This contribution discusses a novel approach viz. polarized proton scattering at angles close to and including 0°, to provide the *complete E1 response in nuclei* up to excitation energies well above the region of the GDR. At proton energies of 200 – 400 MeV the cross sections at small momentum transfers are dominated by isovector spinflip-M1 transitions (the analog of the Gamow-Teller mode) and by Coulomb excitation of non-spinflip E1 transitions [16, 17]. A separation of these two contributions, necessary for an extraction of the E1 response, is achieved by two independent methods: a multipole decomposition analysis of the angular distributions (MDA) and the measurement of polarization transfer (PT) observables.

2. Experiment and analysis

The experimental techniques are described in Ref. [18] and details of the ^{208}Pb measurement in Ref. [19]. A spectrum measured at 0° spectrometer angle is displayed in Fig. 1. Besides strong transition at low excitation energies, in the giant resonance region prominent excitation of the GDR is observed. Because of the excellent energy resolution $\Delta E = 25 - 30$ keV (FWHM) pronounced fine structure is visible, a phenomenon now established as a global feature of giant resonances [20, 21]. The grey line indicates the experimental background mainly due to small-angle scattering in the target It is determined from the focal-plane position distribution in non-dispersive direction achieved by operating the spectrometer optics in an underfocus mode [18].

For an extraction of the cross sections due to M1 and E1 excitations, two independent methods are used. The first one is based on the measurement of the polarization transfer coefficients D_{NN}, D_{SS} and D_{LL}, where N (normal), S (sidewards), and L (longitudinal) define a coordinate system in the frame of the moving proton. A complete set of these observables permits a unique identification of the spinflip character of a transition [22]. At 0°, $D_{NN} = D_{SS}$. This allows the definition of the total spin transfer Σ

$$\Sigma \equiv \frac{3\,(2D_{SS} + D_{LL})}{4}, \qquad (2)$$

which takes values $\Sigma = 1$ and 0 for spinflip and non-spinflip transitions, respectively. Figure 2 presents a decomposition of the total cross sections (top) into $\Delta S = 1$ (middle) and $\Delta S = 0$ (bottom) parts for the ^{208}Pb data. In the giant resonance region $\Sigma \approx 0$ is found as expected from the non-spinflip character of the GDR. in general, the spinflip part of the cross section is small except for the region $E_x \simeq 7 - 9$ MeV, where the spinflip M1 resonance was observed previously [23].

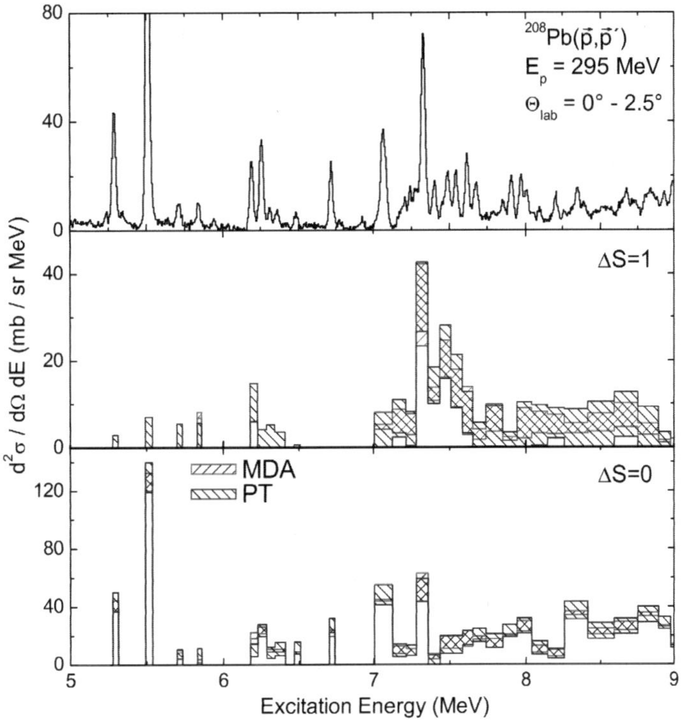

Figure 2. Decomposition of non-spinflip ($\Delta S = 0$) and spinflip ($\Delta S = 1$) cross section parts based on the MDA and PT, respectively, in the excitation energy region $5-9$ MeV. The hatched areas indicate the experimental uncertainties. Excellent agreement between the two completely independent methods is observed.

Alternatively, a decomposition of the cross sections into multipole contributions based on their different angular dependence can be performed. Predictions for the shapes of the angular distributions needed for such an analysis were obtained from calculations with the code DWBA07 [24] using microscopic quasiparticle-phonon model (QPM) wave functions [25] and the Love-Franey effective proton-nucleus interaction [26]. For E1 transitions, the interference of Coulomb and nuclear contributions was taken into account. Although data were available up to 10°, it was decided to limit the angle range to $\Theta \leq 4°$ in order to enhance the selectivity on lower multipoles of interest here at small momentum transfers. This allowed a restriction to transferred angular

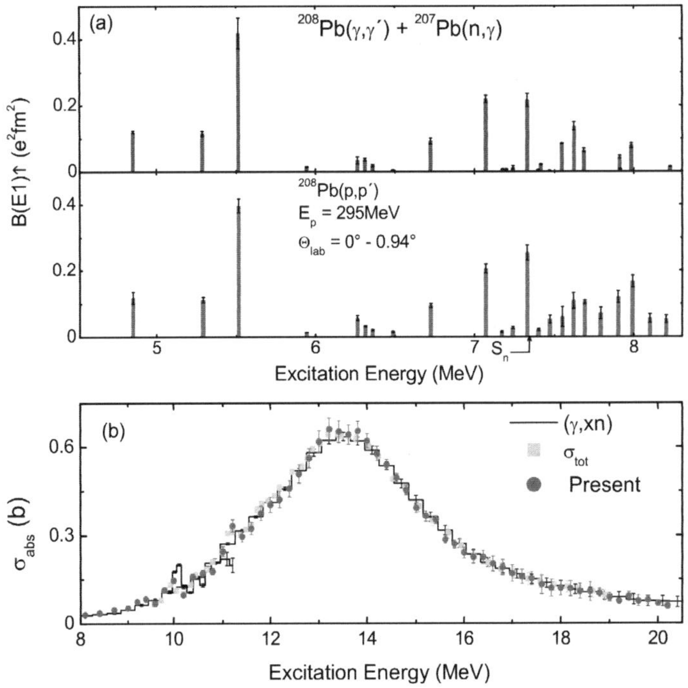

Figure 3. (a) B(E1) strengths in ^{208}Pb in the region $E_x \simeq 4.8 - 8.2$ MeV as deduced from the present work in comparison with (γ,γ') and (n,γ) experiments [25, 29, 30, 31]. (b) Photoabsorption cross sections in the GDR region from the present work compared to (γ, xn) [32] and total photoabsorption [33] measurements.

momenta $\Delta L = 0$ (M1) and $\Delta L = 1$ (E1), while higher multipoles were subsumed in a $\Delta L = 2$ contribution. In order to account for the model dependence of this assumption, the analysis was repeated replacing $\Delta L = 2$ by $\Delta L = 3$. The χ^2-weighted mean of the fits with all possible combinations of theoretical angular distributions provides the final result. In the GDR region the M1 contribution was zero within error bars and was replaced by a phenomenological background describing the data at high excitation energies.

A direct comparison of both methods is possible under the assumption that all contributions $\Delta L > 0$ in the multipole decomposition are of non-spinflip nature. Figure 2 demonstrates excellent agreement up to $E_x = 9$ MeV. At higher excitation energies the comparison is hampered by the contributions of quasifree scattering in the spectrum, which are known to contain a non-negligible spinflip part [27] and cannot be distinguished in the multipole analysis. In any case, both methods agree that $\Delta S = 1$ contributions are very small.

3. E1 strength distribution

It can be schown that under the experimental conditions at very small angles ($\Theta < 1°$) E1 transitions arise purely from Coulomb excitation. This allows an extraction of reduced transition strengths using semiclassical theory [28]. The resulting B(E1) strength up to 8.2 MeV is compared in Fig. 3(a) with an average over all available ^{208}Pb(γ,γ') and ^{207}Pb(γ,n) data (see [25, 29, 30, 31] and references therein). Excellent agreement is obtained up to S_n. The excess strength in the (p,p') data above threshold can be attributed to previously unknown partial neutron decay widths. Figure 3(b) provides a summary of photoabsorption cross section data on ^{208}Pb in the GDR region. Besides the present work, results using positron annihilation in flight [32] and tagged photons [33] are available. Again, very satisfactory agreement of all three measurements is observed.

4. Polarizability

As discussed above, an important quantity, which can be derived from the data, is the electric dipole polarizability. We find $\alpha_D = 18.9(13)$ fm$^3/e^2$ for the E1 strength up to 20 MeV. By taking an average of all available data including excitation energies up to 130 MeV [32, 33], a result with further reduced uncertainty $\alpha_D = 20.1(6)$ fm$^3/e^2$ is obtained. The covariance ellipsoid of the correlation between α_D and the neutron skin thickness r_{skin} in the approach of Ref. [8] is shown in Fig. 4. Only with the present precision for α_D (hatched band) one can constrain the neutron skin thickness to $r_{\text{skin}} = 0.156^{+0.025}_{-0.021}$ fm. The hitherto most precise determinations of this quantity for ^{208}Pb [34, 35] deduced from exotic atoms ($r_{\text{skin}} = 0.18 \pm 0.02$ fm) and hadron scattering ($r_{\text{skin}} = 0.211^{+0.054}_{-0.063}$ fm), respectively, are in excellent agreement with our result based on a totally independent method.

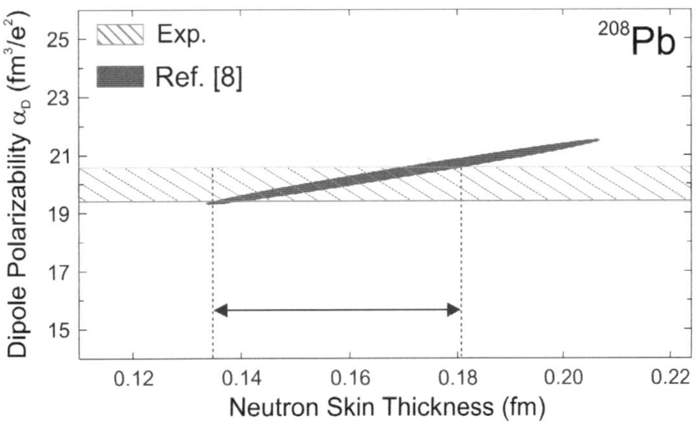

Figure 4. Extraction of the neutron skin in ^{208}Pb based on the correlation between r_{skin} and the dipole polarizability α_D established in Ref. [8]

Recent calculations of neutron matter and neutron star properties [36] in the framework of chiral effective field theory suggest $r_{\text{skin}} = 0.17 \pm 0.03$ fm. The predictions are sensitive to three-nucleon forces, which may be further constrained by the present results. Since the correlation between polarizability, neutron skin thickness and symmetry energy is model-dependent, viz. $r_{\text{skin}} \propto \alpha_D \cdot a_{\text{sym}}$ [37], a systematic study with a variety of EDFs as well as experimental tests in other nuclei would be important.

5. Conclusions and outlook

Polarized proton scattering at very forward angles allows to study, with high resolution, the complete electric dipole response of nuclei from low excitation energies up to the GDR. The E1 strength distribution deduced in a benchmark experiment on ^{208}Pb is in excellent agreement with available data. It provides, however, new information in the region around the neutron emission threshold where all previous experiments had limited accuracy. A precise value for the E1 polarizability can be extracted with important consequences for a determination of the neutron skin and the symmetry energy in neutron-rich nuclei. Although controversially discussed

[8], $r_{\rm skin}$ may independently be derived from a similar correlation with the PDR strength [9, 12], which is accurately determined by the present data as well.

Beyond these results, the experiment also provides information on the spin-M1 resonance in ^{208}Pb. Furthermore, the fine structure of the dipole modes contains information on level densities [21] and characteristic scales [20], giving insight into their dominant damping mechanisms. Systematic studies of the E1 and spin-M1 response in nuclei are currently underway with this new powerful technique.

Acknowledgments
The experiments have been performed in a TU Darmstadt/iThemba LABS/University Kyoto/University Kyushu/University Niigata/University Osaka/RCNP Osaka/University Tohoku/RIKEN Tokyo/CNS University Tokyo/Universidad de Valencia/University of the Witwatersrand collaboration. I am particularly grateful for the contributions of Y. Fujita, A.M. Heilmann, H. Matsubara, I. Poltoratska, V.Yu. Ponomarev, A. Richter and A. Tamii to the results presented. This work was supported by the Deutsche Forschungsgemeinschaft under contracts SFB 634, NE 679/2-2 and NE 679/3-1.

References

[1] Berman B L and Fultz S C 1975 *Rev. Mod. Phys.* **47** 713
[2] Paar N, Vretenar D, Khan E and Colò G 2007 *Rep. Prog. Phys.* **70** 691
[3] Piekarewicz J 2006 *Phys. Rev.* C **73** 044325
[4] Carbone A, Colò G, Bracco A, Cao L-G, Bortignon P F, Camera F and Wieland O 2010 *Phys. Rev.* C **81** 041301
[5] Horowitz C J and Piekarewicz J 2001 *Phys. Rev. Lett.* **86** 5647
[6] http://hallaweb.jlab.org/parity/prex
[7] Bender M, Heenen P-H and Reinhard P-G 2003 *Rev. Mod. Phys.* **75** 121
[8] Reinhard P-G and Nazarewicz W 2010 *Phys. Rev.* C **81**, 051303(R)
[9] Piekarewicz J 2011 *Phys. Rev.* C **83** 034319
[10] Bohigas O, Van Giai N and Vautherin D 1981 *Phys. Lett.* B **102** 105
[11] Adrich P *et al* 2005 *Phys. Rev. Lett.* **95** 132501
[12] Klimkiewicz A *et al* 2007 *Phys. Rev.* C **76** 051603(R)
[13] Wieland O *et al* 2009 *Phys. Rev. Lett.* **102** 092502
[14] Savran D *et al* 2008 *Phys. Rev. Lett.* **100** 232501
[15] Rusev G *et al.* 2008 *Phys. Rev.* C **77** 064321
[16] Frekers D *et al* 1990 *Phys. Lett.* B **244** 178
[17] Heyde K, von Neumann-Cosel P and Richter A 2010 *Rev. Mod. Phys.* **82** 2365
[18] Tamii A *et al* 2009 *Nucl. Instrum. Meth.* A **605** 3
[19] Tamii A *et al* 2011 arXiv:1104.5431
[20] Shevchenko A *et al* 2004 *Phys. Rev. Lett.* **93** 122501
[21] Kalmykov Y *et al* 2006 *Phys. Rev. Lett.* **96** 012502
[22] Suzuki T 2000 *Prog. Theor. Phys.* **321** 859
[23] Laszewski R M, Alarcon R, Dale D S and Hoblit S D 1988 *Phys. Rev. Lett.* **61** 1710
[24] Raynal J 2007 program DWBA07.
[25] Ryezayeva N *et al.* 2002 *Phys. Rev. Lett.* **89** 272502
[26] Franey M A and Love W G 1985 *Phys. Rev.* C **31** 488
[27] Baker F T *et al* 1997 *Phys. Rep.* **289** 235
[28] Bertulani C A and Baur G 1988 *Phys. Rep.* **163** 299
[29] Enders J *et al* 2003 *Nucl. Phys.* A **724** 243
[30] Shizuma T *et al* 2008 *Phys. Rev.* C **78** 061303(R)
[31] Schwengner R *et al* 2010 *Phys. Rev. C* **81** 054315
[32] Veyssiere A, Beil H, Bergère R, Carlos P and Lepretre A 1970 *Nucl. Phys.* A **159** 561
[33] Schelhaas K P *et al* 1988 *Nucl. Phys.* A **489** 189
[34] Friedman E and Gal A 2007 *Phys. Rep.* **452** 89
[35] Zenihiro J *et al* 2010 *Phys. Rev.* C **82** 044611
[36] Hebeler K, Lattimer J M, Pethick C J and Schwenk A 2010 *Phys. Rev. Lett.* **105** 161102
[37] Satuła W, Wyss R A and Rafalski M 2006 *Phys. Rev.* C **74** 011301(R)

Journal of Physics: Conference Series **337** (2012) 012005 doi:10.1088/1742-6596/337/1/012005

Nuclei in the "Island of Stability" of Superheavy Elements

Yuri Oganessian
Flerov Laboratory of Nuclear Reactions (Joint Institute for Nuclear Research)
141980 Dubna, Moscow region, Russia
oganessian@jinr.ru

Abstract. The observation of atomic numbers Z that are 40% larger than that of Bi, the heaviest stable element, is an impressive extension of nuclear survival. Although the super heavy nuclei (SHN) are at the limits of Coulomb stability, shell stabilization lowers the ground-state energy, creates a fission barrier, and thereby enables the SHE to exist. The fundamentals of the modern theory concerning the mass limits of nuclear matter have been experimentally verified.

We are going to consider how big a nucleus may be, what is the limit of atomic nuclei mass and how it is determined.

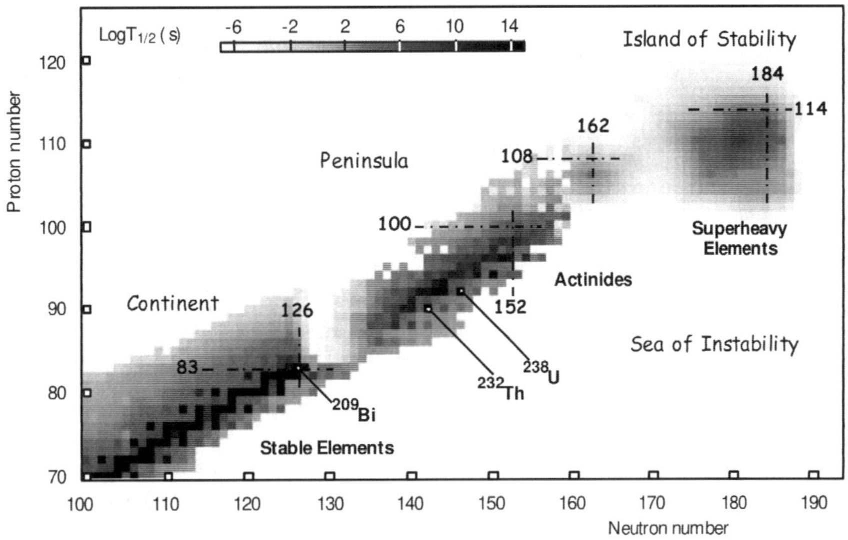

Figure. 1. Map of the nuclides in the region of heavy nuclei. Black squares - isotopes of stable elements. The half-lives of all artificial elements are denoted by different degrees of darkening (upper scale). New nuclear shells Z=108, N=162 and Z=114, N=184 are predicted by the macro-microscopic nuclear theory.

The nuclei shown in Fig.1 having different numbers of protons and neutrons, have different binding energies and, therefore, different decay probabilities (half-lives).

In the region of transuranium elements, a sharp decrease in the stability of nuclei with increasing atomic number is observed. Indeed adding 10 protons and 4 neutrons to ^{238}U lead to a decrease in the period of α-decay of ^{252}No by a factor of 10^{16} with respect to uranium. An even stronger decrease is expected for another type of decay - the spontaneous fission. Here the drop is 23 orders of magnitude! A simple extrapolation into the region of heavier nuclei shows that the limits of existence of elements will be determined by spontaneous fission. When the half-life is reduced to 10^{-14} s, the nuclei will decay long before they obtain its atomic structure. According to a simple extrapolation, this situation arises even for nuclei with atomic numbers 106-108.

Such a limit of atomic nuclei was predicted by N. Bohr and J. A. Wheeler [1], long before the first artificial elements were obtained. The fission of heavy nuclei was simulated by the process of deformation of a drop of charged liquid (the nuclear liquid drop model). According to Bohr and Wheeler, in the case of uranium a potential barrier about 6 MeV high prevents the fission of the nucleus (Fig.2 upper panel). If one adds an energy exceeding 6 MeV to the uranium nucleus, it will split into two fragments within about 10^{-19} sec. However, fission can occur spontaneously, by tunneling through the fission barrier. The half-life of uranium relative to fission is 10^{16} years. Increasing the number of protons in the nucleus (advancing to the transuranium elements) reduces the fission barrier and greatly increases the probability of spontaneous fission. When the fission barrier drops, the nucleus will split into two fragments for about 10^{-19} s. According to the calculations of the liquid drop model, this situation occurs immediately after Z > 100. It should be noted that when, using nuclear reactors, artificial elements up to Fermium (Z = 100) were synthesized, it seemed that the theoretical predictions have been fully confirmed.

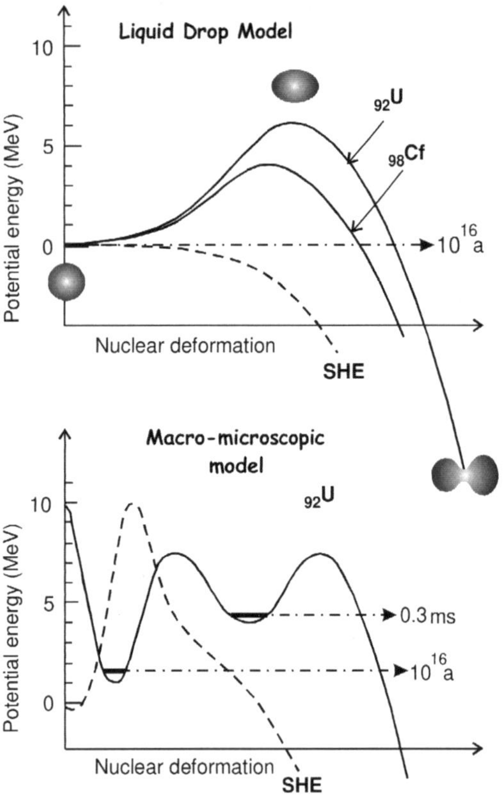

Figure 2 Nuclear potential energy at different deformations. Upper graph show the calculation in the liquid drop model; lower graph – with shell corrections. Dotted curve - the fission barrier of the nucleus with Z= 108, N = 162.

However, in 1962, in our Laboratory in Dubna, an unexpected effect was observed, which caused great doubts in the analogy of nuclear fission and a liquid drop. The isotope of ^{242}Am, obtained in a nuclear reaction, experienced spontaneous fission with two, very different half-lives: 10^{14}a, and 0.014 s [2]. Subsequently, a similar phenomenon was observed in more than 31 nuclei with Z = 92-97 [3]. Definitely, fission in these nuclei occurs from two states - the ground and isomeric states (Fig.2 lower panel). But the shape isomerism of nuclei is completely incompatible with describing the fission process as a split of a liquid drop. In subsequent experiments to measure the heights of fission barriers of nuclei with Z = 80-100, significant differences from the predictions of the liquid drop model of fission were also found.

It is noteworthy that by this time it was already known that nuclei in their ground states differ in form and energy and at certain «magic» numbers of protons and/or neutrons the nuclear binding energy considerably increases. However, it was always assumed that reaching strong deformation on the way to splitting into two fragments, the structure of the nucleus disappears and the fission process can still be considered in the liquid drop approximation. In fact, as was shown in [4], nuclear structure does not disappear with increasing deformation, but is modified and continues to play a significant role in the process of nuclear fission. Calculation of the potential energy of the nucleus at large deformations, taking into account structural effects, explained the experimentally observed high fission barrier of the «doubly magic» nucleus ^{208}Pb (Z = 82, N = 126), as well as the shape isomerism of uranium and transuranic nuclei, as a consequence of the double-humped fission barrier of actinide nuclei.

It was natural to extend this approach to the region of nuclei with Z ≥ 104, where, according to the droplet model, nuclei may not exist. A formal calculation led to unexpected results. It turned out that in the deformed nucleus with mass 270 (Z = 108 and N = 162) a fission barrier occurs, which leads to an increase in its half-life to seconds (instead of 10^{-19} sec). But even higher stability appears in heavier (superheavy) nuclei with proton number Z = 114 and a large number of neutrons N = 184. Here the strong effect of new nuclear shells works, similarly to the shells Z = 82 and N = 126 in the «doubly magic» spherical nucleus ^{208}Pb. Some new calculations indicate that on the map of nuclei-heavyweights form a large enough area, called the «Island of Stability» of superheavy elements (see Fig.1). In the region where the liquid-drop model predicted that the nuclei should decay within 10^{-19} sec, now the half-lives of «long-lived» nuclei at the peak of the island of stability are expected to reach thousands and even millions of years! (see, for example, review [5]).

Hypothetical superheavy elements radically change our ideas about the limits of the material world, and they indicate very significant expansion of the boundaries of existence of atomic nuclei and chemical elements. Here we have a problem how to check this extravagant theoretical hypothesis experimentally.

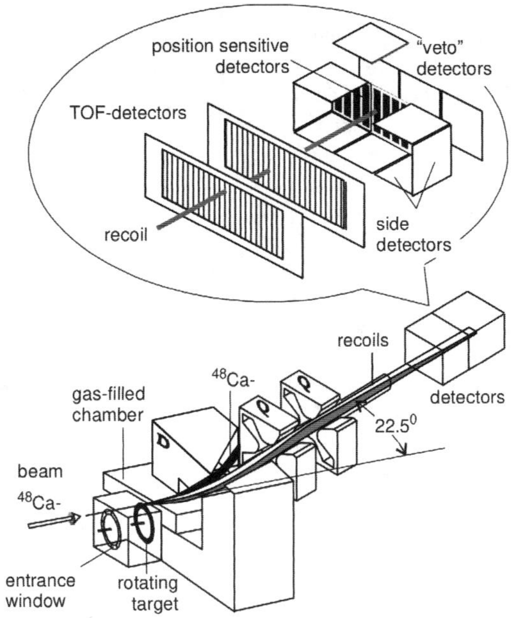

Figure 3. A schematic view of gas-filled recoil-separator, which was used in the experiments on the synthesis of superheavy elements.

The known method of obtaining transuranic elements in successive neutron capture reactions with powerful reactors and even in nuclear explosions, confined to the observation of nuclei only up to ^{257}Fm. In reactions induced by light ions (Z_i = 6-10) it was advance to elements with Z = 106 only. Entry into the area of Z>106 occurred after the discovery the so-called «cold fusion» reactions when the target nuclei - ^{208}Pb or ^{209}Bi and massive projectiles with Z_i = 22-30 [6] were used. In the reactions of «cold fusion» the formed compound nuclei are slightly heated, they are cooled by the emission of one neutron, and therefore have a relatively high survival rate. This was their main advantage, which allowed in the years 1978-1998 to synthesize 6 new elements up to 112 [7].

But unfortunately, as will follow later, this method cannot be used for the synthesis of nuclei located on the «island of stability» of superheavy elements. The problem is that with increasing the mass and charge of the bombarding ion the Coulomb repulsion grows, and thus greatly decreases the probability of formation of compound nuclei with increased charge numbers. There is another fundamental limitation. In cold fusion reactions, due to lack of neutrons, the produced heavy nuclei are outside the predicted «island of stability». Even for the most heavy isotopes of elements 110, 111, and 112 obtained in this way, the half-lives are estimated to be only tens or hundreds of microseconds. It seemed that using more asymmetric masses of the interacting nuclei (the target nucleus is heavier than the mass of projectile) might reduce the strength of the Coulomb repulsion and thus raise the estimated probability of fusion by 4-5 orders of magnitude. But the compound nuclei formed in these reactions are very hot; the losses during their cooling are even greater. Unsuccessful attempts to synthesize superheavy elements in reactions of «hot fusion», undertaken in various laboratories in the years 1977-1985, led to the pessimistic view that the elements heavier than 112 might be quite stable, but it would be practically impossible to produce them.

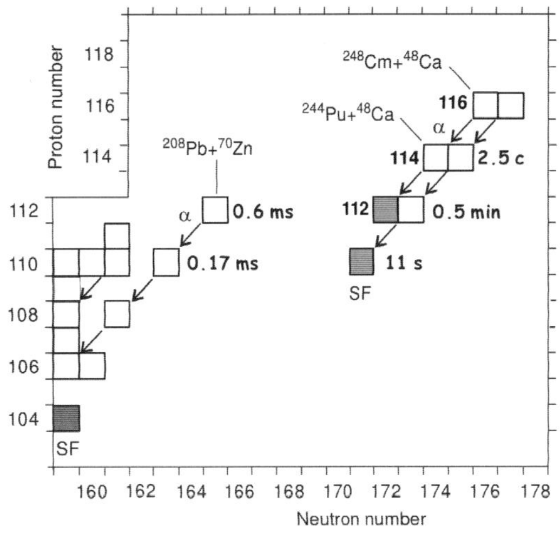

Figure 4. Decay chains of isotopes with Z=114 and 116, synthesized in the fusion reactions with calcium-48 projectiles

In this conclusion, in our opinion, there was, however, an inherent contradiction. If the theoretical predictions were correct, the nuclei near the closed shells should have high fission barriers. In fact, it is this circumstance that causes the enormous stability of superheavy nuclei against spontaneous fission. But the high fission barrier must also prevent *the fission of the heated compound nucleus* and enhance its survival in the process of neutron emission during its cooling. In other words, the nuclei lying in the «island of stability» should fission less strongly and should survive easier. Moreover, as we approach the boundaries of the „Island of Stability», the rise of the formation cross sections of superheavy nuclei would be a direct indication of the inclusion of the effect of the new nuclear shells with Z = 114 (probably 120-122) and N = 184. Based on these assumptions, we chose a different reaction of synthesis of superheavy elements.

Neutron-rich nuclei of SHE can in principle be synthesized if the heaviest actinides: ^{244}Pu, ^{243}Am, ^{248}Cm, ^{249}Bk and ^{251}Cf, obtained in powerful nuclear reactors, were used as targets. As the projectile, we have chosen the rare isotope of calcium with mass number 48. As a result of two years work using an ECR-source and the U-400 accelerator, a stable beam of ^{48}Ca ions was obtained with high intensity: 5.10^{12} -1.10^{13} pps. This allowed conducting long-term experiments (6000 hours/year) over the past 10 years. The registration of rare events of formation and decay of superheavy atoms was performed by the gas-filled recoil separator (Fig.3). In this setup, the nuclei emitted from the target, enter a hydrogen atmosphere (pressure about 1 Torre) and are separated by mass and ionic charge in a magnetic field and focus on the focal plane detector, located at a distance of 4 m from the target. The reaction products pass the distance between the target and the detector for 1 microsecond. The separator has a transmission for evaporation residues about 35-40%; the by-products of the reaction are suppressed by factor of 10^4-10^7 [8]. The focal plane detector registers the energy, position and time when the recoil nucleus stops in it. Then similar information follows concerning the charged particles (alpha particles or fission fragments) emitted in the decay of the implanted nucleus.

We recall that the theoretical predictions of the nuclei within the «Island of Stability» are stable against spontaneous fission. In the process of successive alpha-decay, the daughter nuclei will move away from the closed shells and will approach the boundaries of the island. The likelihood of spontaneous fission increases rapidly and eventually the chain will be interrupted by spontaneous fission. In other words, we expect that the decay of a superheavy nucleus will have the form of a chain (radioactive family), consisting of successive alpha transitions, relatively long in time, terminated by spontaneous fission. Our focal plane detectors have efficiency of 87% for the registration of alpha particles, almost 100% - for a single spontaneous fission fragment, 42% - for both fragments.

The location of the stopped recoil nucleus at the working surface of the detector is determined with an accuracy of better than 0. 2%.

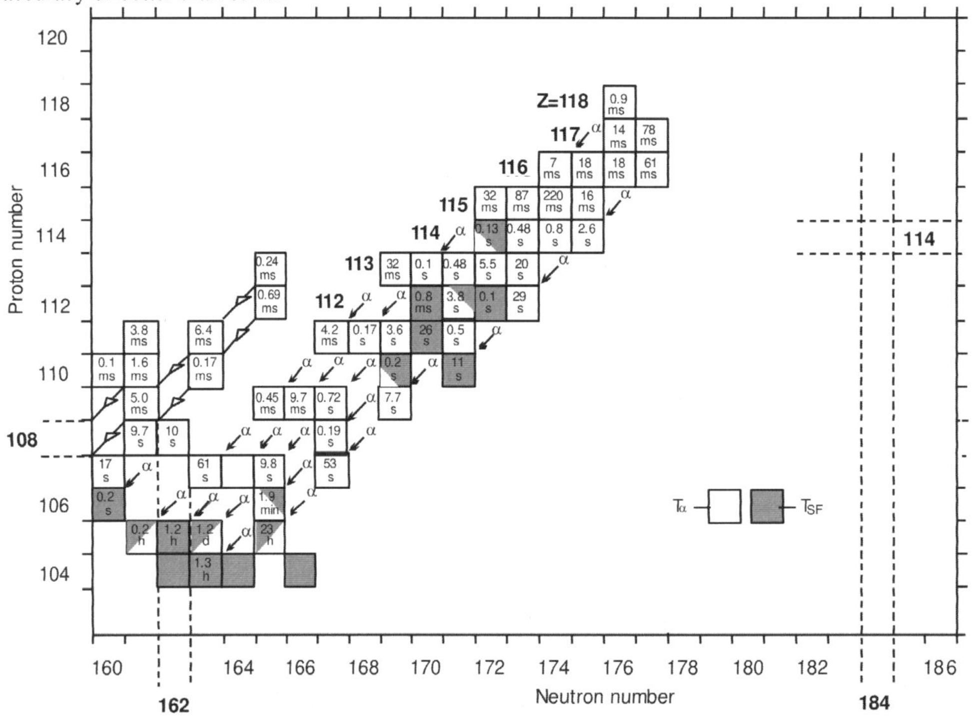

Figure 5. Nuclei with Z=112-118 and products of their radioactive decay - the isotopes of elements 111, obtained in the fusion reactions of the Actinides and ^{48}Ca. The experimental values of half-lives are shown in the squares.

The location of the stopped recoil nucleus at the working surface of the detector is determined with an accuracy of better than 0. 2%.

The coordinates of the follow-up signals can be used to establish the genetic link between the recoil nucleus and its decay. The probability of random coincidences of signals, simulating a correlated decay, can be significantly reduced by switching off the accelerator beam between the registration of the recoil nucleus and its subsequent first alpha-particle decay. For example, the probability of random coincidences in the registration of one event of formation and decay of the nucleus of the isotope of element 117 by the chain $^{293}117\text{-}\alpha{\rightarrow}^{289}115\text{-}\alpha{\rightarrow}^{285}113\text{-}\alpha{\rightarrow}^{281}\text{Rg(SF)}$, is 3.10^{-11} [9].

Just with this setup were synthesized isotopes of superheavy elements with atomic numbers 114 and 116 for the first time [10]. They showed a huge increase in nuclear stability by increasing the number of neutrons in the heavy nucleus (Fig.4). Adding to the previously known nuclei $^{273}110$ and $^{275}112$ 8 neutrons led to an increase in their half-lives by almost 100 000 times!

Our data on the decay characteristics of the isotopes of elements 112, 114 and 116 have been reproduced in 2007-2010 in a number of independent experiments [11-15]. Later, the heaviest element with atomic number 118 [16] was synthesized, followed by the synthesis of elements with odd atomic numbers: 113, 115 and 117 [17,9] in the reactions ^{237}Np, ^{243}Am and ^{249}Bk + ^{48}Ca. The data on the decay characteristics of the isotopes of 6 new superheavy elements, obtained up till now, together with those of the products of their radioactive decay (altogether 47 new nuclei) are shown in Fig.5. They can now be compared with the predictions of the microscopic theory [18].

Such a comparison is shown in Fig.6 for even-Z nuclei, for which the calculation can be done with the greatest accuracy. On the basis of the entire set of experimental data on the properties of superheavy nuclei produced in reactions induced by ^{48}Ca some general conclusions can be made.

1. The largest number of nuclei undergo alpha decay. The experimental energy of the alpha decay of superheavy nuclei differs from the calculated values by no more than 5-7%. This is a good agreement with the calculations of the nuclear macro-microscopic model. The periods of α-decay of the new nuclides were, on the average, up to 10 times higher than predicted by the theory.
2. Spontaneous fission was observed only in 14 isotopes (see Fig.5). They belong to two groups: the isotopes of elements 110-114 with N = 169-172 and the isotopes of lighter elements with Z = 104-106 and N = 160-165. Spontaneously- fissioning nuclei of the first group are located between the closed neutron shells N = 162 and N = 184; they are away from them by 7-12 neutrons. The smallest value of $T_{SF} \approx 10^{-3}$ sec is observed for the even-even nucleus of element 112 with N = 172. As we move away from this area towards an excess or deficit of neutrons, the partial spontaneous fission half-life increases (top graph of Fig.6). For nuclei with Z ≥ 114 and N ≥ 174 the spontaneous fission was not observed. In the second group, the nuclei are located around the deformed shell N = 162; they

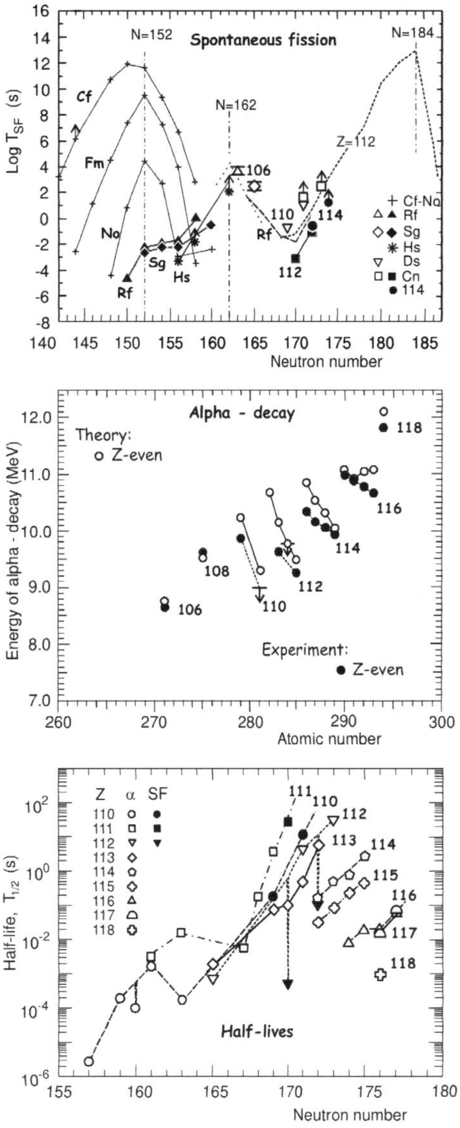

Figure 6. The experimental values of the spontaneous fission half-lives and alpha-decay energies are compared with theoretical expectations. The bottom graph shows the half-lives of superheavy nuclei vs N.

have a relatively high stability. The even-odd nucleus of element 104 with N = 163 has a $T_{SF} \sim 5.10^3$ s. The long-lived odd-Z nuclei, the isotopes of element 105 ($T_{1/2} \approx 10^5$ sec), most probably undergo electron capture. Then the experimentally observed spontaneous fission refers to the decay of even-even isotopes of element 104 with N = 164 and 166. For these nuclei, the so-called fast mode of symmetric fission is expected due to the effect of the nuclear shells Z = 50 and N = 82 in the fission fragments.

Conclusion

Increasing the atomic number of the last stable element - ^{209}Bi ($T_{1/2}$ = 1. 5.10^{19} years) by 40%, we are witnessing an impressive picture of the vitality of atomic nuclei. In the region of extreme Coulomb forces, due to the effect of new shells, the nuclear binding energy in the ground state increases, a fission barrier arises, and strongly increased stability of the heaviest nuclei leads to the existence of a large region of superheavy elements.
The fundamental predictions of the modern theories concerning the mass limits of the atomic nuclei have received their first experimental confirmation.

The experiments were carried out at the U-400 heavy ion cyclotron of the Flerov Laboratory of Nuclear Reactions (FLNR, JINR) in collaboration with LLNL (Livermore, USA), ORNL (Oak Ridge, USA), Research Institute of Atomic Reactors (Dimitrovgrad, Russia) and Vanderbilt University (Nashville, USA).

References

[1] Bohr N and Wheller J A 1939 *Phys. Rev.* **56** 426
[2] Polikanov S M *et al.* 1962 *Soviet Phys.-JETF* **15** 1016
[3] Bjørnholm S and Linn J 1980 *Rev. Mod. Phys.* **52** 725
[4] Brack M *et al.* 1972 *Rev. Mod. Phys.* **44** 320
[5] Sobiczewski A and Pomorski K. 2007 *Prog. Part. Nucl. Phys.* **58**
[6] Oganessian Yu Ts 1975 *Lecture Notes in Physics,* Edt. J. Echlers et al Springer **33** 222
[7] Hofmann S and Munzenberg G 2000 *Rev. Mod. Phys.* **72** 733
[8] Oganessian Yu Ts *et al.* 2000 in Proceedings of the Fourth International Conference on DynamicalAspects of Nuclear Fission (World Scientific, Singapore 334
[9] Oganessian Yu Ts *et al.* 2010 *Phys. Rev.Lett.* **104** 142502
[10] Oganessian Yu Ts *et al.* 2000 *Phys. Rev.*C **62** 041604(R); C**63** 011301(R), (2004) C**70** 064609
[11] Hofmann S *et al.* 2007 *Eur. Phys. J.* A **32** 251
[12] Eichler R *et al.* *Nature* 2007 447 **72**
[13] Stavsetra L *et al.* 2009 *Phys. Rev. Lett.* **103** 132502
[14] Dullmann Ch E *et al.* 2010 *Phys. Rev. Lett.* **104** 252701
[15] Hofmann S *et al.* 2011 *Eur. Phys. J.* to be published
[16] Oganessian Yu Ts *et al.* 2006 *Phys. Rev.* C **74** 044602
[17] Oganessian Yu Ts *et al.* 2004 *Phys. Rev.* C **69** 021601(R) and 2005 C **72** 034611
[18] Muntian I Patyk Z and Sobiczewski 2003 A *Phys. At. Nucl.* **66** 1015

Journal of Physics: Conference Series **337** (2012) 012006 doi:10.1088/1742-6596/337/1/012006

The Origin of Oxygen in the Universe - A new approach to an Old Question

K. E. Rehm

Physics Division, Argonne National Laboratory, 9700 South Cass Av., Argonne, IL, 60439, USA

E-mail: rehm@anl.gov

Abstract. Carbon and oxygen are not only important elements for the existence of life on Earth, but they also play an important role in the evolution of stars towards the end of their life cycle. The formation of ^{12}C through the so-called triple-α reaction is quite well understood. The next step, the formation of ^{16}O through the α capture reaction $^{12}C(\alpha,\gamma)^{16}O$ on the other hand, still has an experimental uncertainty of $\sim$ 30%. Direct measurements of the $^{12}C(\alpha,\gamma)^{16}O$ reaction by detecting either the outgoing γ radiation in a high acceptance Ge-detector array or the residual ^{16}O nuclei in a mass spectrometer do not allow for order-of-magnitude improvements. In this contribution, the possibility of using superheated bubble detectors for a measurement of the time-inverse $^{16}O(\gamma,\alpha)^{12}C$ reaction is being discussed. The first results of a 'proof-of-principle' experiment of the $^{19}F(\gamma,\alpha)^{15}N$ reaction are also being presented.

1. Introduction

The isotopes ^{16}O and ^{12}C are, after ^{1}H and ^{4}He, the third and fourth most abundant nuclei in the visible universe. Most of the carbon and oxygen which we observe today is produced by helium burning in red giant stars. These two elements are not only crucial for all living organisms, but their relative abundances, which are determined by the competition between the triple-α and the $^{12}C(\alpha,\gamma)^{16}O$ reactions, is also an important parameter for the evolution of a massive star at the end of its lifetime during the carbon-, neon-, and oxygen-burning phases [1, 2, 3]. The cross section for the triple-α process is experimentally quite well determined [4]. Our knowledge of the $^{12}C(\alpha,\gamma)^{16}O$ reaction under typical helium burning conditions ($T_9 \sim 0.2$ or $E_{c.m.} \sim 300$ keV), however, is still limited by its small cross section and by the crucial role played by two sub-threshold states in ^{16}O [5].

The history of experiments studying the $^{12}C(\alpha,\gamma)^{16}O$ reaction goes back more than four decades [6, 7]. The magnitude of this cross section, however, is still a hotly debated issue, both experimentally and theoretically, and many recent publications can be found in the literature [8, 9, 10, 11, 12, 13, 14, 15, 16].

Contrary to the triple α reaction which is dominated by the contribution from a resonant state (the so-called Hoyle state) the $^{12}C(\alpha,\gamma)^{16}O$ reaction proceeds mainly through two radiative capture modes to the ^{16}O ground state. One is E1 capture with contributions from the 1^- state at E_x=9.585 MeV (E_r=2.418 MeV) and the sub-threshold 1^- state at E_x=7.117 MeV (E_r=-45 keV). The other is E2 capture which is dominated by the contributions from direct capture and the sub-threshold 2^+ state at E_x=6.917 MeV (E_r=-245 keV). At energies corresponding to the Gamow window for red giant stars ($E_{c.m.}$ ~300 keV), the cross sections are on the order of 10^{-17} b. For that reason, all direct measurements so far were done at higher energies above $E_{c.m.}$ = 890 keV [4]. These data are then extrapolated into the energy region of astrophysical interest using R-matrix theory. Since the higher-energy data are not very sensitive to the contributions from sub-threshold resonances, the published S-factors in the past 30 years range from 1 to 288 keVb for $S_{E1}(300)$ and 7 to 120 keVb for $S_{E2}(300)$ [17].

Two different techniques have been used in the past for direct measurements of the $^{12}C(\alpha,\gamma)^{16}O$ reaction:
• A highly enriched ^{12}C target is bombarded with high intensity α beams and the outgoing γ's are detected in large-acceptance Ge-detector arrays. The measurements performed at the lowest c.m.energies (E_{cm}=890 keV) have used this technique. The main drawback of this technique is the low efficiency that can be achieved for the detection of 8 MeV γ rays.
• A higher detection efficiency can be obtained by studying the reaction in inverse kinematics, i.e. bombarding a ^4He gas target with a ^{12}C beam and detecting the final ^{16}O nucleus in a mass spectrometer. The difficulty with this technique comes from a small ^{16}O beam contamination, which currently restricts the lowest energies to about $E_{cm} \sim$ 1.8 MeV.

Good agreement of the measured cross sections is observed in the energy region studied by both techniques. In addition to these direct techniques, several indirect methods have been investigated in order to improve the reliability of the extrapolations into the astrophysically important energy region. These techniques include data from elastic α scattering on ^{12}C [18, 19], α-transfer reactions to ^{16}O [20], and β-delayed α decay of ^{16}N [22, 23, 21, 11, 24].

In order to discuss the limits and possible improvements for future direct measurements of the $^{12}C(\alpha,\gamma)^{16}O$ reaction, we introduce the concept of a 'modified' luminosity $L' = I \cdot N \cdot \epsilon$, where I is the beam intensity (particles/sec), N the areal target density (target nuclei/cm^2) and ϵ the detection efficiency. When multiplied by the cross section σ, the product $L' \cdot \sigma$ gives the count rate expected in an actual experiment. Sometimes only an order-of-magnitude estimate of the luminosity can be extracted from the information provided in the various publications. The 'modified' luminosities for some of the experiments performed during the last 20 years are shown in Fig. 1. They are typically of the order of 10^{31} $sec^{-1}cm^{-2}$ which, when multiplied by a cross section of 1 pb, translate into an expected count rate of 1 count/day. Improvements by factors of 2-5 are possible by choosing higher beam currents or better Ge detector arrays (perhaps located deep underground). The choice of higher efficiency BaF_2 detectors has also been investigated. However, based on the small cross sections and the relatively small luminosities improvements by orders of magnitude do not seem possible with these techniques in the foreseeable future.

2. Studies of time-inverse Reaction $^{16}O(\gamma,\alpha)^{12}C$ Reaction

For experiments studying astrophysically important proton-capture (p,γ) reactions it has been pointed out [25] that measurements of the time-inverse (γ,p) reactions can yield higher count rates when compared to standard (p,γ) experiments. This is caused by two effects: (a) the possibility to use thicker targets and (b) by a more advantageous phase space factor. This has made Coulomb dissociation studies a competitor to direct (p,γ) experiments. Similar advantages also exist for measurements of (α,γ) and (γ,α) reactions as was pointed out in Ref. [26, 27]. Using the principle of detailed balance, the cross sections for (α,γ) and (γ,α) reactions are (for

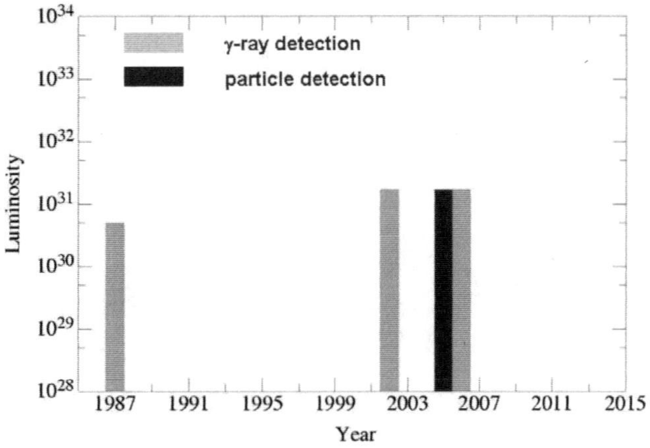

Figure 1. 'Modified' luminosities obtained in some of the direct measurements of the $^{12}C(\alpha,\gamma)^{16}O$ reaction performed during the last 20 years.

the system $^{12}C(\alpha,\gamma)^{16}O$) related by the equation

$$\frac{\sigma(\gamma,\alpha)}{\sigma(\alpha,\gamma)} = \frac{2\mu c^2 E_{cm}(\alpha,\gamma)}{2E_\gamma^2} \tag{1}$$

where μ and E_{cm} are the reduced mass and the c.m. energy in the $^{12}C + \alpha$ system, and E_γ is the corresponding γ energy in the $^{16}O + \gamma$ system (typically 8 MeV for $E_{cm} \sim$1 MeV). For the lowest energies studied so far the enhancement factor $\frac{\sigma(\gamma,\alpha)}{\sigma(\alpha,\gamma)}$ is about 50.

Another advantage for studying the time inverse $^{16}O(\gamma,\alpha)$ reaction originates from the large range of an 8 MeV γ which has a mean absorption length of about 40 cm in water. Thus, in (γ,α) experiments, target thicknesses of tens of g/cm^2 can be used, which results in yields that are higher by about 6 orders of magnitude when compared to direct $^{12}C(\alpha,\gamma)^{16}O$ experiments where target thicknesses of typically 10 μg/cm^2 are used.

The implementation for measuring the time-inverse $^{16}O(\gamma,\alpha)^{12}C$ reaction requires a tunable γ beam with the highest possible intensities and a detector system that can identify the occurrence of a breakup of ^{16}O into ^{12}C and α while being insensitive to γ's at the level of at least 10^{-11}. In Ref. [26] an optical TPC has been discussed. While it can identify the occurrence of a $^{16}O(\gamma,\alpha)^{12}C$ reaction it operates with gases at pressures of $\sim$100 mbar and thus does not make full use of the large range of 8 MeV γ's. In the following we will discuss the advantageous of using a bubble chamber filled with superheated liquids.

3. The use of a bubble chamber for nuclear astrophysics experiments

Bubble chambers were one of the principle detectors for tracking energetic particles in high energy physics experiments. Invented more than 50 years ago, their operational principles are well understood [28]. These type of detectors have recently seen a renaissance in Dark Matter searches and three groups are utilizing superheated liquids in a search of nuclear recoils induced by possible WIMP interactions [29, 30, 31]. The liquids used in these experiments are fluorocarbon compounds which are superheated close to room temperatures. However, water, (the

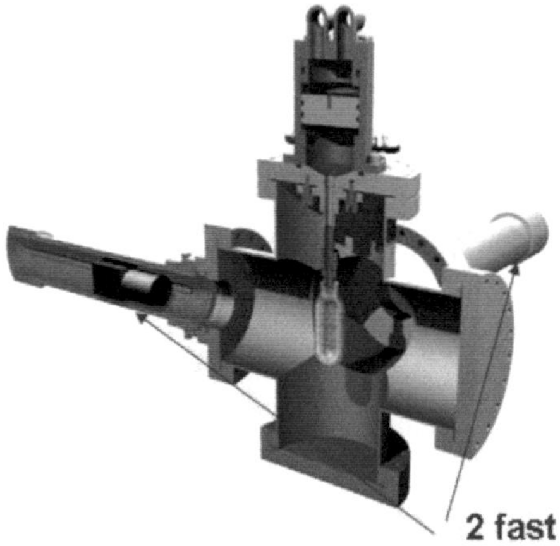

Figure 2. Schematic of the experimental setup used in the first experiment at the HIγS facility as a 'proof-of-principle' study of the ^{19}F$(\gamma,\alpha)^{15}$N reaction. See text for details.

most convenient oxygen-containing liquid), requires temperatures of ~250 C and pressures in excess of 75 atm.

Contrary to bubble chambers used in high energy physics which are superheated only for a short period of time, Dark Matter searches require detectors that are superheated for longer periods (e.g. days). This can be achieved by choosing a glass vessel as a container for the superheated liquid. When located deep-underground, background rates of ~counts/day have been achieved. In order to avoid the high-pressure/temperature conditions required for a water-filled bubble chamber we have therefore, for a first 'proof-of-principle' experiment, studied the time-inverse of the reaction ^{15}N$(\alpha,\gamma)^{19}$F by bombarding fluoro-carbon liquids with γ beams from the HIγS facility at Duke University.

Fig. 2 shows a schematic of the experimental setup. A γ beam from the HIγS facility collimated to a diameter of 1 cm hits a cylindrical glass vessel (diameter ~3 cm) filled with C_4F_{10}. This fluoro-carbon compound is superheated at T~30 C and p~3 atm. If an incident γ causes a ^{19}F$(\gamma,\alpha)^{15}$N reaction the outgoing α-^{15}N pair is stopped in the liquid and acts as a nucleation center producing a bubble. The glass vessel is continously monitored by two CCD cameras taking pictures every 10 ms. If a bubble is detected, the pressure in the vessel is immediately increased which brings the liquid out of the superheated state, thus leading to a quenching of the bubble. After a time delay of typically 1 s, the pressure is again reduced bringing the detector back to the superheated state for the detection of another event. Fig. 3 shows a sequence of photographs taken 10 ms apart showing the growth and the quenching of a bubble produced via the ^{19}F$(\gamma,\alpha)^{15}$N reaction in a superheated C_4F_{10} liquid. The energy of the γ beam was 5.3 MeV.

The normalization of the incident γ flux is achieved by using a high efficiency Ge detector located after the bubble chamber. With this setup, an excitation function of the ^{15}N$(\alpha,\gamma)^{19}$F reaction has been measured in the energy range E_γ=5-6 MeV. The uncertainty of the incident

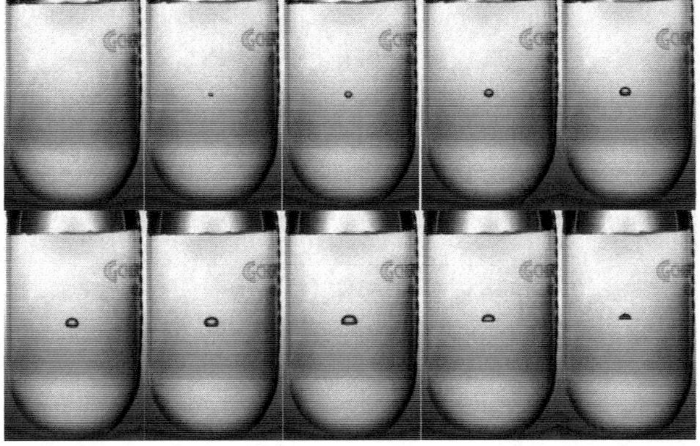

Figure 3. Sequence of photographs taken with one of the CCD cameras at time intervals of 10 ms. When the occurrence of a bubble is detected, the pressure in the vessel is increased, leading to the quenching of the bubble. This pressure increase takes typically 50 msec.

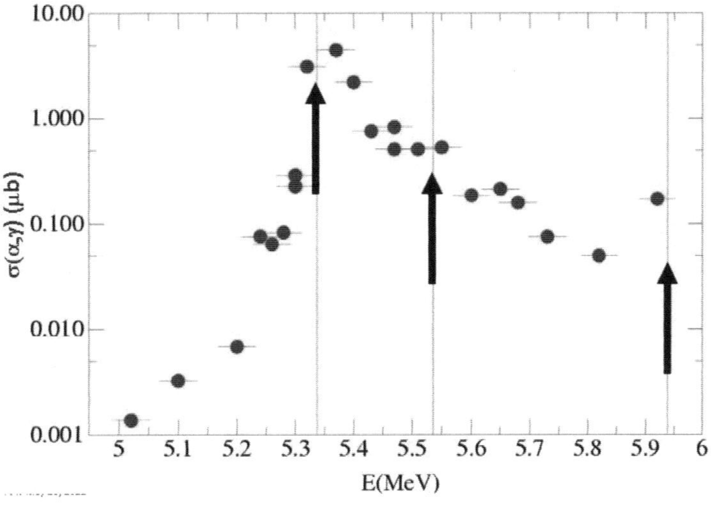

Figure 4. Preliminary cross sections obtained in a 'proof-of-principle' experiment for the $^{15}N(\alpha,\gamma)^{19}F$ reaction. The arrows indicate the location of known resonances in ^{19}F. See text for details.

γ-beam is determined by the energy spread of the electron beam in the storage ring and by the amount of collimation and was about 1% for this experiment. A preliminary excitation function for the $^{15}N(\alpha,\gamma)^{19}F$ reactions is shown in Fig. 4. The cross section scale covers almost four orders of magnitude ranging from 1 nb to 10 μb. The cross sections in Fig. 4 have not been unfolded with the experimental resolution. In order to measure the yield at the maximum of the excitation function, the intensity of the incident γ beam had to be reduced to 5×10^3 γ's/sec. For a 20 cm long vessel filled with water and assuming a beam intensity of 5x10^7 γ's/sec, the calculated count rate for the $^{12}C(\alpha,\gamma)^{16}O$ reaction at a cross section of 1 pb is about $\sim$ 100 events/day. This estimate includes the increase in cross section for the $^{16}O(\gamma,\alpha)^{12}C$ reaction discussed in Section 2. This rate is about a factor of 100 larger than what can be obtained with standard techniques.

The next step involves a similar test using a water-filled bubble chamber. A setup that allows measurements at 250° C and p=75 atm is presently being assembled at Argonne. Contrary to the ^{19}F(γ,α) reaction which used the mono-isotopic ^{19}F, reactions on the target contaminants ^{17,18}O can contribute to the ^{16}O(γ,α) reaction. For this reason water, which is highly enriched in ^{16}O, has been purchased. Similar to the study of the ^{19}F(γ,α) reaction we will start the measurement of the ^{16}O(γ,α) reaction in the vicinity of the known 1^- resonance at E_γ= 9.585 MeV, where several cross sections measurements using other techniques have been performed in the past and will extend the (γ,α) measurements to the lowest energies possible. Reactions induced by cosmic ray neutrons can be eliminated by using an acoustical readout system as done e.g. in Ref. [32].

Acknowledgments

The bubble chamber experiment is a collaboration between Argonne National Laboratory, Fermi National Accelerator Laboratory, University of Illinois, Duke University and the University of North Carolina. I want to thank my collaborators B. Digiovine, C. Ugalde, Roy Holt, D. Henderson(ANL), A. Robinson, A. Sonnenschein (FNAL), N. Sturcchio (UoI), A. Raut, G. Rusev, A. Tochenv (Duke) and A. Champagne (UNC) for their contributions to this experiment. This work was supported by the US Department of Energy, Office of Nuclear Physics under contract No. DE-AC02-06CH11357 and by the NSF JINA Grant. No. PHY0822648.

[1] T. A. Weaver and S. E. Woosley, Phys. Rep. **227**, 65 (1993).
[2] S. E. Woosley and T. A. Weaver, Astrophys. Journal, Suppl., **101**, 181 (1995).
[3] G. Wallerstein *et al.* Rev. Mod. Phys. **69**, 995 (1997)
[4] L. R. Buchmann and C. A. Barnes, Nucl. Phys. **777**, 254 (2006).
[5] (www.tunl.duke.edu/nucldata).
[6] H. M. Loebenstein, D. W. Mingay, H. Winkler, and C. S. Zaidins, Nucl. Phys. A**91**, 481 (1967).
[7] R. J. Jaszcak, J. H. Gibbons, and R. L Macklin, Phys. Rev. C**2**, 63 (1970).
[8] J. W. Hammer *et al.*, Nucl. Phys. A **752**, 514c (2005).
[9] M. Assuncao *et al.*, Phys. Rev. C **73**, 055801 (2006).
[10] A. Belhout *et al.*, Nucl. Phys. A**793**, 178 (2007).
[11] R. H. France III, E. L. Wilds, J. E. McDonald, and M. Gai, Phys. Rev. C **75**, 065802 (2007).
[12] P. Descouvemont, J. Phys. G**35**, 014006 (2008).
[13] C. Matei, C. R. Brune, and T. N. Massey, Phys. Rev. C**78**, 065801 (2008).
[14] F. Strieder, J. Phys. G**35**, 014009 (2008).
[15] P. Tischhauser *et al.*, Phys. Rev. C **79**, 055803 (2009).
[16] H. Makii *et al.*, Phys. Rev. C **80**, 065802 (2009).
[17] R. Kunz *et al.*, Astrophys. J. **567**, 643 (2002).
[18] R. Plaga *et al.*, Nucl. Phys. A**465**, 291 (1987).
[19] P. Tischhauser *et al.*, Phys. Rev. Lett. **88**, 072501 (2002).
[20] C. R. Brune *et al.*, Phys. Rev. Lett. **83**, 4025 (1999).
[21] R. Azuma *et al.*, Phys. Rev. C **50**, 1194 (1994).
[22] L. R. Buchmann *et al.*, Phys. Rev. Lett. **70**, 726 (1993).
[23] Z. Zao *et al.*, Phys. Rev. Lett. **70**, 2066 (1993).
[24] X. D. Tang *et al.*, Phys. Rev. C **81**, 045809 (2010).
[25] G. Baur and H. Rebel, J. Phys.(London) **G20**, 1 (1994).
[26] M. Gai, 21st Winter Workshop on Nuclear Dynamics, arXiv:nucl-ex/o504003v1.
[27] Y. Xu *et al.*, Nucl. Instr. Meth. A **581**, 866 (2007).
[28] F. Seitz, Phys. Fluids **1**,2 (1958).
[29] S. Archambault *et al.*, Phys. Lett. B**682**, 185 (2009).
[30] E. Behnke *et al.*, Phys. Rev. Lett. **106**, 021303 (2011).
[31] M. Felizardo *et al.*, Phys. Rev. Lett. **105**, 211301 (2010).
[32] S. Archambault *et al.*, New J. Phys. **13**, 043006 (2011).

Medium Effects in Reactions with Rare Isotopes

C.A. Bertulani[(1)] **and M. Karakoç**[(1,2)]

[(1)]Department of Physics and Astronomy, Texas A&M University-Commerce, Commerce, TX 75429, USA

[(2)] Department of Physics, Akdeniz University, 07058, Antalya, Turkey

E-mail: carlos_bertulani@tamu-commerce.edu, mesutkarakoc@gmail.com

Abstract. We discuss medium effects in knockout reactions with rare isotopes of weakly-bound nuclei at intermediate energies. We show that the poorly known corrections may lead to sizable modifications of knockout cross sections and momentum dsitributions.

1. Introduction

Most practical studies of the medium corrections of nucleon-nucleon scattering are done by considering the effective two-nucleon interaction in infinite nuclear matter, or G-matrix, as a solution of the Bethe-Goldstone equation

$$\langle \mathbf{k}|G(\mathbf{P},\rho_1,\rho_2)|\mathbf{k}_0\rangle = \langle \mathbf{k}|v_{NN}|\mathbf{k}_0\rangle - \int \frac{d^3k'}{(2\pi)^3}\frac{\langle \mathbf{k}|v_{NN}|\mathbf{k}'\rangle Q(\mathbf{k}',\mathbf{P},\rho_1,\rho_2)\langle \mathbf{k}'|G(\mathbf{P},\rho_1,\rho_2)|\mathbf{k}_0\rangle}{E(\mathbf{P},\mathbf{k}')-E_0-i\epsilon} \tag{1}$$

with $\mathbf{k}_0$, $\mathbf{k}$, and $\mathbf{k}'$ the initial, final, and intermediate relative momenta of the NN pair, $\mathbf{k} = (\mathbf{k}_1 - \mathbf{k}_2)/2$ and $\mathbf{P} = (\mathbf{k}_1 + \mathbf{k}_2)/2$. If energy and momentum is conserved in the binary collision, $\mathbf{P}$ is conserved in magnitude and direction, and the magnitude of $\mathbf{k}$ is conserved. v_{NN} is the nucleon-nucleon potential. E is the energy of the two-nucleon system, and E_0 is the same quantity on-shell. Thus $E(\mathbf{P},\mathbf{k}) = e(\mathbf{P}+\mathbf{k}) + e(\mathbf{P}-\mathbf{k})$, with e the single-particle energy in nuclear matter. It is also implicit in Eq. 1 that the final momenta $\mathbf{k}$ of the NN-pair also lie outside the range of occupied states.

We have performed a study of medium effects in knockout reactions which include different methods to treat medium effects [1]. To test the influence of the medium effects in nucleon knockout reactions, we consider the removal of the $l = 0$ halo neutron of ^{15}C, bound by 1.218 MeV, and the $l = 0$ neutron knockout from ^{34}Ar, bound by 17.06 MeV. The reaction studied is ^{9}Be(^{15}C,^{14}C$_{gs}$). The total cross sections as a function of the bombarding energy are shown in figures 1. The solid curve is obtained with the use of free nucleon-nucleon cross sections. The dashed curve includes the geometrical effects of Pauli blocking. The dashed-dotted curve is the result using the Brueckner theory, and the dotted curve is the phenomenological parametrization of the free cross section.

In figure 2 we plot the longitudinal momentum distributions for the reaction ^{9}Be(^{11}Be,^{10}Be), at 250 MeV/nucleon [1]. The dashed curve is the cross section calculated using the NN cross section from the Brueckner theory and the solid curve is obtained the free cross section. One sees that the momentum distributions are reduced by 10%, about the same as the total cross

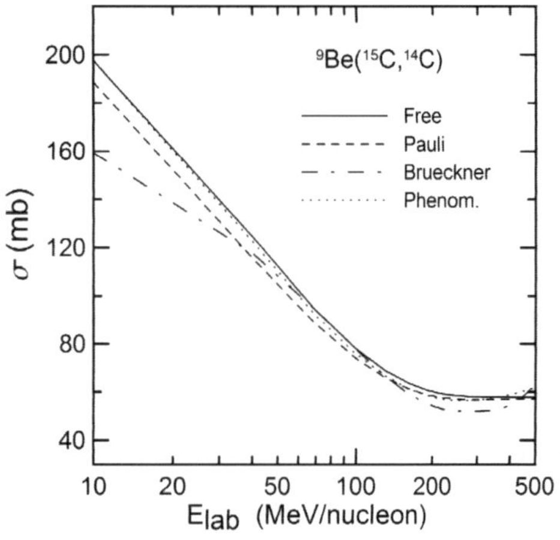

Figure 1. Total knockout cross sections for removing the $l = 0$ halo neutron of ^{15}C, bound by 1.218 MeV, in the reaction ^{9}Be(^{15}C,^{14}C$_{gs}$). The solid curve is obtained with the use of free nucleon-nucleon cross sections. The dashed curve includes the geometrical effects of Pauli blocking. The dashed-dotted curve is the result using the Brueckner theory, and the dotted curve is a phenomenological parametrization.

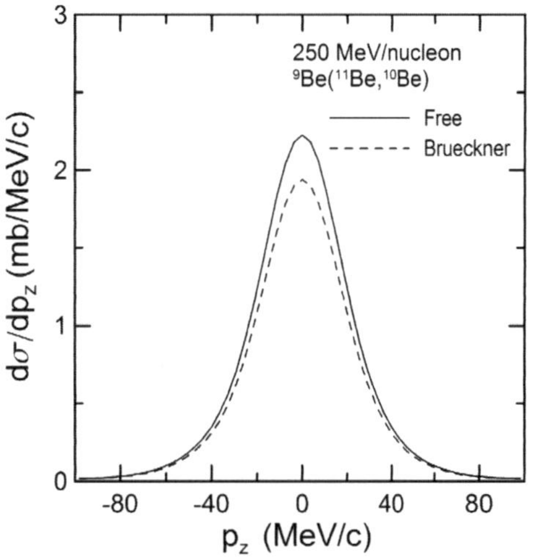

Figure 2. Longitudinal momentum distribution for the residue in the ^{9}Be(^{11}Be,^{10}Be), reaction at 250 MeV/nucleon. The dashed curve is the cross section calculated using the NN cross section from the Brueckner theory and the solid curve is obtained the free cross section.

sections, but the shape remains basically unaltered. If one rescales the dashed curve to match the solid one, the differences in the width are not visible.

Table 1. Spectroscopic factors (C^2S).

Projectile	Residue	*nlj*	C^2S NuShell@MSU	C^2S Ref[4]	Sn(Sp)(MeV)
^{14}O	^{13}O	$0p_{3/2}$	4.97	3.7	23.176
	^{13}N	$0p_{1/2}$	1.83	1.8	4.628
^{36}Ca	^{35}Ca	$1s_{1/2}$	2.16		19.113
	^{35}K	$0d_{3/2}$	3.54		2.559

Based on these results we have developed a new version of the code MOMDIS [2] in order to treat medium effects an Coulomb recoil properly. In the next sections we report a few preliminary results, for reactions which are being planned at RIKEN.

Table 2. Bound state potential parameters.

Projectile	Residue	V_0(MeV)	R_0(fm)	a(fm)
^{14}O	^{13}O	81.46	2.50	0.6
	^{13}N	43.10	3.00	0.8
^{36}Ca	^{35}Ca	63.19	3.93	0.6
	^{35}K	47.63	3.93	0.6

2. Proton and neutron knockout of numerous systems

In this work, ^{9}Be(^{14}O,^{13}O), ^{9}Be(^{14}O,^{13}N) at 50 and 300 AMeV, and ^{9}Be(^{36}Ca,^{35}Ca), ^{9}Be(^{36}Ca,^{35}K) at 70.5 and 300 AMeV cases have been analyzed. The ground state spins for nuclei ^{13}O, ^{13}N, ^{35}Ca and ^{35}K are respectively $3/2^-$, $1/2^-$, $1/2^+$ and $3/2^+$ are taken from Ref. [3], except for ^{35}Ca. The ground state spin for ^{35}Ca is not defined. Nevertheless, we have assumed in our calculations it has $J^\pi = 1/2^+$ ground state spin.

Table 3. Total cross-sections.

Projectile	Residue	E(MeV)	σ_{MOMDIS}(mb)	σ(mb)	σ_{exp}(mb)
^{14}O	^{13}O	57	9.61	51.60	13.4±1.4
		300	16.04	86.10	
	^{13}N	57	29.96	59.21	67±6
		300	31.50	62.26	
^{36}Ca	^{35}Ca	70.5	6.72	15.39	
		300	8.07	18.48	
	^{35}K	70.5	8.96	33.62	
		300	10.56	39.63	

A single nucleon removal cross section σ from the $J^\pi = 0^+$ the ground state (g.s.) of projectile to the g.s. of the knockout residue is given by

$$\sigma = \left(\frac{A}{A-1}\right)^n C^2 S \sigma_{MOMDIS} \tag{2}$$

where σ_{MOMDIS} is calculated using a modified version of the MOMDIS code [2]. The A-dependent term is a center-of-mass correction to the shell-model spectroscopic factors (C^2S) where n is major oscillator shell number ($n = 1$ for sp shell nucleus ^{14}O and $n = 2$ for sd shell nucleus ^{36}Ca) [4, 5].

The spectroscopic factors which are used in the present work are calculated with the shell model code NuShell@MSU [6]. We have calculated C^2S considering the WBT interaction for ^{14}O [7] and USDA interaction for ^{36}Ca [8]. Our calculation is in agreement with Ref. [4] for ^{14}O proton removal case, but the neutron removal spectroscopic factor C^2S for the present calculation is 34% bigger than in Ref. [4].

The bound state wave functions are calculated with Woods-Saxon potentials and the parameters are tabulated in Table 5. The total cross sections obtained with these parameters are shown in Table 6.

Table 4. Separation energies from [3].

Projectile (g.s.)	Residue (g.s.)	nlj	Sp(MeV)
^{9}C, $3/2^-$	^{8}B	$0p_{3/2}$	1.301
^{17}F, $5/2^+$	^{16}O	$0d_{5/2}$	0.600
^{27}P, $1/2^+$	^{26}Si	$1s_{1/2}$	0.860

Table 5. Bound state potential parameters.

Projectile	Residue	V_0(MeV)	R_0(fm)	a(fm)
^{9}C	^{8}B	47.69	2.50	0.6
^{17}F	^{16}O	71.25	2.80	0.6
^{27}P	^{26}Si	19.76	3.30	0.5

Table 6. Nuclear densities.

Nucleus	Model	a(fm)	α(fm)
^{8}B	LDM		
^{16}O	Gaussian[9]	1.833	1.544
^{26}Si	LDM		
^{12}C	Gaussian[9]	1.73	1.38

The ^{14}O proton removal result is in agreement with data from Ref.[4], but for neutron removal case there is a huge difference, probably due to the very big value of $C^2S = 4.97$. This is probably the reason why ^{14}O cases have different bound state potential parameters, which we have searched for best matching to the data of Ref. [4].

References

[1] C.A. Bertulani and C. De Conti, Phys. Rev. C 81, 064603 (2010).
[2] C. Bertulani and A. Gade, Comp. Phys. Comm. 175, 372 (2006).
[3] National Nuclear Data Center (nndc), Brookhaven National Laboratory, available at http://www.nndc.bnl.gov/nudat2/.
[4] C. Louchart, A. Obertelli, A. Boudard, and F. Flavigny, Phys. Rev. C 83, 011601 (2011).
[5] A. Gade, P. Adrich, D. Bazin, M. D. Bowen, B. A. Brown, C. M. Campbell, J. M. Cook, T. Glasmacher, P. G. Hansen, K. Hosier, et al., Phys. Rev. C 77, 044306 (2008).
[6] B. A. Brown and W. Rae (2007), private communication.
[7] E. K. Warburton and B. A. Brown, Phys. Rev. C 46, 923 (1992).
[8] B. A. Brown and B. H. Wildenthal, Annu. Rev. Nucl. Part. Sci. 38, 29 (1988).
[9] H. De Vries, C. W. De Jager, and C. De Vries, Atomic Data and Nuclear Data Tables 36, 495 (1987).

Journal of Physics: Conference Series **337** (2012) 012008 doi:10.1088/1742-6596/337/1/012008

Resonance Scattering to Study Exotic Nuclei at the Limits of Stability

V Z Goldberg[1], B T Roeder[1], G V Rogachev[2], G G Chubarian[1], E D Johnson[2], C Fu[3], A A Alharbi[1], M L Avila[2], A Banu[1], M McCleskey[1], J P Mitchell[2], E Simmons[1], G Tabacaru[1], L Trache[1], R E Tribble[1]

[1]Cyclotron Institute, Texas A&M University, College Station, TX 77843-3366, USA
[2]Department of Physics, Florida State University, Tallahassee, FL 32306-4350, USA
[3]Indiana University, Bloomington, IN 47408, USA

Email: goldberg@comp.tamu.edu

Abstract. We report on the study of a new nuclide, ^{14}F, and on the idea of an experiment to solve the structure mystery of nuclei in the region around ^{9}He.

1. Introduction

Examining nuclear matter under extreme conditions makes the most demanding test of our understanding of nuclear structure. A perfect opportunity is provided by the study of exotic nuclei which are far from the valley of stability. This is because the addition of a single nucleon can change the properties of a light nucleus, and because theoretical calculations, such as *ab-initio* calculations, can only be made for light nuclei at present. ^{14}F and ^{9}He, two exotic nuclei beyond the limits of stability, provide good systems to test the parameters of the nucleon-nucleon (NN) interactions used in these theoretical calculations. In the present work we consider application of the resonance scattering induced by rare isotope beams to obtain experimental data on both borders of nuclear stability.

2. ^{14}F

Fig. 1 illustrates the typical experimental setup for the Thick Target Inverse Kinematics (TTIK) method [1]. In this method, a beam of heavy ions is stopped in the target material and the light recoil product of the elastic scattering reactions (protons in our case) comes out of the target due to much smaller specific energy loss and is detected.

At the Texas A&M University Cyclotron Institute, a 31 MeV/u ^{13}O beam was produced after MARS separation [2] for the ^{14}F experiment. The beam was then slowed down, close to the entrance to the scattering chamber, to minimize the loss of intensity. Stacks of 100 μm polypropylene (C_3H_6) foils (total thickness 1.5 mm) were used to degrade the energy of the beam of ^{13}O from 31 MeV/u to ≈ 11 MeV/u. A scintillating foil was placed before the entrance to the scattering chamber. The light signal from the particles passing through the foil was detected by a pair of photomultiplier tubes (PMTs). The signals from the PMTs were used to monitor the beam, to obtain a "start" signal for time-of-flight (TOF) measurements, and to identify the main beam contaminant, ^{10}C. Our experimental setup was similar to that used in [3,4]. The scattering chamber was filled with methane gas (CH_4) which was used as a safe substitute for hydrogen. The CH_4 was separated from the high vacuum of MARS by a 3 μm Havar foil. However, a few modifications made the target system more "active" than before. A windowless ionization chamber (ICE) was placed in the scattering chamber close to the entrance window (see Fig. 1) to measure the specific energy loss of incoming ions and discriminate protons related with the interaction of the beam with the degrader, which resulted in the destruction of ^{13}O. A pair of quadrant-silicon detector telescopes (QSDs) (the same as in [4]) was also mounted inside smaller windowless ionization chambers 515 mm from the entrance window. Each QSD consisted of four square detectors (12.5 × 12.5 ×1 mm^3) and was followed by a similar veto detector to eliminate high-energy particles that passed through the first QSD. The role of the small ionization chambers was to allow *ΔE–E* analysis for the light ions that are not stopped in the gas and can reach the Si detectors. Further details about the experiment setup and data analysis are given in Ref. [5].

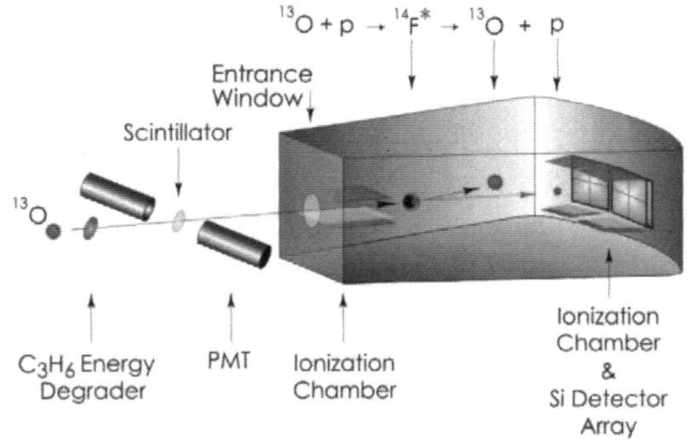

Figure 1. The setup for the ^{14}F experiment. The "gray box" is the scattering chamber. See explanation in the text and Ref. [5].

Fig. 2 presents the excitation functions for ^{13}O +p elastic scattering. A code of the complete R-matrix analysis was used [6]. The results of the analysis are summarized in Table 1. As seen in Table 1, ^{14}F is unstable by 1.56 MeV relative to proton decay, which corresponds to an atomic mass excess of 31 960 ± 50 keV for ^{14}F using the mass tables in [7]. According to the recent *ab-initio* calculations [8], it is expected be unstable by ~3 MeV. While new calculations are needed to specify the necessary corrections to the theoretical approaches, part of the disagreement between the predictions and the present result should be related with the Thomas–Ehrman shift of levels in mirror nuclei. This shift down toward greater stability in proton rich nuclei is the largest (and therefore famous) for *s*-states. The shift is strongly dependent upon the single particle structure of the state. As reported in Table 1, the ground state in ^{14}F has nearly pure shell model structure, and thus the effect of the Thomas–Ehrman shift on the ^{14}F binding energy should be larger than normal.

It is easy to note looking at Fig. 6 that the shell model calculations produce a much more compressed level scheme than the *ab-initio* calculations. The latter are in better agreement with the experimental data. We suppose that this indicates that the residual interactions should be modified in the shell model for a better description of these exotic nuclei.

Table 1. Levels in ^{14}F.

E_R (MeV)[a]	E_{ex} (MeV)[b]	J^π	Γ	Γ/Γ_{sp}
1.56±0.04	0.00	2-	910±100	0.85
2.1±0.17	0.54	1-	~1000	0.6
3.05±0.06	1.49	3-	210±40	0.55
4.35±0.10	2.79	4-	550±100	0.5

[a] Energy above ^{13}O+p threshold.
[b] Excitation Energy in ^{14}F.

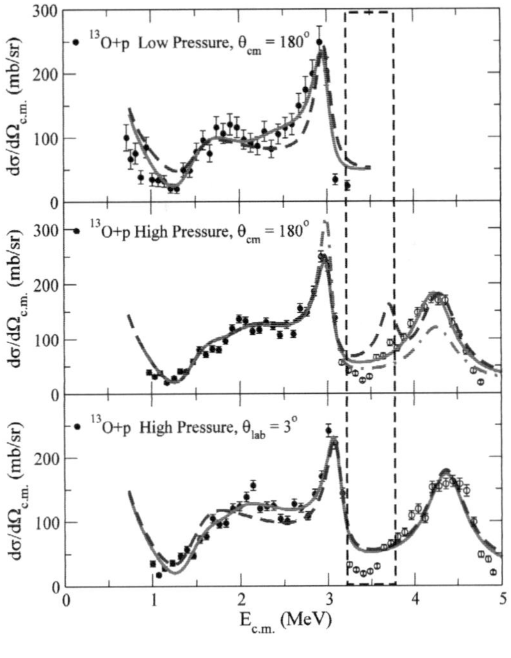

Figure 2. Excitation functions for ^{13}O+p elastic scattering are given in comparison with R-Matrix calculations. The solid line (red) is the best fit calculation following the ^{14}F level scheme as given in Table 1. The dashed box shows the region where the data are distorted because the QSD detector was not fully depleted. **Top Panel.** The dashed line (blue) is a fit with 1^- as the ground state (instead of 2^-). **Middle Panel.** The dashed line (blue) is a calculation with a second hypothetical 2^- state at high energy. The dashed-dot line (green) is a calculation with a 4^- state at 3 MeV (instead of 3^-) and a 3^- state at 4.35 MeV (instead of 4^-). **Bottom Panel.** The dashed line (blue) is the fit without the 1^- first excited state. See text for further discussion.

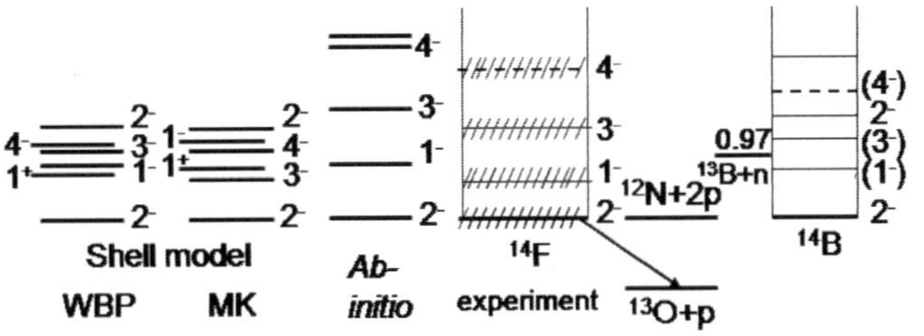

Figure 3. ^{14}F level scheme from this work compared with shell-model calculations, ab-initio calculations [8] and the ^{14}B level scheme. See Ref. [5] for further explanation about these calculations.

3. A mystery in the structure of ^{9}He

The structure of ^{9}He, with its two protons and seven neutrons, could be expected to be simple: two protons fill the *s*-shell, while six neutrons fill the $p_{3/2}$ sub shell and the extra neutron should be in the $p_{1/2}$ shell. Since the time of its first observation in the ^{9}Be(π^+,π^-) reaction [9], a ($1/2^-$) state, unstable by 1.1 MeV to neutron emission, was considered to be the ground state of ^{9}He. The authors compared the excitation energies of the observed levels with theoretical predictions of Ref. [10] and noted that these predictions, based on calculations for nuclei on the line of stability, "appear to be remarkably good for ^{9}He"[9]. Then several high resolution measurements with heavy ion mass transfer reactions [11, 12] brought more exact data. A state of ^{9}He at 1.27 ± 0.10 MeV above the ^{8}He + n threshold with $\Gamma = 0.10 \pm 0.06$ MeV was found. It is remarkable that the width appeared to be more than ten times smaller than what could be expected from "naïve" shell model considerations. The narrow width of the $½^-$ state is direct evidence for its complicated, non shell-model structure, and could be a sign of unusual nuclear structure beyond the neutron dripline.

Then, a surprising result was obtained from neutron-fragment velocity difference measurements in the two-proton knockout reaction ^{9}Be(^{11}Be, ^{8}He + n)X [13]. The MSU group [13] found that the ground-state of ^{9}He was a virtual s-wave state within 0.2 MeV of the ^{8}He + n threshold. Evidently, this result is also in contradiction to the conventional shell structure, and can again be considered as a manifestation of unusual structure near the dripline. Additionally, this result is confusing when one considers the ground state in ^{10}He [14]. New predictions for the ground state of ^{10}He on the basis of new results for ^{9}He gave a much more tightly bound ^{10}He. Finally, the recent availability of a more

Journal of Physics: Conference Series **337** (2012) 012008 doi:10.1088/1742-6596/337/1/012008

intense, rare beam of ^{8}He made the long waited ^{2}H(^{8}He,p)^{9}He experiment feasible [15]. The energy resolution and the counting statistics were poor in [15] due to evident experimental difficulties. Still, the authors claimed that their data exclude existence of the narrow ½$^-$ resonances which have been previously observed [11, 12]. Simultaneously, Ref. [15] supported finding an *s*-state near the threshold for the ^{9}He decay into ^{8}He+n on the basis of the interference pattern between broad l=0 and l=1 structures, but it was noted that "the energy resolution and the quality of measured angular distributions are not sufficient to draw solid conclusions about the exact properties of the *s*-wave contributions".

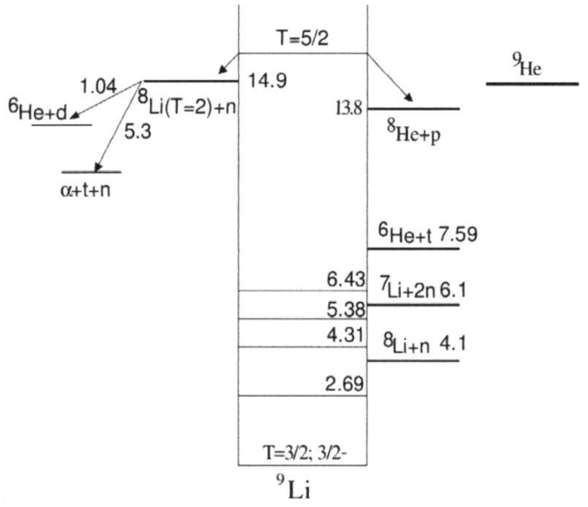

Figure 4. Level scheme of ^{9}Li showing various possible decay channels for excited states.

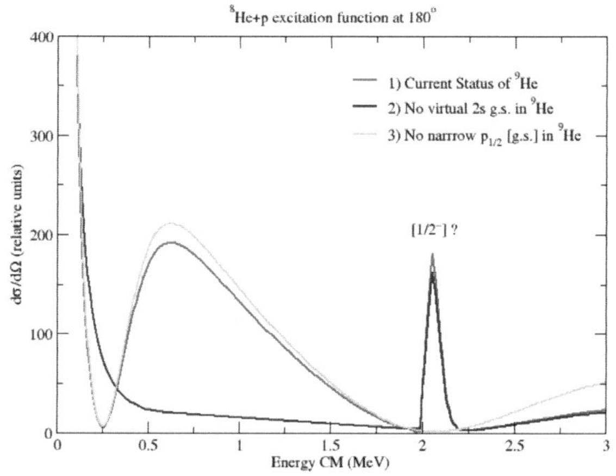

Figure 5. The effect of a possible 2s resonance (analog of the virtual ground state in ^{9}He) on the excitation function of the ^{8}He+p elastic scattering. Also shown is the expectation for a narrow width $p_{1/2}$ first excited state.

Probably the very confusing situation around ^{9,10}He should also be combined with the similar situation for the heavy hydrogen isotope, ^{7}H. The possibility of observing ^{7}H as a sharp resonance was first stimulated by the unusual stability of ^{8}He [16]. The discovery of ^{7}H as a resonance in the ^{12}C(^{8}He,^{13}N)^{7}H reaction was claimed in Ref [17]. However, several more recent attempts to observe ^{7}H in the ^{2}H(^{8}He,^{3}He)^{7}H reaction with an expected higher yield were not successful [18].

Thus, there remain many problems to be resolved in the structure of ^{9}He, and all evident nuclear reactions involving direct measurement of ^{9}He have been tried. However, resonance reactions again can be useful, but this time the study should be related with population of the ^{9}He analog states in ^{9}Li.

In the ^{8}He+p resonance elastic scattering, excitations of ^{9}Li levels (see Fig.4) with T=3/2 or T=5/2 are allowed in accordance with isobaric spin conservation. Strong resonance population of (unknown) T=3/2 resonances at high excitation energy in ^{9}Li is not expected due to the small ratio of the proton width of the initial channel to the total width of the hypothetical T=3/2 resonances. However, only two decay channels are allowed by T conservation for the lowest states with T=5/2. These are the initial channel ^{8}He+p, and its isobaric conjugate, ^{8}Li(T=2)+n (Fig.4). The analog of the "sharp" ½$^-$ state in ^{9}He with T=5/2 [11,12] should be present at an excitation energy of 16.1 MeV in ^{9}Li [19]. As can be seen in Fig. 4, more than 2 MeV of energy interval below the ½$^-$ (T=5/2) state in ^{9}Li are open for the investigation in ^{8}He+p resonance scattering as a result of the Coulomb interaction and the *p-n* mass difference. Also the Coulomb interaction turns the virtual ½$^+$ ground state in ^{9}He [13] into a quasi-stationary state in ^{9}Li.

The first (and only) study of the ^{8}He+p resonance scattering found evidence for rather broad p and d states with T=5/2 at excitation energies above 17 MeV in ^{9}Li [19]. These findings appeared to be in agreement with the recent investigation of ^{9}He spectrum [7]. However, the most interesting region corresponding to the controversial lowest states in ^{9}He could not be investigated in [19]. The on-line beam of ^{8}He [19] separated at FLNR was of high energy and had a large energy spread. The poor parameters of the beam resulted in poor resolution (~400-500 keV) in the region of the possible $\frac{1}{2}^{-}$ sharp state. At present, there are reaccelerated low energy ^{8}He beams at SPIRAL (France) and TRIUMF (Canada), and it is possible to make TTIK measurements at the ^{8}He initial energy of ~4 MeV/A and energy spread of <0.5%. Under these conditions, the energy resolution in the excitation region of ~16 MeV in ^{9}Li will be better than 50 keV, and one should easily observe a resonance peak with 100 keV width. It is well know that the elastic scattering cross sections are highly sensitive to the scattering phase. Therefore the parameters of the *s*-wave scattering near the threshold can be fixed. Fig. 5 illustrates the effect of the 2s virtual state in ^{9}He on the low energy ^{8}He+p resonance scattering at 180^{0} c.m.s. The strong effect of this state is obvious as a result of interference between the nuclear and the Coulomb scattering. We plan to make this measurement in the near future with a setup very similar to that used for the ^{14}F study.

4. Conclusion

In summary, we hope that we have demonstrated the importance of resonance reactions for exciting studies at both borders of nuclear stability.

Acknowledgements

This work was supported by the US Department of Energy under Grant Nos. DE-FG02-93ER40773 and DE-FG52-06NA26207 and by the US National Science Foundation under Grant No. PHY-456463. The authors thank Prof. A.M. Shirokov and Prof. P. Navratil for valuable discussions.

References

[1] Artemov K P *et al.* 1990 *Sov. J. Nucl. Phys.* **52** 406
[2] Tribble R E, Burch R H and Gagliardi C A 1989 *Nucl. Instrum. Methods.* A **285** 441
[3] Goldberg V Z, Chubarian G G, Tabacaru G, Trache L, Tribble R E, Aprahamian A, Rogachev G V, Skorodumov B B, and Tang X D 2004 *Phys. Rev.* C **69** 031302(R)
[4] Fu C *et al.* 2008 *Phys. Rev.* C **77** 064314
[5] Goldberg V Z *et al.* 2010 *Phys. Lett.* B **692** 307
[6] Rogachev G V 1999 Ph. D. thesis, Kurchatov Institute, Moscow
[7] Audi G, Wapstra A H and Thibault C 2003 *Nucl. Phys.* A **729** 337
[8] Maris P, Shirokov A M, and Vary J P 2010 *Phys. Rev.* C **81** 021301(R)
[9] Seth K K, Artuso M, Barlow D, Iversen S, Kaletka M, Nann H, Parker B and Soundranayagam R 1987 *Phys. Rev. Lett.* **58** 1930
[10] Poppelier N A F M, Wood L D and Glaudemans P W M 1985 *Phys. Lett.* B **157** 120
[11] Bohlen H G, Blazevic A, Gebauer B, von Oertzen W, Thummerer S, Kalpakchieva R, Grimes S M and Massey T N 1999 *Prog. Part. Nucl. Phys.* **42** 17
[12] von Oertzen W *et al.* 1995 *Nucl. Phys.* A **588** 129c
[13] Chen L, Blank B, Brown B A, Chartier M, Galonsky A, Hansen P G and Thoennessen M 2001 *Phys. Lett.* B **505** 21
[14] Korsheninnikov A A *et al.* 1994 *Phys. Lett.* B **326** 31
[15] Golovkov M S *et al.* 2007 *Phys. Rev.* C **76** 021605
[16] Baz' A I, Goldansky V I, Goldberg V Z, and Zeldovich Y B 1972 *Light and Intermediate Nuclei Near the Borders of Nucleon Stability* (Moscow: Nauka) [in Russian].
[17] Caamano M *et al.* 2007 *Phys. Rev. Lett.* **99** 062502
[18] Nikolskii E Y *et al.* 2010 *Phys. Rev.* C **81** 064606
[19] Rogachev G V *et al.* 2003 *Phys. Rev.* C **67** 041603(R)

Nuclear Physics in Astrophysics V IOP Publishing
Journal of Physics: Conference Series **337** (2012) 012009 doi:10.1088/1742-6596/337/1/012009

Experimental study of α-induced reactions on ^{64}Zn for the astrophysical γ-process

Gy Gyürky[1], J Farkas[1], Z Halász[1], Zs Fülöp[1], E Somorjai[1], T Szücs[1], P Mohr[1,2], A Wallner[3,4]

[1] Institute of Nuclear Research (ATOMKI), H-4001 Debrecen, POB.51., Hungary
[2] Diakonie-Klinikum, D-74523 Schwäbisch Hall, Germany
[3] Faculty of Physics, University of Vienna, 1090 Wien, Austria
[4] ANSTO, Locked Bag 2001, Kirrawee DC NSW 2232, Australia

E-mail: gyurky@atomki.hu (Gy. Gyürky)

Abstract.
For the synthesis of the heavy, proton rich isotopes in the astrophysical γ-process the precise knowledge of α-induced cross sections is of high importance. We have initiated a comprehensive study of the ^{64}Zn + α system involving the cross section measurement of different reaction channels as well as the elastic scattering at low, astrophysically relevant energies. In this paper the experimental technique and some preliminary results of the ^{64}Zn(α,p)^{67}Ga cross section measurement are presented.

1. Introduction

In various processes of heavy element nucleosynthesis, like e.g. in the astrophysical γ-process [1], the reaction rates are calculated from theoretical cross sections since experimental data are scarce. In this mass region the theoretical cross sections are typically obtained from the Hauser–Feshbach statistical model. In this model the cross section can be described by a formula similar to the Breit–Wigner formula for isolated resonances:

$$\sigma \sim \frac{\Gamma_{\mathrm{in}}\Gamma_{\mathrm{out}}}{\Gamma_{\mathrm{tot}}} \tag{1}$$

where Γ_{in}, Γ_{out} and Γ_{tot} are the widths of the entrance and the exit channels and the total, respectively. In the statistical model, transmission coefficients T_i take the role of the widths Γ_i of the Breit-Wigner formalism. If an α-induced reaction is considered, at low, astrophysically relevant energies Γ_{in} is determined mainly by the Coulomb barrier penetration and can be described by a properly chosen α-nucleus optical potential. Low energy optical potentials show a significant ambiguity, different global optical potentials lead to largely different reaction cross sections when they are used in statistical models (see e.g. [2]). Therefore, the study of the α-nucleus optical potential deserves a special attention.

The optical potential can be studied experimentally by carrying out high precision elastic scattering experiment over a broad angular range [3]. The deviation from the Rutherford cross section can be compared with the predictions of global optical potentials and local potentials can also be constructed. On the other hand, if the elastic scattering cross section measurement

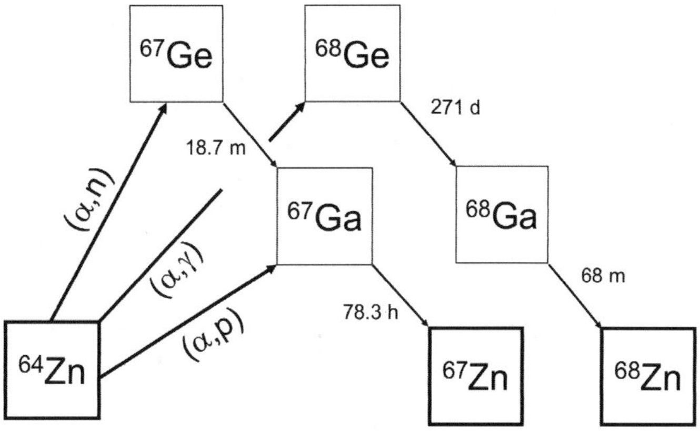

Figure 1. Alpha-induced reactions on ^{64}Zn. The half-lives of the decays of the reaction products are also indicated.

is carried out in the nearly full angular range, it can give direct information on the total cross section since the flux which is missing from the elastic channel can be attributed to the sum of all non-elastic channels [4].

If the cross section of different reaction channels can be measured along with the elastic scattering at the same energies, a more complete description of a given reaction can be achieved. Therefore a research program has been started for the comprehensive study of the ^{64}Zn + α system.

2. Alpha-induced reactions on ^{64}Zn

At a typical γ-process temperature of 3 GK the Gamow window of the ^{64}Zn + α reaction is in between 4 and 6.5 MeV. In this energy range only the ^{64}Zn(α,γ)^{68}Ge and ^{64}Zn(α,p)^{67}Ga reaction channels are open (the first one is exothermic while the latter has a threshold at 4.2 MeV). In order to have substantial deviation from the Rutherford cross section, the elastic scattering, however, may be measured only at higher energies, where the ^{64}Zn(α,n)^{67}Ge channel is also open. Figure 1 shows schematically these reactions. In all three cases the reaction products are radioactive, thus the cross section can be measured with the activation method. The aim of the present work is to measure all three channels as well as the elastic scattering[1]. The first step is the measurement of the ^{64}Zn(α,p)^{67}Ga reaction. Since the product of the ^{64}Zn(α,n)^{67}Ge reaction decays with a short half-life to the product of the ^{64}Zn(α,p)^{67}Ga reaction, and the separation of the two reactions by the decay counting leads to increased uncertainty, the ^{64}Zn(α,p)^{67}Ga cross section has been measured first below the (α,n) threshold located at $E_\alpha = 9.56$ MeV.

3. Experimental procedure

Natural isotopic composition zinc targets have been prepared by evaporating metallic Zn onto thin Al foil backings. The target thickness has been measured by weighing and in some cases checked by Rutherford Backscattering (RBS). Further RBS measurements are in progress for the target characterization. The targets have been irradiated by the α-beam of the cyclotron accelerator of ATOMKI. The beam current was kept below 1 μA in order not to have target deterioration (the target stability was continuously monitored by RBS). In the first experimental campaign the cross section has been measured at nine different α-energies between 6.3 and

[1] Some other channels with multiple particle emission are also open, but the cross sections of these reactions are typically much weaker than that of the studied ones.

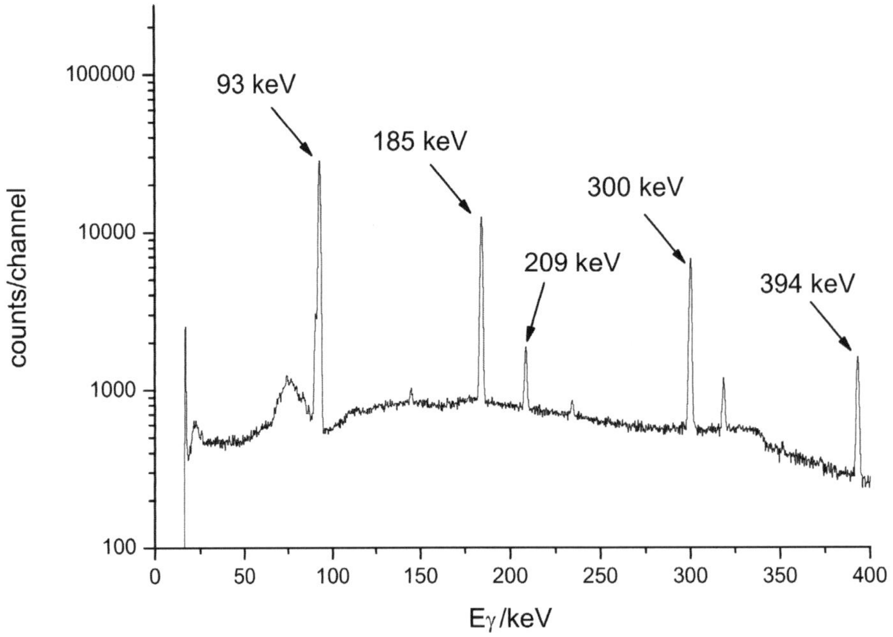

Figure 2. Typical γ-spectrum taken for 14 hours on the target irradiated with 8.6 MeV α-beam. The peaks used for the analysis are indicated by arrows.

9.5 MeV. The irradiations lasted between 5 and 18 hours depending on the energy. The induced activities have been measured by detecting the γ-rays following the electron capture decay of the ^{67}Ga reaction product with a 100 % relative efficiency HPGe detector equipped with 4π shielding. The decay of ^{67}Ga was followed for at least 12 hours by detecting its strongest γ-lines at 93.3, 184.6, 209.0, 300.2 and 393.5 keV. The absolute efficiency of the detector has been measured with several calibrated sources. Figure 2 shows a typical γ-spectrum taken on the target irradiated with 8.6 MeV α-beam. The peaks used for the analysis are indicated by arrows.

4. Results and outlook

The ^{64}Zn(α,p)^{67}Ga cross section has been measured below the (α,n) threshold in the center-of-mass energy range between 5.9 and 8.8 MeV which is at the upper edge and above the relevant Gamow window. In this energy range the cross section varies between 3 and $5{\cdot}10^4$ μb. The preliminary results can be seen in Figure 3 where the cross sections are plotted versus the c.m. energy. Also shown are the calculations carried out by the NON-SMOKER [5] and TALYS [6] codes. In general the agreement between experiment and theory is reasonably good, although stronger deviations can be found towards lower energies. Further measurements, the final analysis of the data and the comparison with theoretical predictions are in progress.

In principle, the ^{64}Zn(α,γ)^{68}Ge reaction can also be studied in the investigated energy range by activation. Some experimental difficulties are, however, encountered. The half-life of ^{68}Ge is relatively long (271 days) making the activity of the irradiated targets rather low. Moreover, the decay of ^{68}Ge is not followed by any γ-emission. Its decay product, ^{68}Ga is also radioactive and emits some γ-rays, but the relative intensity of its strongest γ-line is only 3 %. Thus the γ-detection based activation measurement of the ^{64}Zn(α,γ)^{68}Ge is challenging. Some attempts have been made to observe the ^{68}Ga decay, but only very low statistics could be obtained in

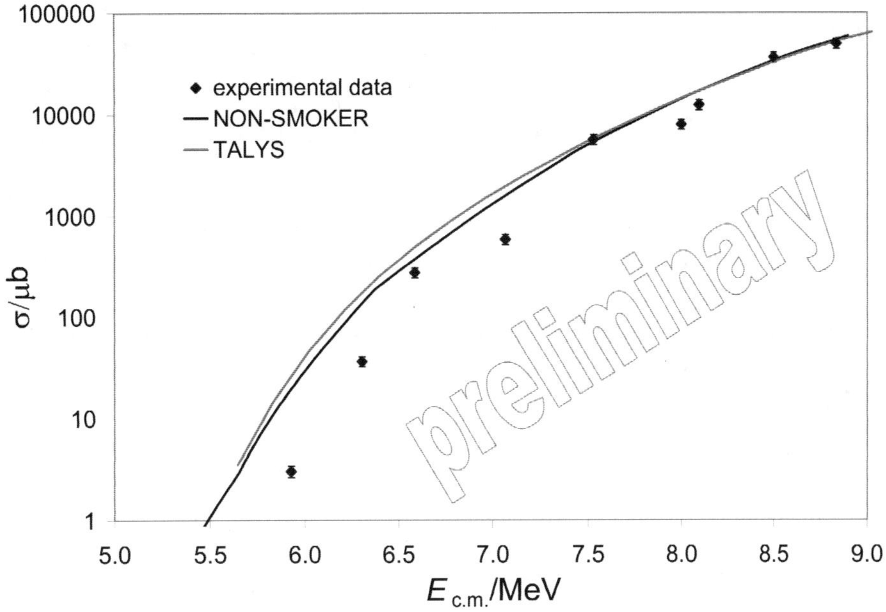

Figure 3. Preliminary results of the ^{64}Zn(α,p)^{67}Ga cross section measurement. The experimental data as well as predictions obtained with the NON-SMOKER and TALYS codes are shown.

several weeks of counting. To circumvent this problem a different method is planned. ^{68}Ge has long enough half-life to be measured with Accelerator Mass Spectrometry (AMS) technique. Some development for such an experiment is in progress at the VERA facility [7] in Vienna, Austria.

Acknowledgments

This work was supported by the European Research Council StG. 203175 and OTKA grants K68801 and NN83261(EuroGENESIS).

References

[1] Arnould M and Goriely S 2003 *Phys. Rep.* **384** 1
[2] Gyürky Gy *et al.* 2010 *J. Phys. G* **37** 115201
[3] Kiss G G *et al.* 2009 *Phys. Rev. C* **80** 045807
[4] Mohr P *et al.* 2010 *Phys. Rev. C* **82** 047601
[5] Rauscher T and Thielemann F-K 2001 *At. Data Nucl. Data Tables* **79** 47, www.nucastro.org
[6] Koning A J, Hilaire S and Duijvestijn M C 2008 *Proceedings of the International Conference on Nuclear Data for Science and Technology - ND2007* 211
[7] Wallner A 2010 *Nucl. Instr. and Meth.* **B268** 1277

Towards an intense radioactive ^{8}Li beam at SARAF Phase I

T Y Hirsh[1,2], D Berkovits[2], M Hass[1], P Jardin[3], A Pichard[3], M L Rappaport[1], Y Shachar[1] and I Silverman[2]

[1]Weizmann Institute of Science, Rehovot, Israel
[2]Soreq NRC, Yavne 81800, Israel
[3]GANIL, Caen, France

tsviki.hirsh@weizmann.ac.il

Abstract. The ^{8}Li(α,n)^{11}B reaction plays an important role in the r-process nucleosynthesis occurring in type II supernovae and in binary neutron stars. At high rates, this reaction can lead, despite the short ^{8}Li half-life, to production of seed nuclei through a chain of reactions, which can then cause the formation of heavier nuclei via (n,γ) reactions. Several measurements of this reaction were carried out in the last few years using fast ^{8}Li beams that are available at several RIB facilities at limited currents. A new production method for light Radioactive Ion Beams (RIB) via fast neutron reactions is conceptually able to provide orders of magnitude higher RIB currents, up to 10^{12} pps. The high efficiency of the two-target production method has been experimentally proven for ^{6}He production, in a similar manner to the ^{8}Li production. We present a unique apparatus for ^{8}Li production and extraction at the Soreq Applied Research Accelerator Facility - SARAF. In 2011, it is expected that SARAF Phase-I will produce high intensity, 5 MeV deuteron beams for limited experiments. By using a LiF neutron converter, neutrons with energies of up to 20 MeV and a total yield of 10^{12} fast n/sec will be available per 1 mA of deuterons. As a secondary production target we are using discs of 65% porous B_4C, mounted inside a high temperature furnace. ^{8}Li, produced via the ^{11}B(n,α)^{8}Li reaction channel, diffuses out of the B_4C target and is ionized in a thin rhenium surface ionizer. The total efficiency of ^{8}Li RIB creation will be measured by means of alpha and beta measurements. Following the experiment at SARAF Phase I, better production and extraction schemes of light RIB, of great interest in various nuclear and astrophysics-related measurements, may become possible.

1. Introduction

In the last few years we have embarked on a program to develop light Radioactive Ion Beams (RIB), to be used in nuclear astrophysics and neutrino physics. In particular, we propose a scheme of order-of-magnitude increase in intensities of short-live isotopes such as ^{6}He and ^{8}Li. This program is motivated by the newly-constructed SARAF accelerator at Soreq, but is also of major importance to leading European facilities such as SPIRAL2 at GANIL as well as ISOLDE and the future β-beam neutrino facility at CERN. One of the main directions is the production of RIB's by secondary neutrons from intense deuteron beams that will be available at SARAF.

Continuing this line of research, we are proceeding now to perform a first optimized production and extraction experiment for ^{8}Li RIB at SARAF. Our approach for light RIB production consists of a two targets system [1]. This method was found to produce ^{8}Li and ^{6}He beams with orders of magnitude more yield than available today. Detailed Monte Carlo simulations indicate that production yields of the order of 10^{12} for ^{8}Li and 10^{13} for ^{6}He are possible with the future SARAF-II 40 MeV 2 mA deuteron beam [1]. However, the RIB release and manipulation efficiencies of this system are not yet fully known [2].

High intensity ^{8}Li beams may used, e.g., to re-measure the ^{8}Li(α,n)^{11}B cross section, which is believed to have a dominant part in some schemes of the r-process following a type II SN explosion [3]. This reaction was previously measured in several experiments, although a large discrepancy has emerged between the respective results. Recently, this discrepancy has been suggested to be due to technical limitations of the detection system in some of the measurements [4]. A new set of measurements, using more intense beams that may achieve better precision, mostly at the lower energy range where the cross section is lower and thus higher statistics is required, may shed a clearer light on the diversity of cross section measurements.

In this paper we present the first experiment that is planned to take place during SARAF phase I experimental period.

2. Two stage irradiation setup

The present scheme consists of a sequential irradiation of two targets. A primary target for deuterons, which is practically a fast neutron converter, and a secondary target that serves as a production target, positioned in the proximity of the primary target to maximize the neutron flux. The effectiveness of this setup stems from the separation of the two most major problems in RIB production, namely, heat removal of the beam power and extraction of the radioisotopes, which are now handled in two separate targets. It is a similar scheme to the one used recently at ISOLDE for the production of ^{6}He [5].

Inside the secondary target, production of ^{6}He, for example, is obtained through a two-reaction channel, ^{9}Be(n,2n)^{8}Be and the actual production reaction of ^{9}Be(n,α)^{6}He. The first reaction effectively doubles the number of neutrons and in addition lowers the neutron energy, resulting in a better fit between the neutron energy and the (n,α) reaction cross section and thus also contributes to an increase in the ^{6}He production.

3. ^{8}Li RIB production apparatus

At phase I of SARAF we are planning to conduct an optimized full ^{8}Li production and extraction experiment. Measurement of the entire system's efficiency would later allow performing a scale-up for SARAF phase II or SPIRAL2 light RIB yields. The experiment is composed of several stages, presented below:

3.1. *The primary target - LiFTiT*

The primary target, *LiFTiT – Lithium Fluoride Thick Target* is essentially an accelerator based high power neutron converter. The fast neutrons are a consequence of high positive Q-Value deuteron induced reactions. By using the 5.2 MeV deuteron beam of SARAF Phase-I, fast neutrons of up to 22 MeV are generated, enabling the possibility to produce high intensity ^{6}He and ^{8}Li beams already in Phase-I.

The material chosen to serve as a target is Lithium-Fluoride, in which both compound elements contribute to the total fast neutron yield due to the high Q-values for the (d,n) and (d,2n) channels in both cases. Calculations preformed using the TALYS code [6] have resulted in $6{\cdot}10^{11}$ n/sec for neutrons with energies higher than 4 MeV, emitted in a directed cone of Ω<90° and for a 2 mA, 5.2 MeV deuteron beam. For instance, a 200 ml BeO target placed in vicinity of this would produce ~10^9 ^{6}He/sec.

LiFTiT consists of a 350 micrometer-thick LiF crystal welded to a copper back that comprises an inner water cooling loop (fig.1). To guarantee an efficient cooling of the copper back during dissipation of the few kW beam power, water is pressurized through thin micro-channels [7]. The water channels, 300 μm in the short dimension, were drilled using spark and wire EDM processes and significantly reduce the chance of a breakdown due to water bubbling.

3.2. The secondary B_4C target

The ^{8}Li isotopes are produced inside a secondary porous target of Boron-Carbide via the $^{11}B(n,\alpha)^8Li$ reaction enabled by energetic neutrons. Fast release rate of the short lived ^{8}Li out of the target is one of the main concerns, and since it proceeds through a diffusive-effusive process, the high melting point and porous structure of the secondary B_4C target (65% of the nominal density and 1-5 μm grain size) are essential.

The secondary target is 80 mm in diameter and 40 mm in length. The large volume is necessary to provide a large solid angle and is also due to the long mean free path energetic neutrons. The B_4C is contained inside a 1500 deg C vacuum furnace, to accelerate the diffusion process. The furnace was designed in a way that only a thin layer of molybdenum heat shield separates the primary and secondary targets to avoid slowing of the neutrons.

The target is capsulated inside an inner smaller Ta cavity which has one small diameter exit. Due to the high temperature, the ^{8}Li atoms released from the B_4C will pass through the small exit port, and reach the surface ionizer which is connected to that port.

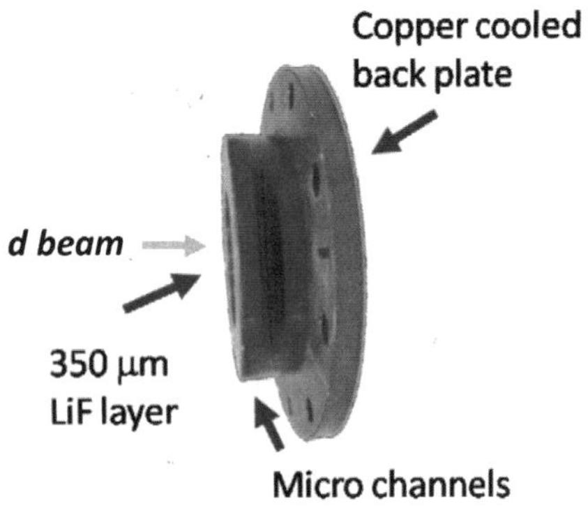

Figure 1 – LiFTiT target

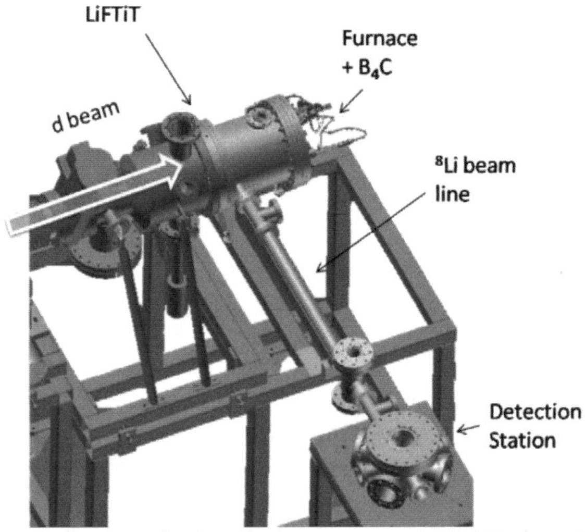

Figure 2 – The irradiation setup at SARAF phase I

3.3. Surface ionizer

A thin sheet of rhenium wrapped as a small diameter tube is used as a surface ionizer, based on the design of SPIRAL2 surface ionizer [8]. The thin Re setup was found to be efficient for ionization of all alkalis [8].

The ionizer is heated to high temperature by means of Joule heating; this process also has a side effect of a small voltage difference which favors the ^{8}Li ions direction towards the extraction electrode. It is assumed that this small applied voltage has much larger effect on the total extraction efficiency than, for example, the ionizer temperature [9].

The entire ionizer and B_4C cage are kept biased to several kV so the extracted ^{8}Li are immediately accelerated towards the detection station (fig.2).

3.4. ^{8}Li decay measurements

Schematically, the detection station is located 1 m from the main targets, leaving enough space to introduce radiation shielding, for reduction of the background levels that might be introduced by the intense neutron source. In between, the RIB is manipulated and shaped to be focused on the detection station. Inside the detection station the ^{8}Li ions are being implanted in a thin 5 μm Ni foil in the center of the detection station that is aligned by 45° to the beam. The ^{8}Li nuclei decay by a beta and

Journal of Physics: Conference Series **337** (2012) 012010 doi:10.1088/1742-6596/337/1/012010

consequently, two alphas due to the spontaneous decay of the ^{8}Be daughter. Two sets of detectors are placed in the vicinity of the foil to measure the resulting alpha and beta radiation, including two silicon detectors and a thick plastic scintillator which are grouped as a dE-E monitor. Even though a high gamma background is expected due to radiation from the neutron source, a major part of it would be discriminated by combining the dE-E telescope and alpha coincidence measurement. For the current setup we expect to obtain sufficient data to allow an efficiency calibration of the entire setup and the various components within.

4. Future Perspectives

Once the efficiency values for the two-stage irradiation setup are measured, the scheme could be easily scaled up to accommodate for the larger-scale experiment with 40 MeV beams. At these energies, the production yields are 3 orders of magnitude higher, intense beams of light RIB, of great interest in various nuclear and astrophysics related measurements, may become feasible. For instance, ^{8}Li and ^{6}He are considered to be two major candidates for the *beta-beam* β^- emitters, for which these isotopes must be produced at sufficiently high intensity as a pre-requisite to the implementation of this concept. The double-target production method, coupled with high intensity fast neutrons sources based on a deuteron LINAC, may achieve a ^{6}He or ^{8}Li sources which are intense enough to be facilitate the *beta-beam* project.

5. Conclusions

During the period of SARAF phase-I we are planning to conduct an optimized production and extraction series of experiments for ^{8}Li RIB. These experiments may provide better understanding of the method and consequently maximize the production and extraction mechanisms for light RIB's. These intense beams may facilitate more precise measurements in astrophysics and various nuclear-physics related measurements.

This work is supported in part by the Pazi Foundation and the FP7 SPIRAL2PP collaboration

References

[1] M. Hass *et al.* J. Physics G**35**, 104042 (2008).
[2] T. Y. Hirsh *et al.* POS(NuFact08)090 (2008)
[3] M. Terasawa *et al.* Ast. J., **562**, (2001) 470-479
[4] M. La Cognata *et al. Ast. J.* **706**, (2009) L251-L255
[5] T. Stora *et al., ISOLDE Newsletter*, Spring (2010) ; to be published in EPL.
[6] http://www.talys.eu
[7] D. Petrich *et al.* Nuc. Ins. Meth. A **596** ,269-275 (2008)
[8] A. Pichard *et al.* Rev Sci Inst. 81(2) 02A908 (2010)
[9] C. Eleon *et al.* Nuc. Ins. Meth. Phys. Res. B, **266** (2008) 4362–4367

Note added in Proof:

The first test run using the LiFTiT production target has taken place after submission of this manuscript; the resulting data is presently being analysed.

Nuclear Physics in Astrophysics V IOP Publishing
Journal of Physics: Conference Series **337** (2012) 012011 doi:10.1088/1742-6596/337/1/012011

Investigation of the CNO-break-out reaction: $^{15}O(2p,\gamma)^{17}Ne$ by the Coulomb dissociation of ^{17}Ne

Justyna Marganiec
ExtreMe Matter Institute EMMI, GSI Darmstadt, Germany
The Joint Institute for Nuclear Astrophysics JINA, Notre Dame, IN, USA
E-mail: j.marganiec@gsi.de

Thomas Aumann
Kernreaktionen und Nukleare Astrophysik, GSI Darmstadt, Germany
Institut für Kernphysik, TU Darmstadt, Germany

Michael Heil
Kernreaktionen und Nukleare Astrophysik, GSI Darmstadt, Germany

Ralf Plag
Kernreaktionen und Nukleare Astrophysik, GSI Darmstadt, Germany
Goethe-Universität, Frankfurt am Main, Germany

Felix Wamers
Kernreaktionen und Nukleare Astrophysik, GSI Darmstadt, Germany
Institut für Kernphysik, TU Darmstadt, Germany

for the LAND-R^{3}B collaboration

Abstract. By the Coulomb dissociation of ^{17}Ne, the time-reversed reaction $^{15}O(2p,\gamma)^{17}Ne$ has been investigated. This reaction might play an important role in the *rp* process, as a bypass of the ^{15}O waiting point. The secondary ^{17}Ne ion beam of 500 AMeV has been dissociated on a Pb target, and the reaction products have been recorded with the LAND-R^{3}B experimental setup (GSI). This experiment allows to determine the Coulomb dissociation cross section σ_{Coul}, which can be converted into a photo-absorption cross section σ_{photo}, and a radiative-capture cross section σ_{cap} for the $^{15}O(2p,\gamma)^{17}Ne$ reaction. Additionally, informations about the structure of the ^{17}Ne nucleus, a possible two-proton halo, may be obtained. The analysis is still in progress.

1. Introduction

At high temperature and density conditions, the CNO cycle is linked with the *rp* process by the α capture reaction on the ^{15}O nucleus. By the *rp* process, which is a sequence of proton captures and β^+ decays, the initial CNO material can be processed towards heavier nuclei. ^{15}O is a waiting point nucleus, which hampers the reaction flow between the CNO cycle and the FeNi-mass region. The $^{15}O(2p,\gamma)^{17}Ne$ reaction can bridge this waiting point. The three-body radiative capture can proceed sequentially [1] or directly from the three-body continuum [2]. It has been suggested that the reaction rate can be enhanced by a few orders of magnitude by taking into account the three-body continuum [2]. In order to verify these calculations, the $^{15}O(2p,\gamma)^{17}Ne$ reaction has to be investigated and the radiative capture cross section has to be determined.

This experiment is also interesting for nuclear structure. The proton dripline nucleus ^{17}Ne is a promising candidate for a two-proton halo, because of a comparatively small $2p$ separation energy (S_{2p} = 970 keV) [3]. The mixture of the d^2 and s^2 configurations of the two protons outside the ^{15}O core in the ^{17}Ne ground state is still unknown. In the case of large s^2 weights of proton configurations the ^{17}Ne is a halo nucleus, while in the case of large d^2 weights the nucleus ^{17}Ne has no halo structure (Fig. 1).

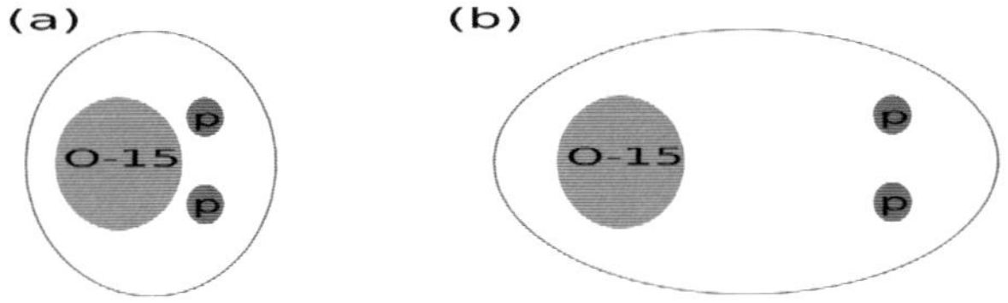

Figure 1. Schematic representation of the structure of ^{17}Ne, if: (a) - the d^2 component dominant; (b) - the s^2 component dominant.

2. Coulomb dissociation method

The Coulomb dissociation method is an indirect technique, which is useful in case of very small cross sections, unstable nuclei and three particles in an entrance channel. In this method, instead of the radiative-capture process, the time-reversed process is studied, and the Coulomb field of a heavy nucleus is used as a source of photons [4] (Fig. 2).

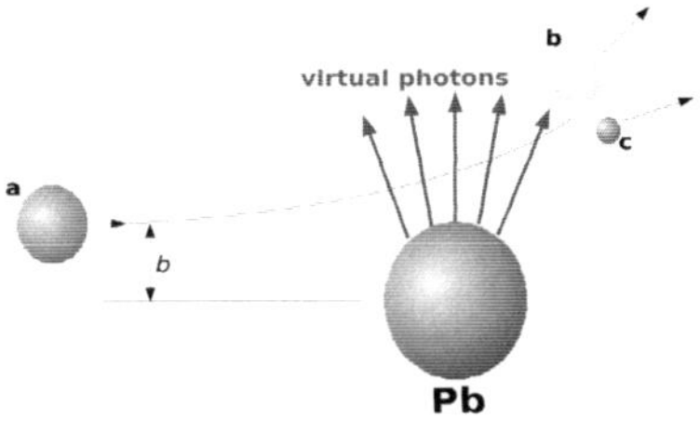

Figure 2. Coulomb dissociation in the field of a Pb target. (b - impact parameter)

Using virtual photon-theory, a Coulomb dissociation cross section σ_{Coul} can be converted into a photo-absorpion cross section σ_{photo}, which can then be converted into a radiative-capture cross section σ_{cap} with the help of the detailed-balance theorem.

3. Experiment and preliminary results

The primary ^{20}Ne beam from an ion source was preaccelerated in the universal linear accelerator (UNILAC) and injected into the heavy-ion synchrotron (SIS) at GSI in Darmstadt, Germany. The accelerated primary beam was directed to a beryllium production target and the secondary ^{17}Ne beam was produced via nuclear fragmentation. The reaction products were separated by means of the fragment separator (FRS) [5], where dipole magnets filter out all species except those with a specific A/Z ratio ($B\rho = \frac{p}{Q} \propto \frac{A}{Z}\beta\gamma$). The secondary beam of ^{17}Ne, with an energy

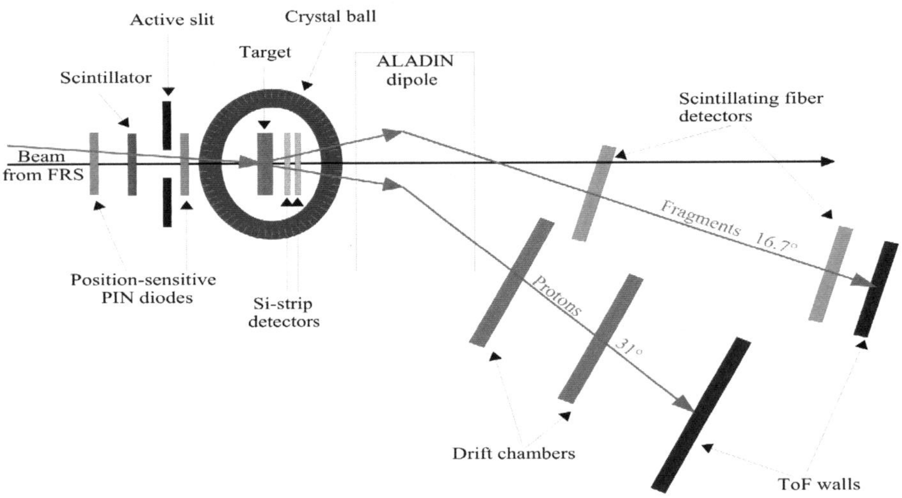

Figure 3. Experimental setup.

$E = 500$ AMeV, was identified in the experimental area (Fig 3), by means of energy-loss and position measurements with position-sensitive pin diodes and time-of-flight measurements.

Dissociation of ^{17}Ne nuclei was induced by a secondary Pb target (200 mg/cm^2). The reaction products have been identified using two Si-strip detectors placed before the large-gap dipole magnet (ALADIN). After the magnet, two scintillating-fibre detectors and a two-layer Time-of-Flight (ToF) wall have been used to detect the fragments. The protons have been recorded by two drift chambers and a large two-layer ToF wall [6]. By means of time-of-flight and energy-loss measurements, all reaction products have been identified (Fig. 4).

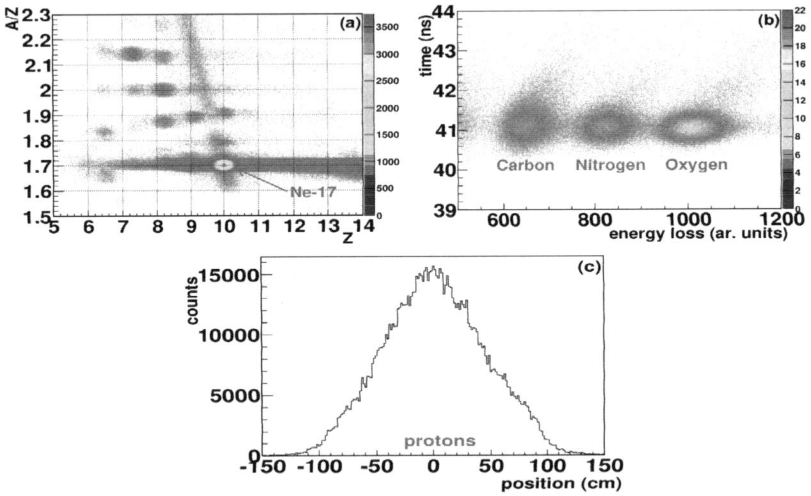

Figure 4. (a) - The identification of incoming beam nuclei; (b) - the identification of outgoing fragments; (c) - detected protons in a large two-layer ToF wall.

To obtain masses and momenta of the reaction products a tracking procedure was utilized. The mass specra of oxygen are given in Fig. 5.

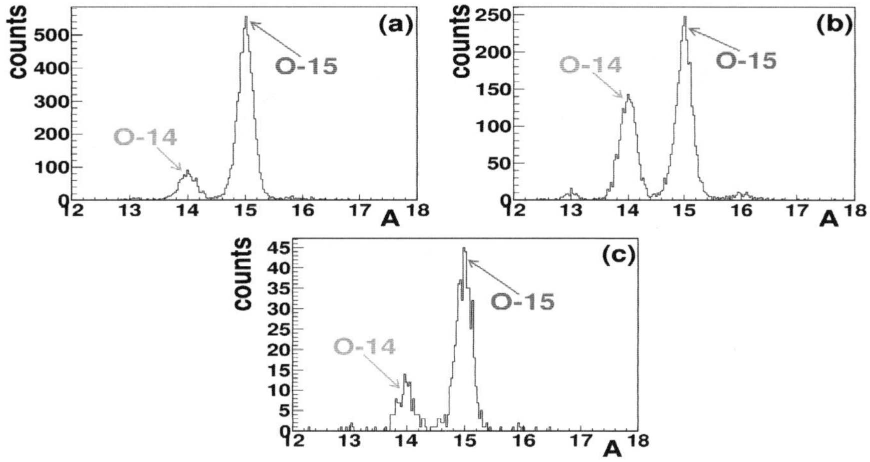

Figure 5. The oxygen mass spectra after tracking. (a) - for data with the Pb target; (b) - for data with the C target (to subtract the nuclear contribution); (c) - for data without any target (to subtract the background).

The measured Coulomb dissociation cross section is given by the formula:

$$\sigma_{Coul} = p_{Pb}\left(\frac{M_{Pb}}{d_{Pb}N_{Av}}\right) - p_C\left(\alpha\frac{M_C}{d_C N_{Av}}\right) - p_{empty}\left(\frac{M_{Pb}}{d_{Pb}N_{Av}} - \alpha\frac{M_C}{d_C N_{Av}}\right), \tag{1}$$

where p is the interaction probability for a target, M - the molar mass of target material [g/mol], d - the target thickness [g/cm^2], N_{Av} - Avogadro's number [mol^{-1}] and α is a scaling factor between Pb and C targets.

Using this formula the preliminary exclusive Coulomb dissociation cross section has been determined. The value is $\sigma_{Coul} = 253 \pm 36$ mb (only statistical uncertainy).

4. Summary

The secondary ^{17}Ne beam has been produced by fragmentation reactions of ^{20}Ne, and the ^{15}O$(2p,\gamma)^{17}$Ne reaction has been studied. The incoming beam and outgoing reaction products have been identified, and the preliminary exclusive Coulomb dissociation cross section has been obtained $\sigma_{Coul} = 253 \pm 36$ mb (only statistical uncertainy). The next steps of the analysis will be the calculations of the photoabsorption and the radiative capture cross sections, which are not only relevant for the rp process but also are of interest with regard to the two-proton halo structure of ^{17}Ne. The analysis is still in progress.

5. Acknowledgments

This project was supperted by the German Federal Ministry for Education and Research (BMBF), EU(EURONS), ExtreMe Matter Institute EMMI, GSI Darmstadt and FIAS Frankfurt Institute for Advanced Studies.

References

[1] Görres J *et al.* 1995 *Phys. Rev. C* **51** 392
[2] Grigorenko L V, Zhukov M V 2005 *Phys. Rev. C* **72** 015803
[3] Kanungo R *et al.* 2003 *Phys. Lett. B* **571** 21
[4] Baur G and Bertulani C A 1986 *Nucl. Phys.* **A458** 188
[5] Geissel H *et al.* 1992 *Nucl. Instrum. Meth.* **B 70** 286
[6] http://www-linux.gsi.de/~rplag/land02/

Nuclear Physics in Astrophysics V IOP Publishing
Journal of Physics: Conference Series **337** (2012) 012012 doi:10.1088/1742-6596/337/1/012012

Anomalous properties of neutron resonances in Pt isotopes

M Krtička[1], P E Koehler[2], F Bečvář[1], J A Harvey[2], K H Guber[3]

[1]Charles University, Faculty of Mathematics and Physics, Prague, Czech Republic
[2] Physics Division, Oak Ridge National Laboratory, Oak Ridge, USA
[3] Reactor and Nuclear Systems Division, Oak Ridge National Laboratory, Oak Ridge, USA

E-mail: `krticka@ipnp.troja.mff.cuni.cz`

Abstract. We obtained an unprecedentedly large number of s-wave neutron widths and total radiation widths through $\mathcal{R}$-matrix analysis of neutron cross-section measurements on enriched Pt samples. Careful analysis of resonance parameters of even-even Pt targets was performed and a brief summary of findings is given.

1. Introduction

Neutron resonance parameters remain some of the most important information for testing the validity of predictions of the random matrix theory (RMT) [1], even more than half a century after these data served as the original stimulus for its creation. Today, RMT pervades the physics of almost all complex systems. In nuclear physics, predictions of RMT are expected to correctly describe fluctuation properties of nuclear levels at relatively high excitation such as near the neutron threshold in medium-weight and heavy nuclei.

Predictions of RMT implicitly yield that spacing of neutron resonances with the same spin and parity J^π follow so-called Wigner distribution [2] and reduced neutron widths, $\Gamma^0_{\lambda n}$, of s-wave resonances λ follow a χ^2_ν distribution with one degree of freedom, $\nu = 1$, which is known as the Porter-Thomas distribution (PTD) [3]. Today, the general consensus is that data on resonance spacing and $\Gamma^0_{\lambda n}$ agree with these expectations. However, there are problems with both the data and analysis techniques used in probably the best test of the RMT predictions to date [4]. In addition, a few deviations from the PTD have been reported in the past, see references in [5]. Therefore, it is still worthwhile to perform new tests of predictions of RMT, especially considering that measurement and analysis techniques have improved considerably since the 1970's when most of the data used in previous tests were taken.

To make reliable conclusions regarding the validity of the RMT predictions, it is important that the data set be as pure, complete, and large as possible. Perennial problems with neutron resonance data have been (i) contamination of s- by p-wave resonances and/or resonances of neighboring isotopes, (ii) obtaining enough resonances with known J^π to perform statistically meaningful tests, and (iii) missed resonances due to finite experimental threshold.

For our recent data on Pt isotopes measured at the Oak Ridge Electron Linear Accelerator (ORELA), problem (i) was minimized because Pt is near the peak of the s- and the minimum of the p-wave neutron strength functions ($S_0/S_1 \approx 10$), and because we made high resolution cross-section measurements on natural Pt as well as four samples enriched in ^{192,194,195,196}Pt.

Solving problem (ii) has been addressed in our case by very good sensitivity for observation of resonances up to neutron energies, E_{n}, of several keV. If only even-even targets are considered in the analysis, all their s-wave resonances have $J^{\pi} = 1/2^{+}$, so that difficulty (ii) with the resonance spin assignment also disappears. Solving all three problems is facilitated by a novel approach of using an artificially imposed, energy-dependent threshold in the analysis.

2. Experiment and data reduction

Data from measurement of neutron capture and total cross sections for ^{192,194,195,196,nat}Pt at the ORELA facility were exploited. During the experiment, cross sections were measured in the energy range from 10 eV up to several hundred keV.

Details of the measurements can be found in Ref. [6]. The ORELA was operated at a pulse rate of 525 Hz, a pulse width of 8 ns, and a power of 7-8 kW. Capture measurements were made at a source-to-sample distance of 40.12 m with a pair of C_6D_6 detectors using the pulse-height-weighting technique, and were normalized via the saturated 4.9-eV resonance in the ^{197}Au(n,γ) reaction. Total neutron cross sections were measured, at the same time on a separate flight path, via transmission using a ^{6}Li-loaded glass scintillator at a source-to-detector distance of 79.83 m.

The multilevel, multichannel, $\mathcal{R}$-matrix code SAMMY [7] was used to fit both our transmission and capture data and extract resonance parameters. Orbital angular momenta up to and including p-waves ($\ell_{\mathrm{n}} = 1$) were included in the analysis. Resonance energies E_{λ}, total radiation widths $\Gamma_{\lambda\gamma}$ and observables $g_J\Gamma_{\lambda\mathrm{n}} = g_J\Gamma^0_{\lambda\mathrm{n}}\sqrt{E_{\lambda}/1\ \mathrm{eV}}$, where g_J is the statistical factor for resonances with spin J, were used in subsequent analyses described below. For even-even targets, s-wave resonances have $g_J \equiv g_{1/2} = 1$, and hence, $g_J\Gamma_{\lambda\mathrm{n}} = \Gamma_{\lambda\mathrm{n}}$. Overall, we determined parameters for 159, 413, 423, 258, and 11 resonances in energy range up to 5.0, 16.0, 7.5, 16.0, and 5.0 keV in ^{192,194,195,196,198}Pt+n systems, respectively.

An asymmetrical shape in the transmission data could be used to assign $\ell_{\mathrm{n}} = 0$ resonances [6, 8], see Fig. 1. However, there remained many weak resonances, most of which are p wave, for which we could not unambiguously determine the ℓ_{n} value. As shown below, the potential problem posed by these resonances can be surmounted.

3. Analyses of resonance parameters

The need to use a threshold on observables $g_J\Gamma_{\lambda\mathrm{n}}$, to guard against possible systematic errors due to instrumental effects and p-wave contamination, was realized very early on [3] in tests of theoretical predictions. Use of a E_{n}-dependent s-wave threshold of the form $T_0 = a_0 E_{\mathrm{n}}^{3/2}$, where a_0 is a constant factor, is a key improvement in our method compared to previous analyses for at least three reasons. First, p-wave contamination is eliminated equally effectively at all energies. Second, as shown in Fig. 1, our instrumental threshold follows very closely this same energy dependence, being at the same time substantially lower than T_0; thus, possible diffusiveness of the instrumental threshold can be surmounted equally effectively at all energies. Third, statistical precision of any analysis is maximized by allowing the largest number of s-wave resonances to be included.

Results from the analysis of the distribution of $g_J\Gamma^0_{\lambda\mathrm{n}}$ in even-even Pt targets were already published [8]. Using the maximum-likelihood (ML) method and assuming that the distribution of reduced neutron widths is governed by a χ^2_{ν} distribution with unknown number of degrees of freedom it was found that best estimates of ν were $\widehat{\nu}_{\mathrm{exp}} = 0.57^{+0.16}_{-0.15}$, $0.47^{+0.19}_{-0.18}$, and $0.60^{+0.28}_{-0.26}$ in ^{192,194,196}Pt, respectively. This indicated a deviation from the PTD. But, as pointed out in [8], it is premature to draw from these values of $\widehat{\nu}_{\mathrm{exp}}$ a reliable conclusion.

To check the veracity of the ML results for ^{192,194}Pt, we undertook additional analysis. First, fixing an expectation value $\mathrm{E}[\Gamma^0_{\lambda\mathrm{n}}]$ of $\Gamma^0_{\lambda\mathrm{n}}$, we drew from the PTD and a uniform distribution of energies E_{λ} a random sample, consisting of n_0 pairs $[E_{\lambda}, \Gamma^0_{\lambda\mathrm{n}}]$, where n_0 is

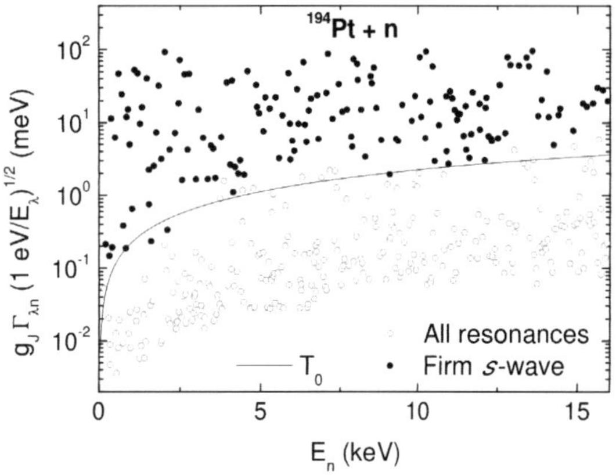

Figure 1. Energy-reduced values of $g_J \Gamma_{\lambda\text{n}}$ for resonances of ^{194}Pt. Full-circles represent resonances to which we assigned $\ell_\text{n} = 0$. For the meaning of the threshold T_0 see the text.

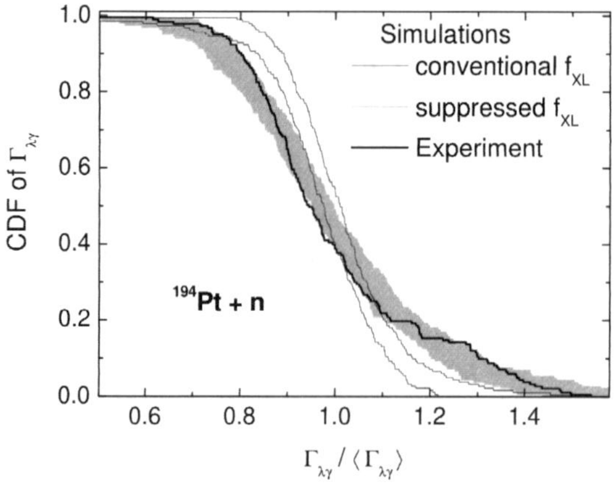

Figure 2. Empirical cumulative distribution functions (CDFs) of widths $\Gamma_{\lambda\gamma}$ – see the main text. Here, $\langle\Gamma_{\lambda\gamma}\rangle$ stands for averaging over the finite sample of $\Gamma_{\lambda\gamma}$'s.

the number of resonances above T_0. Then, with the aid of the ML analysis, we estimated ν for this sample. From a large number of samples for each value of $\text{E}[\Gamma^0_{\lambda\text{n}}]$ we constructed the empirical cumulative distribution function (CDF) of estimates $\widehat{\nu}$ and determined a probability $p_\nu = P(\widehat{\nu} > \widehat{\nu}_\text{exp} | \, \text{E}[\Gamma^0_{\lambda\text{n}}])$ that *a simulated* value $\widehat{\nu}$ is higher than the ML estimate $\widehat{\nu}_\text{exp}$ deduced from *experimental* values of $\Gamma^0_{\lambda\text{n}}$. Given the value of $\text{E}[\Gamma^0_{\lambda\text{n}}]$, the probability p_ν represents the statistical significance at which the validity of the PTD can be rejected. Values of p_ν were very high ($\approx 99.5\%$ in cases of ^{192,194}Pt for the most probable values of $\text{E}[\Gamma^0_{\lambda\text{n}}]$), but varied considerably with $\text{E}[\Gamma^0_{\lambda\text{n}}]$. Therefore, we undertook further analysis to impose limits on $\text{E}[\Gamma^0_{\lambda\text{n}}]$. Simple statistics (named as Z and Z^2 in [8]) similar to von Mises, Kolmogorov-Smirnov, or Anderson-Darling ones were adopted for this purpose; see [8] for details. Combining results from p_ν, Z, and Z^2 for ^{192,194}Pt we arrived at the conclusion that PTD can be rejected with a statistical significance of at least 99.997%. Our ^{196}Pt data were not used in this analysis as the smaller number of observed resonances resulted in reduced statistical precision.

Our data allow also analysis of distribution of total radiation widths, $\Gamma_{\lambda\gamma}$, from individual resonances. Unfortunately, fluctuation properties of $\Gamma_{\lambda\gamma}$ are much more complicated than those of $\Gamma^0_{\lambda\text{n}}$ as the widths $\Gamma_{\lambda\gamma}$ are given by

$$\Gamma_{\lambda\gamma} = \sum_f \sum_{XL} \frac{f_{XL}(E_\gamma)\xi^2_{XLf} E_\gamma^{(2L+1)}}{\rho(E, J, \pi)}, \tag{1}$$

where E_γ is γ-ray energy, $\rho(E, J, \pi)$ is the density of resonances (at excitation energy E) with spin J and parity π, and $f_{XL}(E_\gamma)$ is radiative strength function for multipolarity XL. The sum runs over all final levels f allowed by selection rules. Based on the predictions of RMT it is believed that ξ^2_{XLf} is the random number from the PTD. The empirical CDF of measured widths $\Gamma_{\lambda\gamma}$ for ^{194}Pt+n system is shown in Fig. 2, see the thick solid line.

As each width $\Gamma_{\lambda\gamma}$ is a sum of many terms, there is no simple prediction on distribution of $\Gamma_{\lambda\gamma}$. To check the compatibility of experimental distribution of $\Gamma_{\lambda\gamma}$ with RMT predictions we simulated a large number of $\Gamma_{\lambda\gamma}$ values using the DICEBOX [9] code for different models of $f_{XL}(E_\gamma)$ and $\rho(E, J, \pi)$. Statistics similar to Z and Z^2 used in analysis of distribution of $\Gamma^0_{\lambda\text{n}}$ [8] were then used to check the agreement between experimental data and simulations.

Unfortunately, the distribution of $\Gamma_{\lambda\gamma}$ depends on adopted models of $f_{XL}(E_\gamma)$ and $\rho(E, J, \pi)$. Employing the conventional models for these quantities [10] we found that the distribution of the simulated widths $\Gamma_{\lambda\gamma}$ is markedly narrower compared to that of the widths deduced from our ^{192,194,196}Pt measurements. This disagreement makes it possible to reject the validity of the conventional models for $f_{XL}(E_\gamma)$ and $\rho(E, J, \pi)$ [10] with high statistical confidence. The degree of disagreement is illustrated in Fig. 2, where the corridor delimited by thin curves represents the region covered by 10^3 simulated CDFs for statistical variables $\Gamma_{\lambda\gamma}$.

There are several ways to reconcile the simulated CDF with the experimental one while maintaining the statistical approach: (i) random numbers ξ^2_{XLf} could be sampled from a distribution which differs from the PTD – a χ^2_ν distribution with $\nu \sim 0.2 - 0.5$ instead of $\nu = 1$ is likely to be needed if this is the sole modification; (ii) $f_{XL}(E_\gamma)$ could be strongly suppressed for small E_γ – predictions with suppressed $f_{XL}(E_\gamma)$ similar to that reproducing γ-ray spectra from (n,γ) reactions on ^{197}Au [11] is shown in Fig. 2, see the the corridor in gray; (iii) there could be abrupt changes in level density as a function of excitation energy E. As significant decrease of level density with E at some interval of E would be needed we guess that possibility (iii) is very unlikely.

Finally, using our data on ^{194}Pt we tested the validity of the multivariate distribution (MVD) of resonance energies E_λ predicted by the RMT. However, contrary to RMT predictions, we assumed that $\Gamma^0_{\lambda\mathrm{n}}$ fluctuate according to χ^2_ν with $\nu = 0.47$, in conformity what we found before [8]. As it is impossible to identify J^π of weak resonances, a fraction of s-wave resonances must be controllably treated as missing resonances. Keeping this in mind, we simulated a large number of sequences of resonance energies E_λ of ^{194}Pt together with the corresponding widths $\Gamma^0_{\lambda\mathrm{n}}$. As a next step, by imposing the condition of threshold T_0, from each such sequence we extracted a sequence of energies of the above-threshold resonances. Then, the MVD of the resulting energy sequences was compared with the sequence of energies E_λ of experimentally observed above-threshold resonances. Using a statistic similar to Z^2, from this comparison we found that our experimental data on resonance energies can reject the validity of the RMT-based MVD of energies E_λ only on a relatively low statistical significance level of about 95%. Nevertheless, we stress that our knowledge about the *true* distribution of $\Gamma^0_{\lambda\mathrm{n}}$ is limited. For this reason the results of this outlined analysis should be taken with care.

Acknowledgement

This work was supported by the U.S. DOE under Contract No. DEAC05-00OR22725 with UT-Battelle, LLC, and by Czech Research Plans MSM-021620859 and INGO-LA08015.

References

[1] Guhr T *et al.*, 1998, *Phys. Rep.* **299**, 189
[2] Balian R, and Bloch C, 1970, *Ann. Phys. (N.Y.)* **60** 401
[3] Porter CE and Thomas GE, 1956, *Phys. Rev.* **104**, 483
[4] Haq, RU, Pandey A, and Bohigas O, 1982, *Phys. Rev. Lett.* **48** 1086; Bohigas O, Haq RU, and Pandey A, 1983, in *Nuclear Data for Science and Technology*, edited by K. H. Bockhoff (D. Reidel, Dodrecht), p. 809
[5] Koehler PE, 2010, *EPJ Web of Conferences* **2**, 05001
[6] Koehler PE *et al.*, 2002, *J. Nucl. Sci. and Tech.*, **Suppl. 2**, 546
[7] Larson NM, 2008, Oak Ridge National Laboratory Technical Report No. ORNL/TM-9179/R8
[8] Koehler PE, Bečvář F, Krtička M, Harvey JA, Guber KH, 2010, *Phys. Rev. Lett.* **105** 072502
[9] Bečvář F, 1998, *Nucl. Instrum. Methods Phys. Res. A* **417**, 434
[10] Capote R *et al.*, 2009, *Nucl. Data Sheets* **110**, 3107
[11] Krtička M *et al.*, 2006, *AIP Conference Proceedings* **831**, 481

Journal of Physics: Conference Series **337** (2012) 012013 doi:10.1088/1742-6596/337/1/012013

Self-consistent theory of stellar electron capture rates

N. Paar, D. Vretenar

Physics Department, Faculty of Science, University of Zagreb, Croatia

Y. F. Niu

State Key Laboratory of Nuclear Physics and Technology, School of Physics, Peking University, Beijing 100871, China

J. Meng

State Key Laboratory of Nuclear Physics and Technology, School of Physics, Peking University, Beijing 100871, China
School of Physics and Nuclear Energy Engineering, Beihang University, Beijing 100191, China
Department of Physics, University of Stellenbosch, Stellenbosch 7602, South Africa

E-mail: npaar@phy.hr

Abstract. Recently a novel theory framework has been established for description of electron capture rates at temperatures and densities in stellar environment, based on the relativistic energy density functional. The model includes finite temperature relativistic mean field to determine single-particle basis and the corresponding thermal occupation factors of target nucleus, while relevant charge-exchange excitations are described by the self-consistent finite temperature relativistic random phase approximation (FTRRPA). The calculated electron capture rates for ^{54,56}Fe are shown in comparison with other advanced theory approaches.

1. Introduction

Electron capture on nuclei plays an important role in the evolution of core-collapse supernova [1]. During the precollapse phase of the star evolution, it reduces the number of electrons available for the electron degeneracy pressure support , while beta decay goes in the opposite direction. At the same time, the neutrinos produced by the electron capture freely escape from the star at densities $\leq 10^{11}$ g/cm^3, removing energy and entropy from the core[1, 2]. For the initial electron-to-baryon ratio $Y_e \approx 0.5$, β^- decay processes can be effectively hindered by the electron degeneracy, but become competitive when the nuclei become more neutron-rich.

At lower central stellar densities ($\leq 10^{10}$ g/cm^3) and temperature range 300 - 800 keV, electron capture occurs mainly on nuclei in the mass range $A \sim 60$ [1]. Since the electron chemical potential becomes of the same order of magnitude as the nuclear Q-value, the respective rates are sensitive to the detailed structure of Gamow-Teller (GT) transition strength. However, for higher densities and temperature, nuclei with mass numbers $A > 65$ become quite abundant, the electron chemical potential is noticeably larger than the Q-value, thus the electron capture rates are mainly determined by the overall GT strength and its centroid energy. When reaching even higher densities ($\rho > 10^{11}$ g/cm^3), the electron chemical potential becomes larger than ≈ 20

MeV. Therefore, in addition to the GT strength, forbidden transitions should also be taken into account in modeling the electron capture rates at higher densities [1, 2].

Various approaches have so far been employed in calculations of stellar electron capture rates, often strongly dependent on the empirical data. The first tabulation of the rates for nuclei between $A = 21$ and 60 has been accomplished by Fuller, Fowler, and Newman (FFN), based on the independent particle model for allowed transitions, supplemented by the empirical values of the Gamow-Teller(GT) matrix elements [3]. Within the shell model Monte Carlo method (SMMC) applied for fp-shell nuclei, the thermal effects and GT transitions have been calculated in a microscopic way [4]. The large-scale shell-model diagonalization (LSSM) approach provided tabulation of weak interaction rates for more than 100 nuclei in the mass range $45 \leq A \leq 65$ [5]. In order to overcome the limitations of the shell model, in the study of nuclei beyond fp-shell, a hybrid model of the SMMC and random phase approximation (RPA) has been employed for large number of nuclei involved in stellar core collapse[6, 7]. The electron capture rates have also been studied with the proton-neutron quasiparticle RPA (QRPA) employing the Nilsson model and separable GT forces [8]. More recently, thermal QRPA (TQRPA) has been introduced, based on the Woods-Saxon potential and separable multipole and spin-multipole particle-hole interactions, with temperature taken into account by the thermofield dynamics [9].

Most of currently available models of the stellar electron capture rates have to be adjusted to the experimental data for the structure properties of involved nuclei. Therefore, extrapolations toward target nuclei away from the valley of stability are rather unreliable. Feasible systematic description of nuclear ground state properties and excitations over the nuclide chart currently can be provided only by the nuclear energy density functionals. Recently two self-consistent frameworks have been introduced for modeling the electron capture cross sections and rates, based on *i)* Skyrme functionals [10] and *ii)* relativistic energy density functionals [11]. In this proceedings we give an overview of recent progress in self-consistent modeling of the stellar electron capture rates in the relativistic framework [11].

2. Electron capture rates based on relativistic energy density functional

The relativistic density functional theory, realized in terms of the relativistic mean field (RMF) and the relativistic Hartree Bogoliubov (RHB) model, has achieved great success in the description of ground-state properties and excitations in nuclei all over the periodic table, including those far away from the stability line [12, 13, 14, 15]. For the study of astrophysical processes, temperature effects have recently been taken into account in the self-consistent relativistic RPA , which has been used to investigate excitations without charge exchange, in particular the low-energy monopole and dipole response in nuclei at finite temperature[16].

Since the electron capture on nuclei involves transitions with charge exchange, recently the self-consistent finite temperature relativistic random phase approximation (FTRRPA) has been extended for the implementation in the charge-exchange channel [11]. The properties of the nuclear initial state, i.e. single nucleon basis and the corresponding thermal occupation probabilities, are obtained within the RMF at finite temperature, formulated using the effective Lagrangian with medium-dependent meson-nucleon couplings with DD-ME2 parameterization [17]. The FTRRPA equations are derived using the single-nucleon basis of the RMF at finite temperature [16]. In a self-consistent approach, the residual interaction terms in the FTRRPA matrix are obtained as the derivative of the single particle Dirac Hamiltonian from the same Lagrangian as in the RMF framework. In the relativistic RPA, the configuration space includes not only proton-neutron pairs in the Fermi sea but also pairs formed from the fully or partially occupied states in the Fermi sea and the empty negative-energy states from the Dirac sea. The spin-isospin-dependent residual interaction terms are generated by the exchange of π and ρ mesons. For the ρ-meson density-dependent coupling strength we choose the same functional form used in the effective interaction in the RMF [17]. The particle-hole residual interaction includes thermal occupation probabilities for protons and neutrons obtained from the self-conistent RMF calculations at finite temperature.

The electron capture rate is evaluated using the cross section and distribution of electrons $f(E_e, \mu_e, T)$ at given temperature [11]. In stellar conditions present in the collapsing core of supernova, electrons are well described by the Fermi-Dirac distribution, with temperature T and electron chemical potential μ_e[7]. As an example, in Fig. 1 we show the calculated electron capture rates for ^{54,56}Fe for different conditions associated with the initial phase of the core-collapse supernova.

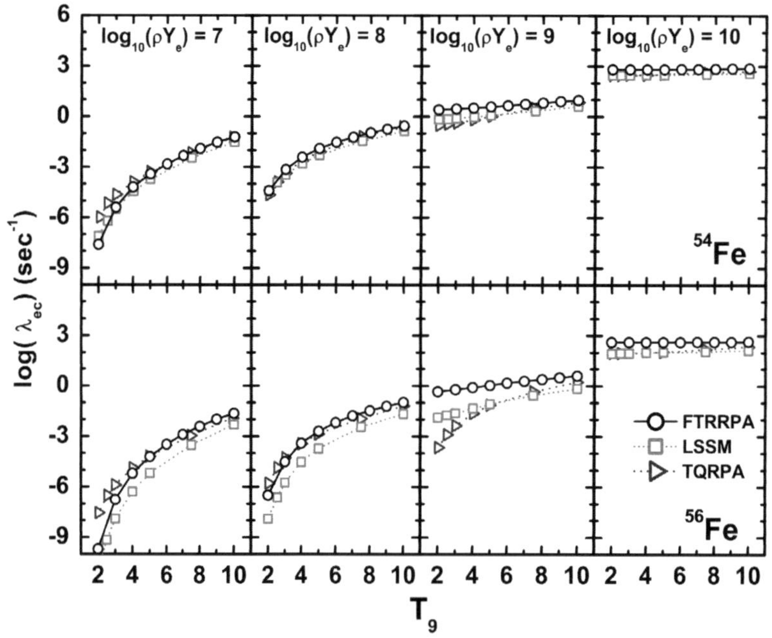

Figure 1. Electron capture rates as functions of temperature T for ^{54,56}Fe at selected densities ρYe (g cm^{-3}), calculated with the FTRRPA based on DD-ME2 interaction, in comparison with the LSSM [5] and TQRPA [9] results. Taken from Ref. [11].

The FTRRPA rates are displayed as functions of temperature T_9 ($T_9 = 10^9$ K) for ^{54,56}Fe and selected densities ρY_e (g cm^{-3}), in comparison with the results from studies based on LSSM and TQRPA[5, 9]. In general, the electron capture rates increase with temperature and electron density. For high electron densities, the rates increase slower, and in the case of $\rho Y_e = 10^{10}$ g/cm^3 the temperature dependence almost vanishes. At high electron densities, when the electron chemical potential is not small, high-energy electrons are able to excite most or even all the GT$^+$ transitions even at lower temperatures. For ^{54}Fe, at lower densities, e.g. $\rho Y_e = 10^7, 10^8$ g/cm^3, the FTRRPA agrees with the shell model calculations very well. At higher densities, e.g. $\rho Y_e = 10^9$ g/cm^3, with electron chemical potential about 5 MeV that is similar as the threshold energy, at lower temperatures the FTRRPA yields a higher rate than the shell model. One can understand this difference in the fragmentation of the shell model GT$^+$ strength over the energy range $0 \sim 10$ MeV[2].

When the electrons at lower temperature excite all the GT$^+$ strength at $E \simeq 2$ MeV in FTRRPA, only a fraction of the shell model strength could be excited [11]. Since part of the shell model GT$^+$ strength is located at higher energies than in the FTRRPA, the resulting rate is smaller than in the FTRRPA case. When the electron density becomes even higher, e.g. $\rho Y_e = 10^{10}$ g/cm^3 with chemical potential about 11 MeV, the high-energy electrons at lower temperatures are gradually sufficient enough to excite all the transition strength in the shell model approach as well, thus the discrepancy with the FTRRPA reduces. However, as the GT$^+$ centroid energy of the FTRRPA is lower than that of the shell model, in the former case the rates are somewhat lower. For ^{56}Fe, at densities $\rho Y_e = 10^7, 10^8$ g/cm^3 the FTRRPA results agree better with the TQRPA, while the LSSM gives lower rates. At higher densities, the FTRRPA yields a more similar trend with the shell model than the TQRPA, but the values are still above the shell model results. In general, the discrepancy between the FTRRPA and the shell model rates is larger in ^{56}Fe than ^{54}Fe. The main reason lies in the difference between the GT$^+$ centroid energy between the two models, that is larger in ^{56}Fe [11].

3. Conclusion

Recently a novel self-consistent theory framework has been introduced for modeling the electron capture rates in the initial phase of supernova core collapse, based on the relativistic energy density functional [16]. It is realized in terms of the finite temperature RMF and FTRRPA, allowing inclusion of all relevant charge-exchange transitions induced in the electron capture process. The electron capture rates have been studied for ^{54,56}Fe target nuclei in different stellar environments characteristic for the core collapse supernova, and compared with the shell model and the TQRPA. It is shown that the FTRRPA results in similar trend of temperature dependence as other approaches, but the overall values are somewhat larger, due to differences in the structure of GT$^+$ strength distributions. It is important to note that the finite temperature RMF and FTRRPA provide universal framework for description of stellar weak-interaction processes at finite temperature in a self-consistent way. This approach is especially valuable for those nuclei not accessible to the shell model diagonalization approach, and it is advantageous in comparison to other models which necessitate input from the experimental data on nuclei under consideration, which may not always be available.

Acknowledgments

This work is supported by the Chinese-Croatian project 'Nuclear structure and astrophysical applications", State 973 Program 2007CB815000, the NSF of China under Grants No. 10775004, No. 10947013, and No. 10975008, the Unity through Knowledge Fund (UKF Grant No. 17/08) and MZOS - project 1191005-1010 in Croatia.

References

[1] Janka H-Th, Langanke K, Marek A, Martínez-Pinedo G, and Müller B 2007 *Phys. Rep.* **44** 38
[2] Langanke K. and Martínez-Pinedo G 2003 *Rev. Mod. Phys.* **75** 819
[3] Fuller GM, Fowler WA, and Newman MJ 1980 *Astrophys. J. Suppl. Ser.* **42** 447
[4] Dean D, Langanke K, Chatterjee L, Radha P, and Strayer M 1998 *Phys. Rev.* C **58** 536
[5] Langanke K and Martínez-Pinedo G 2001 *At. Data Nucl. Data Tables* **79** 1
[6] Langanke K, Kolbe E, and Dean D 2001 *Phys. Rev.* C **63** 1
[7] Juodagalvis, Langanke K, Hix WR, Martinez-Pinedo G, and Sampaio JM 2010 *Nucl. Phys.* A **848** 454
[8] Nabi JU and Klapdor-Kleingrothaus HV 1999 *At. Data Nucl. Data Tables* **71** 149
[9] Dzhioev AA, Vdovin AI, Ponomarev VYu, et al. 2010 *Phys. Rev.* C **81** 015804
[10] Paar N, Colò G, Khan E, and Vretenar D 2009 *Phys. Rev.* C **80** 055801
[11] Niu YF, Paar N, Vretenar D, and Meng J 2011 *Phys. Rev.* C **83** 045807
[12] Ring P 1996 *Prog. Part. Nucl. Phys.* **37** 193
[13] Vretenar D, Afanasjev AV, Lalazissis GA, and Ring P 2005 *Phys. Rep.* **409** 101
[14] Meng J, Toki H, Zhou SG, Zhang SQ, Long WH, and Geng LS 2006 *Prog. Part. Nucl. Phys.* **57** 470
[15] Paar N, Vretenar D, Khan E, and Colò G 2007 *Rep. Prog. Phys.* **70** 691
[16] Niu YF, Paar N, Vretenar D, and Meng J 2009 *Phys. Lett.* B **681** 315
[17] Lalazissis GA, Nikšić T, Vretenar D, and Ring P 2005 *Phys. Rev.* C **71** 024312

Journal of Physics: Conference Series **337** (2012) 012014 doi:10.1088/1742-6596/337/1/012014

Equation of State of Symmetric And Asymmetric Nuclear Matter At Various Densities And Temperatures

S Shlomo

Cyclotron Institute, Texas A&M University, College Station, TX 77843-3366, USA

Email: shlomo@comp.tamu.edu

Abstract. The equation of state (EOS) of nuclear matter (NM) is an important ingredient in the study of properties of nuclei at and far from stability, of structures and evolution of compact astrophysical objects, such as neutron stars and core-collapse supernovae, and of heavy-ion collisions (HIC). We will review the current status of the EOS of NM, as deduced from theoretical and experimental studies of structure and reactions of nuclei.

1. Introduction

The saturation point of the equation of state (EOS), i.e., the energy as a function of matter density, for the symmetric nuclear matter (SNM) at zero temperature (T = 0) is well determined from ground state properties of nuclei, such as binding energies and central matter densities, by extrapolation to infinite nuclear matter (NM). To extend our knowledge of the EOS beyond the saturation point of the SNM, an accurate value of the NM incompressibility coefficient K, which is directly related to the curvature of the EOS, is needed. An accurate value of the symmetry energy coefficient, J, is needed for the EOS of asymmetric NM. It is well known that the energies of compression modes in nuclei, such as the isoscalar giant monopole resonance (ISGMR) and the isoscalar giant dipole resonance (ISGDR), are sensitive to the value of K and the energies of the isovector giant dipole resonance (IVGDR) are sensitive to J. We consider, in particular, the issue of determining the parameters of the effective nucleon-nucleon interaction from properties of finite nuclei within the Hartree-Fock (HF) approach. We will present results of fully self-consistent HF-based random-phase-approximation (RPA) for the ISGMR and compare with experimental data.

A proper description of nuclei away from stability and the structure and evolution of compact astrophysical objects requires knowledge of the EOS at extreme conditions of density (low and high), isospin asymmetry and temperature. Due to the important effect of correlations the mean-field approach is inadequate. Studies of multifragmentation of hot nuclear systems, created in high energy proton induced reactions or in intermediate energy heavy-ion collisions (HIC), have been employed to determine the EOS of nuclear matter at extreme conditions. It is common to adopt the freeze-out assumption and/or the coalescence model to reconstruct the properties of the dense excited matter using the fragment yields. We will discuss important effects that should be accounted for in the analysis of experimental data and comment on the values of T, matter density and the symmetry energy density at low density determined from HIC.

2. EOS at zero temperature

The development of a modern and more realistic nuclear energy density functional (EDF) [1] for accurate predictions of properties of nuclei is the subject of enhanced activity, since it is very important for the study of properties of rare nuclei with unusual neutron-to-proton ratios that are difficult to produce experimentally and likely to exhibit interesting new phenomena associated with isospin, clusterization and the continuum. Since the pioneering work of Brink and Vautherin [2], continuous efforts have been made to readjust the parameters of the Skyrme-type effective nucleon-nucleon (NN) interaction to better reproduce experimental data [3]. We have recently considered the Skyrme EDF [4] associated with the standard parameterization of Skyrme type interactions [5]:

$$V_{12} = t_0(1 + x_0 P_{12}^{\sigma})\delta(\vec{r}_1 - \vec{r}_2) + \frac{1}{2}t_1(1 + x_1 P_{12}^{\sigma})[\overleftarrow{k}_{12}^2\delta(\vec{r}_1 - \vec{r}_2) + \delta(\vec{r}_1 - \vec{r}_2)\vec{k}_{12}^2]$$
$$+ t_2(1 + x_2 P_{12}^{\sigma})\overleftarrow{k}_{12}\delta(\vec{r}_1 - \vec{r}_2)\vec{k}_{12} + \frac{1}{6}t_3(1 + x_3 P_{12}^{\sigma})\rho^{\alpha}\left(\frac{\vec{r}_1 + \vec{r}_2}{2}\right)\delta(\vec{r}_1 - \vec{r}_2) + \quad (1)$$
$$iW_0\overleftarrow{k}_{12}\delta(\vec{r}_1 - \vec{r}_2)(\vec{\sigma}_1 + \vec{\sigma}_2) \times \vec{k}_{12}.$$

where t_i, x_i, α , and W_0 are the parameters of the interaction, P_{12}^{σ} is the spin exchange operator, $\vec{\sigma}_i$ is the Pauli spin operator, $\vec{k}_{12} = -i(\vec{\nabla}_1 - \vec{\nabla}_2)/2$, and $\overleftarrow{k}_{12} = -i(\overleftarrow{\nabla}_1 - \overleftarrow{\nabla}_2)/2$. Here, the right and left arrows indicate that the momentum operators act on the right and on the left, respectively. We determined within the HF approximation a new and more realistic Skyrme interaction (named KDE0) by fitting [4] a set of extensive data on binding energies, "bare" single particle energies, charge root mean square (rms) radii, and rms radii of valence nucleon density distribution of nuclei. We have included in the fit, for the first time, the data on the constraint energies of the isoscalar giant monopole resonances (ISGMR) of nuclei and imposed additional constraints, such as a non-negative value for the slope of the symmetry energy density at high nuclear matter (NM) density (at three times the saturation density of NM) and the Landau stability constraints on nuclear matter. We have implemented, for the first time, the simulated annealing method (SAM) together with an advanced least square method to search for the global minima. In Table 1 we present the results of fully self-consistent HF-based RPA for the centroid energies of the ISGMR (see Refs. [6,7]) for several nuclei, using the KDE0 [4], SK255 [6], and SGII [8] Skyrme interactions and the results obtained within the relativistic mean-field (RMF)-based RPA obtained for the NL3 interaction [9] and compare with the experimental data of Refs. [10,11]. Considering the results for the SK255 and NL3, we conclude that it is possible to build *bona fide* Skyrme forces with K close to that obtained in relativistic models. We have also found that (see the review in Ref. [12]) $K_{nm} = 240 \pm 20$ MeV. The uncertainty of 20 MeV is mainly due to the uncertainty in the value of *J*.

Table 1. Fully self-consistent RPA results for the centroid energies of the ISGMR for interactions with various values of K and J (in MeV), compared with experimental data.

Nucleus	$\omega_1 - \omega_2$	Expt.	NL3	SK255	SGII	KDE0
^{90}Zr	0-60		18.7	18.90	17.89	18.03
	10-35	17.81± 0.30		18.85	17.87	17.98
^{116}Sn	0-60		17.1	17.31	16.36	16.58
	10-35	15.85± 0.20		17.33	16.38	16.61
^{144}Sm	0-60		16.1	16.21	15.26	15.46
	10-35	15.40± 0.40		16.19	15.22	15.44
^{208}Pb	0-60		14.2	14.34	13.57	13.79
	10-35	13.96± 0.20		14.38	13.58	13.84
K (MeV)			272	255	215	229
J (MeV)			37.4	37.4	26.8	33.0

3. Temperature and density dependence of EOS from heavy ion collisions

We point out that the decay of highly excited nuclear matter produced in intermediate energy heavy-ion collisions is a complex dynamic process. During the collision a dense and hot nuclear matter is created which then expands and fragments. A simple approach for describing the process is the freeze-out concept, in which one assumes that the expanding nuclear matter reaches thermal equilibrium. With decreasing density, the reaction rates decrease and the equilibration process is suppressed. At this point the thermal and chemical equilibriums are frozen out and the reaction product distribution is identical to the cluster distribution at the freeze-out point. In a macrocanonical ensemble, the statistical properties of the system can be evaluated using the grand partition sum

$$\mathscr{Z} = \sum_{\{f\}} e^{(\mu_n N_f + \mu_p Z_f)/T} Q_f, \quad Q_f = \exp(-F_f/T). \tag{2}$$

In Eq. (1), N_f, Z_f, Q_f and F_f are the total neutron number, the total charge, the partition function, and the free energy for a given event f, respectively. The Lagrange multipliers T, μ_n and μ_p are determined by the corresponding conservation laws. Using (2) one finds [13] for the relative yield of fragments s the expression

$$\frac{\bar{n}_s}{\bar{n}_n^{N_s}\,\bar{n}_p^{Z_s}} = \frac{1}{2}\left(\frac{1+\kappa}{2\chi}\right)^{A_s-1} A_s^{3/2}\left(\lambda_T^3\right)^{A_s-1} Q_s^{*}\, e^{-(B_s+E_C(s))/T}. \tag{3}$$

Here, $\bar{n}_s$ is the average density of clusters s with $A_s=N_s+Z_s$ nucleons, $\chi = V'/V_0$ is the hindrance factor, λ_T is the thermal nucleon wave-length, Q_s^* is the internal partition function, and B_s and $E_C(s)$ are the binding energy and the Wigner-Seitz Coulomb energy corrections, respectively.

A key element in the investigation of the nuclear multifragmentation phenomena is the derivation of the temperature T at the freeze-out point, which can be simply obtained from (3) by considering ratios of the yields of fragments [14]. In particular, one finds that the dependence of the excitation energy of the disassembling system on T, i.e., the caloric curve, shows irregularities which may be interpreted as a possible liquid-gas phase transition.

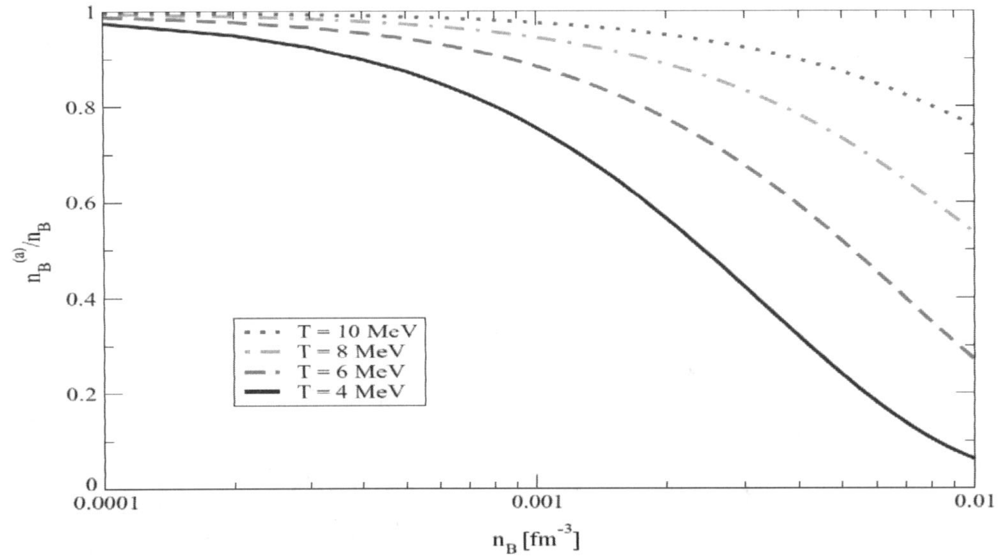

Figure 1. The ratio between the baryon density determined without ($n_B^{(a)}$) and with (n_B) medium effect.

In the analysis of the experimental data of fragment yields we have considered the effects of:

(i) The long range Coulomb interactions among fragments in the freeze-out volume [15]. Here we employ the Wigner-Seitz approximation;

(ii) The radial collective flow [16]. An expanding system, in a strict thermodynamic sense, is not in equilibrium. However, if the time scale involved in the expansion is much larger compared to the equilibration times in the expanding complex, i.e., the flow velocity is quite small compared to the average nucleonic velocity, the assumption of thermodynamic equilibrium may not be inappropriate;

(iii) The post emission decay (secondary decay) processes of the fragments emitted from the the freeze-out surface [17];

(iv) The effect of the medium on the binding energies of clusters [18]; and,

(v) The effect of clusterization on the symmetry energy density [19];

We have demonstrated that: (i) Coulomb effects are important when considering ratios of yields of isotones; (ii) The flow affects the freeze-out volume; (iii) Taking into account the post emission decay one extracts the same freeze-out temperature using different thermometers (double yields ratios); (iv) Although, at low density, the temperature calculated from given yields changes only modestly if medium effects are taken into account, larger discrepancies are observed when the nucleon densities are determined from measured yields, see Figure 1 (taken from Ref. [18]); (v). Due to clusterization, at low density nuclear matter, the symmetry energy is much larger than that predicted by mean-field models (see Ref. [19]).

Acknowledgements

This work was supported by the US Department of Energy under Grant Nos. DE-FG02-93ER40773. We thank V. M. Kolomietz, J. De, A. Kolomiets, B.K. Agrawal, T. Sil, G. Roepke and J. B. Natowitz for discussions and collaboration on work reported here.

References

[1] Kohn W 1999 *Rev. Mod. Phys.* **71** 1253
[2] Vautherin D and D. M. Brink D M 1972 *Phys. Rev.* C **5**, 626
[3] Bender M, Heenen P H and Reinhard P-G 2003*Rev. Mod. Phys.* **75**, 121
[4] Agrawal B K, Shlomo S and Au V Kim 2005 *Phys. Rev.* C **72**, 014310
[5] Chabanat E, Bonche P, Haensel P, Meyer J and Schaefer R 1997 *Nucl. Phys.* A **627**, 710
[6] Agrawal B K, Shlomo S, and Au V Kim 2003 *Phys. Rev.* C **68**, 031304R
[7] Sil T, Shlomo s, Agrawal B K and Reinhardt P-G 2006 *Phys. Rev.* C **73**, 034316
[8] Giai N V and Sagawa H 1981 *Phys. Lett.* B **106**, 379
[9] Lalalazissis G A, Konig J and Ring P 1997 *Phys. Rev.* C **55** 540
[10] Youngblood D H *et al.* 2004 *Phys. Rev.* C **69** 034315
[11] Youngblood D H *et al.* 2004 *Phys. Rev.* C **69** 054312
[12] Shlomo S, Kolomietz V M and Colo G 2006 *Eur. Phys. J.* A **30**, 23
[13] Shlomo S and Kolomietz V M 2005 Rep. Prog. Phys. **68** 1
[14] Albergo S, Costa S, Costanzo E and Rubbino A 1985 Nuovo Cimento **A89** 1
[15] Kolomiets A, Kolomietz V M and Shlomo S 1997 Phys. Rev. C **55** 1376
[16] Shlomo S, De J N and Kolomiets A 1997 Phys. Rev. C **55** R2155
[17] Kolomiets A et al. 1996 Phys. Rev. C **54** R472
[18] Shlomo S, Ropke G, Natowitz J B, Quin L, Hagel K, Wada R and Bonasera A 2009 Phys. Rev. C **79** 034604
[19] Natowitz J B, Ropke G, Typel S, Blaschke D, Bonasera A, Hagel K, Klahn T, Kowalski S, Qin L, Shlomo S, Wada R and Walter H H 2010 Phys. Rev. Lett. **104** 202501

Nuclear Physics in Astrophysics V IOP Publishing
Journal of Physics: Conference Series **337** (2012) 012015 doi:10.1088/1742-6596/337/1/012015

Astrophysical(α, γ) reaction in inverse kinematics; Electron screening effect in the beta-decay

P Ujić[1,2], F de Oliveira[1], A Lagoyannis[3], T J Mertzimekis[3], S Harissopulos[3], P Demetriou[3], A Spyrou[4], C Stodel[1], MG Saint-Laurent[1], O Kamalou[1], A Lefebvre-Schuhl[5], S Grevy[1], L Caceres[1], M Lewitowicz[1], M A Amthor[1], L Perot[6], A Coc[5], V Tatischeff[5], J Kiener[5], O Sorlin[1], A Lepailleur[1], M Assié[1], B Bastin[1], L Achouri[7], R Borcea[8], C Borcea[8], M Stanoiu[8], G de France[1], E Clement[1], A Pautrat[9], A Buta[8], L Gaudefroy[10], V Meot[10] and I Deloncle[5]

[1] GANIL, CEA/DSM-CNRS/IN2P3, Caen, France
[2] VINCA Institute, University of Belgrade, Belgrade, Serbia
[3] Tandem Accelerator Lab., Inst. of Nuclear Physics, NCSR "Demokritos", Athens, Greece
[4] NSCL/MSU, East Lansing, Michigan, USA
[5] CSNSM, CNRS/IN2P3 and Université Paris Sud XI, F-91405 Orsay Campus, France
[6] IPNO-IN2P3-CNRS, Université Paris-Sud, Orsay, France
[7] LPC Caen, ENSICAEN, Université de Caen, CNRS/IN2P3, Caen, France
[8] IFIN-HH, Bucharest, Romania
[9] CRISMAT, Caen, France
[10] CEA/DAM Bruyères-le-Châtel, France

E-mail: ujic@vinca.rs

Abstract. The abundance calculations of the p-nuclei produced in explosive stellar sites rely on the Hauser-Feshbach (HF) theory with the alpha-article optical model potential (α-OMP) one of its major ingredients. To date, most of the (α, γ) cross sections measured show that HF calculations can be wrong by a factor of ten or more especially when phenomenological α-OMP are employed. To investigate the relevant uncertainties entering the HF calculations and furthermore develop global microscopic α-OMPs, systematic (α, γ) cross-section measurements are necessary. This led us to perform a feasibility study of (α, γ) measurements in inverse kinematics that will allow us to employ also radioactive beams in the future. Hence, the ^{4}He(^{78}Kr,γ)^{82}Sr reaction was studied using the LISE3 spectrometer to separate the ^{82}Sr recoils from the primary ^{78}Kr beam. Although an excellent rejection factor $> 10^{10}$ was achieved, the position of the ions of interest was unexpectedly masked by a secondary beam of high intensity. Given these, new setup improvements are proposed to remove the pollutant ions.

Recently, many experiments were conducted in order to study the influence of the environment (especially in a metallic material) on the decay probability of radioactive nuclei. Additionally, hydrogen-like fusion reactions were performed indicating a change in the cross-section due to the influence of the Coulomb field screening induced by quasi-free electrons in metals. This was explained by the Debye screening model which treats metallic electrons within Maxwell-Boltzmann statistics. We measured the decay rate of ^{19}O in metallic, insulating and superconducting environments whereas the electrons in the superconductors should obey the Bose-Einstein statistics. The decay rate measurement was supported by a branching ratios measurement. We found that the effect on the decay rate, if any, is less than the 0.1%, far below the theoretical predictions.

Journal of Physics: Conference Series **337** (2012) 012015 doi:10.1088/1742-6596/337/1/012015

1. Introduction

Abundance calculations of p nuclei make an use of the nuclear statistical model for the calculation of the reaction rates needed for the solution of an extended reaction network involving over than 20000 nuclear reactions on $\approx$ 2000 nuclei [1]. The majority of calculations have to rely on cross-section predicted by the Hauser-Feshbach (HF) theory. Some of the very few (α, γ) data show that the reaction rates calculated using phenomenological alpha-particle optical potentials can be wrong by a factor of ten or more. These uncertainties might be reduced by systematic cross-section measurements of α-particle capture reactions at relevant energies, mainly in the A=80-140 and A=170-200 region [2]. The (α, γ) measurements at sub-Coulomb energies in inverse kinematics using state-of-the-art detectors combined with velocity filters could be the best solution. In the present work we report on the ^{78}Kr$(\alpha, \gamma)^{82}$Sr reaction, the first feasibility study of an α-particle capture reaction in inverse kinematics at energies relevant to the p process, performed on the Wien Filter (WF) of the LISE3 spectrometer at GANIL.

In the second part of this article we will describe the electron screening effect in the beta-decay. It is possible to calculate the influence of the screening effect (SE) on the beta decay rate by considering Fermi golden rule. Many light elements fusion experiments were conducted where metals were used as a host for hydrogen, deuterium or helium and it was noticed in many of these experiments that the SE is larger than the maximal theoretical limit. These findings have been confirmed by different groups [3, 4, 5] and also for some beta decaying nuclei (e.g. [6]). It was found that the high SE occurs in general in metals. The electrons in metals obeys the Pauli principle of exclusions and thus obey the Fermi distribution. However, bosons obey the Bose-Einstein distribution and as the electrons in superconductors (SC) are organized as pairs called "Cooper pairs" which are bosons, it was decided to examine the SE in superconductors. Actually, Stoppini [7] predicted that the Cooper pairs could induce high SE.

2. Experimental setup of the ^{78}Kr$(\alpha, \gamma)^{82}$Sr reaction in inverse kinematics

The ^{78}Kr$(\alpha, \gamma)^{82}$Sr reaction was investigated at an energy of 1.76 MeV/u using the LISE3 spectrometer installed at GANIL. The target used at GANIL consisted of $\approx$ 1.3$\times$ 10^{17} atoms/cm^2 of Helium implanted into a 50 μg/cm^2 Al foil. The expected cross-section for the reaction was ($\approx$ 100 μb) and the difference between the velocity of the ^{78}Kr beam and the compound nucleus ^{82}Sr is $\sim$5% only. Such a small velocity difference is a real challenge for the separation of the recoiling nuclei and the beam. The 12 meters long block of Wien Filter of the LISE spectrometer was used for the separation of the ^{82}Sr nuclei from the primary stable beam ^{78}Kr^{+8}. The advantage of this device is the fact that the selection of the nuclei of interest is based on the velocity and not on the charge state of the ions. The intensity of the primary beam was 7.8$\times 10^9$ pps. The main part of the primary beam was stopped on the "61W" slits located in the center of the WF and the rest was stopped on the "62" slits located at the end of the spectrometer, as shown in Figure 1. In this setup with closed slits the velocity acceptance was 0.55 % Extensive ion optics simulations were necessary, where it was necessary to have a fine tunning of the quadrupoles as well. For these purposes, the ZGOUBI code version 5.1.0. was used.

3. Results and conclusions of the ^{78}Kr$(\alpha, \gamma)^{82}$Sr reaction in inverse kinematics

The analysis of the data has shown that the measured rejection factor of the primary beam is very good, i.e. higher than 10^{10}. However, we have observed particles at $\sim$84 MeV which cannot be explained by energy loss and scattering in the target, neither by the scattering elsewhere in the separator. The SRIM simulation has shown that the energy and the counting rate of the locus at 84 MeV cannot be explained by the straggling in the target. Since the target is very thin, the most probable reason for this energy discrepancy is microscopic dust deposited on the target. The dust particles of diameter from 10 μm down to 1 μm are present in the atmosphere

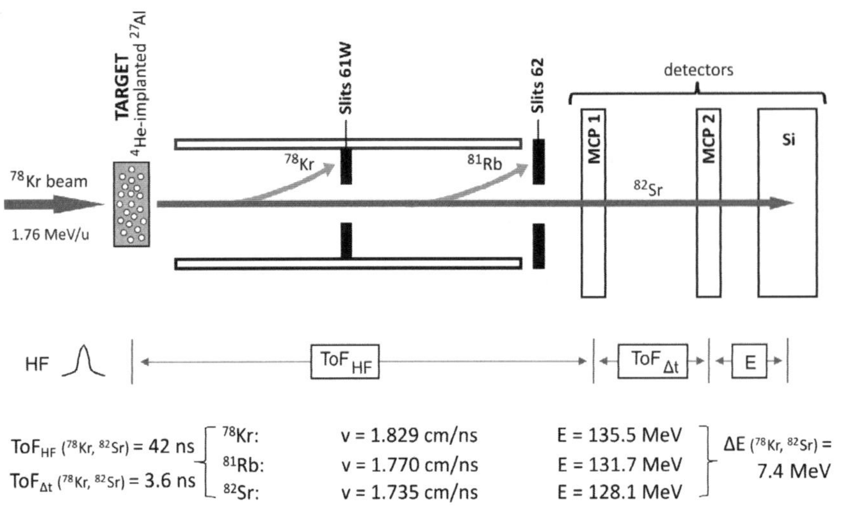

Figure 1. Experimental setup employed ^{78}Kr(α,γ) reaction in inverse kinematics

with density of 1 cm^{-3}. We used simplified model of a spherical PVC dust particle of 2.5 μm in diameter deposited on the surface of the target. If such particle is placed at the beam spot on the target, it can produce the energy seen at the locus 84 MeV and regarding its effective surface, it explains the measured count rate of the locus 84 MeV.

We didn't succeed to separate any events attributed to the ^{82}Sr, mainly due to the high presence of the background at our zone of interest coming from the locus 84 MeV. However, we can conclude that the obtained rejection factor was quite good - higher than 10^{10}. Regarding the high risk of the scattering on the dust deposited on the solid target, the use of the gas windowless target is indispensable. Considering the achieved rejection factor of the primary beam and the possible use of the windowless gas target, this type of experiment is promising.

4. Experimental setup of the half-life measurement in superconcuctive environment

As a host superconductor we used a niobium as it has convenient critical temperature of $\approx$9.2 K, which allows us to use simple cooling system based on liquid helium. Additionally, the niobium can support high irradiation by heavy ions and still retain the superconductive attributes [8]. We measured the half-lives of ^{19}O and ^{19}Ne, and the branching ratio of ^{19}O, since each branch has its own Q value and the same screening energy will have different influence on different branches. It was intended to measure the decay curve repeating the implantation-measurement cycles until we achieve enough statistics to obtain the precision of the measurement of 0.1%. The implantation period was chosen to last for approximately two duration of two half-lives of ^{19}O (^{19}Ne), while the measurement time was about 20 half-lives of given nucleus. In order to lower down the possible systematic influences on the measurement it was decided to measure in one by one hour cycles. Namely, we measured several implantation-measure cycles during one hour in a superconducting (SC) phase of the Nb, then we would change the temperature and measure several implantation-measure cycles during one hour in a metallic phase of Nb.

5. Results and conclusions of the half-life measurement in superconcuctive environment

The obtained results are shown on the table 1. Although, all measured half-lives and branching ratios are practically within the one sigma error bars, we can notice certain systematic behavior.

Table 1. The measured change of the half-life and branching ratio of ^{19}O in different environments and temperatures [9]. The uncertainty of the estimated half-lives shows the statistical uncertainty and the uncertainty due to the correction on the ^{19}Ne contamination.

Material and phase	Temp. [K]	BR of 197/1554	BR of 197/4377	Half-life ^{19}O [s]	Half-life ^{19}Ne [s]
Nb SC	4	3.611 ± 0.015	3000 ± 200	$26.454 \pm 0.015 \pm 0.007$	17.231 ± 0.007
Nb metal	16	3.607 ± 0.015	2700 ± 170	$26.427 \pm 0.016 \pm 0.007$	17.236 ± 0.006
Nb SC	4	3.595 ± 0.017	3100 ± 200	$26.472 \pm 0.015 \pm 0.007$	17.231 ± 0.008
Nb metal	90	3.582 ± 0.017	2800 ± 200	$26.461 \pm 0.016 \pm 0.007$	17.237 ± 0.009
Teflon	300	3.606 ± 0.038	4300 ± 800	$26.502 \pm 0.085 \pm 0.008$	N/A

If we compare each measurement of ^{19}O at superconducting phase and its corresponding measurement in metallic phase we can see that the branching ratio and the half-life is always higher in the superconducting environment. We have the same situation for the ^{19}Ne with opposite change between superconducting and metallic phase since ^{19}Ne decays by β^+ and ^{19}O decays by β^-. This behavior is exactly what should be expected if the screening effect on the decay probability would exist. Thus, we decided to combine all these measurement – the branching ratio and half-life of oxygen and the half-life of neon and to estimate the average effect. The average screening effect in superconducting phase in comparison to metallic phase was 1.1% ± 0.9% of the effect predicted by the screening model in superconductors. These changes would correspond to the screening energy of 880 eV for the oxygen and 360 eV for the neon. However, we also realized that method used in previous publications, to calculate a screening energy cannot be true, since both parent and daughter nucleus are subjected to the screening and this has to be included in the estimation of the Q-value:

$$Q_{scr} = \left[M(A,Z)c^2 - U_{scr}(Z)\right] - \left[M_{-1}(A,Z \pm 1)c^2 - U_{scr}(Z \pm 1)\right] - m_e c^2 = Q_\beta - U_{scr}(Z=1),$$

where $U_{scr} = \frac{1}{4\pi\epsilon_0}\frac{Z_e Z e^2}{R_{scr}}$, where $Z_e = 1$ is the charge of electron 1 and Z is the atomic number of the considered nucleus. This equation means that the screening potential is independent on the atomic number of the considered nucleus and it is always equal to the screening potential on the hydrogen nuclei for the given screening radius. This cast strong doubts about the efficiency of the screening effect on the change of the probability of the beta-decay.

6. Acknowledgements

The work partially supported by FP7/REGPOT/LIBRA Project (Grant 230123) and IN2P3.

References

[1] Arnould M and Goriely S 2003 *Phys. Rep.* **384** 1–84
[2] Demetriou P, Grama C and Goriely S 2002 *Nucl. Phys. A* **707** 253
[3] Czerski K, Huke A, Biller A, Heide P, Hoeft M and Ruprecht G 2001 *EPL* **54** 449
[4] Kasagi J, Yuki H, Baba T, Noda T, Ohtsuki T and Lipson A 2002 *JPSJ* **71** 2881
[5] Raiola F, Gang L, Bonomo C, Gy
"urky G, Aliotta M, Becker H, Bonetti R, Broggini C, Corvisiero P and et al D 2004 *EPJA* **19**(2) 283
[6] Limata B, Raiola F, Wang B, Yan S, Becker H, D'Onofrio A, Gialanella L, Roca V, Rolfs C, Romano M *et al.* 2006 *EPJA* **28** 251
[7] Stoppini G 1991 *Nuovo Cimento* **13** 1181
[8] Neum
"uller H, Lehmann M, Ischenko G, Klaum
"unzer S and Adrian H 1977 *Phys. Lett. A* **62** 446
[9] Ujić P et al., to be submitted

Nuclear Physics in Astrophysics V IOP Publishing
Journal of Physics: Conference Series **337** (2012) 012016 doi:10.1088/1742-6596/337/1/012016

Upper Limit on the molecular resonance strengths in the $^{12}C+^{12}C$ fusion reaction

X. D. Tang[1], X. Fang[1], B.Bucher[1], H. Esbensen[2], C. L. Jiang[2], K. E. Rehm[2], and C. J. Lin[3]

[1]Department of Physics, University of Notre Dame, Notre Dame, IN 46556, USA
[2]Physics Division, Argonne National Lab, Argonne, IL 60439, USA
[3]China Institute of Atomic Energy, P. O. Box 275(10), Beijing 102413, P. R. China

Abstract. Molecular resonances complicate the prediction of the carbon fusion rate at the energies of astrophysical relevance. By comparing the cross sections of three carbon isotope fusion reactions, $^{12}C+^{12}C$, $^{12}C+^{13}C$ and $^{13}C+^{13}C$, an empirical relationship among the three systems is found. With this relationship and the Equivalent Square Well (ESW) model, for the first time, the resonant cross sections in the $^{12}C+^{12}C$ fusion reactions are quantitatively described. Consequently, an upper limit on the molecular resonance strengths in $^{12}C+^{12}C$ fusion reactions is established. With this limit, we claim that the strong resonances at 1.5 MeV and 2.14 MeV are not realistic. This upper limit posts an additional constraint for the superburst models.

1. Introduction

In 1960 Almqvist, Kuehner and Bromely discovered several resonances in collisions between ^{12}C nuclei. For at least three energies, $E_{c.m.}$=5.68, 6.00 and 6.32 MeV, they observed resonances in the yield of collisional byproducts: p, α, n and γ. These resonances have characteristic widths of about 100 keV. The resonances were identified as signatures of the formation of nuclear molecules[1, 2]. In the following years, as measurements were pushed down towards lower energies, the discoveries of such resonances continued as far as the lowest measured energies. For instance, the most recent published measurement of the $^{12}C+^{12}C$ fusion reported a strong resonance at $E_{c.m.}$=2.14 MeV[3].

Apart from the interesting molecular resonances, the $^{12}C+^{12}C$ fusion reaction plays a crucial role in a number of important astrophysical scenarios[4]. The important energy range spans from 1 MeV to 3 MeV in the center of mass frame, which is only partially covered by the measurements of the $^{12}C+^{12}C$ fusion reaction. An extrapolation is the only resource available to obtain the fusion reaction rate for the astrophysical applications. The currently adopted reaction rate is established based on the modified S factor S*(E)[5], which is defined as,

$$S^*(E) = \sigma(E)Ee^{\frac{87.21}{\sqrt{E}}+0.46E}. \tag{1}$$

An averaged S* factor of 3×10^{16} MeV*b was obtained by fitting the data measured by Patterson[5], Spinka[6] and Becker[7]. This averaged value was extrapolated down to lower energies with the simple assumption that the averaged modified S factor is constant at sub-barrier energies[5, 8]. At present, there is nothing known about the exact energies and strengths of these resonances in the energy region of astrophysical relevance. Therefore, the current carbon fusion rate based on the averaged S* factor is highly uncertain [9].

This uncertainty leaves an ambiguous space in astrophysical models and prevents us from a precise understanding of the underlying physics. For example, superbursts are long, energetic, and rare thermonuclear flashes on accreting neutron stars in low-mass X-ray binaries[10]. These bursts are considered to be triggered by the unstable ^{12}C burning in the ash left over from the rp-process on the surface of a neutron star. The ash is heated up with the heat generated in the crust of a neutron star and ignites the carbon in the ash. However, with the currently adopted carbon fusion rate, superburst modes fail to explain the ignition of carbon at the column depth inferred from observations [11]. Another difficulty is that the amount of ^{12}C in the ash does not seem to be sufficient to trigger the superburst. Inspired by the strong resonance observed at $E_{c.m.}$=2.14 MeV, a hypothetical resonance at $E_{c.m.}$=1.5 MeV with a similar strength was proposed to enhance the fusion reaction rate at $T=0.5\times10^9$ K by a factor of over two orders of magnitude and thereby alleviate the discrepancy between superburst models and observations[10].

2. An Empirical relationship among the carbon isotope fusion reactions

In contrast to the striking resonances in the ^{12}C+^{12}C fusion reaction, other carbon isotope fusion reactions, such as ^{12}C+^{13}C and ^{13}C+^{13}C, behave more regularly. Only minor resonance features have been observed in these two systems [12, 13]. The shape of the fusion cross sections at sub-barrier energies is primarily dominated by the Coulomb barrier penetration effect. To remove this effect and reveal more details of nuclear interaction, the cross sections of all three carbon fusion systems are converted into the S* factors using eq.1. The advantage of using this conversion over the traditional astrophysical S factor is that the cross section ratios among the three systems are preserved.

The experimental S* factor for all three carbon isotope fusion systems are shown in Fig.1. There are several important features in these carbon isotope fusion systems:

(i) The ^{12}C+^{12}C cross sections are bound from above by the cross sections of the other two carbon isotope fusion systems.
(ii) The major resonant cross sections of ^{12}C+^{12}C (E_r=3.08, 4.28, 4.92, 5.67, 5.96 and 6.26 MeV) match remarkably well with the other two carbon isotope fusion cross sections, especially ^{13}C+^{13}C.
(iii) Overall, the ^{12}C+^{13}C cross sections are the highest among the three carbon isotope fusion systems. The deviation between ^{12}C+^{13}C and ^{13}C+^{13}C occurs in the energy of 3.5 to 5 MeV (excluding the 4.25 MeV resonance in ^{12}C+^{13}C) and is less than 30%.
(iv) The 4.25 MeV resonance in ^{12}C+^{12}C is also present in ^{12}C+^{13}C.

In the ^{12}C+^{13}C and ^{13}C+^{13}C fusion reactions, the valence neutron of ^{13}C introduces an additional degree of freedom for excitation, besides the ^{12}C core excitation. Despite this complication, the maximum deviation among the ^{12}C+^{12}C peak cross sections and the ^{12}C+^{13}C and ^{13}C+^{13}C fusion cross sections is less than 30%. Based on this strong correlation, we claim:

(i) In the ^{12}C+^{12}C fusion reaction, all the major resonances with maxima matching the ^{12}C+^{13}C and ^{13}C+^{13}C cross sections should have the same reaction mechanism;
(ii) This reaction mechanism should also be present in the fusion of ^{12}C+^{13}C and ^{13}C+^{13}C;
(iii) The effect arising from the valence neutron in ^{13}C is not strong compared to this reaction mechanism but it is strong enough to make the resonances disappear in the fusion of ^{12}C+^{13}C and ^{13}C+^{13}C.

One possible mechanism is the coupling effect between the incident elastic channel and the excited channels proposed by Imanishi and others[15, 2]. In the Nogami-Imanishi model, the

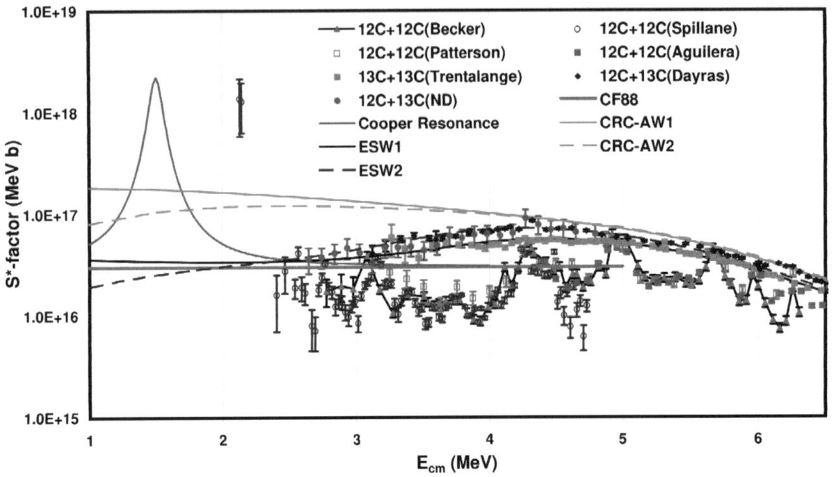

Figure 1. The experimental S* factors of three carbon isotope fusion reactions, $^{12}C+^{12}C$[3, 5, 7, 14], $^{12}C+^{13}C$ from this work (green points) and Ref.[12] and $^{13}C+^{13}C$[13]. The averaged S* factor of $^{12}C+^{12}C$(S*=3×10^{16} MeVb), labled as CF88(the red line), is recommended based on the data of Becker, Patterson and Spinka (not shown here)[8]. There are some data with large error bars in Spillane's data. In this paper, only those with relative errors less than 70% are shown. The hypothetical resonance proposed by Copper *et al.* is also shown. The explanations for *CRC–AW*1,2 and *ESW*1,2 are available in the section 3.

effect of the coupling to the $^{12}C(2^+$, 4.44 MeV) state is crucial for the formation of the $^{12}C+^{12}C$ molecular resonances. The coupling effect leading to the resonances only takes place when the energy matches the condition of the molecular states[15]. In the other two carbon fusion systems, such a coupling effect should also exist. The presence of valence neutron significantly increases the level density of compound nuclei[16]. As a result, the matching condition becomes valid at any energies and the coupling effect can take place at any energy. This may be the reason for the observed correlations among these three carbon fusion systems.

3. Upper limit on the resonant strengths

The correlations among the three carbon isotope systems provide a possibility to establish an upper limit for the important $^{12}C+^{12}C$ reaction. If all the major resonances in $^{12}C+^{12}C$, including those to be discovered at lower energies, are driven by the same reaction mechanism, one can expect to establish an extrapolation based on the resonant cross sections observed at higher energies. To test the predictive power of extrapolating models and achieve a more reliable extrapolation within the energy range of 1 to 3 MeV, it is necessary to push the measurements of the carbon isotope fusion reactions towards lower energies.

The recent study of $^{12}C+^{13}C$ at Notre Dame provides a great opportunity to test the predictive power of extrapolating models. In this experiment, the lowest measured cross section has been pushed down from 1 μb [12] to 20 nb [17]. The dominant component in the error bars is the systematic uncertainty of the theoretical branching ratio which is estimated as 20% [12]. Two different models have been applied to test their predictive power at low energies. The first one is the standard coupled-Reaction-Channels calculation[19] using the the Akyüz-Winther potential (CRC-AW). In the calculation, the $^{13}C(3/2^-$, 3.684 MeV) state is included. The other is the Equivalent Square Well (ESW) potential, which has been chosen traditionally to model the sub-barrier fusion cross sections and provide extrapolation within the energy ranges of astrophysical interest[9]. Using the new $^{12}C+^{13}C$ data at lower energies, it is demonstrated that CRC-AW calculation (labled as $CRC\text{–}AW2$ in Fig. 1) over-predicts the fusion cross sections by a factor of 3.3 at $E_{c.m.}$=2.640 MeV. This observation agrees with the hindrance study[18]. The ESW extrapolation taken from Ref.[12] (labeled as $ESW2$ in Fig. 1) proves its superior prediction power by matching the new data at the energies below 3.1 MeV within their quoted error bars.

After the validation, these two models are used in $^{12}C+^{12}C$ to establish the upper limit for the resonant cross sections. The ESW model extrapolation, labeled as $ESW1$ in Fig. 1, is obtained by fitting those major peak cross sections (E_r=3.08, 4.28, 4.92, 5.67, 5.96 and 6.26 MeV) which match the $^{12}C+^{13}C$ and $^{13}C+^{13}C$ fusion cross sections. Even though the fit is based on the peak cross sections above 3 MeV, the extrapolation to lower energies provide an upper bound on most of Spillane's data as well, except the strong resonance observed at 2.14 MeV which is about a factor of 40 higher than the ESW extrapolation. The preliminary result from the $^{12}C+^{12}C$ experiment at Naples[20] found that $S^*=(1.4\pm3.3)\times10^{16}$ MeV b at 2.15 MeV, which disagrees with the strong resonance found in Ref.[3] but agrees well with our upper limit. Another safer upper limit can be estimated using CRC-AW which has been proven to over-predict the fusion cross sections at extreme sub-barrier energies. In this calculation, the coupling to the $^{12}C(2^+$,4.44 MeV) state is included. The result is shown as $CRC\text{–}AW1$ in Fig. 1. The two resonances at 1.5 MeV and 2.14 MeV are about a factor of 10 higher than the CRC-AW calculation.

Even though we do not know the exact location of the potential resonance in the unexplored energy range, with the established upper limit we can claim the real $^{12}C+^{12}C$ fusion rate is much less than Cooper *et al.* proposed. Taking the over-predicted S* factors calculated with CRC-AW, the real rate should not be more than a factor of 6 stronger than the currently adopted fusion rate based on the S* factor in the energy range of 1 to 3 MeV. The upper limit established with the ESW fit, the best fit we have achieved, is only 20% higher than the currently adopted rate in the range of 1 to 2.5 MeV. As shown in Fig. 1, the averaged S* factor of Spillane's data tends to suggest a smaller average than the currently adopted value. Therefore, the factor of more than 40 enhancement proposed in Ref.[10] is not realistic. The upper limit establish in this work posts an additional constraint on the current superburst models. Since the carbon burning rate cannot be as high as Cooper *et al.* proposed, it is likely that some unknown physics process is not included in the current superburst models.

References

[1] Almqvist, Bromley and Kuehner, Phys. Rev. Lett. **V4**, (1960) 515
[2] W. Greiner, *Nuclear Molecules* (World Scientific 1994)
[3] T. Spillane *et al.*, Phys. Rev. Lett. **98** (2007) 122501
[4] L. R. Gasques *et al.*, Phys. Rev. C **76**, (2007) 035802
[5] J.R.Patterson and H.Winkler and C.S.Zaidins, Ap. J. **157** (1969) 367
[6] H.Spinka and H.Winkler, Nucl. Phys. A**233** (1974) 456
[7] H.W.Becker, K.U.Kettner, C.Rolfs, and H.P.Trautvetter, Z. Phys. A**303** (1981) 305
[8] G. R. Caughlan and W. A. Fowler, Atomic Data And Nuclear Data Tables **40** (1988)283
[9] C. A. Barnes and S. Trentalange and S. C. Wu, "Heavy Ion Reactions in Nuclear Astrophysics," *Treatise on Heavy-Ion Sceince* (Plenum preesslm, New York 1985) Volume **6**
[10] R. L. Cooper, A. W. Steiner, and E. F. Brown, Ap. J. **702**, (2009)660
[11] L. Keek *et al.*, Astro & Astrophys., (2008) 479
[12] R. A. Dayras, R. G.Stokstad, Z.E.Switkowski, and R.M.Wieland, Nucl. Phys. A**265**, (1975) 153
[13] S. Trentalange, S. C. Wu, J. L. Osborne, and C. A. Barnes, Nucl. Phys. A**483** (1988) 406
[14] E. F. Aguilera *et al.*, Phys. Rev. C **73**, (2006) 064601
[15] B. Imanishi, Phys. Lett. B**27**, (1968) 267
[16] B. Heusch *et al.*, Phys. Rev. C **23**, (1981) 1527
[17] M. Notani *et al.*, Nucl.Phys. A **834**, (2010) 192c
[18] S. Misicu and H. Esbensen, Phys. Rev. C **75**, (2007) 034606
[19] K. Hagino, N. Rowley, and A.T. Kruppa, Comput. Phys. Comm **123** (1999) 143
[20] J. Zickefoose, U. Conn. Thesis (2010)

Journal of Physics: Conference Series **337** (2012) 012017 doi:10.1088/1742-6596/337/1/012017

Bare nucleus S(E) factor of the ^{2}H(d,p)^{3}H and ^{2}H(d,n)^{3}He reactions via the Trojan Horse Method

A. Tumino[1,2], C. Spitaleri[1], A.M. Mukhamedzhanov[3], S. Typel[4,5], M. Aliotta[6], V. Burjan[7], M. Gimenez del Santo[8], G.G. Kiss[1,9], V. Kroha[7], Z. Hons[7], M. La Cognata[1], L. Lamia[1], J. Mrazek[7], R.G. Pizzone[1], S. Piskor[7], G.G. Rapisarda[1], S. Romano[1], M.L. Sergi[1], R. Spartà[1]

[1]Laboratori Nazionali del Sud - INFN and DMFCI Università di Catania, Catania, Italy
[2]Università degli Studi di Enna "Kore", Enna, Italy
[3]Cyclotron Institute Texas A&M University - College Station, Texas, USA
[4]Excellence Cluster Universe - Technische Universität München, Garching, Germany
[5]GSI Helmholtzzentrum für Schwerionenforschung GmbH - Theorie Darmstadt Germany
[6]School of Physics and Astronomy - University of Edinburgh; SUPA, UK
[7]Nuclear Physics Institute of ASCR - Rez near Prague, Czech Republic
[8]Departamento de Fisica Nuclear - Universitade de Sao Paulo, Sao Paulo, Brasil
[9]ATOMKI - Debrecen, Hungary

E-mail: tumino@lns.infn.it

Abstract. The Trojan Horse Method was applied for the first time to the ^{2}H(d,p)^{3}H and ^{2}H(d,n)^{3}He reactions by measuring the ^{2}H(^{3}He,p^{3}H)^{1}H and ^{2}H(^{3}He,n^{3}He)^{1}H processes in quasi free kinematics. The ^{3}He+d experiment was performed at 18 MeV, corresponding the a d-d energy range from 1.5 MeV down to 2 keV. This range overlaps with the relevant region for Standard Big Bang Nucleosynthesis as well as with the thermal energies of future fusion reactors and deuterium burning in the Pre Main Sequence phase of stellar evolution. This is the first pioneering experiment in quasi free regime where the charged spectator is detected. Both the energy dependence and the absolute value of the bare nucleus S(E) factors have been extracted for the first time. They deviate by more than 15% from available direct data with new S(0) values of 57.4±1.8 MeVb for ^{3}H+p and 60.1±1.9 MeVb for ^{3}He+n. None of the existing fitting curves is able to provide the correct slope of the new data in the full range, thus calling for a revision of the theoretical description. This has consequences in the calculation of the reaction rates with more than a 25% increase at the temperatures of future fusion reactors.

1. Introduction

The ^{2}H(d,p)^{3}H and ^{2}H(d,n)^{3}He reactions at low energies are important in pure and applied physics. These reactions are part of the Standard Big Bang Nucleosynthesis (SBBN) network, synthesizing D, ^{3}He, ^{4}He and ^{7}Li in the early universe. Before the WMAP accurate results, SBBN was the only way to evaluate the baryonic density in the Universe (η), by comparing observed and calculated light-element abundances. Since measuring primordial abundances has improved, the interest in the reaction rates as input parameters for accurate consistency tests of the SBBN model has increased again. This applies in particular to the deuterium, the best baryometer, as its primordial abundance decreases rapidly with η and its later chemical evolution is simple, being essentially destroyed in stars [2]. The uncertainty of η inferred from deuterium is contributed equally by the observational error in the deuterium abundance and by the uncertainties in the cross sections of reactions belonging to the deuterium SBBN network [1]. Among them, the ^{2}H(d,p)^{3}H and ^{2}H(d,n)^{3}He processes are the most important and still critical points at the SBBN relevant energies ($d-d$ relative energies E_{dd} from 50 to 350 keV). All published measurements for both channels have been mainly focused either on the lower E_{dd} region [3, 4, 5, 6] or at E_{dd} above 1 MeV [7]. An accurate calculation of the reaction rate needs the cross section to be known from E_{dd}=30 keV to at least 1 MeV. All the reaction rates quoted so far rely either on polynomial [8, 9] or in the best case on R-Matrix fits [10], whose different slopes bring inside the calculations deviations of more than 15%. In a recent work [11] the total cross section for both ^{2}H(d,p)^{3}H and ^{2}H(d,n)^{3}He reactions was measured at E_{dd} from 55 to 325 keV in steps of about 50 keV. Although uncertainties of better than 4% are quoted, there are some concerns about the procedure followed to calculate the reaction rates. For this reason, a new experimental campaign was recommended in order to provide new data in the SBBN energy range of interest. Moreover, the importance of the ultra-low energy region below 30 keV is twofold: it overlaps with the burning temperatures (0-10 keV) of deuterium in Pre Main Sequence (PMS) stars and corresponds to the thermal energy range of future fusion power plants. However, the measured cross section below 10 keV, available only for the ^{2}H(d,p)^{3}H channel with a gas target [5], has shown a clear enhancement attributed to electron screening. The deduced electron screening potential of U_e= 25±5 eV is significantly larger than the adiabatic limit (14 eV) provided by atomic physics. Unfortunately, nuclear reactions in the laboratory are affected by a different mechanism of electron screening than in a plasma [12], either astrophysical or laboratory fusion with inertial confinement. Hence, the laboratory screening effects of the bound electrons should be removed from the data to assess the reaction rate correctly. This makes the knowledge of the bare nucleus cross section essential for several purposes, from basic to applied research. Usually, a simple extrapolation of available unscreened higher energy data [13] is taken as an approximation of the bare nucleus cross section, and compared to measured low-energy data to extract the electron screening potential. However, this approach can lead to large uncertainties. One can consider that available theoretical extrapolations of the bare nucleus ^{2}H(d,p)^{3}H and ^{2}H(d,n)^{3}He cross sections provide quite different S(0) values wich vary by 10-15%. A valid alternative approach is represented by the Trojan Horse Method (THM) ([14, 15, 16, 17] and references therein), that provides at present the only way to measure the energy dependence of the bare nucleus cross section down to the relevant ultra-low energies, overcoming the main problems of direct measurements [17, 18]. The THM has already been applied several times to reactions connected with fundamental astrophysical problems ([14, 15, 16] and references therein). In the present paper we report on a novel investigation of the deuterium depletion reactions ^{2}H(d,p)^{3}H and ^{2}H(d,n)^{3}He [19] throughout the energy range relevant for pure and applied physics by means of the THM. The method was applied by choosing the ^{2}H(^{3}He,p^{3}H)^{1}H and ^{2}H(^{3}He,n^{3}He)^{1}H reactions in QF kinematics with ^{3}He as the Trojan horse. For the first time in the application of the THM, we have detected the particle that acts as a spectator to both ^{2}H(d,n)^{3}He and ^{2}H(d,p)^{3}H binary processes, namely the proton. This technique was mandatory for the ^{3}He+n fusion channel to avoid the limitation of standard neutron detectors. However, it was well suited also for the ^{3}H+p channel, preventing unwanted QF events from target break-up to be detected. A detailed account of the measurement and related data analysis is given in [19]. Here, we briefly report on the main analysis details and final results.

2. Data analysis and results

After the channel selection, accomplished as reported in [19], the occurrence and the dominance of the QF mechanism in both cases was indicated by the consistency within 3% of the shape of the experimental momentum distribution deduced in PWIA (see [14, 15, 16] for details) with the theoretical one [19]. According to previous R-matrix calculations [10], it is necessary to include the $l = 0$ and $l = 1$ components for both ^{2}H(d,p)^{3}H and ^{2}H(d,n)^{3}He reactions in the determination of the $S(E)$ factor throughout the investigated E_{cm} region. With this information, data analysis follows the standard procedure as reported in [14, 15, 16] and references therein. With the deduced scaling ratio of the s- and p-wave contributions, the $S(E)$ factors for both reactions were then extracted as a function of E_{cm} in steps of 20 keV, with a statistical error of 4%. The normalization of the S(E) factors to direct data has to be carried out in a region where electron screening effects are negligible (E_{cm} ≥15 keV). However, available data sets below 200 keV for both ^{2}H(d,p)^{3}H and ^{2}H(d,n)^{3}He reactions present different accuracies and large scatters, probably due to large systematic errors in some of those measurements [8]. Thus, a weighted normalization was carried out to direct data of [3, 4, 5, 6, 7, 11], from E_{cm}= 1.5 MeV down to 15 keV, each with their quoted total errors (both statistical and systematic combined in quadrature). The weighted normalization procedure [21] clearly favours the most accurate direct data, some of which have total uncertainties better than 1% [5, 7]. A total of 43 THM data points were used in the energy regions where direct data are available. The procedure leads to an overall normalization error of about 1%, obtained from the standard error combination in quadrature. The $S(E)$ factors of the ^{2}H(d,p)^{3}H and ^{2}H(d,n)^{3}He reactions are shown in the upper part of Figs. 1 and 2, respectively. TH data are shown as black full dots, with uncertainties accounting for statistical and normalization errors. Direct data from [3, 4, 5, 6, 7] are shown as colored symbols (the legend in both figures associates a symbol and a color to each ref.). The very regular trend of TH data contrasts with that of direct data taken as a whole, both in energy dependence and absolute values, with deviations of more than 15%. The polynomial/R-matrix fits to direct data, usually taken as reference, are shown as green [8], blue [9] and yellow [10] dashed lines. They barely overlap with most of the direct low-energy data, in particular in the case of the ^{3}He+n channel, probably due to their larger scatter. Conversely, the green line seems to agree with low-energy TH data, while for E_{cm} >200 keV they seem to be better reproduced by the blue line. However, none of them correctly reproduces the slope of the THM S(E)-factor in the entire energy region investigated, thus calling for a revision of the previous theoretical descriptions based on these new high quality data. We provide a new parameterization of the THM S(E) factor as a sum of the theoretical $l = 0$ and $l = 1$ contributions (red and blue solid lines, respectively), with relative weight fixed from the fit to the measured three-body coincidence yield. This gives the total S(E) factor shown as a black solid line in each figure. For a better assessment and comparison, the bottom panels of Figs. 1 and 2 show the residual scattering in the direct data about this curve divided by the weighted dispersion σ (1.82 keVb for the ^{3}H+p and 4.24 keVb for the ^{3}He+n channel), as reported in [9] for a one parameter fit or renormalization where systematics dominate. The dashed horizontal lines represent the 1-sigma error bars. These plots help to visualize the trends of the deviation from the normalized theoretical S(E) factor for each of the direct data sets. The THM parameterizations of the S(E) factors lead to new values of S(0) = 57.4±1.8 keVb for ^{3}H+p and 60.1±1.9 keVb for ^{3}He+n, with uncertainties including the 1% normalization error and 3% coming from the theory, combined in quadrature. The insets in the upper panel of both Figs. 1 and 2 help to compare these values with previous ones usually taken as reference [8, 9, 10]. The deviations are of 15%-20%. A comparison between the THM S(0) factors to the ^{2}H(d,p)^{3}H direct data below 15 keV provides further insight into the electron screening effect. Low-energy direct data at 14.95 keV from [5] were first normalized to the THM S_{bare}(E) (black solid line) and then fitted with the screening function [23] leaving U_e as free parameter. This provides a value of $U_e = 13.2 \pm 1.8$ eV, not exceeding the adiabatic limit (14 eV) for a molecular deuteron target (gas target), but covering it with its uncertainty. Further improvements in the precision of direct low-energy data would help to pin point the electron screening potential.

This work was supported in part by the Italian MIUR under Grant (RBFR082838).

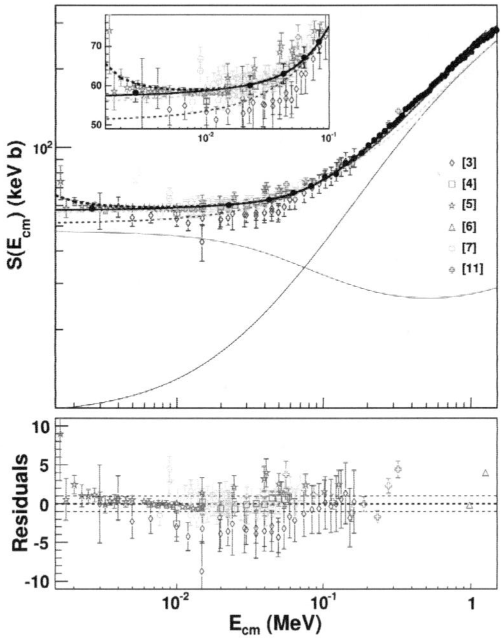

Figure 1. (Color online) $S(E)$ factor for the ^{2}H(d,p)^{3}H reaction: black dots=THM results; colored symbols=direct data. See text for the description of the lines. Bottom panel: residuals of direct data to the black solid line, with dashed lines indicating the 1-sigma error bars.

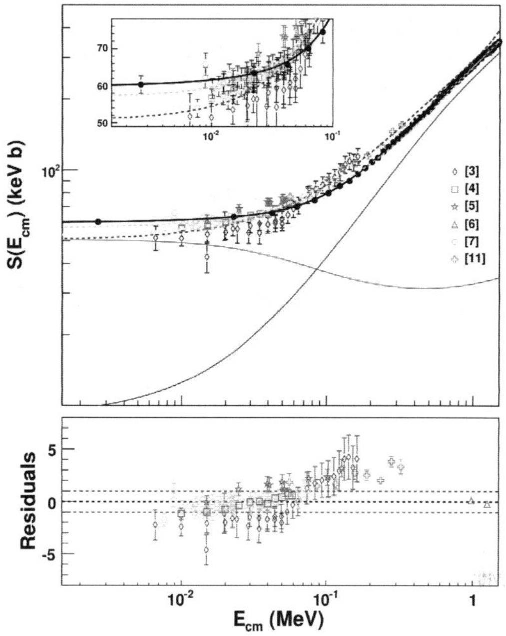

Figure 2. (Color online) $S(E)$ factors of the ^{2}H(d,n)^{3}He reaction: black dots=THM data; colored symbols=direct data. Lines have the same meaning as in Fig.1. Same meaning also for the bottom panel.

References

[1] G. Fiorentini *et al.*, Phys. Rev. D58 (1998) 063506
[2] D. Tytler *et al.*, Phys. Scr. T85 (2000) 12
[3] A. Krauss *et al.*, Nucl. Phys. A465 (1987) 150
[4] R.E. Brown and N. Jarmie, Phys. Rev. C41 (1990) 1391
[5] U. Greife *et al.*, Z. Phys. A351 (1995) 107
[6] The First Res. Group, Chin. J. Nucl. Phys. 9 (1985) 723
[7] R.L. Schulte *et al.*, Nucl. Phys. A192 (1972) 609
[8] C. Angulo *et al.*, Nucl. Phys. A656 (1999) 3
[9] R.H. Cyburt, Phys. Rev. D70 (2004) 023505
[10] P. Descouvemont *et al.*, At. Data Nucl. Data Tab. 88 (2004) 203
[11] D.S. Leonard *et al.*, Phys. Rev. C73 (2006) 045801
[12] G. Shaviv and N. Shaviv, Astroph. J. 529 (2000) 1054
[13] H.S. Bosch and G.M. Hale, Nucl. Fusion 32 (1992) 611
[14] C. Spitaleri *et al.*, Phys. Rev. C63 (2001) 005801
[15] A. Tumino *et al.*, Phys. Rev. C67 (2003) 065803
[16] M. La Cognata *et al.*, Astrophysical J. 708 (2010) 796
[17] A. Tumino *et al.*, Phys. Rev. Lett. 98 (2007) 252502
[18] A.M. Mukhamedzhanov *et al.* J. Phys. G: Nucl. Part. Phys. 35 (2008) 014016
[19] A. Tumino *et al.* Phys. Lett. B 700 (2011) 111 and Phys. Lett. B (2011) doi:10.1016/j.physletb.2011.10.056
[20] S. Typel and G. Baur, Annals of Physics 203 (2003) 228
[21] F.C. Barker, Nucl. Phys. A707 (2002) 277
[22] S. Kimura and A. Bonasera, Phys. Rev. C 76 (2007) 031602(R)
[23] H.J. Assenbaum *et al.*, Z. Phys. A327 (1987) 461
[24] G. Bluege *et al.*, Phys. Rev. C 36 (1987) 21

Decay of $^{48\text{-}50}$Ar isotopes

L. Weissman[1]
Soreq Nuclear Research Center, Yavne 81800, Israel
U. Bergmann, J. Cederkall, L. Fraile, S. Franchoo, H.O.U. Fynbo, T. Fritioff, U. Koster and ISOLDE collaboration
ISOLDE, CERN, CH-1211 Geneva 23, Switzerland
O. Arnd, I. Dillman, O. Hallmann, K.-L. Kratz, B. Pfeiffer, A. Wohr
Institut fur Kernchemie, Universitat Mainz,Germany
L. Gaudefroy, O. Sorlin
GANIL, BP 55027, F-14076 Caen Cedex 5, France

Abstract: Information on β-decay properties of neutron-rich $^{48\text{-}50}$Ar was obtained at the ISOLDE mass-separator facility at CERN using isobaric selectivity. This was achieved by a combination of a plasma-ion source with a cooled transfer line and subsequent mass-separation. Normally, argon beams cannot be mass-separated from intense multi-charged symmetric fission krypton and xenon. Several techniques were used successfully in order to overcome this problem. Implication of the obtained information for a better understanding of the origin of the ^{48}Ca/^{46}Ca isotopic anomaly discovered in inclusions from the Allende meteorite is discussed.

1. Introduction

Knowledge of the nuclear properties from the Z~20 and N≤28 mass region of the nuclear chart is important for understanding one of long persisting astrophysical problems, the Ca/Ti/Cr isotopic anomalies observed in some meteoritic inclusions [1,2]. The decay properties of neutron-rich argon isotopes are the least known compared to other nuclei relevant for clarification of the ^{48}Ca/^{46}Ca meteorite anomaly. As the ^{48}Ar nucleus is, most likely, the main progenitor of the ^{48}Ca isotope the lack of information on its decay properties limits considerably the predictive power of the astrophysical network calculations [3]. These nuclei are also of great interest of nuclear structure physics due to unexpected observation of collectivity and N=28 shell erosion in that region [4]. Therefore any new experimental information on neutron-rich argon isotopes is highly desirable.

Sufficient yields of the neutron-rich argon nuclei can be obtained at ISOL facilities as ISOLDE (CERN). However ,an experimental problem significantly hinders such measurements. The radioactive argon isotopes extracted and mass-separated from a FEBIAD plasma ion sources with cooled transfer line [5] could not be mass-separated from multiply charged krypton and xenon symmetric fission noble gas products that have the same A/q ratio. This results in an overwhelming doubly-charged krypton and triply-charged xenon backgrounds for the singly-charged argon beams [6]. In the case of ^{48}Ar isotope, which is of main importance, the signal-to-noise ratio is especially unfavorable.

In several short experiments performed during 2002-2004 years we have tested several techniques in order to overcome this problem and study decay of neutron-rich argons:

1. Separating doubly-charged argon beams rather than single-charge ones significantly reduced multicharged krypton and xenon background. However, the overwhelming singly charged neon background presents the problem for the case of even argon isotopes like ^{48}Ar [6]. In the latter case use of a detector setup with a beta telescope allows one to clean greatly the signal from neon decay.

[1] Corresponding author: E-mail: weissman@soreq.gov.il

2. Comparison of the γ-spectra collected from the direct proton bombardment of an actinide target to those arising from the use of the neutron converter [7] results in an unambiguous identification of the γ-rays originating from decay of the argon isotopes [8].

3. Finally, a mixture of mass-separated radioactive noble gases was injected into a PHOENIX type Electron Cyclotron Resonance Ion Source (ECRIS) that has been installed in recent years at an ISOLDE beam line for testing of charge-breeding techniques. Modification of the charge-state distributions of the different elements with consequent mass-separation allows for a significant improvement in the quality of the argon signal [9] .

In this paper we briefly describe the experiment and summarize results of decay spectroscopy studies of neutron-rich ^{48,49,50}Ar isotopes obtained by the techniques listed above. The results confirm the picture where the ^{48}Ar is the main progenitor of ^{48}Ca.

2. Experimental Results

Format limitations of this contribution do not allow for detailed description of the experimental setup, experimental procedure and the results. The experimental details as target or neutron converter irradiation conditions, setting of the ion source and the mass-separator, the detector setups and the calibration procedures can be found in [Wei04, Fri06]. Very briefly: the ions were extracted by a 60 kV voltage, mass-separated and implanted into an aluminized tape, which allowed to remove the long-lived daughter activity away from the detector setup.

The implantation point was placed close to Kapton windows, and was surrounded by thin plastic beta detectors. Two Ge detectors of 75 % and 65 % relative efficiency were placed next to the implantation point for detection of γ-radiation. The time and energy signals were collected by a VME-based data acquisition system. Typical time cycle included opening the mass-separator gate for some period after a proton pulse and its closure till the next pulse for study of decay curves. The time cycle was optimized for each particular case.

2.1 ^{50}Ar

The most neutron-rich ^{50}Ar isotope, presented the easiest case as it did not suffer from multicharged background. Two β-gated γ-spectra were collected for the mass-separator A/q value of 50 (singly charged ^{50}Ar beam) for direct proton bombardment of the uranium target and for bombardment of the neutron converter. A number of unknown γ-transitions, 93, 128, 172, 770, 3425 and 3596 keV, were observed in the "direct bombardment" spectrum. In addition the 3084 keV transition from ^{49}Ca decay was also observed. These transitions were not seen in the "neutron converter" spectrum indicating that the transitions of interest belong to the decay of ^{50}Ar nucleus and its daughters. Application of gates on the strongest γ-transitions indicated coincidences between 3425 and 128, 172 keV lines. A correlation was also established between 93 and a weak, 770 keV transition (both transitions belongs to the ^{49}Ar decay (see below)). The transitions of interest exhibited the similar half-lives within the experimental uncertainties, ensuring the assumption that they belong to the same ^{50} Ar decay. The averaging of the half-lives obtained from gating a trigger spectrum by six the γ-transitions yielded 106(6) ms. The implantation rate of ^{50}Ar was estimated from intensities of these transitions as 11 atoms per μC of protons on the target. The neutron emission probability in the ^{50}Ar decay was estimated to be 37(7) %.

2.2 ^{49}Ar

Two β-gated γ-spectra collected for the mass-separator A/q value of 49 (singly charged ^{49}Ar beam) for direct proton bombardment of the uranium target and for bombardment of the neutron converter. Number of transitions, 93, 278, 450, 770, 1240 and 1300 keV, are observed in the direct bombardment spectrum and are not seen in the background spectra. Moreover the 143 keV transitions in the direct bombardment spectrum are stronger than corresponding transitions in the neutron converter spectrum (144 keV from ^{98}Rb). The 780, 3084, 3832 keV transitions from the decay of A=49 and A=48 descendant nuclei are also seen in the direct bombardment spectrum. Application of gates on the strongest γ-transitions indicated the coincidences between the 93 keV line and 770, 1240, 1300 keV lines, as well as the coincidence between the 278 keV and 450 keV lines. No coincidences were established with the 143 keV transition. The decay half-lives were obtained by fitting the trigger time spectra gated by the γ-transitions of interest. The averaged half-live obtained with six transitions is 236(8) ms. The neutron emission probability in the ^{49}Ar decay was estimated to be 29(6) %. The

Journal of Physics: Conference Series **337** (2012) 012018 doi:10.1088/1742-6596/337/1/012018

implantation rate of ^{49}Ar was estimated from intensities of these transitions as 80 atoms per μC of protons on the target.

2.3 ^{48}Ar

2.3.1 Measurement with the β-γ detector setup

A few short measurements were made at A/q=48 settings of the mass-separator (singly-charged ^{48}Ar beam). Numerous transitions from A=96 and A=144 chains resulted to overwhelming background and high dead time. A weak the 3832 keV transition from ^{48}K decay is the only indication of the presence of the ^{48}Ar isotope in the beam. In addition only the unknown 3196 keV transition, which did not appear in the neutron-converter spectrum, can be singled out as the one belonging to the ^{48}Ar decay. The half-life obtained by gating the trigger signal by the transition is 381(35) ms, which is not consistent with any decays of M=96 and M=144 chains. Applying a coincidence gate on the 3196 keV transition yielded correlation to the 143 keV γ-line.

2.3.2 Measurement with the β-telescope setup

Another measurement at A/q=24 separator settings (doubly-charged ^{48}Ar) was performed with a detector setup utilizing an implantation tape, a β-telescope and two Ge detectors. The detailed description of that set-up will be given in [10]. The beam was dominated by very intense singly-charged ^{24}Ne ($T_{1/2}$=3.38 min), however, the large difference in the ^{24}Ne and ^{48}Ar decay Q-values was utilized for improvement of the signal-to-noise ratio. Applying of the high-energy gate (E_{β}>2.5 MeV) on the time stamp signal resulted in the significantly different decay curve comparing to the ungated time spectrum. Fitting of the decay curve yielded ^{48}Ar half-life of 412(19) ms. This value is consistent with the result recently reported by the GANIL group, 475(40) ms [11]. The fitting range was insensitive to the contribution from the decay of the longer-lived $^{47,48}K$ daughters. Thus it was not possible to determine from the measurement the neutron emission probability. Only the 143 keV γ-ray was observed after gating the Ge detector spectrum by the high-energy betas.

2.3.3 Measurement with the secondary ECR ion source.

A short separate measurement was performed using the secondary ECR Ion source (ECRIS) source. This ion source was installed at an ISOLDE beam line to study possibility of charge-breeding of the radioactive beams [12]. The ISOLDE beam at the mass-separator setting of A/q=48 beam was injected into the ECRIS. The separator beam gate was open for 1 second after each proton pulse. The ions were ejected from the ECRIS after 80 ms charge-breeding time. The highly charge ions were ejected from the source, mass-separated by the secondary 102 degree mass separator and implanted into aluminized tape surrounded by a β-γ detector setup. The main experimental problems were associated with X-rays from the ECR source and the background from the radioactive neutral atoms diffused into the source, and stable beams of closed magnetic rigidity [9]. The mass-separated $^{48}Ar^{9+}$ beam charge state corresponded to the best signal-to-noise ratio. The background taken at the same separator and proton cycle conditions but with closed beam gated (X-rays from the ECRIS and breeding of radioactive neutral) was subtracted from the spectrum. During implantation the tape was moved only once in every 3 minutes. One can observe from the spectrum both 143 and 3196 keV γ-transitions, along with 2013.5 and 3832 keV transitions from the decay of the longer-lived ^{47}K and ^{48}K daughter products. A relatively weak signal from A=96 chain is seen, while no activity originating from the A=144 chain was observed at all. The averaging of the half-lives obtained by gating the trigger spectrum by 143 and 3196 keV γ-transitions yields 430(70) ms, which is in agreement with the previous results. The neutron emission probability in the ^{48}Ar decay was estimated from the intensities of the 2013.5 and 3832 keV transitions and was found to be to be 38(6) %.

3. Discussion

The frame of this proceeding does not allow one neither detailed presentation of the experimental results, nor extended comparison with the results of the QRPA and large-scale shell model calculations. This will be done in the upcoming publication [10]. Here we discuss only integration of the obtained the gross-decay properties of the $^{47-50}Ar$ isotopes into the astrophysical picture which attempts to explain the high $^{48}Ca/^{46}Ca$ anomaly in the EK-1-4-1 inclusions of the Allende meteorite.

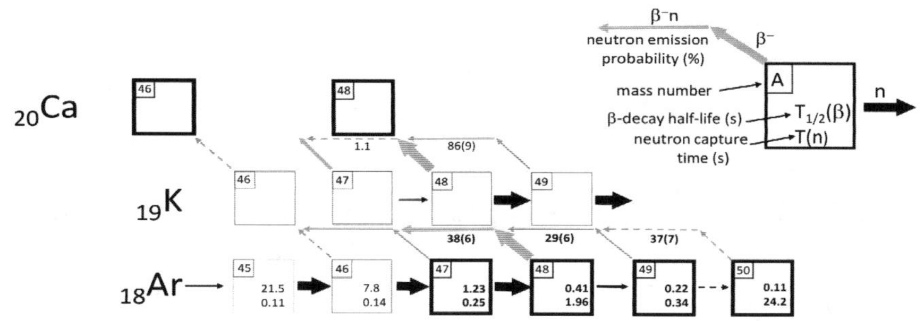

Figure. 1 Population of the ^{48,46}Ca isotopes as result of a weak r-process.

A weak r-process is an astrophysical scenario which could explain the overproduction of ^{48}Ca [3]. The process takes place at the moderate neutron densities and temperature. The competition between neutron capture and beta-decay determines the weak r-process path. The weak process path calculated for neutron density of $4{\bullet}10^{19}$ n/cm^3 and temperature of $0.8T_9$ [3] is presented in Fig. 1. At these conditions the calcium isotopes are mostly produced by beta-decay of their argon progenitors. The argon chain is populated via the ^{45}Cl waiting point, which decay half-life was found to be much shorter than expected due to erosion of the N=28 gap for S and Cl isotopes [13]. The ^{46}Ar half-life is much longer than the neutron capture probability (Fig. 1). Therefore this isotope and, hence ^{46}Ca, will be weakly populated as result of the process. For the ^{48}Ar isotope the decay half-life becomes significantly shorter than the neutron capture time ensure the waiting point character and its strong population. About two thirds of the produced ^{48}Ar population becomes ^{48}Ca. The rest ends up as ^{47}Ti after beta-delayed neutron emission in the ^{48}Ar decay. Any population of ^{49}Ar isotope eventually ends up as ^{48}Ca due to relatively high P_n-value of the ^{49}Ar decay and very high P_n-value of the ^{49}K decay.

In conclusion, we have performed a study of beta-decay of neutron-rich isotopes. Several experimental techniques were used in order to overcome the experimental problems associated with the ISOL method. The results favor the assumption that the origin of the ^{48}Ca/^{46}Ca anomaly is due to the weak r-process. The obtained information on the decay of the argon isotopes together with the recent measurements of the neutron capture rates on ^{44,46}Ar [14] can be implemented now in the new generation of astrophysical network calculations in order to narrow down the astrophysical conditions at which overproduction of ^{48}Ca takes place.

4. References

[1] Lee T, Papanastassiou D A and Wasserburg G J 1978 *Astrophys .J.* **220** L.21
[2] Neiderer F R, 1980 *Astrophys. J.* **240** L73
[3] Kratz K L *et al.* 2001 *Mem. Soc. Astron. Italiana* **72** 453
[4] Glasmacher T *Annu. Rev. Nucl. Part. Sci.* 1998 **48** 1
[5] Sundell S, Ravn H *Nucl. Instr. and Meth* 1992 B **70** 160
[6] Weissman L *et al.* 2003 *Phys. Rev. C* **67** 054314
[7] Catherall R, Lettry J, Gilardoni S, Koster U, *Nucl. Instr. and Meth.* 2003 **204** 235
[8] Weissman L *et al.* 2004 *Phys. Rev. C* **70** 024304
[9] Fritioff T *et al.* 2006 *Nucl. Instr. and Meth. A* **556** 31
[10] Weissman L. *et al.*, in preparation.
[11] Grevy S *et al* . 2004 *Phys.Lett. B* **594** 252
[12] Delahaye P. *et al*, 2006 *Rev. Sci. Instr.* **77** 03B105
[13] Sorlin O *et a.l*, 1993 *Phys Rev C* **47** 2941
[14] Gaudefroy L et al., 2006 *Eur. Phys. J.* **27** 309

Journal of Physics: Conference Series **337** (2012) 012019 doi:10.1088/1742-6596/337/1/012019

On a possible observation of "dihelion" in dissociation of relativistic ^{9}C nuclei

D A Artemenkov, N K Kornegrutsa, D O Krivenkov, R Stanoeva and P I Zarubin
V I Veksler and A M Baldin Laboratory of High Energy Physics, JINR, Dubna, Russia

E-mail: zarubin@lhe.jinr.ru

Abstract. Nuclear track emulsion is exposed to 1.2A ^{9}C GeV nuclei. Pairs of 2^{3}He nuclei having unusually narrow opening angles are observed in channel $^9C \rightarrow 3^3He$ pointing to the possible 2^{3}He resonance near the production threshold.

1. Introduction

The BECQUEREL Collaboration explores the peripheral dissociation of light relativistic nuclei in nuclear emulsion [1]. The structural features of a projectile nucleus are most clearly manifested in the coherent dissociation of nuclei not accompanied by the formation of the target nucleus fragments and mesons (or in "white" stars). Nucleon clustering in stable and radioactive Be, B, C and N isotopes are studied in this approach with a complete observation of the fragments at a record angular resolution [2-7]. Macro photos of such interactions are assembled in the collection [1].

In particular, nuclear track emulsion is irradiated in a ^{9}C nucleus beam produced at the JINR Nuclotron by the fragmentation of 1.2 A GeV ^{12}C nuclei [8]. Dominance of the ^{9}C isotope in the secondary beam is confirmed by emulsion measurements of the ionization of the secondary beam nuclei, features of the charge topology of their fragmentation, as well as momentum measurements of the accompanying the ^{3}He nuclei. As is established, the ^{8}B + p, ^{7}Be + 2p, 2He + 2p and He + 2H + 2p channels are leading ones in the ^{9}C coherent dissociation [7]. The two latter are related to the ^{7}Be core dissociation.

In addition, population of the three ^{3}He nucleus state is observed in the ^{9}C coherent dissociation with 14% probability. Its origin can be caused by a virtual regrouping of a neutron from the ^{4}He cluster to form ^{3}He clusters as well as the presence of the 3^{3}He component in the ^{9}C ground state. In the second variant, the probability of the 3^{3}He ensemble production points to the significance of this deeply bound configuration in the wave function of the ^{9}C ground state.The mechanism of the ^{9}C coherent dissociation in the channels with nucleon separation and the 3^{3}He channel is a nuclear diffractive interaction which is established on the basis of measurements of the total transverse momentum transferred to the fragment ensemble.

2. Narrow pairs in coherent dissociation $^9C \rightarrow 3^3He$

Few pairs of ^{3}He nuclei with opening angles Θ_{2He} less than 10^{-2} rad are detected in the $^9C \rightarrow 3^3He$ channel. The macro photo of one identified event is shown in figure 1. This observation indicates to a possible existence of a narrow 2^{3}He resonant state with the decay energy near the 2^{3}He mass threshold (or "dihelion"). In the same way the formation of ^{8}Be nuclei is reliably manifested in the production of ^{4}He pairs with extremely small opening angles in the relativistic dissociation of $^9Be \rightarrow 2^4He$ [3] and $^{10}C \rightarrow 2^4He + 2p$ [5,6]. Significant probability of coherent dissociation $^9C \rightarrow 3^3He$ makes it an efficient source for search for an analog of the unbound ^{8}Be nucleus among the ^{3}He pairs. In what follows, a rather unexpected and potentially important feature of the spectrum Θ_{2He} of ^{3}He pairs produced in the dissociation of ^{9}C [7] and ^{8}B [4] nuclei, is given.

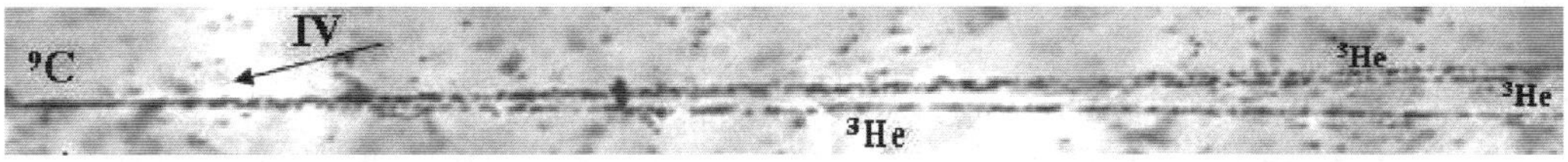

Figure 1. Macro photo of "white" star of ^{9}C dissociation to 3 ^{3}He nuclei in nuclear track emulsion; the interaction vertex **IV** is shown by the arrow.

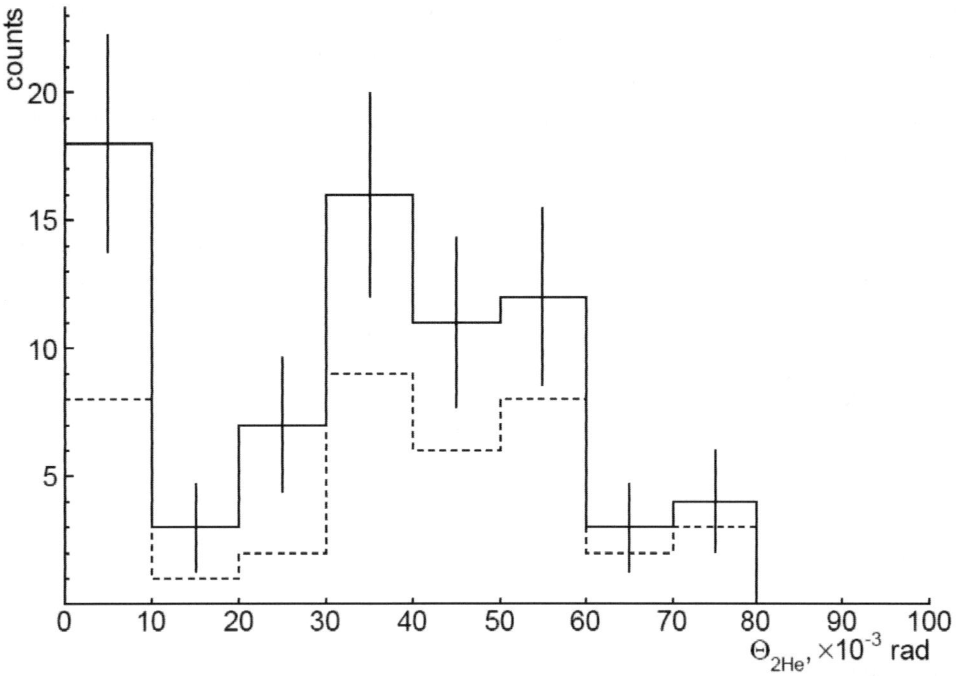

Figure 2. Total distribution of opening angles Θ_{2He} between the relativistic He fragments in the "white" stars $^9C \rightarrow 3^3He$ and in events $^8B \rightarrow 2He + H$ with the formation of target nucleus fragments or meson; dotted line indicates the "white" stars contribution.

In figure 2 the dotted line shows the distribution Θ_{2He} for "white" stars $^9C \rightarrow 3^3He$. Its main part, corresponding to 30 pairs, is described by a Gaussian distribution with mean value $<\Theta_{2He}> = (46 \pm 3) \cdot 10^{-3}$ rad (RMS $16 \cdot 10^{-3}$ rad). In addition, thanks to excellent spatial resolution eight 2^3He pairs within $\Theta_{2He} < 10^{-2}$ rad are reliably observed. These pairs form a special group with a mean value $<\Theta(2^3He)> = (6 \pm 1) \cdot 10^{-3}$ rad and RMS 3·10-3, which is obviously beyond the previous description.These values correspond to the average relative energy $<Q(2^3He)> = (142 \pm 35)$ keV (RMS 100 keV). The parameter $Q(2^3He)$ was defined as the difference between the invariant mass of the pair and the double ^{3}He mass assuming that the fragments conserve the ^{9}C momentum per nucleon.

3. Narrow pairs in peripheral dissociation $^8B \rightarrow 2He + H$

Additional search for the resonance 2^3He is carried our in the events of 1.2A GeV peripheral dissociation $^8B \rightarrow 2He + H$. In this case, inelastic interactions with target nucleus fragments or produced mesons are selected in order to enhance the "dihelion" effect. This condition provides the selection of interactions with knocking of a neutron out of the ^{4}He cluster in the ^{8}B nucleus. Thus, the resulting distribution $\Theta(2^3He)$ in the figure also includes a separate group of narrow pairs with $<\Theta(2^3He)> = (4.5 \pm 0.5) \cdot 10^{-3}$ rad (RMS $1.5 \cdot 10^{-3}$ rad), corresponding to the case of "white" stars $^9C \rightarrow 3^3He$.

The total distribution for both the (figure 2) makes the indication to the existence of the 2^3He resonance more reliable. Moreover, the question arises about the nature of the broad peak with a maximum of $\Theta(2^3He)$ about $(40 - 50) \cdot 10^{-3}$ rad. It is possible that in this $\Theta(2^3He)$ region the 2^3He system shows its similarity with the first excited 2^+ state of the ^{8}Be nucleus.

4. Conclusions

Obviously, the "dihelion" indication should be reviewed using a significantly larger statistics.The BECQUEREL Collaboration performed irradiation of nuclear track emulsion in a mixed beam of 1.2A GeV ^{12}N, ^{10}C and ^{7}Be nuclei [5]. Thus, there are new opportunities with regard to the issue of "dihelion" based on the analysis of the found 400 "non-white" stars ^{7}Be $\rightarrow$ 2^{3}He with knocking out of a neutron and the formation of fragments of target nuclei or mesons, as in the case of ^{8}B $\rightarrow$ 2He + H.

However, it is possible that the "dihelion" formation is due to the presence of the 2^{3}He component in the ^{9}C and ^{8}B structures. In principle, in a lighter ^{7}Be nucleus such a component can be suppressed, this means that the "dihelion" formation can be suppressed as well. Therefore, it is important to search for the 2^{3}He resonance with high statistics exactly in low energy ^{9}C and ^{8}B beams. At the same time, pointing to the existence of "dihelion", our observation motivates the search for a mirror state of a pair of nuclei ^{3}H - "ditriton".

Acknowledgments

This work was supported by grants from 96-1596423, 02-02-164-12a, 03-02-16134, 03-02-17079, 04-02-17151. 04-02-16593, and 09-02-9126 CT-a RFBR. The authors thank S. P. Kharlamov (FIAN, Moscow), and A. I. Malakhov (JINR, Dubna) for discussions and friendly support.

References

[1] The BECQUEREL Project 2011 http://becquerel.jinr.ru
[2] Peresadko N G et al. 2007 *Phys. At. Nucl.* **70**, 1226 (Preprint arXiv:nucl-ex/0605014)
[3] Artemenkov D A, Krivenkov D O, Shchedrina T V, Stanoeva R and Zarubin P I 2008 *Few Body Syst.* **44**, 273
[4] Stanoeva R et al. 2009 *Phys. At. Nucl.* **72**, 690 (Preprint arXiv:0906.4220)
[5] Kattabekov R R, Mamatkulov K Z et al. 2010 *Phys. At. Nucl.* **73**, 2110 (Preprint arXiv:1104.5320)
[6] Artemenkov D A, Alikulov S S, Kattabekov R R, Mamatkulov K Z, Kornegrutsa N K, Krivenkov D O and Zarubin P I 2011 *Few Body Syst.* **50**, 259 (Preprint arXiv: 1105.2438)
[7] Krivenkov D O et al. 2010 *Phys. At. Nucl.* **73**, 2103 (Preprint arXiv:1104.2439)
[8] Rukoyatkin P A et al. 2008 *Eur. Phys. J. ST* **162**, 267

Beta decay measurements from ^{6}He using an electrostatic ion beam trap

O.Aviv[1], S. Vaintraub[1,2], T. Hirsh[1,2], A. Dhal[1], M. L. Rappaport[1], D. Melnik[2], O. Heber[1], D. Schwalm, D. Zajfman, K. Blaum[3], M. Hass[1]

[1]Weizmann Institute of Science, Rehovot 76100, Israel
[2]Soreq Nuclear Research Center, Yavne 81800, Israel
[3]Max-Planck-Institut für Kernphysik, D-69117 Heidelberg, Germany

E-mail: ofer.aviv@weizmann.ac.il

Abstract. We plan to establish the ground work of a novel experimental setup that will enable precision measurements of β−ν correlation from ^{6}He using a unique method which incorporates a radioactive ion beam, ion trapping, ion bunching, and a radiation detection system. For the production of the ^{6}He radioisotopes we plan to use neutron-induced reactions and an electron ion beam trap (EBIT) for ionization. The $^{6}He^{+}$ radioisotopes will be stored in an electrostatic ion beam trap (EIBT), commonly used in atomic and molecular physics. The entire apparatus will be built at the Weizmann Institute. In the following we present the method, the present status of the setup and future plans.

1. Introduction

New physics beyond the standard model became one of the major interests in recent years. Trapped radioactive atoms and ions present an excellent laboratory for the study of fundamental interactions and symmetries, e.g. detecting the minute possible deviations to the angular correlation between a neutrino and an electron in the beta-decay process [1]. The advantage in using radioactive isotopes is that once the isotopes are produced, a more or less table top experiment can be performed which reduce considerably the required resources compared to large colliders as LHC. In recent years there have been several experiments with trapped radioisotopes for beta-neutrino correlation measurements: The work of the LBL group with ^{21}Na [3] using a magneto optical trap (MOT), the work of the TRINAT group with ^{38m}K [3] using a MOT, and the work of the GANIL group with ^{6}He using a Paul trap [4]. Here we describe a novel technique for beta decay studies using an Electrostatic Ion Beam Trap (EIBT) [4, 5]. The advantage of an EIBT in the current context is the ability to have the ions as a beam which enhance the accuracy of the ion kinematics as well as a large free field region that simplified the detection configuration. Another major advantage of the EIBT is that it can be used to study practically any radioactive isotope, unlike other existing setups (e.g. MOT). Moreover, the long storage times (few seconds) are ideal for studying ^{6}He having a half-life of τ=807 ms. ^{6}He decays via beta-minus to ^{6}Li and to an anti- neutrino. The β−ν angular distribution for allowed beta decay from an un-polarized nucleus is given by [7]:

$$dW \propto E_e P_e (E-E_0)^2 \left(1 + a_{\beta\nu} \frac{\vec{P}_e \cdot \vec{P}_\nu}{E_e E_\nu}\right) \quad (1)$$

Journal of Physics: Conference Series **337** (2012) 012020 doi:10.1088/1742-6596/337/1/012020

Here E_0 is the energy released in the decay, being 3.5 MeV for ^{6}He, and E_e, P_e, E_ν, P_ν are the energies and momenta of the electron and neutrino. For a pure Gamow-Teller decay as in the case of ^{6}He, the expected angular coefficient is $a_{\beta\nu}=-1/3$. Any deviation from the prediction of -1/3 may indicate physics beyond the standard model.

2. Experimental Setup

The scheme of the experimental setup is shown in Fig. 1. d+t neutron generator (produced by NSD-FUSION, Germany) produces fast (14.1 MeV) neutrons which hit a hot (1500 K) BeO target located inside a furnace. As a result ^{6}He are produced via ^{9}Be(n,α)^{6}He reaction. The ^{6}He diffuse into an EBIT source (produced by DREEBIT, Germany) [8] where they are ionized, accumulated, and bunched. The ion bunch is then accelerated (~7keV), steered, focused and injected into the trap chamber where beta decay studies are performed. The beam-line is maintained at a pressure of 10^{-7} -10^{-8} Torr, and the trap chamber is pumped from below by a cryogenic pump to ~10^{-10} Torr.

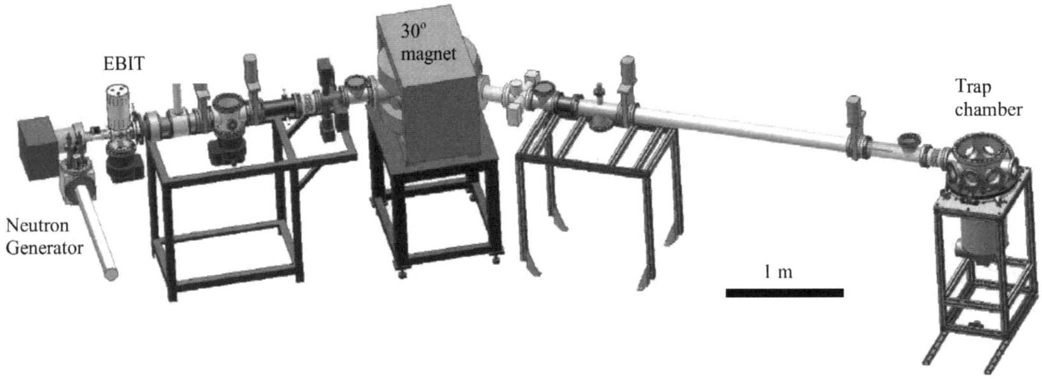

Figure 1. Experimental scheme of WIRED (Weizmann Institute Radioactive Electrostatic Device). See text for details.

2.1. The EIBT – Principles of operation

The EIBT acts similarly to an optical resonator, here ions bounce back and forth between electrostatic mirrors [4]. Fig. 2 shows a scheme of the trap chamber. The mirror electrodes as well as the detectors are mounted on a precision optical table. Each mirror consists of 8 electrodes which produces a retarding field that reflects the beam along its path and focus it on the lateral direction. The innermost electrodes are grounded to create a field-free region. In the region the ion beam has a well-defined kinetic energy and direction in space. Upon injection the entrance mirror is grounded, once the beam fills the entire trap, the potential on the entrance mirror are raised quickly (~100 ns), thus the ion beam oscillates back and forth between the two mirrors. By applying a small RF voltage of a few volts on one of the mirror electrodes, the ion beam can be bunched [9], creating a small moving radioactive source. A pickup ring electrode located inside the trap is used to determine the radioactive source bunch position at any given moment. While stored the beam of ^{6}He$^+$ can decay, the electron and recoil products are measured in coincidence using a set of dedicated detectors which are located in the field free region of the trap. The position and energy of the electron are detected on a silicon strip detector backed by a scintillator. The position and time of flight (and therefore there energy) of the recoil Li^{++} are measured on a position sensitive multi-channel-plate (MCP) detector situated along the beam line. Following the decay, shakeoff electrons emitted from the daughter recoil will increase its charge state. This happens in the field-free region where the recoils have ~7 keV. Therefore we do not expect any major effect on our results.

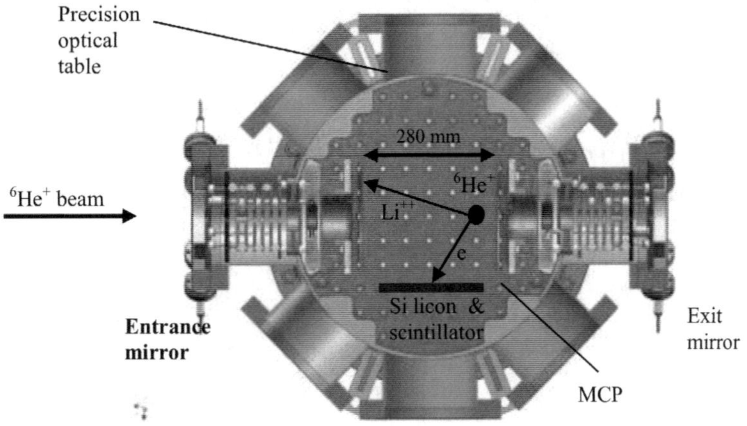

Figure 2. Scheme of the EIBT chamber. See text for details.

The MCP has an annular hole which allows the stable trajectories of the stored ion bunch. Thus the momentum of the unmeasured neutrino can be reconstructed and the beta-neutrino angular distribution determined. A discussion concerning beta decay of ^{6}He in an EIBT using Monte Carlo simulations can be found in [10].

3. Status and future plans

To date, much of the setup has been installed. Production of ^{6}He is expected by mid of 2012. Trapping tests with a ^{4}He beam are expected by 9/2011. Preliminarily measurements with stored ^{6}He are expected by 2013. At a later stage, we envisage taking advantage of uniquely high yields of radioactive nuclei that will become available at the new and modern SARAF accelerator at Soreq [11].

4. Acknowledgments

This work has been supported by Israel Science Foundation, the Minerva Program of the Max-Planck Society, and by the research grant from the Estate of David Turner.

References

[1] J. Behr and G. Gwinner, J. Phys. G: Nucl. Part. Phys. **36** 033101 (2009).
[2] N. Scielzo, S. J. Freedman, B.K. Fujikawa, and P.A. Vetter, Phys. Rev. Lett. **93**, 102501 (2004).
[3] A. Gorelov *et al.*, Phys. Rev. Lett. **94**, 142501 (2005).
[4] X. Fiechard *et al.*, Phys. Rev. Lett. **101**, 212504 (2008).
[5] D. Zajfman, O. Heber et al., Phys. Rev. A **55**, 1577 (1998).
[6] L. H. Andersen, O. Heber, D. Zajfman, J. Phys. B: At. Mol. Opt. Phys. **37** R57 (2004).
[7] J. D. Jackson, S. B. Trieman, H. W. Wyld, Phys. Rev. **106**, 517 (1957).
[8] H. J. Kluge, K. Blaum, F. Herfurth, W. Quint, Physica Scripta, **104**, 167-177 (2003).
[9] H. B. Pedersen *et al.*, Phys. Rev. A, **65**, 042703 (2002).
[10] S. Vaintraub *et al.*, Journal of Physics Conference Series, in press.
[11] T. Hirsh *et al.*, Journal of Physics Conference Series, in press.

Macroscopic properties of neutron stars including deformation

M Urbanec[1], E Běták[1,2] and Z. Stuchlík[1]

[1] Inst. of Physics, Fac. Philos. and Sci., Silesian Univ., 74601 Opava, Czech Rep.
[2] Inst. of Physics, Slovak Acad. Sci., 84511 Bratislava, Slovakia

E-mail: martin.urbanec@fpf.slu.cz, betak@savba.sk, zdenek.stuchlik@fpf.slu.cz

Abstract. We used the equations of state of dense nuclear matter to construct the macroscopic properties of neutron stars and test them using available observational results. The Dirac-Brueckner-Hartree-Fock mean field calculations approximated by their parameterized form are the basis of our calculations. We calculated the central pressure and density and correspondingly the possible radius and mass both with and without allowance for hyperons first, and compared these results with recent astronomical observations, and, finally, we included effect of deformation and rotation.

1. Introduction

Neutron stars play important role in different areas of modern physics. We devote our attention on the possibility that phenomena related to neutron stars can be used as tests of equation of state (EoS) of asymmetric nuclear matter with allowance for hyperons.

A wide spectrum of different equations of state of nuclear matter and their applications to astrophysical problems has been reported in the literature (see, e.g., [1–7]). All these EoS yield (nearly) the same properties close to the standard nuclear density ($\rho_N \approx 0.16$ nucleon/fm^3 $\approx 2.7 \times 10^{14}$ g/cm^3), but when one is far off this value, s/he has to rely more on underlying principles than on possible experimental verification of predicted physical observables.

Our calculations use the Dirac-Brueckner-Hartree-Fock (DBHF) mean field approach [8, 9] in its parameterized form [10, 11] which reproduces the full nuclear matter calculations[12–14].

Frequently reported and/or calculated quantities of neutron stars are their mass M and radius R, but sometimes also quantities like the radiation radius R_∞, ratio of baryon to gravitational mass and others [1, 4, 7, 15–28].

2. Equations of state of neutron star matter

We follow the DBHF mean field approach (see e.g. [8, 9, 29]), which allows to consider various compositions of the neutron star matter, and also inclusion of non-nucleonic degrees of freedom. The full mean-field DBHF calculations of nuclear matter [12–14] have been parameterized [10]. We choose three parameterizations, which were shown to yield the best fits to the well-known properties of nuclear matter, namely: HA corresponding to the best RMF fit to results of Huber *et al* [13]; LA to those of Lee *et al* [14]; MA to the results of [12], respectively. For details as well as for the values of coupling constants for the individual sets of parameterization, we refer

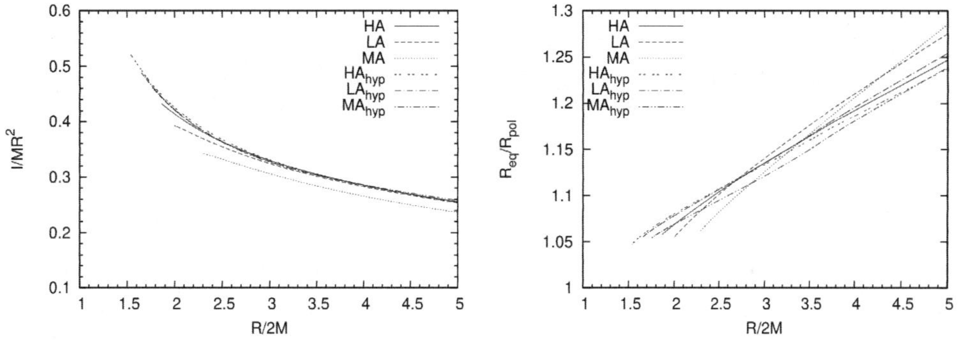

Figure 1. Inertia moment *(left)* and the ratio of the equatorial to the polar radii *(right)* vs. compactness for three parameterizations of the equation of state, both with and without hyperons.

to the original paper [10]. With the inclusion of hyperons, the full Lagrangian density becomes extended [11].

3. Other ingredients of the model

The matter in neutron stars is in β-equilibrium, which determines the particle fractions at each density.

Whereas the nuclear EoS are the dominant input for the calculations in the high-density region, namely $\rho \geq 10^{14}$ g/cm^3. For lower densities, the EoS used are: Feynman-Metropolis-Teller EoS [30] for 7.9 g/cm$^3 \leq \rho \leq 10^4$ g/cm^3 where matter consists of e$^-$ and $^{56}_{26}$Fe; Baym-Pethick-Sutherland EoS [31] for 10^4 g/cm$^3 \leq \rho \leq 4.3 \times 10^{11}$ g/cm^3 with Coulomb lattice energy corrections; and Baym-Bethe-Pethick one [32] for 4.3 g/cm$^3 \times 10^{11} \leq \rho \leq 10^{14}$ g/cm^3.

The hydrostatic equilibrium is given by the Tolman-Oppenheimer-Volkoff equation relating the pressure $P(r)$ and the energy density $\rho(r)$ [33, 34],

$$\frac{\mathrm{d}P}{\mathrm{d}r} = -(\rho + P)\frac{m(r) + 4\pi r^3 P}{r(r - 2m(r))}. \tag{1}$$

We refer here to our previous papers on the properties of neutron stars calculated in this way [25, 37], and concentrate ourselves on the neutron star deformation and rotation.

The centrifugal force in a star depends on the local spin frequency $\bar{\omega}$ measured in a local inertial reference frame. To the lowest order, $\bar{\omega}$ satisfies differential equation (see, e.g., [35])

$$\frac{1}{r^4}\frac{\mathrm{d}}{\mathrm{d}r}\left(r^4 \mathrm{e}^{-\Phi-\lambda}\frac{\mathrm{d}\bar{\omega}}{\mathrm{d}r}\right) + \frac{4}{r}\left(\frac{\mathrm{d}}{\mathrm{d}r}\mathrm{e}^{-\Phi-\lambda}\right)\bar{\omega} = 0, \tag{2}$$

where Φ and λ are metric functions of a non-rotating star.

The total moment of inertia $I = J/\Omega$, where J is the total stellar angular momentum, is given by $I = \frac{8}{3}\pi\int_0^R r^4(\rho + P/c^2)\frac{\bar{\omega}}{\Omega}\exp(-\Phi - \lambda)\mathrm{d}r$ [35]. In the Newtonian limit, $\bar{\omega} = \Omega$, $\lambda = \Phi = 0$, $P \ll \rho c^2$, what gives $I_{\mathrm{Newton}} = \frac{8}{3}\pi\int_0^R r^4 \rho \mathrm{d}r$.

The resulting inertia moment vs. compactness for six parameterizations of the equation of state (see [25, 37]) and the ratio of the equatorial to the polar radii vs. compactness, as calculated for the rotation frequency 716 Hz (the most rapidly rotating star) are presented in Fig. 1. We see that all calculated curves lie very close one to each other (especially in the case of the inertia moment), what suggests that these quantities are generally very weakly dependent on the exact form of the equation of state, even if hyperons are included.

4. Conclusions

We calculated the properties of neutron stars stemming from the parametrized DBHF approach. In addition to usual quantities, like mass, radius, etc., we calculated also the neutron star deformation and moment of inertia vs. compactness. The nonzero deformation, which was obtained also for small compactness, suggests that the deformation is a common feature for these stellar objects and that it will be possible to study creation of gravitational waves.

Acknowledgments

The authors are grateful to J. Kotulič Bunta and Š. Gmuca for the availability of their computer codes. The work has been supported by the Czech grants MSM 4781305903 (EB and ZS) and LC 06014 (MU) and by the VEGA grant 2/0029/10 (EB).

References

[1] Lattimer J M and Prakash M 2001 *Astrophys. J.* **550** 426
[2] Haensel P, Zdunik J-L and Douchin F 2002 *Astron. and Astrophys.* **385** 301
[3] Rikovska Stone J, Guichon P A M, Matevosyan H H and Thomas A W 2003 *Phys. Rev. C* **68** 034324
[4] Lattimer J M and Prakash M 2007 *Phys. Rep.* **442** 109
[5] Weber F, Negreiros R and Rosenfeld P 2007 Neutron star properties and the equation of state of superdense matter arXiv:0705.2708
[6] Burgio C 2008 *J. Phys. G* **35** 014048
[7] Steiner H, Lattimer J and Brown E 2010 *Astrophys. J.* **722** 33
[8] Walecka J D 2004 *Theoretical Nuclear and Subnuclear Physics, 2nd Edition* (London and Singapore: Imper. Coll and World Sci.)
[9] Weber F 1999 *Pulsars as Astrophysical Laboratories for Nuclear and Particle Physics* (Bristol: IoP Publishing)
[10] Kotulič Bunta J and Gmuca Š 2003 *Phys. Rev. C* **68** 054318
[11] Kotulič Bunta J and Gmuca Š 2003 *Phys. Rev. C* **70** 054309
[12] Li G Q, Machleidt R and Brockmann R 1992 *Phys. Rev. C* **45** 2782
[13] Huber H, Weber F and Weigel M K 1995 *Phys. Rev. C* **51** 1790
[14] Lee C-H, Kuo T T S, Li G Q and Brown G E 1998 *Phys. Rev. C* **57** 3488
[15] Walter F and Matthews L 1997 *Nature* **1997** 358
[16] Pons J A, Walter F M, Lattimer J M, Prakash M, Neuhäuser R and An P 2002 *Astrophys. J.* **564** 981
[17] Bejger M and Haensel P 2002 *Astron. and Astrophys.* **396** 917
[18] Bejger M, Bulik T and Haensel P 2002 *Monthly Not. Royal Astron. Soc.* **364** 635
[19] Trümper J, Burwitz V, Haberl F and Zavlin V E 2004 *Nucl. Phys. B Proc. Suppl.* **132** 560
[20] Podsiadlowski P, Dewi J D M, Lesaffre P, Miller J C, Newton W G and Stone J R 2005 *Monthly Not. Royal Astron. Soc.* **361** 1243
[21] Klähn T *et al* 2006 *Phys. Rev. C* **74** 035802
[22] van Kerkwijk M and Kaplan D 2007 *Astrophys. J. Suppl.* **308** 191
[23] Nice D, Stairs I and Kasian L 2008 *AIP Conference Proc.* **983** 453
[24] Stuchlík Z, Török G, Hledík S and Urbanec M 2009 *Class. Quant. Gravity* **26** 035003
[25] Urbanec M, Běták E and Stuchlík Z 2010 *Acta Astron.* **60** 149
[26] Demorest P B, Pennucci T, Ransom S M, Roberts M S and Hessels J W T 2010 *Nature* **467** 1081
[27] Güver T, Wroblewski P, Camarota L and Özel F 2010 *Astrophys. J.* **719** 1807
[28] Gandolfi S, Carlson J and Reddy S 2011 *The maximum mass and radius of neutron stars and the nuclear symmetry energy* arXiv:1101.1921
[29] Krastev P G and Sammarruca F 2006 *Phys. Rev. C* **74** 025808
[30] Feynman R, Metropolis N and Teller E 1949 *Phys. Rev.* **75** 1561
[31] Baym G, Pethick C and Sutherland P 1971 *Astrophys. J.* **170** 299
[32] Baym G, Bethe H and Pethick C 1971 *Nucl. Phys. A* **175** 225
[33] Oppenheimer J R and Volkoff G M 1939 *Phys. Rev.* **55** 374
[34] Tolman R C 1939 *Phys. Rev.* **55** 364
[35] Haensel P, Potekhin A Y and Yakovlev D 2007 *Neutron Stars 1* (Heidelberg: Springer)
[36] Neuhäuser R, Hambaryan V V, Tetzlaff N, Hohle M N and Eisenbeiss T 2011 *Proc. Sci.* **NIC XI** 037
[37] Urbanec M, Běták E and Stuchlík Z 2010 in *Proc. FINUSTAR-3 Conf., Rhodos 2010 (in press)*

Journal of Physics: Conference Series **337** (2012) 012022 doi:10.1088/1742-6596/337/1/012022

Study of α-Decay Properties of Superheavy Nuclei

Andreea-Ioana Budaca[1] **and Ion Silisteanu**

Horia Hulubei National Institute of Physics and Nuclear Engineering, Bucharest-Magurele, RO-077125, Romania, EU

E-mail: [1] `abudaca@theory.nipne.ro`

Abstract. The α-decay is considered from the viewpoint of the many body features of internal nuclear motion and the theory of resonance reactions. The α-half-lives are derived from clustering and scattering amplitudes given by self-consistent nuclear models for the nuclear shell structure and reaction dynamics. Calculations are performed for superheavy nuclei with Z=102-120 using: only the measured E_α values, microscopic (shell model) or macroscopic (one body) cluster formation amplitudes and, resonance scattering amplitudes. Theoretical results for α-half-lives are compared to data and empirical estimates. We prove that the Brown systematics ($logT_\alpha$ (s) vs. $Z_d^{0.6} \cdot Q_\alpha^{-1/2}$, where Q_α (MeV) is the effective decay energy, and Z_d is the charge number of daughter nucleus) of current decay data is very useful in the analysis and interpretation of data and the prediction of new results. The accuracy of available experimental half-lives is discussed.

1. Introduction

In case of the α-decay of a single resonance state k into a set n of channels, the partial half-live is: $T_n^k = ln2 \cdot \hbar/\Gamma_n^k$, the partial decay width being expressed [1, 2] as:

$$\Gamma_n^k = 2\pi \left| \frac{\int_{r_{\min}}^{r_{\max}} I_n^k(r) u_n^0(r) dr}{\int_{r_{\min}}^{r_{\max}} I_n^k(r) u_n^k(r) dr} \right|^2, \tag{1}$$

where $I_n^k(r)$ is the particle (cluster) formation amplitude (FA)

$$I_n^k(r) = r\langle \Psi_k \mid \mathcal{A}\{[\Phi_D(\eta_1)\Phi_\alpha(\eta_2)Y_{lm}(\hat{r})]_n\}\rangle, \tag{2}$$

and $u_n^k(r)$, $u_n^0(r)$ are the solutions of the systems of differential equations

$$\left[\frac{\hbar^2}{2m}\left(\frac{d^2}{dr^2} - \frac{l(l+1)}{r^2}\right) - V_{nn}(r) + Q_n\right]\left[\begin{array}{c} u_n^0(r) \\ u_n^k(r) \end{array}\right] + \sum_{m \neq n} V_{nm}(r)\left[\begin{array}{c} u_m^0(r) \\ u_m^k(r) \end{array}\right] = \left[\begin{array}{c} 0 \\ I_n^k(r) \end{array}\right], \tag{3}$$

where V_{nm} is the matrix elements of α-daughter nucleus potential.

2. Methods

Shell model α-formation amplitude (SM) Following [4] we use in Eq.(2) the w.f. $\mid \Psi_k^{SM}(r_i) >=$ det $\parallel \psi_{nlj}^{SM}(r_i) \parallel$, i=1,A and $\mid \Phi_D^{SM} >=$ det $\parallel \psi_{nlj}^{SM}(\eta_{1i}) \parallel$, i=1,A-4 , and the α-particle wave function is chosen as: $\Phi_\alpha(\eta_2) = (2/(1/2)!)^{3/2}(\beta/\pi)^{9/4}(\rho_1^2 + \rho_2^2 + \rho_3^2)(4\pi)^{-3/2}\chi_{00}(s_1 s_2)\chi_{00}(s_3 s_4)$, where the

α-particle oscillator parameter is $\beta = 0.484\ fm^{-2}$, χ_{00} is the singlet spin function and the internal spatial coordinates $\rho_1 = (r_1 - r_2)/\sqrt{2}$, $\rho_2 = (r_3 - r_4)/\sqrt{2}$, $\rho_3 = (r_1 + r_2 - r_3 - r_4)/2$, are connected to the individual coordinates r_i of four nucleons. Replacing these functions in Eq.(2) we get

$$I_n^{k[SM]}(r) = r\langle \Psi_k^{SM}(r_i) \mid \mathcal{A}\left\{\left[\Phi_D^{SM}(\eta_1)\Phi_p(\eta_2)Y_{lm}(\hat{r})\right]_n\right\}\rangle. \quad (4)$$

We used s.p. configurations to α-half-lives ($\rightarrow \log T_\alpha^{SM}$) Z=102-120$\rightarrow$p (s.p. proton states): $1i_{13/2}, 2f_{7/2}, 2f_{5/2}, 3p_{3/2}, 3p_{1/2}$; N=150-178$\rightarrow$n (s.p. neutron states): $2g_{9/2}, 2g_{7/2}, 3d_{5/2}, 3d_{3/2}, 4s_{1/2}$.
Resonance formation amplitude (RFA) The RFA was defined [2, 3] by means only of the eigenvalues and eigenfunctions. The asymptotic overlap integral or the RFA $I_n^k(r \rightarrow \infty) \longrightarrow I_n^{res}(r) = u_n^0(r) \simeq \bar{G}_n(r)$, is represented by the normalized irregular Coulomb w.f. $\bar{G}_n$.
Empirical approximations:
Viola-Seaborg formula [5] (**VS**): $\log T_\alpha(s) = (aZ_d + b){Q_\alpha}^{-1/2} + (cZ_d + d) + h_{e-o}$, with the parameters a, b, c, d and even-odd corrections h_{e-o} [8] for different (Z, N) parities.
Tagepera-Nurmia formula [6] (**TN**): $\log T_\alpha(years) = 1.615(Z_d{Q_\alpha}^{-1/2} - Z_d^{2/3}) - 27.815 + h_{e-o}^{(y)}$.
Brown formula [7] (**B**): $\log T_\alpha(s) = 9.54Z_d^{(0.6)} \cdot {Q_\alpha}^{-1/2} - 51.37 + h_{e-o}$.
The effective decay energy includes screening corrections $E_{sc} = 6.53 \cdot 10^{-5} \cdot Z_d^{7/5} - 8.0 \cdot 10^{-5} \cdot Z_d^{2/5}$: $Q_\alpha(MeV) = (A/(A-4)) \cdot E_\alpha^{exp} + E_{sc}$.

3. Results

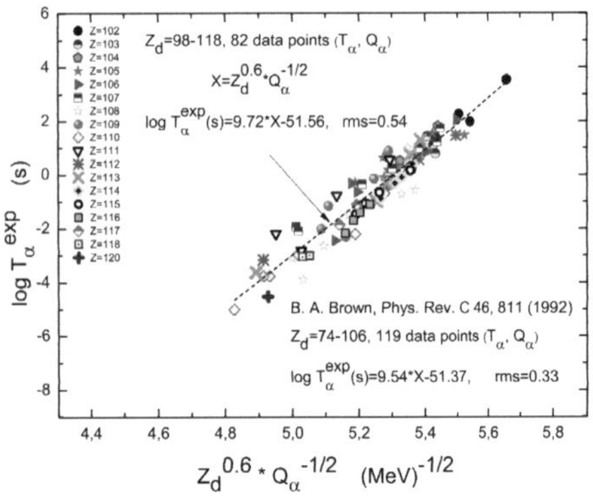

Figure 1. The values of $\log T_\alpha^{exp}$ (s) [9–11] vs. $Z_d^{0.6}{Q_\alpha}^{-1/2}$ and the best fit line of the data.

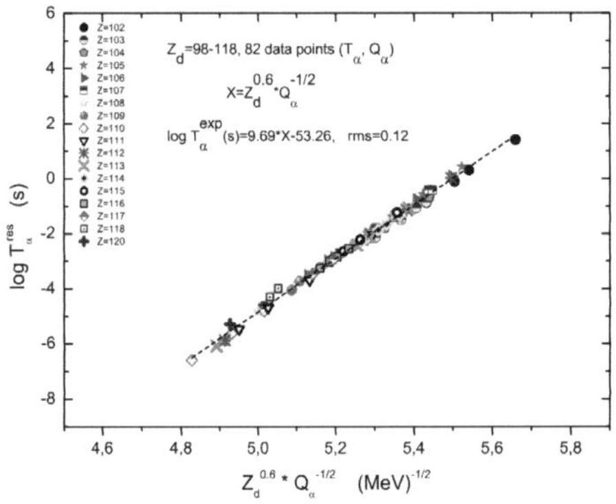

Figure 2. The values of $\log T_\alpha^{res}$ (s) vs. $Z_d^{0.6}.{Q_\alpha}^{-1/2}$ and the best fit line of these values.

- The rms errors in T_α^{exp} are larger by one order magnitude than for calculated ones (Table 1). The ratio $T_\alpha^{res}/T_\alpha^{SM}$ is compatible with the finite size corrections of nucleons to the o.b. rate [3]. The ratio $T_\alpha^{SM}/T_\alpha^{VS} \approx 1.3 - 1.4$ is almost constant.
- The scatter and the values of $\log T_\alpha^{exp}$ (Fig. 1) can not be reproduced by $\log T_\alpha^{res}$ without h_{e-o} (Fig. 2). By adding h_{e-o} corrections to $\log T_\alpha^{SM}$ (s) values (Fig. 4), the agreement with experimental data is improved and basic trends in systematics of data (Fig. 3) are well reproduced.
- A key problem in the understanding of the α-decay and determining the half lives is especially the importance of nuclear structure in the exit channels. The role of "finite sizes", screening and even-odd corrections has been stressed.
- We hope these results will provide some guidance for using the resonant-particle spectroscopy (RPS) and the recoil decay tagging (RDT) techniques in detailed α-decay studies of SHN.

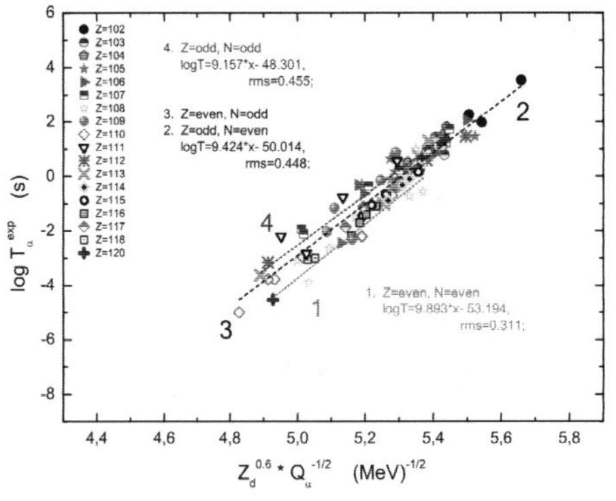

Figure 3. The same as in Fig. 2 but the fit lines correspond to h_{e-o} corrections.

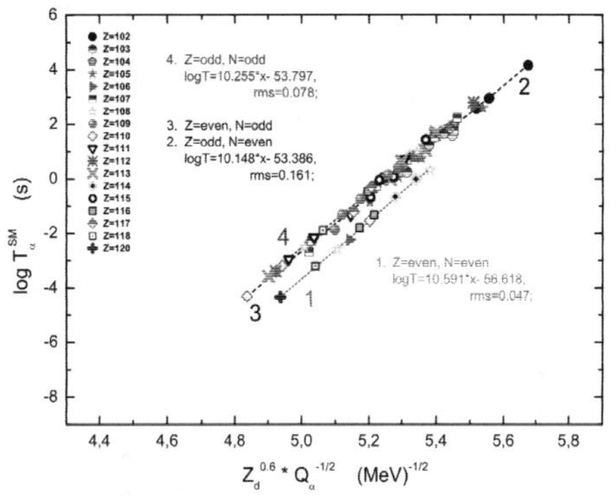

Figure 4. The values of $\log T_\alpha^{SM} + h_{e-o}$ vs. $Z_d^{0.6} Q_\alpha^{-1/2}$ and the best fit lines of these values.

Table 1. The best fit lines of experimental and calculated results together with the root mean square (rms) errors included.

Z-N	$\log T^{exp}$	rms	$\log T^{B}$	rms
e-e	$\mathbf{9.893 \cdot x - 53.194}$	0.311	$\mathbf{9.001 \cdot x - 48.347}$	0.006
e-o,o-e	$\mathbf{9.424 \cdot x - 50.014}$	0.448	$\mathbf{8.965 \cdot x - 47.193}$	0.140
o-o	$\mathbf{9.157 \cdot x - 48.301}$	0.455	$\mathbf{8.989 \cdot x - 47.170}$	0.003
	$\log T^{res} + h_{e-o}$	rms	$\log T^{TN}$	rms
e-e	$\mathbf{9.552 \cdot x - 52.521}$	0.083	$\mathbf{10.865 \cdot x - 58.980}$	0.126
e-o, o-e	$\mathbf{9.742 \cdot x - 52.592}$	0.200	$\mathbf{10.552 \cdot x - 56.323}$	0.164
o-o	$\mathbf{9.866 \cdot x - 53.100}$	0.101	$\mathbf{10.629 \cdot x - 56.574}$	0.073
	$\log T^{SM} + h_{e-o}$	rms	$\log T^{VS}$	rms
e-e	$\mathbf{10.591 \cdot x - 56.618}$	0.047	$\mathbf{10.527 \cdot x - 56.408}$	0.044
e-o, o-e	$\mathbf{10.146 \cdot x - 53.386}$	0.161	$\mathbf{10.083 \cdot x - 53.181}$	0.159
o-o	$\mathbf{10.255 \cdot x - 53.797}$	0.078	$\mathbf{10.192 \cdot x - 53.591}$	0.076

References

[1] Silişteanu I, Budaca A I and Silişteanu A O 2010 *Rom. J. Phys.* **55** 10881110
[2] Silişteanu I, Scheid W and Săndulescu A 2001 *Nucl. Phys.* A **679** 317
[3] Silişteanu I, Scheid W 1995, *Phys. Rev.* C **51** 2023
[4] Mang H J 1960 *Phys. Rev.* **119** 1069
[5] Viola V E and Seaborg G T 1966 *J. Inorg. Nucl. Chem.* **28** 741
[6] Tagepera R and Nurmia M 1961 *Ann. Acad. Sci. Fenn. Serie* A **78**
[7] Brown B A 1992 *Phys. Rev.* C **46** 811
[8] Sobiczewski A, Patyk Z and Cwiok S 1989 *Phys. Lett.* B **224** 1
[9] Oganessian Yu Ts, Abdulin F Sh and Bailey P D *et al.* 2010 *Phys. Rev. Lett.* **104** 0142502
[10] Oganessian Yu Ts 2006 *Pure. Appl. Chem.* **78** 889
[11] Hofmann S and Münzenberg G 2000 *Rev. Mod. Phys.* **72**, 733

Nuclear Physics in Astrophysics V IOP Publishing
Journal of Physics: Conference Series **337** (2012) 012023 doi:10.1088/1742-6596/337/1/012023

Gamma-ray strength at low energies using relativistic QRPA with exact coupling to the continuum.

I. Daoutidis and S. Goriely

Institut d'Astronomie et d'Astrophysique, ULB, CP226, 1050 Bruxelles, Belgium

E-mail: `idaoutid@ulb.ac.be`, `sgoriely@ulb.ac.be`

Abstract. Continuum-quasiparticle random-phase Approximation (CQRPA) within the relativistic point-coupling model with density-dependent coupling constants is applied to investigate collective excitations in spherical nuclei. In particular we study the impact of the exact continuum on the giant-dipole and pygmy resonance of several Sn isotopes as well as the radiative neutron capture rates of importance for astrophysical calculations.

1. Introduction

The study of exotic nuclei within the relativistic density functional theory has gained considerable interest in recent years, because these models are based on Lorentz invariance, connecting in a consistent way the spin and spatial degrees of freedom of the nucleus and providing thus a relatively simple phenomenological description for many nuclear properties all over the periodic table. The same functionals can be used to investigate not only the static properties of these systems but also their dynamics, such as collective excitations and nuclear response functions. However, while the static properties can be extracted directly from the relativistic mean field theory, the collective ones can be studied only under the small amplitude variations of the density around the static value, which is done by solving the random-phase approximation (RPA).

This work is devoted to an investigation of relativistic quasiparticle RPA (QRPA) models with an exact treatment of the coupling to the continuum. In addition to this continuum QRPA (CQRPA), the most common spectral representation is also considered for comparison. In both cases, we use a point-coupling Lagrangian which is capable of reproducing a wide range of experimental data [2]. The BCS model with constant pairing force is also applied, in order to treat the pairing correlations of the open-shell nuclei [3].

The basic idea behind the RPA approach is that under the influence of an external field oscillating with the frequency ω the nucleus is excited. The cross section of this process is proportional to the strength function:

$$S(\omega) = -\frac{1}{\pi} Im \int d^3r d^3r' \, f^*_{ext}(\mathbf{r}) R(\mathbf{r}, \mathbf{r}'; \omega) f_{ext}(\mathbf{r}'), \tag{1}$$

where $f_{ext}(\mathbf{r})$ is the external operator inducing the reaction. The response function of the nucleus $R(\mathbf{r}, \mathbf{r}'; \omega)$ has been modified in order to incorporate the effect of the exact coupling to the continuum (See [2, 3] for details). The main advantage of this approach is that the

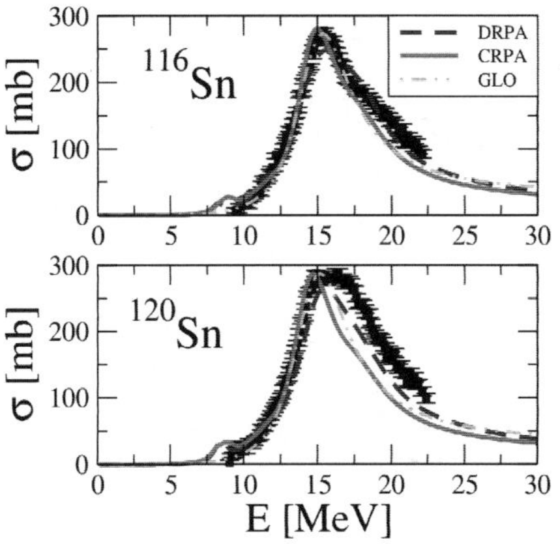

Figure 1. Photoabsorption cross section for the stable isotopes ^{116,118,120,124}Sn. The red solid line represents the CQRPA calculations, while the blue dashed line shows the conventional DQRPA results. Also shown is the GLO formula, as recommended by Ref. [6]. The experimental results are taken from [7].

large number of antiparticle-hole pairs which is required in conventional methods is taken into account effectively and without any numerical cost. Hence one can reduce the numerical effort up to one order of magnitude, as compared to conventional RPA approaches without loosing in predictive power. In addition, the continuum RPA gives automatically information about the escape width, which can be a significant portion of the total width of the multipole giant resonances.

2. Applications

The Isovector Giant Dipole resonance (IVGDR) is the most well studied collective excitation and the first to have been observed experimentally. An external electromagnetic field causes protons and neutrons to oscillate in opposite phases with respect to each other and leading to a pronounced resonance peak in the photoabsorption cross section.

In Fig. 1 we show the excitation energy of the IVGDR strength for several Sn isotopes, using both the continuum and the conventional approach, the latter being referred here as discrete QRPA (DQRPA) model. We see that the agreement with the experimental observations is excellent. The point-coupling force PC-F1 [1] is used for the interaction, while an additional energy-dependent width is considered to account for the spreading width of the resonance.

As expected, the pygmy dipole resonance (PDR) appears at energies around $E \sim 8$ MeV. At such an energy, the PDR may lie below the neutron separation energy and become undetectable through (γ,n) experiments. However, recent experiments based on the (^{3}He,^{3}He'γ) and (^{3}He,$\alpha\gamma$) probes [4, 5] has provided valuable information about the low-lying γ-ray strength below the neutron threshold. In particular, the γ-ray strength of the three even-A isotopes ^{116,118,122}Sn has recently been determined, making these nuclei particularly important in the study of the pygmy mode. In Fig. 2 we compare the recent experimental data with the relativistic QRPA predictions which appear to agree fairly well. Both the CQRPA and DQRPA calculations show a well pronounced PDR around 8 MeV, though the shape of the PDR slightly differs.

Finally, it has been shown that the low-lying E1 strength, close to or below the particle separation energies can affect the neutron capture rates of relevance for r-process nucleosynthesis [9]. For that reason, we have applied the continuum QRPA cross sections to Hauser-Feschbach reaction rate calculations to investigate the impact of the exact coupling to the continuum. We

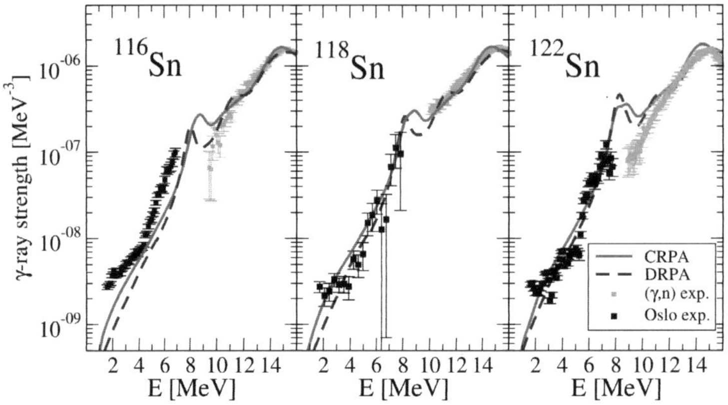

Figure 2. Comparison of the CQRPA (solid line) and DQRPA (dashed line) E1 strength of ^{116,118,122}Sn with available experimental data. The low-energy data are taken from [4] while the (γ,n) data are from [7]. For ^{122}Sn the data are from [5] and [8].

find that for stable Sn isotopes, CQRPA gives similar results to conventional QRPA approaches, but significant discrepancies are found for heavier isotopes with $A > 132$. In particular, with the CQRPA model, reaction rates are larger by a factor of four larger, as compared to those obtained with the non-relativistic HFB+QRPA [9] and by a factor of 10 to 100 times with respect to those obtained with the Generalized Lorentzian (GLO) model [10], as shown in Fig. 2.

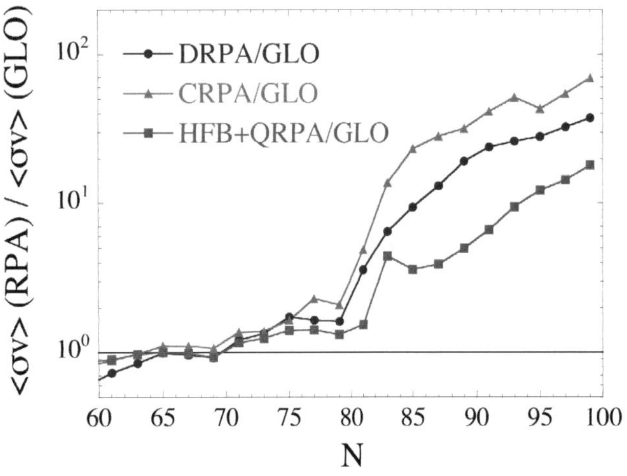

Figure 3. Ratio of the radiative neutron capture rates obtained with RPA approaches with respect to those obtained with the GLO model [10]. The circles and triangles correspond to the DRPA and CRPA cases, respectively, as studied in the present paper, while the squares correspond to the non-relativistic HFB+QRPA strength of Ref. [9]. The rates are calculated for odd-N Sn targets at a temperature of 10^9 K.

References

[1] T. Bürvenich, D. Madland, J. Maruhn, and P.-G. Reinhard, *Phys. Rev.* **C65**, 044308 (2002).
[2] J. Daoutidis and P. Ring *Phys. Rev.* **C80**, 024309 (2009).
[3] J. Daoutidis and P. Ring *Phys. Rev.* **C83**, 044303 (2011).
[4] H. K. Toft et.al., Phys. Rev. **C81**, 064311 (2010).
[5] H. K. Toft et.al., Phys. Rev. **C83**, 044320 (2011).
[6] T. Belgya, *et al.*, *Handbook for calculations of nuclear reaction data, RIPL-2*, IAEA-TECDOC-1506
[7] A. Lepretre et.al., Nucl. Phys. **A219**, 39 (1974).
[8] V. V. Varlamov, Yad. Konst. **1**, 52 (1993).
[9] S. Goriely, E. Khan, and M. Samyn, *Nucl. Phys.* **A739**, 331 (2004).
[10] J. Kopecky and M. Uhl, *Phys. Rev.* **C41**, 1941 (1990).

Effects of Tensor Correlations on the Positive Parity States of the Even-Even Nuclei in the sd Shell

by

J. O. Fiase* and F. Gbaorun

Department of Physics, Benue State University,
P. M. B. 102119, Makurdi, Benue State,
Nigeria.

Abstract

Effects of tensor correlations on the positive parity states of ^{20}Ne nucleus are examined. Two-body nuclear matrix elements are obtained by the lowest order constrained variational technique with and without tensor correlations. The matrix elements calculated are used as input into the NUSHELL shell model code to calculate the energy spectrum of ^{20}Ne nucleus with and without tensor correlations. We have found that the effects of tensor correlations is to open up the calculated energy spectrum and provide reasonable agreement with experiment, whereas the energy spectrum calculated without tensor correlations is compressed and provides significant disagreement with experimental data. This feature is consistent with our observations for nuclei in the range $18 \leq A \leq 40$

The results presented here re-emphasize the strong evidence found in nuclear structure calculations that tensor correlations are very important in nuclei and their presence cannot be ignored.

*** fiasejo@yahoo.com**

1.0 Introduction.

The non-relativistic many nucleon Hamiltonian in the rest-frame of the nucleus may be written as [1]

$$H = \sum_{i>j} \left[p_{ij}^2 / M(A) + V_{ij} \right], \tag{1}$$

where $\underline{p_{ij}}$ is the relative momentum, $\underline{p_{ij}} = \frac{1}{\sqrt{2}}\left(\underline{p_i} - \underline{p_j}\right)$, $M(A) \approx m_N A$ is the total mass of the nucleons, while V_{ij} is taken here to be the Nijmegen potential [2]. An effective Hamiltonian may be defined which takes the form [3,4]:

$$H_{eff} = \sum_{i>j} f_{ij}\left(\frac{p_{ij}^2}{M} + V_{ij}\right) f_{ij}, \tag{2}$$

where the f_{ij} are two-body correlation operators. In earlier studies regarding nuclear matter and finite nuclei [4,5], the two-body correlation functions were found to have several features. The most important of these features was found to be the tensor correlations, hence, the two-body correlation functions were parameterized in the form [6,7]

$$\begin{aligned} r_{ij} < r_c \quad &: \quad f_{ij} = 0 \\ r_{ij} > r_c \quad &: \quad f_{ij} = f(r_{ij}) \sum_{\lambda} \left(1 + \alpha_\lambda(A) S_{ij}\right), \end{aligned} \tag{3}$$

where the sum λ is over all reaction channels and $f(r_{ij})$ is parameterized as:

$$f(r_{ij}) = 1 - \exp[-\beta(r_{ij} - r_c)^2]. \quad (4)$$

In eq. (4), $r_c = 0.25$ fm and $\beta = 25\,\mathrm{fm}^{-2}$ which represents the short-range repulsion of the bare NN interaction, S_{ij} is the usual tensor operator; $\alpha_\lambda(A)$ is the strength of the tensor correlations which is measured by the condition:

2. Results.

The matrix elements of eq. (2) were calculated in a harmonic oscillator basis. There are 63 two-body matrix elements comprising 28 diagonal and 35 non-diagonal ones.

In Fig. 1, we present the T = 1 and T = 0 non - diagonal two-body matrix elements for the A = 20 system for the present calculation and those of the fitted interaction of Wildenthal [8]. Fig. 1 is the result when the strength of the tensor correlations is switched on to take the value of 0.0550. This gives an impressive agreement with the Wildenthal [8] interaction. Though not shown, there is significant disagreement when tensor correlations is switched off.

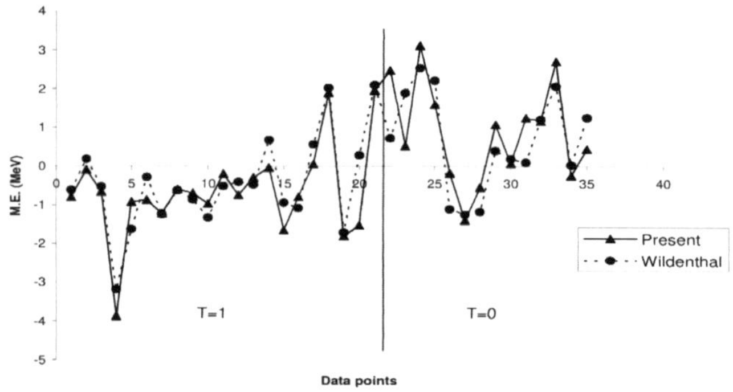

Fig. 1. T = 1 and T = 0 non-diagonal matrix elements for the present calculation and Wildenthal's with $\alpha_\lambda(A) = 0.0550$, $A = 20$, and $\hbar\omega = 13$. Doted line represents Wildenthal's matrix elements while the solid line represents the present calculation.

In Fig. 2 we present the experimental spectrum of ^{20}Ne nucleus together with the present calculation with and without tensor correlations. It is found that the first five experimental energy levels are well reproduced when tensor correlations are switched on. On the other hand the same levels are also reproduced but the spectrum is somehow compressed when tensor correlations are switched off. We also present the spectrum using the wildenthal [8] interaction for comparison. This feature is found in all the nuclei in the range $18 \le A \le 40$.

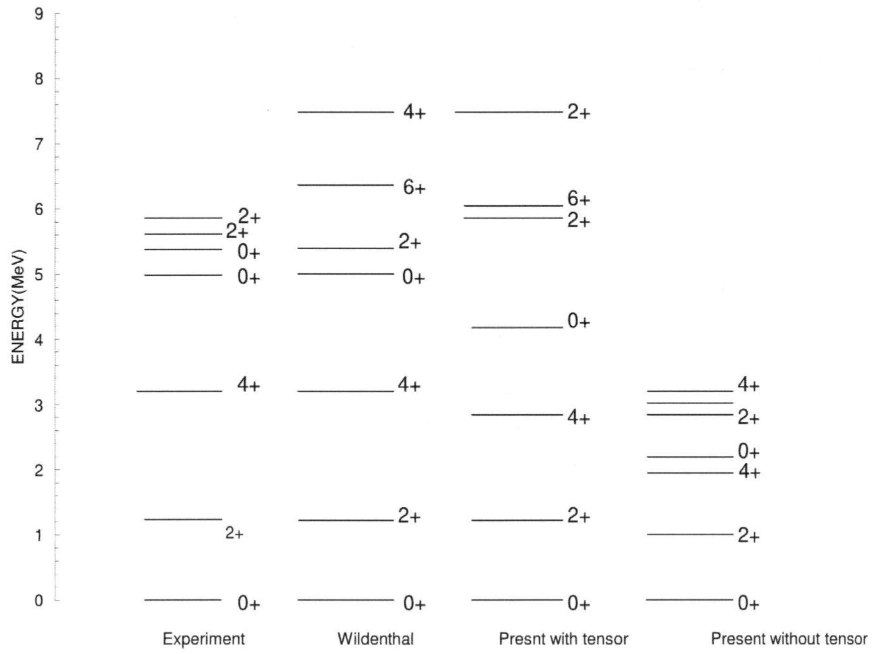

Fig 2. ^{20}Ne Energy Spectrum. $\hbar\omega = 13$, $\alpha = 0.0550$ (with tensor), $\alpha = 0.0000$ (without tensor)

5. Summary and Conclusion.

In this work we have studied the effects of the tensor correlations on the spectrum of ^{20}Ne nucleus. We have found that the strength of the tensor correlations is very important in obtaining reasonable agreement with experiment. Indeed we have found that when the strength of tensor correlations is switched off, the calculated energy spectrum either compress very severely or disagrees significantly with experiment. This feature is found in all the nuclei in the range $18 \leq A \leq 40$. We expect to find the same kind of trend on measurable quantities when studying astrophysical quantities where tensor correlations are involved.

References

[1] J.M. Irvine, G.S. Mani, V.F.E. Pucknell, M. Vallieres, and F. Yazici, Ann. Phys. (N.Y.) **50**, 411 (1976).
[2] V.G.J. Stoks, R.A.M. Klamp, C.P.F. Terheggen, J.J. De Swart: Phys. Rev. **C49**, 2950 (1994)
[3] J. M. Irvine, The Nuclear Matter Saturation Problem, Annales UMCS (Lublin, Poland) Section AAA XLII (1987).
[4] J.O. Fiase, K.R.S. Devan, A. Hosaka: Phys. Rev **C66**, 014004(2004)
[5] J. Fiase, A. Hamoudi, J.M. Irvin, F. Yazici: J. Phys. G. Nucl. Phys. **14**, 27 (1988)
[6] J.M. Irvine, Prog. Part. Nucl. Phys. **5**, 1 (1980)
[7] J. Fiase, Phys. Rev. C **63**, 037303 (2001)
[8] Wildenthal H 1984 Prog. Part. Nucl. Phys. **11** 5

Journal of Physics: Conference Series **337** (2012) 012025 doi:10.1088/1742-6596/337/1/012025

Status of the ACCULINNA-2 project at FLNR

A.S. Fomichev[a], G.M. Ter-Akopian[a], A.A. Bezbakh[a], V. Chudoba[a], A.V. Daniel[a], M.S. Golovkov[a], A.V. Gorshkov[a], V.A. Gorshkov[a], L.V. Grigorenko[a,d], G. Kaminski[a,b,*], S.A. Krupko[a], Yu.Ts. Oganessian[a], Y.L. Parfenova[a], S.I. Sidorchuk[a], R.S. Slepnev[a], L. Standylo[a,c], S.V. Stepantsov[a], R. Wolski[a,b], S.N. Ershov[b], V.K. Lukyanov[b], B.V. Danilin[d], A.A. Korsheninnikov[d], V.Z. Goldberg[e], M. Pfützner[f], I.G. Mukha[g], H. Simon[g], O.B. Tarasov[h], N.K. Timofeyuk[i], M.V. Zhukov[j]

[a]Joint Institute for Nuclear Research, Dubna, Russia

[b]Institute of Nuclear Physics PAN, Krakow, Poland

[c]Institute of Nuclear Problems, Warsaw, Poland

[d]RRC The Kurchatov Institute, Moscow, Russia

[e]Cyclotron Institute, Texas A&M University, College Station, USA

[f]Institute of Experimental Physics, Warsaw University, Warsaw, Poland

[g]Gesellschaft für Schwerionenforschung mbH, Darmstadt, Germany

[h]NSCL, Michigan State University, East Lansing, Michigan, USA

[i]Department of Physics, University of Surrey, Guildford, UK

[j]Fundamental Physics, Chalmers University of Technology, Göteborg, Sweden

*E-mail: Grzegorz.Kaminski@ifj.edu.pl

Abstract. The project of a new and more powerful in-flight fragment separator ACCULINNA-2 at U-400M cyclotron in FLNR, JINR planned to build in addition to the existing separator ACCULINNA is presented. The new separator will provide high intensity RIBs in the lowest energy range (5÷50 MeV/nucleon) which is attainable for in-flight separators. The possibilities for the astrophysics studies at the proposed device are presented. ACCULINNA–2 separator is planned to be constructed in the years 2010-2015. The current status of the project is reported.

1. Introduction

An important part of the planned third generation of Dubna Radioactive Ion Beams (DRIBs) facilities [1] at the Flerov Laboratory of Nuclear Reactions of the Joint Institute for Nuclear Research (FLNR, JINR) in Dubna is the construction of the in-flight fragment separator ACCULINNA-2 [2,3] as an upgrade of the existing separator ACCULINNA [4]. The project of the new separator should be realized in parallel with continuous operation of ACCULINNA fragment separator and should gradually replace the latter after commissioning. The new facility is expected to be a more universal and powerful instrument providing wider range of high quality light exotic beams for nuclear physics studies. One of the key points of the project is to significantly increase the intensities of radioactive beams at the FLNR. The separator, together with the beam diagnostics system, should become a standard instrumentation for the laboratory as well as external users. Construction of the facility is planned in the years 2011-2015.

2. Anticipated scientific agenda

The proposed ACCULINNA-2 project is focused on the experimental study of nuclear properties far from the stability valley in the area of light nuclei. The investigation of nuclei far from β-stability valley and even beyond the nuclear stability lines is important for understanding the properties of nuclear matter under extreme conditions. Major physical topics related to the study of the phenomena observed in light exotic nuclei to be considered by means of the future facility are: nuclear reactions, nuclear structure and astrophysical applications. Each of them has a preferable set of methods by which its problems could be best analyzed. The scientific agenda of the ACCULINNA-2 separator considers a wide range of physical methods to study the properties of the exotic nuclear systems which should be accessible with this instrument. A strong emphasis is put on the possibility of the astrophysics studies. It is very important to carry out focused, dedicated investigations of some special cases difficult for the experimental study but crucial for the stellar burning cycles. Most nuclear reactions in stars involve short-lived proton-rich and neutron-rich nuclei. Moreover, nuclear astrophysics benefits in several ways from the investigation of unstable nuclei where the major aspect is new information on the nuclear matter properties and relevant nuclear reactions. A group of adequate methods as possibilities for the program of astrophysical studies with radioactive beams within the frame of the proposed project and details of the research program are given in the letter of intent [2].

3. Outline of ACCULINNA-2 separator

Figure 1 shows a schematic layout of two-stage separation ion-optical system of the separator. The first stage is represented by an energy loss achromat, the second one is a 15-meter straight beam line with beam diagnostics and detector array. Moreover a RF-kicker allowing to significantly improve the RIB purity (especially in case of the proton rich nuclei) is planed to be installed in the second part. The new separator will provide RIBs in the broad range of energies 6÷60 AMeV – the lowest energy

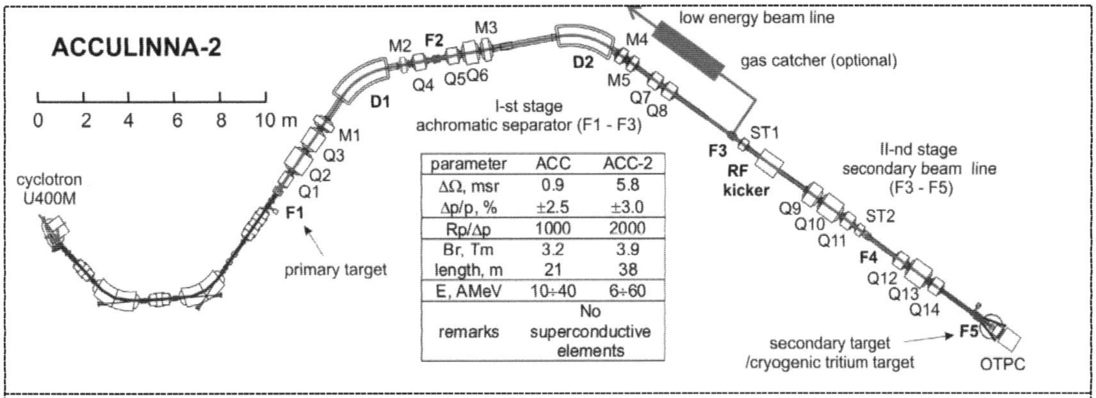

parameter	ACC	ACC-2
ΔΩ, msr	0.9	5.8
Δp/p, %	±2.5	±3.0
Rp/Δp	1000	2000
Br, Tm	3.2	3.9
length, m	21	38
E, AMeV	10÷40	6÷60
remarks		No superconductive elements

Figure 1. A schematic layout of the ACCULINNA-2 fragment separator. F_n denote focuses, dipoles, quadrupole lenses, miltipoles, steerers are indicated by D_n, Q_n, M_n, ST_n, respectively.

range which is attainable for in-flight separators. The solid angle acceptance and the energy spread are 5.8 msr and 6%, respectively. These issues together with cyclotron upgrade will allow to increase the RIB intensities by a factor of 10-15 as compared to the intensities achieved at the ACCULINNA beam line [4]. The expected RIBs intensities at the ACCULINNA-2 separator are listed in Ref. [2].

The new separator will be also equipped with a unique cryogenic tritium target with record characteristics in the world. The combination of the cryogenic target system [5] with a standard instrumentation (silicon shell, gamma and neutron arrays, etc) should be a distinctive feature of this facility giving unique possibilities for the reaction and nuclear structure studies carried out in vicinity and beyond the neutron drip-line. Moreover a novel type of the ionization chamber with optical readout designed for the rare decay studies, the so-called Optical Time Projection Chamber (OTPC) [6] is already being constructed in collaboration with Warsaw University. This device was originally developed to study two proton radioactivity [6] and allows a full kinematics reconstruction of decay event. The results of the experimental studies with OTPC at the beams delivered by the ACCULINNA separator demonstrated that such type of detector offers a unique possibility to study the decay and the structure of both proton and neutron-rich dripline nuclei [7]. Even though the low-cost separator cannot compete with new large in-flight separators in the sense of "crude power", but taking advantages of a unique instrumentation combined with the low intermediate energy secondary beams, the results of the experimental studies are expected to be competitive for many years.

4. Status of the projects

The realization of the project is divided into two main phases: the first in the years 2011-2013 - construction and installation of the achromatic separator beam line and the second phase in the years 2011-2015 - is the developments of the beam diagnostic, control and detection systems, testing and first experiments running. The tentative timetable of the project realization is given in Table 1.

Table 1. The tentative timetable of the ACCULINNA-2 project realization.

	2011	2012	2013	2014	2015
			Separator		
Contract for delivery of the main elements of the separator					
Simulation of the experiments & experimental tests at the existing ACCULINA-1 beam line					
First run of the new separator, tuning, first experiments running (F1-F3)					
Radiation shield & gas cell					
			DAC		
Prototypes / final variant					

Acknowledgments

The authors are grateful to Profs. S.N. Dmitriev, A.G. Popeko and G.G. Gulbekian for overall support of the project.

References

[1] http://159.93.28.88/flnr/dribs.html; http://159.93.28.88/dribs/publ.html
[2] Fomichev A S et al. 2008 *JINR Commun. E13-2008-168, Dubna*
[3] Fomichev A S et al., 2010 *Acta Phys. Pol B* **41** 475-480
[4] Rodin A M et al. 2003 *Nucl. Instrum. and Methods B* **204** 114-118
[5] Yukhimchuk et al. A A 2003 *Nucl. Instrum. and Methods A* 513 439-447
[6] Miernik K et al. 2007 *Nucl. Instrum. Methods A* **581** 194-197
[7] Mianowski S et al. 2010 *Acta Phys. Pol B* **41** 449-456

Nuclear ingredients for cross section calculation of exotic nuclei

S. Goriely[1], S. Hilaire[2], A.J. Koning[3] and M. Girod[2]

[1] Institut d'Astronomie et d'Astrophysique, Universit Libre de Bruxelles, Campus de la Plaine CP 226, 1050 Brussels, Belgium
[2] CEA, DAM, DIF, 91297 Arpajon, France
[3] Nuclear Research and Consultancy Group, P.O. Box 25, 1755 ZG Petten, The Netherlands

E-mail: sgoriely@astro.ulb.ac.be

Abstract. The increasing need for cross sections far from the valley of stability, especially for applications such as nuclear astrophysics, poses a challenge for nuclear reaction models. So far, predictions of cross sections have relied on more or less phenomenological approaches, depending on parameters adjusted to available experimental data or deduced from systematic relations. While such predictions are expected to be reliable for nuclei not too far from the experimentally known regions, it is clearly preferable to use more fundamental approaches, based on sound physical bases, when dealing with very exotic nuclei. Thanks to the high computer power available today, all major ingredients required to model a nuclear reaction can now be (and have been) microscopically (or semi-microscopically) determined starting from the information provided by an effective nucleon-nucleon interaction. All these microscopic ingredients have been included in the latest version of the TALYS nuclear reaction code (http://www.talys.eu/).

1. Introduction

The calculation of nuclear reaction cross sections involves several models such as the optical model, the pre-equilibrium model and the compound nucleus model connected one with another in a nuclear reaction code. These models rely on parameters (or ingredients) which are usually adjusted to reproduce experimental data. However, when one has to deal with nuclear reactions involving exotic targets located far from the valley of stability, the lack of experimental information makes it necessary to extrapolate the parameters on which the reaction models are based. One has then to face a situation where (1) the predictive power and robustness of the nuclear reactions models is crucial and (2) the extrapolations of the models' ingredients have to be trusted. For the nuclear reactions models, the situation is rather positive today since several years of efforts have provided the community with robust and well tested nuclear reactions codes, such as TALYS [1] or EMPIRE [2] for instance, in which the modern nuclear reaction models are coherently implemented. The same holds for phenomenological ingredients which have been used since the infancy of nuclear physics, and have proven their ability in fitting available experimental nuclear data. These ingredients have been assembled in libraries, among which the most famous one is certainly the Reference Input Parameter Library (RIPL) developed for nearly 15 years under the coordination of the IAEA [3]. However, these phenomenological approaches are often based on simple empirical approximations to represent the complexity of nuclear physics from a universal point of view, and it is therefore quite clear that their extrapolations are likely to be more uncertain than those based on sound physical bases usually referred to as microscopic approaches. This is the reasons why several years of efforts have been devoted to derive all the ingredients required by nuclear reaction models from theoretical approaches based on first principles of physics and to include them in a nuclear reaction code. In our case this code is the TALYS code [1], and within the RIPL project (Phase 3), almost all microscopic ingredients implemented in TALYS have also been made available to the nuclear physics community. The present contribution summarizes the various microscopic models developed and implemented in the TALYS code and discusses some future improvements.

2. From phenomenological to microscopic approaches

For astrophysics, but also in other fields, a large amount of nuclear data need to be extrapolated far away from experimentally known regions. In these situations, two major features of the nuclear theories have to be considered. The first one is the *accuracy* of a model. In most applications, this criterion has been the main, if not the unique, one for selecting a model. The second one is the *reliability* of the predictions. A physically sound model based on first principles and that is as close as possible to a microscopic description of the nuclear systems is expected to provide the best possible reliability of extrapolations far away from experimentally known regions. Of course, the accuracy of such microscopic models in reproducing experimental data may be poorer than the one obtained from more phenomenological models in which enough free parameters can guarantee a satisfactory agreement with the data at the expense of the quality of the input physics, and consequently of the reliability.

The most fundamental quantity concerns the nuclear structure properties. These include the nuclear mass, the deformation, the nuclear radius, ... of the nucleus in its ground state as well as in its excited states. A series of Hartree-Fock-Bogolyubov (HFB) mass formulas have recently been constructed. Such HFB calculations are based on a conventional Skyrme force with a density dependent or independent pairing interaction treated in the full Bogolyubov framework with restoration of broken symmetries. Among the various mass model, the BSk14 [4] and BSk21 [5] interactions have been implemented in TALYS. Their respective rms deviations with respect to all measured nuclear masses amount to 0.729 and 0.577 MeV. While the latter gives a much higher accuracy and has been constrained to many more observables, the former has also been used to coherently determined additional nuclear inputs, such as nuclear level density and fission properties. For this reason, BSk14 remains one of the favored interactions for reaction calculations and to estimate the various nuclear ingredients of relevance, i.e (i) the optical model potential (OMP), (ii) the nuclear level densities, (iii) the γ-ray strength functions, and (iv) the fission properties, as discussed below

Starting with the OMP, we have at our disposal the so-called semi-microscopic JLMB approach [7] which essentially requests as inputs the nuclear matter densities. Such an approach is clearly less accurate than phenomenological or local OMP which have been fitted for years to reproduce experimental data in the valley of stability. Yet, since it is not based on a free parameter adjustment, but rather on symmetric nuclear matter properties within the Brückner-Hartree-Fock approach, its predictive power far from the experimentally known regions is increased. The only required inputs of this OMP are the proton and neutron radial matter densities which can be and have been tabulated and stored in databases [1, 3, 6].

Another crucial quantity required, in particular for the statistical decay of the compound nucleus, is the nuclear level density. As for the OMP, a large collection of more or less refined analytical expressions are available in the literature. However these analytical models suffer from severe limitations due to the crude approximations on which they rely. For this reason, we have developed an alternative method to go beyond these severe restrictions. It consists in using a microscopic combinatorial model essentially based on HFB nuclear structure properties. This approach provides nuclear level densities which can compete with the historically employed global phenomenological approaches and simultaneously account for non statistical features [8]. Systematic calculations of nuclear level densities have been performed and made available in a table format. These tables can be used, but also adjusted if need be to fit experimental reaction cross sections (more details can be found in Refs [6, 9, 10]).

A decay channel which always have to be considered in competition with the light particle emission is the photon emission. For this specific channel, the γ-ray strength functions are the counterpart of the particle transmission coefficients obtained from the OMP. Again, a large number of analytical models are available in the literature (see Ref. [3] for an extensive review) reflecting the lack of knowledge that we have about this ingredient in particular for low photons' emission energies. The γ-ray strength function has also been determined from microscopic approaches such as the Quasiparticle Random Phase Approximation for instance. Within this framework, the E1 γ-ray strength functions have been systematically and successfully determined [11] and made available in a table format [3, 6].

Last but not least is the fission channel which is probably the most complex process to model. Phenomenological approaches lead to accurate description of experimental fission cross sections, but at the price of the adjustment of several tens of free parameters whose fine tuning is particularly subtle and makes it almost impossible to make any predictions. Microscopic fission barriers can also be determined coherently with all the previously mentioned ingredients. As pointed out in Ref. [4], the global model of HFB barriers obtained with the BSk14 Skyrme force reproduces the emprical barriers [3] of nuclei with $Z \geq 88$ with an rms deviation of about 0.65 MeV. It remains difficult nowadays to improve such an accuracy with global models, even on the basis of macroscopic-type approaches. Such an accuracy is clearly not sufficient for practical applications, but with a simple normalization, a fair description of experimental fission cross section can be achieved [10]. Such an approach has also the advantage of avoiding the simple picture of double-humped inverted parabolic shapes (whose height and width are freely adjusted among other parameters) and of providing fission paths (along with nuclear level densities at the saddle points) on the same footing for for both measured and unknown nuclei.

3. What's coming next

All the previously mentioned microscopic ingredients have been tabulated either directly within the TALYS nuclear reaction code or in publicly available libraries [3, 6]. Yet, they have all been derived on the basis of a Skyrme-type effective interaction. A future improvement would consist in replacing this interaction by a finite-range interaction of the Gogny type with an explicit and microscopic account of collective correlations beyond mean field. Work along this line has already started improving significantly the ability of the Gogny HFB model to reproduce experimental masses [12]. Thanks to this new interaction, we plan to derive all the microscopic ingredients within a coherent framework.

References

[1] Koning A J, Hilaire S and Duijvestijn M 2008 in *Nucl. Data for Science and Technology*, (EDP Sciences, eds Bersillon et al.) p 211 (see also http://www.talys.eu)
[2] Herman M, Capote R, Carlson B V, Oblozinsky P, Sin M, et al. 2007 *Nucl. Data Sheets* **108** 2655
[3] Capote R, Herman M, Oblozinsky P, Young P G, Goriely S, et al. 2009 *Nucl. Data Sheets* **110** 3107
[4] Goriely S, Samyn M and Pearson J M 2007 *Phys. Rev.* C **75** 064312
[5] Goriely S, Chamel N and Pearson J M 2010 *Phys. Rev.* C **82** 035804
[6] The Brussels Library 2011: *http://www.astro.ulb.ac.be/bruslib*
[7] Bauge E, Delaroche J-P and Girod M 2001 *Phys. Rev.* C **63** 024607
[8] Goriely S, Hilaire S and Koning A J 2008 *Phys. Rev.* C **78** 064307 (see also Goriely et al., these proceedings)
[9] Koning A J, Hilaire S and Goriely S 2008 *Nucl. Phys.* A **810** 13
[10] Goriely S, Hilaire S, Koning A J, Sin M and Capote R 2009 *Phys. Rev.* C **79** 024612
[11] Goriely S, Khan E and Samyn M 2004 *Nucl. Phys.* A **739** 331
[12] Goriely S, Hilaire S, Girod M and Péru S 2009 *Phys. Rev. Lett.* **102** 242501

Latest development of the combinatorial model of nuclear level densities

S. Goriely[1], S. Hilaire[2] and M. Girod[2]

[1] Institut d'Astronomie et d'Astrophysique, Universit Libre de Bruxelles, Campus de la Plaine CP 226, 1050 Brussels, Belgium

[2] CEA, DAM, DIF, 91297 Arpajon, France

E-mail: sgoriely@astro.ulb.ac.be

Abstract. The combinatorial model of nuclear level densities has now reached a level of accuracy comparable to that of the best global analytical expressions without suffering from the limits imposed by the statistical hypothesis on which the latter expressions rely. In particular, it provides, naturally, non-Gaussian spin distribution as well as non-equipartition of parities which are known to have an impact on cross section predictions at low energies [1, 2, 3]. Our previous global models developed in Refs. [1, 2] suffered from deficiencies, in particular in the way the collective effects - both vibrational and rotational - were treated. We have recently improved this treatment using simultaneously the single-particle levels and collective properties predicted by a newly derived Gogny interaction [4], therefore enabling a microscopic description of energy-dependent shell, pairing and deformation effects. In addition for deformed nuclei, the transition to sphericity is coherently taken into account on the basis of a temperature-dependent Hartree-Fock calculation which provides at each temperature the structure properties needed to build the level densities. This new method is described and shown to give promising results with respect to available experimental data.

1. Introduction

Global microscopic models of nuclear level densities (NLD) have been developed for the last decades [2] but are almost never used for practical applications, because of their lack of accuracy in reproducing experimental data or because they do not offer the same flexibility as do highly parameterized analytical expressions. We have therefore developed a combinatorial approach and demonstrated that it can clearly compete with the statistical ones in the global reproduction of experimental data [1, 2] . In addition, we have shown that such tabulated NLD can be normalized, if need be, to perform experimental fits of cross sections [5, 6]. Such an approach provides the energy, spin and parity dependence of NLD, and, at low energies, describes the non-statistical limit which, by definition cannot be described by the traditional statistical formulas though it can have a significant impact on cross section predictions (e.g [3]). However, our combinatorial method still includes phenomenological ingredients that could hamper its microscopic nature, and consequently its predictive power.

Our combinatorial method has been extensively described in Refs. [1, 2, 7] and we just summarize here the main features. It consists in using the single-particle level scheme and pairing properties obtained from axially symmetric Hartree-Fock-Bogoliubov (HFB) method to construct incoherent particle-hole (ph) state densities as functions of the excitation energy, the spin projection and the parity. These incoherent ph state densities are then folded with vibrational excited states to yield total state densities [2, 7]. Level densities are then obtained by constructing rotational bands if any (i.e. if the nucleus is deformed) or using the classical expression relating state and level densities for a spherical nucleus [2]. Although it is well established that at increasing energies, a deformed nucleus becomes spherical, this effect has not been taken microscopically into account up to now.

2. Improvements to the combinatorial NLD model

If the combinatorial level densities we had obtained (and tabulated) so far provide quite good results when compared to available experimental data [1, 2] or for cross section estimates [6], it still suffers from several approximations, as (i) the determination of the vibrational energy levels and (ii) the evolution of the collective enhancement at increasing excitation energies. Both approximations have been improved in the present study.

For the prediction of vibrational energy levels, the five-dimensional collective Hamiltonian based on the Gogny interaction which has proven its ability to estimate systematically the properties of the quadrupole vibrational levels [8] is now followed to replace the previously used systematics. Theoretical predictions of the octupole mode can be obtained using the Quasiparticle Random Phase Approximation [9], but such a systematic study has not been performed yet. For coherence, the D1M Gogny force [4] (i.e the first Gogny-type interaction capable of reproducing measured masses with a high degree of accuracy) is adopted to extract systematically all nuclear ingredients of relevance in NLD calculations.

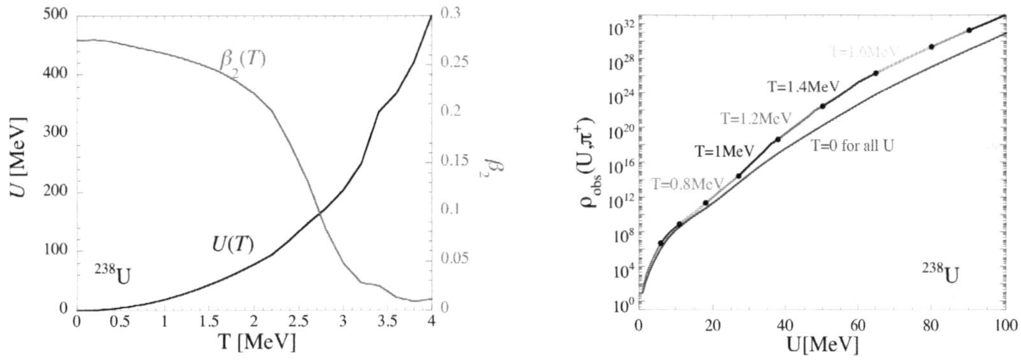

Figure 1. *Left*: Excitation energy U and and quadrupole deformation β_2 as a function of the nuclear temperature T for ^{238}U. *Right*: Total NLD for positive parities of ^{238}U calculated for several temperatures connected smoothly thanks to the procedure described in the text. The points not affected by the smoothing procedure and the $T = 0$ level density are also shown.

Concerning the expected transition from deformed to spherical shape at increasing excitation energies, the temperature-dependent HFB approach is now used, following the method described in Ref. [10]. The calculations have been performed with the same D1M interaction [4] and are based on the minimization of the free energy $F = E - TS$ (where E is the binding energy, T the temperature and S the entropy). The approach followed here consists in determining, at a selected temperature T_i (corresponding to a given excitation energy U_i) the most probable deformation β_i through an unconstrained HFBCS calculation minimizing F. The corresponding evolution of the excitation energy U and quadrupole deformation β_2 with increasing temperatures are illustrated in Fig. 2 for ^{238}U. The well-known approximative relation $U \propto T^2$ as well as the deformed to spherical shape transition can be observed. To estimate the

NLD at an excitation energy U_i and thus at the corresponding deformation β_i, the combinatorial method described above is used with the corresponding single-particle level scheme and pairing properties at the temperature T_i. The NLD are sensitive to the corresponding T-dependent nuclear inputs and display discontinuities stemming from the non-continuous disappearance of shell and pairing effects due to the finite set of temperatures considered. Indeed, the NLD determined at a given temperature is only valid at the corresponding excitation energy. However, for practical reasons, we consider the NLD at T_i to be valid over the interval $[U_i, U_{i+1}]$. To suppress the discontinuities, an differential energy shift is applied to the NLD above U_i (up to U_{i+1}) in such a way as to recover the NLD calculated at U_{i+1} on the basis of the HFB ingredients obtained at T_{i+1}. With this treatment, the discontinuity in the T-dependent NLD are smoothed out, as shown in Fig. 2. The T-dependent NLD is found to be significantly different from

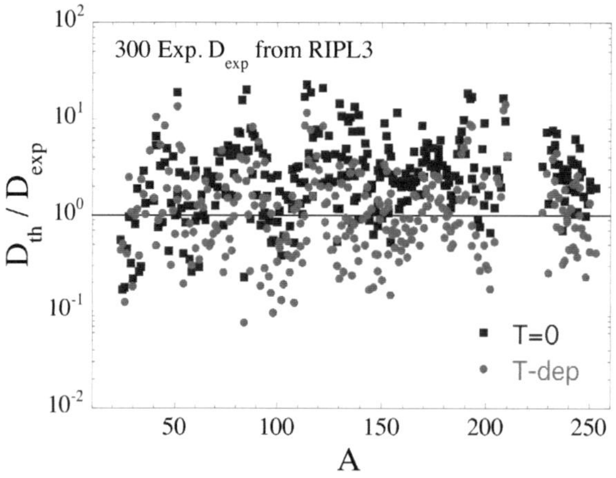

Figure 2. Ratio of HFB plus combinatorial (D_{th}) to the experimental (D_{exp}) s-wave neutron resonance spacings compiled in [11] with the temperature dependent treatment (red circles) and using the HFB ingredients obtained for $T = 0$ (blue squares).

that traditionally obtained at $T = 0$ (Fig. 2). Such a temperature effect may also affect the s-wave spacings at the neutron separation energy. To check that, we have applied the previously described method to all the nuclei for which experimental s-wave spacings are available. In terms of the rms deviation on all the experimental spacings [11], we find $f_{rms} = 3.6$ without accounting for the temperature dependence, and $f_{rms} = 2.7$ otherwise (for the definition of f_{rms}, see Ref. [2]). It should also be mentioned that contrary to what has been done in Ref. [2], no contribution from hexadecapole vibrational phonons are now included for nuclei with $Z < 85$.

References

[1] Hilaire S and Goriely S 2006, *Nucl. Phys.* A **779** 63
[2] Goriely S, Hilaire S and Koning A J 2008 *Phys. Rev.* C **78** **(2008)** 064307 and references therein
[3] Goko S, Utsunomiya H, Goriely S, Makinaga A, Kaihori T, et al. 2006 *Phys. Rev. Lett.* **96** 192501
[4] Goriely S, Hilaire S, Girod M and Péru S 2009 *Phys. Rev. Lett.* **102** 242501
[5] Koning A J, Hilaire S and Goriely S 2008 *Nucl. Phys.* A **810** 13
[6] Goriely S, Hilaire S, Koning A J, Sin M and Capote R 2009 *Phys. Rev.* C **79** 024612
[7] Hilaire S, Delaroche J-P and Girod M 2001 *Eur. Phys. J.* A **12** 169
[8] Delaroche J-P, Girod M, Libert J, Goutte H, Hilaire S, et al. 2010 *Phys. Rev.* C **81** 014303
[9] Péru S, Gosselin G, Martini M, Dupuis M, Hilaire S and Devaux J-C 2011 *Phys. Rev.* C **83** 014314
[10] Martin V, Egido J L and Robledo L M 2008 *Phys. Rev.* C **68** 034327
[11] Capote R, Herman M, Oblozinsky P, Young P G, Goriely S, et al. 2009 *Nucl. Data Sheets* **110** 3107

Journal of Physics: Conference Series **337** (2012) 012028 doi:10.1088/1742-6596/337/1/012028

Multiphonon nuclear response in ^{16}O: A microscopic treatment equivalent to shell model

Knapp F[23], [1], Bianco D[1], Lo Iudice N[1], Andreozzi F[1], Porrino A[1]

[1]Dipartimento di Scienze Fisiche, Università di Napoli Federico II, and INFN, Sezione di Napoli, Napoli, Italy

[2] INFN, Sezione di Napoli, Napoli, Italy

Abstract. We have reformulated an equation of motion method for generating iteratively a basis of multiphonon states built of Tamm-Dancoff phonons. Represented in such a basis, the Hamiltonian matrix has a simple structure and can be easily brought to diagonal form. The method is here adopted to compute the energy levels and the electromagnetic responses in ^{16}O, chosen as testing ground because of its complex structure.

1. Introduction

The increasing evidence of multiphonon collective modes at low-energy [1], and of multiple giant resonance at high energy [2] requires theoretical studies which go beyond Tamm-Dancoff (TDA) or random-phase approximations (RPA). An extension of these, basically harmonic, approximations, is also needed in order to describe the widths and damping of giant resonances.

The existing extensions, known as second RPA (SRPA), derive equations where the RPA particle-hole (ph) are coupled to the $2ph$ configurations . We mention here a recent calculation using a Skyrme force [3] and a relativistic RPA (RRPA) plus phonon coupling (PC) approach applied to the giant dipole resonance (GDR) and to the pygmy giant resonance in spherical open shell nuclei [4].

All the above approaches include up to $2ph$ configurations and, therefore, are more suited to describe the fragmentation of collective modes. Moreover, they rely on the quasiboson approximation, valid in the small amplitude vibrational limit.

Here we report of a reformulation of an equation of motion multiphonon approach (EMPM) [5, 6] which generates iteratively a multiphonon basis built of Tamm-Dancoff phonons and solves the eigenvalue problem in such a basis. The method is basically free of approximations and, therefore, equivalent to shell model.

In order to illustrate its potential we solved the eigenvalue problem for ^{16}O in a space which includes up to three phonons. The isovector $E1$ response will be discussed.

2. The method

We consider a two-body Hamiltonian H in the second quantized form

$$H = \sum_i \epsilon_i a_i^\dagger a_i + \frac{1}{4} \sum_{ijkl} V_{ijkl}\, a_i^\dagger a_j^\dagger a_l a_k \,, \tag{1}$$

[3] Present address: Faculty of Mathematics and Physics, Charles University, Prague, Czech Republic

where ϵ_i are the single-particle energies, V_{ijkl} the antisymmetrized matrix elements of the nucleon-nucleon interaction, $a_i^\dagger$ (a_i) the creation (annihilation) particle operators with respect to the physical vacuum.

We intend to construct n-phonon states $|n;\beta>$ having the following structure

$$|n;\beta>=\sum_{\gamma\lambda} C_{\lambda\gamma}^{(\beta)} O_\lambda^\dagger \; |n-1;\gamma>, \tag{2}$$

where $|n-1;\gamma>$ is a $(n-1)$-phonon state upon which acts the particle-hole $(p-h)$ Tamm-Dancoff phonon operator

$$O_\lambda^\dagger = \sum_{ph} c_\lambda(ph) a_p^\dagger a_{\bar{h}} \tag{3}$$

and $a_p^\dagger$ $(a_{\bar{h}})$ creates a particle (hole) with respect to the ph vacuum $\mid 0\rangle$.

The starting point of the procedure consists in writing the equations of motion

$$< n,\beta|[H,O_\lambda^\dagger]|n-1,\alpha>= \left(E_\beta^{(n)} - E_\alpha^{(n-1)}\right) < n,\beta|O_\lambda^\dagger|n-1,\alpha> . \tag{4}$$

After expanding the commutator and making several manipulations [7], we get, for the n-phonon subspace, the eigenvalue equation

$$A\mathcal{D}C \;=\; E\mathcal{D}C, \tag{5}$$

where $\mathcal{D}(\lambda\alpha,\lambda'\alpha') =< n-1,\alpha' \mid O_{\lambda'} O_\lambda^\dagger \mid n-1,\alpha>$ is the metric matrix and

$$A(\lambda\alpha,\lambda'\alpha') = (E_\lambda + E_\alpha)\delta_{\lambda\lambda'}\delta_{\alpha\alpha'} + \mathcal{V}_{\lambda\alpha,\lambda'\alpha'}. \tag{6}$$

Here, E_λ and E_α are the energies of the TDA phonon $|\lambda>$ and the $n-1$-phonon state $|(n-1),\alpha>$. They interact through a potential of the form

$$\mathcal{V}_{\lambda\alpha,\lambda'\alpha'} = \sum_{rstq} \rho_{\alpha\alpha'}^{(n-1)}(rs) V_{trqs} \rho_{\lambda\lambda'}(tq), \tag{7}$$

where $\rho_{\alpha\alpha'}^{(n)}(rs) =< n,\alpha'|a_r^\dagger a_s|n,\alpha>$ are either particle $(r=p, s=p')$ or hole $(r=h, s=h')$ one-body densities.

All quantities are given by recursive formulas, which allows to solve the eigenvalue Eq. (5) iteratively once the TDA phonons are generated. The redundant states are eliminated by the procedure outlined in [5, 6], based on the Clolesky decomposition method.

A set of orthonormal multiphonon states $\{|0>, |n=1,\lambda>, \ldots |n,\alpha> \ldots\}$ is thus generated. In each n-phonon block, the Hamiltonian is diagonal. Only subspaces differing by at most two-phonons are coupled. The matrix elements $< n',\beta|H|n,\alpha>$ $(n' = n+1, n+2)$ are given by simple recursive formulas so that it is a simple matter to diagonalize the Hamiltonian in the full space.

3. A numerical testground: ^{16}O

We have chosen ^{16}O as testing ground for our method. We used realistic Hamiltonians and adopt either a Nilsson or Hartree-Fock single particle basis. All the shells up to the $N=3$ (p,f) have been include in order to construct the TDA phonons.

A full calculation up to three phonons is too lengthy due to the huge number of the three-phonon states. We have therefore truncated such a subspace by including the three phonon space up to the energy $\simeq 45$ MeV.

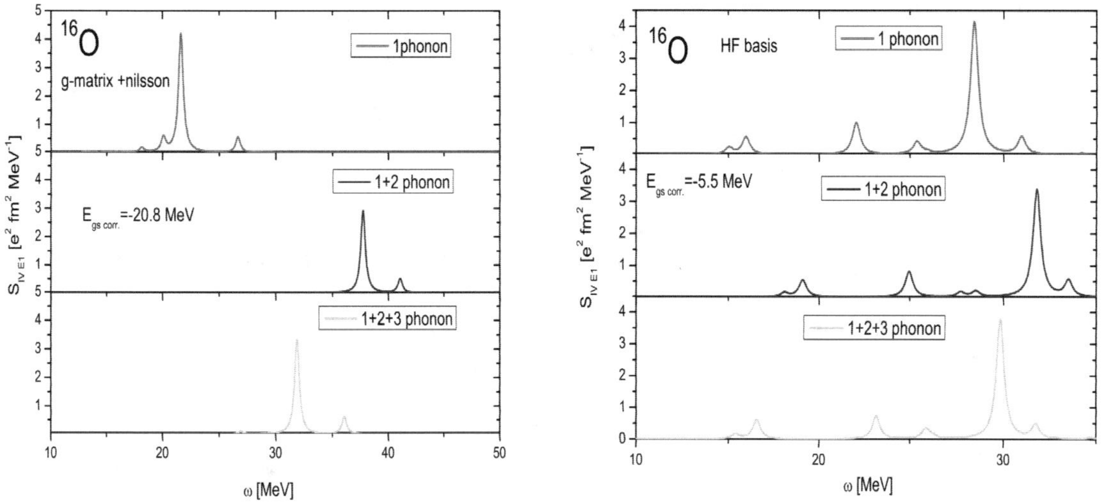

Figure 1. Isovector $E1$ strength distribution in ^{16}O.

The $E1$ strength distribution computed in either Nilsson or HF basis is plotted in Fig. 1. The fragmentation is practically not affected by the number of phonons. It is, instead, strongly sensitive to the basis adopted. In fact, it increases considerably in going from Nilsson to HF.

The position of the peaks depend on both the basis and the number of phonons. In the Nilsson basis, the TDA main peak is at ~ 22 MeV, consistently with experiments. In HF, it is about 7 MeV higher.

If we include the states up to two-phonons, the peaks are shifted upward by 5 MeV in the HF basis and by ~ 18 MeV in Nilsson. This is simply due to the depression of the ground state induced by the coupling between the ph vacuum and the two-phonon subspace. The ground state is depressed by 20.8 MeV in Nilsson case and 5.5 MeV in HF basis. Indeed, the coupling between states differing by two-phonons is the most effective one. One should, therefore, expect a corresponding shift for the excited states once the three-phonons are included.

Indeed, the coupling between one and three phonons shifts downward the peaks but not not enough. The main peak, in fact, is at $\simeq$30 MeV in both bases, too high with respect to the experimental value ~ 21 MeV. It is to be said, however, that only a fraction of the three phonon states have been included. We expect that, if the number of three-phonon states is incremented, the one- to three-phonon coupling would be more effective and, eventually, would bring the peaks close to the experimental ones.

References

[1] For a review Kneissl M, Pietralla N and Zilges A 2006 *J. Phys. G: Nucl. Part. Phys.* **32** R217
[2] Auman T, Bortignon P F, Hemling H 1998 *Annu. Rev. Nucl. Part. Sci.* **48** 351
[3] Gambacurta D, Grasso M and Catara F 2010, Phys. Rev. C **81** 054312
[4] Litvinova E, Ring P and Tselyaev V 2010, Phys. Rev. Lett. **105** 022502
[5] Andreozzi F, Knapp F, Lo Iudice N, Porrino A, and Kvasil J 2007 *Phys. Rev.* C **75** 044312
[6] Andreozzi F, Knapp F, Lo Iudice N, Porrino A, and Kvasil J 2008 *Phys. Rev.* C **78** 054308
[7] Bianco D, Knapp F, Lo Iudice N, Andreozzi F, Porrino A, to be submitted for publication.

Journal of Physics: Conference Series **337** (2012) 012029 doi:10.1088/1742-6596/337/1/012029

Investigating the variation of elastic alpha scattering cross sections in the A ≈ 100 region

G. G. Kiss[1], P. Mohr[1,2], Zs. Fülöp[1], Gy. Gyürky[1], Z. Elekes[1], J. Farkas[1], E. Somorjai[1], D. Galaviz[3], R. T. Güray[4], N. Özkan[4], C. Yalçin[4], J. Görres[5]

[1] Institute of Nuclear Research (ATOMKI), H-4001 Debrecen, POB.51., Hungary
[2] Diakonie-Klinikum, D-74523 Schwäbisch Hall, Germany
[3]Centro de Física Nuclear da Universidade de Lisboa, 1649-003, Lisbon, Portugal
[4]Kocaeli University, Department of Physics, TR-41380 Umuttepe, Kocaeli, Turkey
[5]University of Notre Dame, Notre Dame, Indiana 46556, USA

E-mail: ggkiss@atomki.hu

Abstract. The elastic scattering cross sections for the reactions ^{110,116}Cd(α,α)^{110,116}Cd at energies above and below the Coulomb barrier are measured to provide a sensitive test for the alpha-nucleus optical potential parameter sets. The potentials under study are a basic prerequisite for the prediction of α-induced reaction cross sections, e.g. for the calculation of stellar reaction rates in the astrophysical p- or γ-process. Moreover, the variation of the elastic alpha scattering cross sections along the Z = 48 isotopic and N = 62 isotonic chain is investigated by the study of the ratios of the of ^{110,116}Cd(α,α)^{110,116}Cd scattering cross sections at $E_{c.m.}$ ≈ 15.6 and 18.8 MeV and the ratio of the ^{110}Cd(α,α)^{110}Cd and ^{112}Sn(α,α)^{112}Sn reaction cross sections at $E_{c.m.}$ ≈ 18.8 MeV, respectively. These ratios are sensitive probes for the alpha-nucleus optical potential parameterizations.

1. Motivation

In several astrophysical applications — such as modeling the nucleosynthesis in explosive scenarios like γ process — the reaction rates are taken from statistical model calculations [1]. These calculations utilize global alpha-nucleus optical potential (which combines a Coulomb term with a complex nuclear potential, composed of real and imaginary parts) parameter sets. Considerable efforts have been devoted in recent years to improve the knowledge of the alpha-nucleus optical potential [2, 3, 4, 5, 6, 7]. The extensive use of the statistical model calculations requires experimental tests for these parameter sets. One possible way of testing the different potential parameterizations is to carry out alpha elastic scattering experiments and compare the measured angular distributions to the corresponding predictions from the global potential parameterizations.

Furthermore, a global alpha-nucleus optical potential must be able not only to provide a correct prediction for the elastic scattering cross section but also to describe the variation of the angular distributions along isotopic and isotonic chains. This is especially important for explosive nucleosynthesis scenarios where reaction rates on large number of reactions involving thousands of nuclei have to be taken into account. The reliability of the potential parameterizations can be tested by measuring the elastic scattering cross sections on several nuclei along isotopic and isotonic chains and use the ratio of the Rutherford normalized cross sections as a very sensitive observable for the quality of α-nucleus potentials.

2. Experimental approach

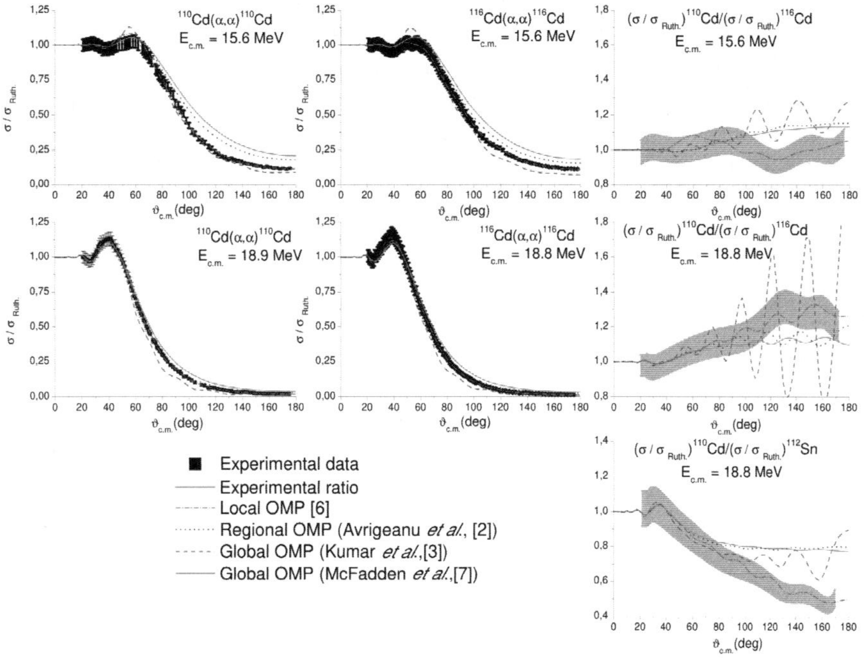

Figure 1. Rutherford-normalized elastic scattering cross sections of the ^{110,116}Cd$(\alpha,\alpha)^{110,116}$Cd reactions at $E_{c.m.} \approx 15.6$ and 18.8 MeV and experimental ratio (gray area with taken into account the experimental uncertainties) of the measured scattering cross sections at $\approx$ 15.6 MeV and 18.9 MeV versus the angle in center-of-mass frame. The cross sections of the ^{112}Sn$(\alpha,\alpha)^{112}$Sn are taken from [4]. The lines correspond to the predictions using a local[7], a regional [2] and global [3, 8] optical model parameter sets (OMP).

The experiment was carried out at the cyclotron laboratory of ATOMKI, Debrecen (further details on the experimental procedure can be found in [7]). The targets were produced by evaporating highly enriched ($\geq 95\%$) ^{110,116}Cd onto thin carbon foils ($\approx 20\ \mu$g/cm^2). The target thickness was determined via alpha particle energy loss measurement using radioactive sources and was found to be approximately 200 μg/cm^2. The energies of the alpha beam were 16.14 and 19.46 MeV with typical beam currents of 150-200 pnA. Six ion implanted silicon detectors with active areas of 50 mm^2 mounted on two rotatable turntables were used to measure the angular distributions. In addition, two detectors were mounted at a larger distance on the wall of the scattering chamber at fixed angles $\vartheta=\pm15°$ left and right to the beam axis. These detectors were used as monitor detectors during the whole experiment to normalize the measured angular distribution and to determine the precise position of the beam on the target.

3. Results and discussion

Complete angular distributions between 20° and 175° were measured at energies of $E_{lab.}$ = 16.14 and 19.46 MeV in 1° (20° $\leq \vartheta \leq$ 100°) and 2.5° (100° $\leq \vartheta \leq$ 175°) steps. On the left side of Figure 1. the Rutherford-normalized cross sections of the $^{110,116}Cd(\alpha,\alpha)^{110,116}Cd$ reactions can be seen. Whereas the Rutherford normalized cross sections cover only about two orders of magnitude between the highest (forward angles at $E_{lab.}$ = 16.14 MeV) and the lowest cross sections (backward angles at $E_{lab.}$ = 19.46 MeV), the underlying cross sections cover more than four orders of magnitude. Over this huge range of cross sections almost the same accuracy of about 4-5% total uncertainty could be achieved. This error is mainly caused by the uncertainty of the determination of the scattering angle in the forward region and from the statistical uncertainty in the backward region.

The behavior of the elastic alpha scattering cross sections along the Cd isotopic (Z=48) chain at $E_{c.m.} \approx$ 15.6 and 18.8 MeV and along the N=62 isotonic chain at $E_{c.m.} \approx$ 18.8 MeV is investigated in this work. Although there are small differences in the center-of-mass energies ($\leq$ 120 keV), the ratios of the Rutherford normalized cross sections are well defined because the dominating $1/E^2$ dependence of the scattering cross section is taken into account during Rutherford normalization. Therefore, the ratios of Rutherford normalized cross sections are a very sensitive test for local, regional and global alpha-nucleus potential parameterizations.

It is found that the ratio of the normalized scattering cross sections shows an oscillation pattern at backward angles (right side of Fig. 1). The large number of experimental points and the low uncertainties on all data sets provide a unique probe to understand the evolution of the alpha nucleus potential along the Cd isotopic chain. In the present work the local potential derived in [7] and the following open access regional and global alpha nucleus optical potential parameterizations are considered: the recent regional potential of Avrigeanu *et al.* [2], the global potential of Kumar *et al.* [3] and the global potential of McFadden and Satchler [8]. No regional or global parameterization can describe correctly the amplitude and the phase of the oscillation pattern of the experimental data at backward angles. This fact indicates that the available theoretical alpha nucleus optical potential parameterizations have to be further improved.

4. Acknowledgments

This work was supported by the EUROGENESIS research program, by OTKA (NN83261, K068801), by the European Research Council (grant agreement no. 203175) and by the Joint Institute for Nuclear Astrophysics (NSF grant PHY0822648). G. G. K. and D. G. acknowledge the support of the Spanish CICYT under the project FPA2005-02379 and MEC Consolider project CSD2007-00042. D. Galaviz acknowledges support from Juan de la Cierva grant (Spanish Ministry of Science). This work was also supported by the Scientific and Technological Research Council of Turkey (TUBITAK) Grant No: 108T508 (TBAG1001) and 109T585 (under the EUROGENESIS research program).

References

[1] T. Rauscher and F.-K. Thielemann, At. Data Nucl. Data Tables **79**, 47 (2001).
[2] M. Avrigeanu *et al.*, At. Data Nucl. Data Tables **95**, 501 (2009).
[3] A. Kumar *et al.*, Nucl. Phys. A**776**, 105 (2006).
[4] D. Galaviz *et al.*, Phys. Rev. C **71**, 065802 (2005).
[5] G. G. Kiss *et al.*, Phys. Rev. C **80**, 045807 (2009).
[6] P. Mohr *et al.*, Phys. Rev. C **82**, 047601 (2010).
[7] G. G. Kiss *et al.*, Phys. Rev. C accepted
[8] L. McFadden and G. R. Satchler, Nucl. Phys. **84**, 177 (1966).

Investigation of α-nuclear potential families from elastic scattering experiments [1]

A. Ornelas[1], D. Galaviz[1], Zs. Fülöp[2], Gy. Gyürky[2], G. Kiss[2], Z. Máté[2], P. Mohr[2,3], T. Rauscher[4], E. Somorjai[2], K. Sonnabend[5] and A. Zilges[6]

[1] Centro de Física Nuclear, University of Lisbon, 1649-003 Lisbon, Portugal
[2] ATOMKI, H-4001 Debrecen, POB. 51, Hungary
[3] Diakonie-Klinikum, D-74523 Schwäbisch Hall, Germany
[4] Departement für Physik und Astronomie, Universität Basel, CH-4056 Basel, Switzerland
[5] Institut für Angewandte Physik, Goethe-University Frankfurt, D-60438 Frankfurt, Germany
[6] Institut für Kernphysik, Universität zu Köln, D-50937, Köln, Germany

E-mail: ornelas@cii.fc.ul.pt

Abstract.
In this work we present the continuation of the reported analysis [1] of the experimentally measured angular distributions of the reaction ^{106}Cd$(\alpha,\alpha)^{106}$Cd at several different energies around the Coulomb barrier. The difficulties that arise in the study of ^{106}Cd-α-nuclear potential and the so called Family Problem are addressed.

1. Introduction

The 35 stable elements located on the proton-rich side of the valley of β-stability that cannot be explained in the framework of slow and fast neutron capture are the so called the *p*-nuclei. These very low-abundant nuclei present one of the most interesting puzzles in nuclear astrophysics. One of the most accepted mechanisms for the synthesis of the *p*-nuclei is based on photon-disintegration reactions on neutron-rich seed nuclei [2, 3]. Possible scenarios for such nucleosynthesis are the C, O and Ne layers of a Type II SN [2].

2. The Family Problem in ^{106}Cd

The sensitivity of nuclear reaction network calculations to the nuclear physics input has been addressed [3, 4] with particular emphasis to the uncertainties related to the α-nuclear potentials in the heavy mass region (A>150). The sensitivity of α-nuclear potentials at high energies (far above the Coulomb barrier) has been extensively studied in the past (see for instance [5, 6]). The present report concentrates on the α-nuclear potential of the system α-^{106}Cd, starting from the 14 families of the potential previously obtained [1] from the analysis of elastic scattered angular distributions measured at energies around the Coulomb barrier [7, 8]. The analysis was performed within the framework of the Optical Model, considering parameterizations of the Woods-Saxon form for both real and imaginary parts of the nuclear potential.

[1] Work supported by FCT grant PTDC/FIS/103902/2008, ERC StG No. 203175, OTKA NN83261 (EuroGENESIS) and OTKA K68801

3. Description of existing α-induced reaction data

Considering as our starting point the 14 families of the real nuclear potential [1], a modified parameterization with an increased surface Woods-Saxon was adopted for the imaginary potential. The astrophysical S-factor for the processes ^{106}Cd$(\alpha,\gamma)^{110}$Sn and ^{106}Cd$(\alpha$,n$)^{109}$Sn [9], was determined for each of the considered potential families using the NON-SMOKER^{WEB} [10] application. The default settings for neutron and proton potential [11], default nuclear level density [12] and theoretical masses [13] were used in the calculations.

The results are shown in Figure 1 together with the experimental data from [9]. At this stage, our evaluation of the potential families is limited to these two reaction channels, since they are primarily dependent on the α-particle width at the considered energies [14].

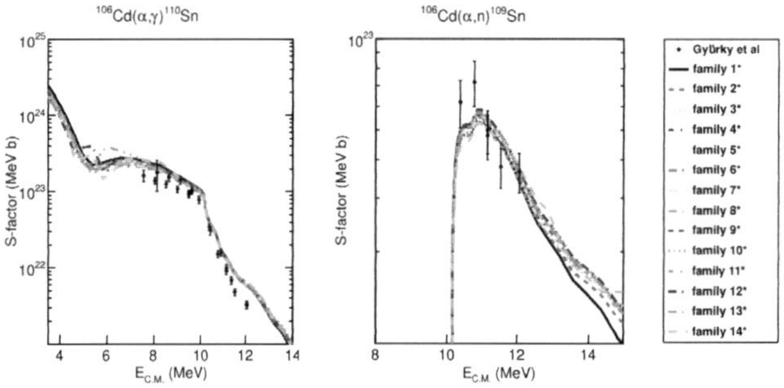

Figure 1. Astrophysical S-factor of the ^{106}Cd$(\alpha,\gamma)^{110}$Sn and ^{106}Cd$(\alpha$,n$)^{109}$Sn reactions [9] together with the results obtained from the 14 different potential families obtained in this study.

As it can be seen in the figures, both the (α,γ) and $(\alpha$,n$)$ processes are well reproduced by almost all of the considered potential families. In the case of the (α,γ) reaction, almost all families present a local minimum at the threshold energy for the $(\alpha$,p$)$ process, which needs to be further investigated. Any of these potentials could be thus used to calculate the reaction cross section at the energy range relevant for p-process calculations: T_9=2-3, with α-energies corresponding to 5.4 and 8.1 MeV [14].

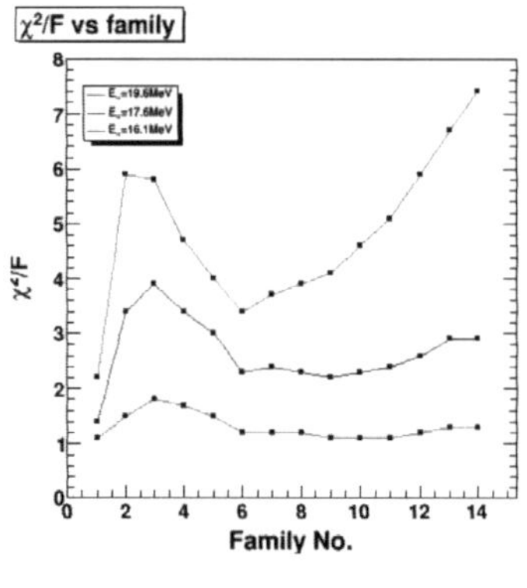

Figure 2. Normalized χ^2 obtained from the analysis of the elastic scattering data by all potential families considered in this work.

The results shown in Figure 2 provide a summary of the description of the different families of the experimental data of the ^{106}Cd$(\alpha,\alpha)^{106}$Cd at energies around the Coulomb barrier, presenting the normalized χ^2 obtained for each of the measured energies. A local minimum is observed around family number 6*, with another minima located on family 1*. The nature of these minima, as well as the particularities of these two families need to be further studied before any definite conclusions can be drawn.

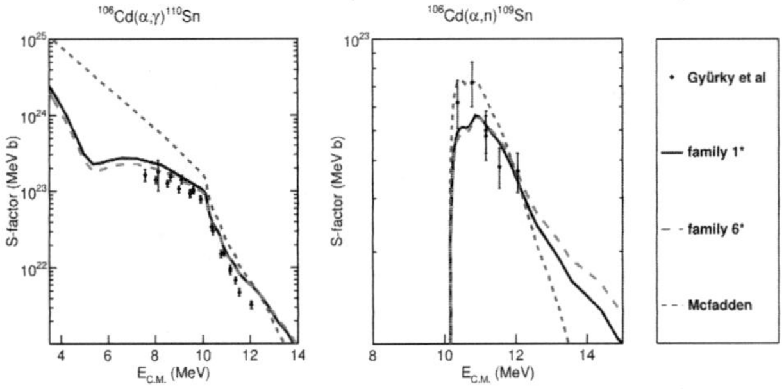

Figure 3. Astrophysical S-factor of the ^{106}Cd$(\alpha,\gamma)^{110}$Sn and ^{106}Cd$(\alpha$,n$)^{109}$Sn reactions [9] together with the results obtained from families 1*, 6* and from the global potential of [15].

4. Conclusions

At this stage, we compare the two selected families with the standard global α nuclear potential from [15]. The results are shown in Figure 3. Analysis and comparison to further global and local α nuclear potentials will be the topic of a dedicated paper. The results from this work, combined with those previously presented [1], highlight the processes that are more sensitive to each part of the considered nuclear potential. While the real part of the potential presents a clear sensitivity to elastic scattering data, induced α-particle capture reactions show a higher sensitivity to the imaginary part of the potential. When analyzing both processes in a coupled way, a minimum of χ^2/F in the description of the elastic scattering data appears. The reaction channels (α,n) and up to some extent (α,γ) at low energies, and elastic scattering reaction data at energies above the Coulomb barrier are necessary to achieve a full description of the potential.

5. Bibliography

[1] Ornelas A, Galaviz D, Fülöp Z, Gyürky G, Kiss G, Máté Z, Mohr P, Rauscher T, Somorjai E, Sonnabend K and Zilges A 2010 *PoS* **(NIC XI)** 241
[2] Arnould M and Goriely S 2003 *Phys. Rep.* **384** 1
[3] Rauscher T 2006 *Phys. Rev. C* **73** 015804
[4] Rapp W, Görres J, Wiescher M, Schatz H and Käppeler F 2006 *Astrophys. J.* **653** 474
[5] Batty C J, Friedman E, Gils H J and Rebel H 1989 *Advances in Nuclear Physics* **19** 1
[6] Atzrott U, Mohr P, Abele H, Hillenmayer C and Staudt G 1996 *Phys. Rev. C* **53** 1336
[7] Galaviz D 2004 *Systematic study of α-nucleus potentials for neutron-deficient nuclei and its astrophysical applications* Ph.D. thesis Technische Universität Darmstadt
[8] Kiss G, Zs F, Gyürky G, Máté Z, Somorjai E, Galaviz D, Kretschmer A, Sonnabend K and Zilges A 2006 *European Physics Journal* **27** 197
[9] Gyürky G, Kiss G G, Elekes Z, Fülöp Z, Somorjai E, Palumbo A, Gorres J, Lee H Y, Rapp W, Wiescher M, Özkan N, Guray R T, Efe G and Rauscher T 2006 *Phys. Rev. C* **74** 025805
[10] Rauscher T code NON-SMOKER^{WEB}, version v5.8.1w URL `http://nucastro.org/websmoker.html`
[11] Jeukenne J P, Lejeune A and Mahaux C 1977 *Phys. Rev. C* **15**(1) 10–29
[12] Rauscher T, Thielemann F K and Kratz K L 1997 *Phys. Rev. C* **56**(3) 1613–1625
[13] Moller P, Nix J, Myers W and Swiatecki W 1995 *Atomic Data and Nuclear Data Tables* **59** 185 – 381 ISSN 0092-640X
[14] Rauscher T 2010 *Phys. Rev. C* **81**(4) 045807
[15] McFadden L and Satchler G R 1966 *Nucl. Phys.* **84** 177

Journal of Physics: Conference Series **337** (2012) 012031 doi:10.1088/1742-6596/337/1/012031

A mathematical structure for nuclei

M. Fabre de la Ripelle

Particle Physics Department, Weizmann Institute of Science, Rehovot, Israël *

Abstract

The Shell Model was accepted even though it still lakes theoretical foundation [1]. A mathematical structure is proposed to give an answer to this statement and establish a needed foundation for this Model.

The nuclear Independent Particle Model (IPM) has been accepted for sixty years despite « no clear demonstration was ever given that the Shell Model is a good approximation » [1].

In the IPM the coordinates of each nucleon are associated with a wave function in a potential well then coordinates are not left over to define the free motion of the center of mass. Other defects make the IPM questionable [2].

"Among all central potential wells the harmonic oscillator potential occupies a special position : it yields the order of single nucleon orbits which, together with a strong spin-orbit interaction give rise to the observed shells in nuclei" [1]

Schematically, avoiding parameters, the wave function of a harmonic oscillator is a solution of $(-\Delta + r^2 - E)\Psi(r,\Omega) = 0$ where r is the radial coordinate in polar coordinates $\vec{x}(r,\Omega)$, Δ is a Laplace operator and E an eigen value.
In the physical 3-dimensional space $r = r_i$ is the radial coordinate of the particle i (coordinates $\vec{x}_i(r_i,\omega_i)$). For many body-systems $r^2 = \sum_1^A r_i^2$ for A identical particles.

In any case this sum of one-body harmonic oscillator potentials is invariant under rotation in the whole $3\times A$-dimensional space. It is the only one for which it is the case.

Let be $V(r)$ a potential invariant under rotation in the whole space, the wave function is the product $\Psi(r,\Omega) = Y_{[L]}(\Omega) u_{[L]}(r) / r^{(D-1)/2}$ of a harmonic polynomial on the unit sphere in the $3\times A$- dimensional space and a radial function $u_{[L]}(r)$ solution of

$$(1)\ \left\{ -\frac{d^2}{dr^2} + \frac{\mathscr{L}(\mathscr{L}+1)}{r^2} + V(r) - E \right\} u(r) = 0$$

where $\mathscr{L} = L + (D-3)/2$, $D = 3A$, $r^2 = 2\sum_1^A (\vec{x}_i - \vec{X}_{c.m})^2$.

L is the degree of the harmonic polynomial $r^L Y_{[L]}(\Omega)$ while $[L]$ is the set of the $3A-1$ quantum numbers associated with the $3A-1$ angular coordinates (Ω).

For ground states the degree L is the smallest available.

L=0 for Bosons. For Fermions the antisymmetry is provided by a Slater determinant

$$D_{[L]}(\vec{x}) = \left\| t_j^i s_j^i \Phi_{n_j,\ell_j}(r_i) r_i^{\ell_j} Y_{\ell_j}^{m_j}(\omega_i) \right\|$$

where i and j refer to the rows and columns respectively, while t and s to isospin and spin respectively.

Journal of Physics: Conference Series **337** (2012) 012031 doi:10.1088/1742-6596/337/1/012031

In the IPM Φ is a radial wave in the well, while in the Hyperspherical Model (HM) $\Phi_{n_j,\ell_j}(r_i) = r_i^{2n_j}$ in such a way that $D_{[L]}(\vec{x})$ becomes an antisymmetric polynomial of degree $L = \sum_1^A (2n_i + \ell_i)$ homogeneous in the coordinates $\vec{x}_i$.

A ground state is obtained when $L = L_m$ is minimum.

Any symmetrical operator seeking to decrease the degree of homogeneous polynomial applied to $D_{[L_m]}(\vec{x})$ gives zero. Therefore $D_{[L_m]}(\vec{x})$ is a translationnaly invariant harmonic polynomial. It is the lowest order polynomial occurring in a harmonic polynomial expansion of the wave function. It generates the Magic Numbers [2] and it complies with the properties of analytical functions, like wave functions, to be expandable in harmonic polynomials. Since the "state" $D_{[L_m]}(\vec{x})$ is translationnaly invariant it can be written in terms of Jacobi coordinates. In order to calculate the two-body correlations a reference pair must be chosen. Let it be *(i, j)* and D_{ij} the Slater determinant $D_{[L_m]}(\vec{x})$ for this choice.

A sum of two-body potentials $V(\vec{r}_{ij})$ generates a sum of two-body amplitudes $P(\vec{r}_{ij}, r)$ which are a solution of [2]

$$(2)\quad (T - E)D_{ij}P(\vec{r}_{ij}, r) + V(\vec{r}_{ij})D_{ij}\sum_{k,\ell} P(\vec{r}_{k\ell}, r) = 0$$

where T is the kinetic energy operator, E the total energy while $V(\vec{r}_{ij})$ depends on spin and isospin for nuclei. A sum of eq. (2) over all pairs *(i, j)* reproduces the Schrödinger equation.

Let us set $P(\vec{r}_{ij}, r) = P_\lambda(\vec{r}_{ij}, r)u(r)/r^{(D-1)/2}$ where P_λ is adiabatically r-dependent, premultiply eq.(2) by D_{ij} and then integrate over all variables except r_{ij}. It yields two-one variable second order differential equations [3, 4] which can be easily solved numerically by using the usual algorithm for integrating an IPM radial Schrödinger equation.

All details can be found in [3, 4] and references therein.

When the potential is hypercentral (r dependence only) the solution of eq. (2) can be exactly calculated. For an infinite hypercentral well were $V(r) = 0$ in eq.(1) and $u(r) = 0$ for $r \geq R$, the kinetic energy for identical fermions in ground state for large A systems is 12% smaller than the one given by the Fermi-gas model with the same r. m. s. radius a [5].

It exibit the difference between the Fermi-gas IPM and the Hyperspherical Model where more freedom is available since with the constraint $\sum_1^A (\vec{x}_i - \vec{X}_{cm})^2 < R^2 \approx Aa^2$ one particle can extend far away of the center of mass.

For testing the validity and accuracy of our numerical algorithm a comparison of the binding energy of A-bosons systems with Afnan-Tang S3 potential obtained first with a Hyperspherical Harmonic Expansion Method (HHEM) using up to fifty coupled equations [6] and then from eq.(2) is shown in the following table.

Journal of Physics: Conference Series **337** (2012) 012031 doi:10.1088/1742-6596/337/1/012031

A	4	6	8	10	12	16	40
HHEM	27.23	111.17	247.41	430.70			
HM	27.25	111.16	247.41	430.68	657.57	1235.45	7826
r.m.s radius	1.44	1.34	1.33	1.33	1.34	1.35	1.42

Other comparisons for ^{6}Li, ^{12}C, ^{16}O and ^{40}Ca can be found in ref. [2].

The Argonne 14 potential does not bind without correlations. It underbids ^{4}He [7]. A Variational Monte Carlo (VMC) calculation was performed with a V6 version of this potential completed by an Urbana VII three-body component [8].

Another exemple of the capability of the HM is provided by a calculation of eq.(2) with a V4 version of Arg. 14 potential where the triplet –even Afnan-Tang S_1 (ATS_1) potential fitting $^{3}S_1$ phase-shifts and the deuteron energy is substituted for the V^{3+} + tensor of Arg. 14.

The binding energies are shown for ^{4}He, ^{16}O and ^{40}Ca with and without the Coulomb potential on the next table.

	^{4}He	^{16}O	^{40}Ca
HM without Coulomb	30.7	128.55	335.9
HM with Coulomb	29.8	112.9	251.43
VMC with Coulomb	30.48	112.8	

The Arg.14 potential reduced to a V_4 potential with a triplet even component without tensor force but fitting the experimental $^{3}S_1$ phase shift simulate the effect of introducing a three-body component in a V6 version of V14.

The HM without Coulomb is in fair agreement with the experimental binding energies. The charge radius of 1.61 fm given by the HM for ^{4}He is in agreement with the experimental Model Independent value.

* Visiting scientist
e-mail address : mineko13@hotmail.com

[1] I. Talmi in Contemporary Concepts in Physics: Simple Models of Complex Nuclei, Vol. 7, Harwwood Academic Publishers, 1993.

[2] M. Fabre de la Ripelle, Nucl. Phys., A 839 (2010) 11-41

[3] M. Fabre de la Ripelle, Lecture Notes in Physics, 273 (1986), Springer-Verlag

[4] M. Fabre de la Ripelle, S. Sofianos, R. M. Adam, Ann. Phys. (NY) 316 (2005) 107

[5] M. Fabre de la Ripelle, J. Navarro, Ann. Phys. (NY) 123 (1979) 185

[6] N. Barnea, M. Viviani, Phys. Rev., C61, 034003

[7] N. Barnea, W. Leidemann, G. Orlandini, Nucl. Phys., A 693 (2001) 565

[8] S. C. Pieper, R. B. Wiringa, V. R. Pandharipande, Phys. Rev., C46 (1992) 1741

Nuclear Physics in Astrophysics V IOP Publishing
Journal of Physics: Conference Series **337** (2012) 012032 doi:10.1088/1742-6596/337/1/012032

A possible underground accelerator in the Dresden Felsenkeller

T Szücs[1,2], D Bemmerer[1], T Cowan[1,3] and K Zuber[3]

[1]Helmholtz-Zentrum Dresden-Rossendorf, Germany
[2]Institute of Nuclear Research (ATOMKI), Debrecen, Hungary
[3]Technische Universität Dresden, Germany

E-mail: tszucs@atomki.hu

E-mail: d.bemmerer@hzdr.de

Abstract. Typical energies in an astrophysical scenario lie far below the Coulomb barrier. Therefore the cross section is tiny, and in an in-beam experiment the experimental counting rate in γ detectors is low. Often it is dominated by cosmic-ray induced background, even if suitable anticoincidence shields are applied. This problem can be overcome by placing the accelerator deep underground, where a sufficiently thick rock overburden attenuates cosmic rays. Based on the success of the LUNA 400 kV underground accelerator in Italy, several higher-energy underground accelerators are in the planning phase worldwide. All of them should be hosted in laboratories shielded by over 1000 m of rock, where cosmic-ray muons are negligible for the purposes of nuclear physics experiments. A preliminary background study indicates that a combined approach, using a shallow underground laboratory below 47 m of rock and an active shield to suppress surviving muons, results in a background level that is similar to deep underground sites.

The modeling of nuclear processes in astrophysical scenarios such as the Big Bang, stars, and stellar explosions requires a host of nuclear physics informations as input parameters, including precise cross sections of radiative capture reactions [1, 2, 3]. These reactions have generally been well-studied for energies above the Coulomb barrier. However, the astrophysically relevant energies lie far lower, and extrapolations are fraught with uncertainty. Therefore, it is desirable to measure the cross section directly at the energy of astrophysical relevance, or at least close to it. At these energies, the Coulomb barrier has to be penetrated by quantum mechanical tunneling, so the cross section values are very low. In turn, this results in very low experimental yields, so that the laboratory background counting rate in a detector becomes a limiting factor.

This difficulty can be overcome by placing the entire laboratory in an underground setting. Rock overburdens that are a few meters thick suffice to suppress the nucleonic part of the cosmic-ray induced background to negligible levels [4]. However, at these depths there remains a flux of cosmic-ray induced muons that is only slowly attenuated [5]. This remaining muon flux is troublesome because it leads to tertiary neutrons that are produced inside the shielding or the detector itself [6] and which are exceedingly difficult to suppress with active or passive shields.

In order for the background to become negligible for the purpose of nuclear astrophysics experiments, depths of about 1000 m rock are required. The only underground accelerator worldwide, the LUNA 400 kV machine, is located in the Gran Sasso laboratory in Italy, below 1400 m rock. It has been used with great success to study solar fusion and big bang nuclear reactions [8].

However, a combined approach is conceivable where (a) a somewhat thinner rock overburden is used to reduce tertiary neutrons to a level that is comparable to the ambient neutron flux in Gran Sasso [7] and (b) the direct effects of secondary muons are vetoed by an active shield detector. In order to test this idea, the γ-ray background in one and the same actively vetoed high-purity germanium (HPGe) detector, a so-called "traveling" detector, has been studied at different sites: at the Earth's surface, shallow underground at the Felsenkeller facility in Dresden/Germany below 47 m of rock [9], and deep underground at Gran Sasso.

The Felsenkeller facility is sited in the Plauenscher Grund former quarry in Dresden. In the 1850's, a tunnel system was dug into the rock, in order to make ice cellars for a nearby, now decommissioned brewery. Since 1982, the Felsenkeller low-level counting facility is installed in one of the nine tunnels. It is presently operated by Verein für Kernverfahrenstechnik und Analytik Rossendorf (VKTA) and used mainly for analytics. The adjacent rock consists of hornblende monzonite, with a relatively high ^{232}Th and ^{238}U content. The low level counting facility hosts two shielded laboratories called Messkammer 1 and 2 (MK1 and MK2). The data reported here have been taken in MK2, which was constructed in 1995 and has 25 cm thick walls of iron and steel, in part pre-1945.

For the background study, a EURISYS Clover-type detector [10] has been used. It consists of four coaxial n-type HPGe detectors arranged like a four-leaf clover. The detectors are 71 mm long and have a tapered circular area of originally 51 mm diameter. The spacing between the crystals is only 0.2 mm, leading to a closely packed geometry. The virtual large detector formed by the four crystals in addback mode has 122% relative efficiency. In the present study, the Clover detector is always used in conjunction with a surrounding BGO scintillator. For the addback mode data, the BGO can act as a Compton suppression veto.

At Felsenkeller the data have been taken in the heavy shielded MK2, but no heavy shield have been applied during the surface and Gran Sasso runs. Therefore, no direct comparison can be made in the low energy region. However, the effect of the escape-suppression shield is most visible at $E_\gamma > 5.5$ MeV, where both the background from radioisotopes and α emitters in the crystal do not play a role any more (fig. 1). At LUNA the data was recorded during the stops of the running experiments. Therefore, the dynamic range of the detector have been set for the

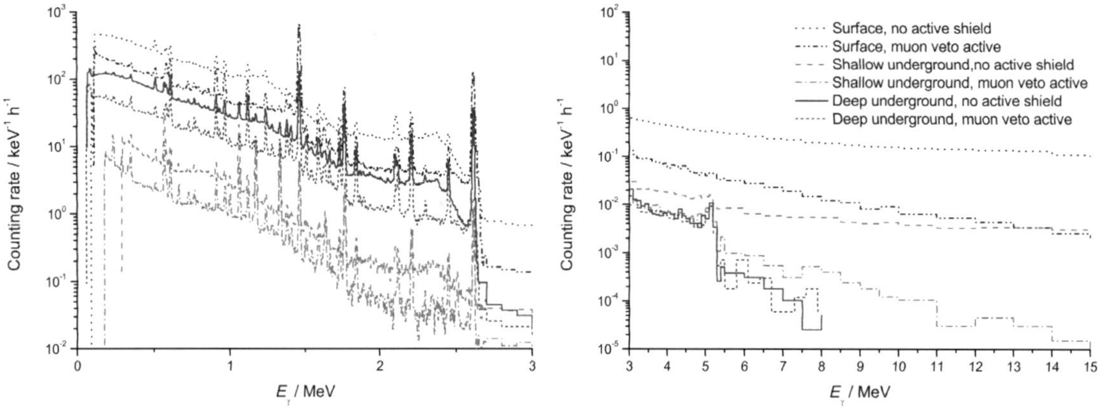

Figure 1. (Color online) γ-ray background in a Clover HPGe detector: Earth's surface, free-running (black dotted line) and with active muon veto (black dot-dot-dashed line). Shallow underground, free-running (red dashed line) and with active muon veto (red dot-dashed line). Deep underground [12, limited to 0-8 MeV], free-running (blue solid line) and with active muon veto (blue short dashed line).

experiment, and the spectra ended at 8 MeV. The recorded energy region is, however, relevant to a number of radiative-capture reactions of interest for astrophysics. For example, for the $^{14}N(p,\gamma)^{15}O$ reaction controlling the rate of the CNO cycles, at astrophysically relevant energies the three major transitions include γ-lines between 6 and 8 MeV γ-energy. In this energy range, the escape-suppressed counting rate of one and the same detector is only a factor 3 worse in Felsenkeller when compared to LUNA.

The science case for a higher-energy accelerator underground requires to move beyond the Sun and study reactions with higher-lying Gamow peaks that are inaccessible at the LUNA 0.4 MV facility. With a 2 MV underground accelerator the following reactions would be opened up for study:

- The (α,γ) reactions of stellar helium burning,
- the $^{13}C(\alpha,n)^{16}O$ and $^{22}Ne(\alpha,n)^{25}Mg$ reactions providing the neutrons for the slow neutron capture process, the so-called astrophysical s-process,
- ^{12}C-induced reactions for stellar carbon burning, and
- selected supernova-related reactions, e.g. $^{40}Ca(\alpha,\gamma)^{44}Ti$

Presently, projects for underground accelerators of 2-4 MV acceleration potential are under discussion at Gran Sasso/Italy [11], Canfranc/Spain, Boulby/UK, and Homestake Mine/US. All these sites have rock overburdens of 1000 m of more and are expected to show background characteristics that are equal to or better than the known background at LUNA [12, and references therein].

However, constructing and maintaining an accelerator laboratory in any of these places requires considerable effort and lead-up time. Therefore, it is conceivable that a staged approach would accelerate progress: In an initial stage, the ease of access to a shallow-underground site like Felsenkeller would be exploited to gain low-energy cross-section data, while deep-underground facilities would still complete their preparatory phase. Subsequently, the even lower background there would be used to pursue the most promising projects, taking advantage of the experience gained shallow underground.

As it has been recognized by the community that it may be necessary to construct more than one underground accelerator (see white paper at `http://www.hzdr.de/felsenkeller`), the call for one or more underground accelerators has been highlighted in the NuPECC Long Range Plan 2010 (`http://www.nupecc.org`).

Acknowledgments

This work has been supported in part by the Herbert Quandt Foundation, DFG (BE 4100/2-1), and EuroGENESIS.

References

[1] Rolfs C and Rodney W 1988 *Cauldrons in the Cosmos* (Chicago: The University of Chicago Press)
[2] Iliadis C 2007 *Nuclear physics of Stars* (Weinheim: WILEY-VCH Verlag Gmbh & Co. KGaA)
[3] Boyd R N 2008 *An introduction to Nuclear Astrophysics* (Chicago: The University of Chicago Press)
[4] Heusser G 1995 *Annu. Rev. Nucl. Part. Sci.* **45** 543-90
[5] Formaggio J A and Martoff C 2004 *Annual Review of Nuclear and Particle Science* **54** 361-412
[6] da Silva A *et al.* 1995 *Nucl. Inst. Meth.* A **354** 553
[7] Belli P *et al.* 1989 *Nuovo Cimento A* **101**, 959-966
[8] Broggini C, Bemmerer D, Guglielmetti A, and Menegazzo R 2010 *Annu. Rev. Nucl. Part. Sci.* **60**, 53-73
[9] Köhler M *et al.* 2009 *Appl. Radiat. Isot.* **67**, 736-740
[10] Duchêne G *et al.* 1999 *Nuclear Instruments and Methods* A **432** 90-110
[11] Formicola A *et al.*, these proceedings
[12] Szücs T *et al.* 2010 *European Physical Journal* A **44** 513-9

Journal of Physics: Conference Series **337** (2012) 012033 doi:10.1088/1742-6596/337/1/012033

The KADoNiS databases - progress and future plans

T Szücs[1], I Dillmann[2,3], R Plag[2,4] and Zs Fülöp[1]

[1]Institute of Nuclear Research (ATOMKI), Debrecen, Hungary

[2]GSI Helmholtzzentrum fr Schwerionenforschung GmbH, Darmstadt, Germany

[3]II. Physikalisches Institut, Justus-Liebig-Universität, Giessen, Germany

[4]Institut für Angewandte Physik, Goethe-Universität, Frankfurt am Main, Germany

E-mail: `tszucs@atomki.hu`

Abstract. The KADoNiS (**K**arlsruhe **A**strophysical **D**atabase **o**f **N**ucleosynthesis **i**n **S**tars) project is an online database (`www.kadonis.org`) for cross sections relevant for s-process and the p-process nucleosynthesis. Recently, the p-process part of the KADoNiS database has been extended, and now includes almost all available experimental data from (p,γ), (p,n), (p,α), (α,γ), (α,n) and (α,p) reactions in or close to the respective Gamow window.

1. Introduction

The nuclei heavier than iron are produced by the three processes called s-, r-, and p-process [1]. The first two involve neutron capture reactions and β-decays. The main difference is the neutron density and therefore the time scale of the neutron capture. For the s-process the capture time scale is much longer (one capture per 100−1000 years) than the β-decay half-lives, therefore the reaction flow is running along the valley of stability up to ^{209}Bi. Here the next neutron capture and β-decay lead to the α-instable ^{210}Po, forcing the material to loop back to ^{206}Pb. The two sites of the s-process are low mass thermally pulsing asymptotic giant branch starts (TP-AGB) for the main s-process ($A\geq90$) [2] and massive stars for the weak s-process (A=60−90) [3].

In the r-process the neutron density is so high ($\gg10^{20}$ cm^{-3}), that subsequent neutron captures drive the material far away from the line of stability close to the neutron drip-line. When the neutron density drops and the temperature decreases ("freeze out"), the material decays back to stability via long β-decay chains. Such high neutron densities and temperatures requires an explosive stellar environment. The exact r-process scenario is still under discussion, but the presently favoured candidates are core collapse supernovae [4] and merging neutron stars [5].

There are about 35 isotopes between ^{74}Se and ^{196}Hg on the proton-rich side of the valley of stability, which cannot be produced via neutron capture reactions. These so-called "p-nuclei" are responsible for only less than 1% of the total solar abundance in mass range heavier than iron, and are mainly produced in a mechanism called "γ-process" [6]. It proceeds by photon-induced reactions starting from pre-existing s- and r-process seed nuclei. First (γ,n) reactions drive the material towards the neutron-deficient region, before charged-particle emitting reactions like (γ,α) and (γ,p) become more probable and divert the reaction flow towards lower masses. Recent simulations revealed that proton-induced reactions are important in the lower mass region, whereas α-induced reactions are important in the higher mass region [7].

However, astrophysical network calculations cannot fully reproduce the observed abundances of all p-nuclei within one scenario [8] and call for additional processes for single isotopes. These reaction network calculations involve thousands of stable and unstable isotopes and tens of thousands of reactions [9]. Only the minority of these reactions are known experimentally, whereas the cross sections for the largest fraction have to be inferred from statistical model calculations.

The fact that self-consistent γ-process studies have problems in synthesizing p-nuclei in the mass regions A <124 and 150< A <165 may result from difficulties related to the astrophysical models as well as from systematic uncertainties of the nuclear physics input. Therefore improvements of nuclear reaction cross sections are crucial for further progress in γ-process models, either by directly replacing theoretical predictions by experimental data or by testing and improving the reliability of statistical models, if the relevant energy range (the so called Gamow window) is not accessible by experiments. For this effort the build-up of an user friendly database from the existing experimental data is needed. The current status and the future aspects of such a database are presented here.

2. Past and present of the KADoNiS database

The first version of the "Karlsruhe Astrophysical Database of Nucleosynthesis in Stars" (KADoNiS, `www.kadonis.org`) database was released in 2005 and was an updated sequel to the previous Bao et al. [10] compilations for (n,γ) cross sections relevant to Big Bang and s-process nucleosynthesis. These compilations contained mainly Maxwellian averaged (n,γ) cross sections (MACS) relevant to the s-process, and KADoNiS provided until now there were three updates [11], [12]. The current version v0.3 contains evaluated Maxwellian-Averaged (n,γ) Cross Section (MACS) at kT=30 keV for 357 datasets between ^{1}H and ^{210}Po, out of which 77 are semi-empirical estimates because only theoretical data exists. The present average experimental uncertainty for all experimental reaction rates at kT=30 keV is ±6.5% (Fig. 1). For the s-only isotopes the uncertainty is ±3.8% (Fig. 2), while the aim is to reach the 1% level. The s-process database contains also the previous recommended values (KADoNiS history), a list of experimental and theoretical values, MACS for energies between kT=5 and 100 keV, reactions rates vs. kT, and Stellar Enhancement Factors (SEF).

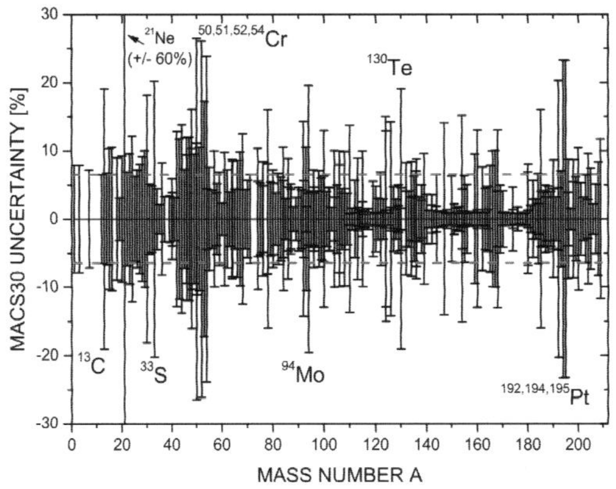

Figure 1. Uncertainties at kT= 30 keV of all experimental datasets. The average value is ±6.5%.

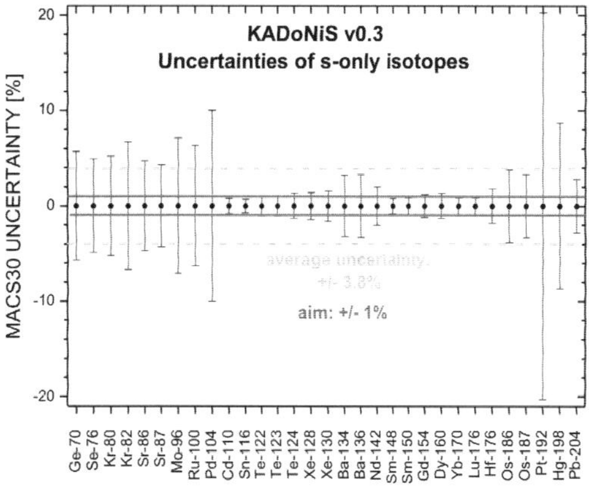

Figure 2. Uncertainties at kT= 30 keV of s-only isotopes. The average value is ±3.8%.

3. The new p-process database

The aim of the p-process database is the collection of all available experimental cross sections in or close to the respective Gamow window and the comparison with various theoretical models. In 2006 a first draft for a p-process database within the KADoNiS project was launched, but it lasted until 2010 until this part was improved in collaboration with the ATOMKI Debrecen and the updated p-process database was published [13].

The basis of the new p-process compilation was the EXFOR database [14], which has the advantage that it contains (almost) all available experimental cross sections and is regularly updated. We scanned it for experimental data from (p,γ), (p,n), (p,α), (α,γ), (α,n), and (α,p) reactions for targets heavier than ^{70}Ge and energies down to 1.5× the upper end of the respective Gamow window for $T=$ 3 GK. This upper cut-off energy was arbitrarily chosen because most available data was measured above the astrophysically relevant Gamow windows, which lie (for $T=$ 3 GK) at $E=$ 3–6 MeV for proton-induced and between $E=$ 7–14 MeV for α-induced reactions.

In the KADoNiS data sheet of each isotope an interactive chart of nuclides is shown, where the nuclei with experimental cross sections in the database are highlighted. After selection of the isotope and the reaction, the data sheet provides the calculated Gamow window (using approximations as discussed in Ref. [15]) for T=2–3 GK for the respective reaction. It is planned to replace these approximated location of the Gamow window with the more accurate numerical solutions provided by Ref. [16]. One can also select one or more datasets and plot these with an online plotter and obtain tabulated datasets. The tabulated datasets contain the center-of-mass energy of the reaction in MeV with uncertainties (if available), the cross sections (in μbarn) and the experimental uncertainties (if available), and then the respective calculated astrophysical S-factor (in units of MeV barn).

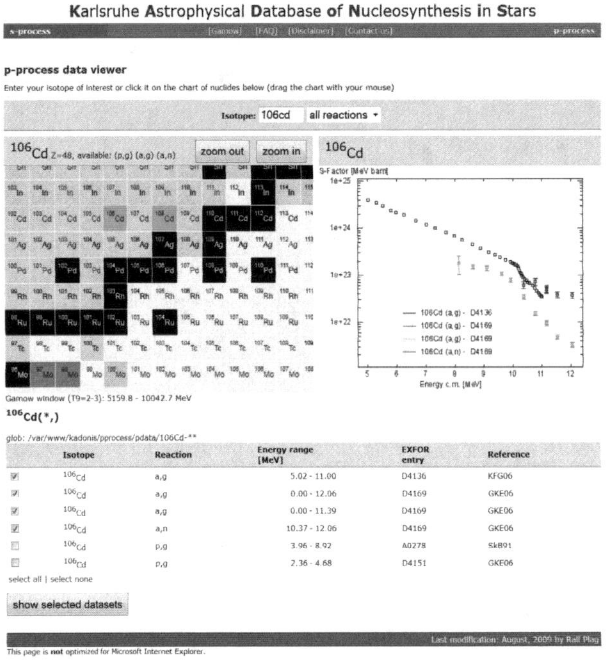

Figure 3. Screenshot of the KADoNiS p-process database.

The experimental basis for (p,γ), (p,n), (p,α), (α,γ), (α,n), and (α,p) cross sections are very different with respect to the energy window for p-process nucleosynthesis. The criteria for the KADoNiS p-process database are fulfilling 34 (p,γ), 15 (α,γ), 70 (p,n), 42 (α,n), and 8 (α,p) reactions. No (p,α) cross section exists within our defined upper cut-off energy, whereas for the inverse (α,p) channel eight reactions are available (^{92,94}Zr, ^{92,98}Mo, ^{96}Ru, ^{106}Cd, and ^{112,124}Sn). However, both reactions play no or only a negligible role in present γ-process models. The most information over the whole mass region can be found for the (α,n) and (p,n) reactions. A lot of effort with respect to astrophysical measurements in the last 15 years was put in determining radiative capture cross sections, which can be related to the photo-induced reactions by detailed balance. For (p,γ) reactions up to now 34 datasets are available, most of them in the lower mass region where their influence was found to be largest [7].

The situation for (α,γ) is much worse, and it was stated by many authors before that more effort should be put in measuring this type of reaction with heavier targets. Up to end of 2009 only 15 isotopes had been measured (and published): ^{70}Ge, ^{91}Zr, ^{96}Ru, ^{106}Cd, ^{107}Ag, ^{113,115}In, ^{112,118}Sn, ^{127}I, ^{136}Xe, ^{139}La, ^{144,154}Sm, and ^{197}Au. All of these measurements do not reach the lower end of the Gamow window, and in the higher mass range, where they have real impact for the network calculations, none of them reach the upper end of the Gamow window.

4. Further plans

The implementation of theoretical data is in progress, which helps to compare the reliability of different theoretical calculations. For each reaction we will give recommended reaction rates and derive fit parameters, which can easily be included in reaction libraries. The database will also be updated regularly (every 1 to 2 years) with new published data. The next step are then paper publications for both databases, still in 2011 for the s-process database, and in 2012 for the p-process database.

Acknowledgments

T.Sz. is supported by EUROGENESIS OTKA-NN83261. I.D. is supported by the Helmholtz association via the Young Investigators project VH-NG-627. R.P. is supported by the Helmholtz association via the Young Investigators project VH-NG-327. Zs.F. is supported by OTKA K68809.

References

[1] Burbidge M E, Burbidge G R, Fowler W A and Hoyle F 1957 *Rev. Mod. Phys.* **29** 547-650
[2] Busso M, Gallino R and Wasserburg G J 1999 *Annu. Rev. Astron. Astrophys.* **37** 239-309
[3] The L-S, El Eid M and Meyer B 2000 *Astrophys. J.* **533** 998-1015
[4] Qian Y-Z 2000 *Astrophys. J. Lett.* **534** L67-70
[5] Freiburghaus C, Rosswog S and Thielemann F-K 1999 *Astrophys. J.* **525** L121-4
[6] Woosley S E and Howard W M 1978 *The Astrophysical Journal Supplement* **36** 285-304
[7] Rapp W, Görres J, Wiescher M, Schatz H and Käppeler F 2006 *Astrophys. J.* **653** 474-89
[8] Arnould M and Goriely S 2003 *Physics Reports* **384** 1-84
[9] Rayet M *et al.* 1990 *Astronomy and Astrophysics* **227** 271-81
[10] Bao Z Y, Käppeler F 2000 *Atomic Data and Nuclear Data Tables* **36** 411-51; and Bao Z Y, Beer H, Käppeler F, Voss F, Wisshak K and Rauscher T 2000 *Atomic Data and Nuclear Data Tables* **76** 70-154.
[11] Dillmann I, Heil M, Käppeler F, Plag R, Rauscher T and Thielemann F-K 2006 *AIP Conf. Proc.* **819** 123-7
[12] Dillmann I, Plag R, Käppeler F, Rauscher T 2010 KADoNiS v0.3 - The third update of the "Karlsruhe Astrophysical Database of Nucleosynthesis in Stars" *EFNUDAT Fast Neutrons, Proc. of the "Scientific Workshop on Neutron Measurements, Theory and Applications"* (April 2009 Geel/Belgium) ed F-J Hambsch (Luxembourg: Publications Office of the European Union) p 55-8
[13] Szücs T, Dillmann I, Plag R, Fülöp Zs 2010 *Proc. of Science* **PoS(NIC XI)247**
[14] `http://www-nds.iaea.org/exfor/exfor.htm`
[15] Rolfs, C E and Rodney W S 1988 *Cauldrons in the cosmos: Nuclear astrophysics*, (Chicago: The University of Chicago Press)
[16] Rauscher T 2010 *Phys Rev.* C **81** 045807

Nuclear Physics in Astrophysics V IOP Publishing
Journal of Physics: Conference Series **337** (2012) 012034 doi:10.1088/1742-6596/337/1/012034

2^+ level densities in ^{40}Ca extracted from high energy-resolution (p,p′) experiments

I Usman[1,2], Z Buthelezi[1], J Carter[2], G R J Cooper[3], R W Fearick[4], S V Förtsch[1], H Fujita[2], Y Kalmykov[5], P von Neumann-Cosel[5], R Neveling[1], I Poltoratska[5], A Richter[5,6], A Shevchenko[5], E Sideras-Haddad[2], F D Smit[1] and J Wambach[5]

[1] iThemba LABS, PO Box 722, Somerset West 7129, South Africa
[2] School of Physics, University of the Witwatersrand, Johannesburg 2050, South Africa
[3] School of Earth Sciences, University of the Witwatersrand, Johannesburg 2050, South Africa
[4] Department of Physics, University of Cape Town, Rondebosch 7700, South Africa
[5] Institut für Kernphysik, Technische Universität Darmstadt, D-64289, Darmstadt, Germany
[6] ECT*, Villa Tambosi, I-38123 Villazzano (Trento), Italy

E-mail: usman@tlabs.ac.za

Abstract. Level densities of 2^+ states in ^{40}Ca have been extracted from the high energy-resolution (p,p′) experimental data by means of a fluctuation analysis. Two different methods of background determination have been considered, namely, Quasi-free calculations and Discrete Wavelet Transform (DWT) technique. The results are then compared to theoretical models based on phenomenological and microscopic approaches.

Level density is one of the classical and fundamental quantities in the description of many-body systems. This is particularly so for reaction cross-sections [1] needed for many practical applications in astrophysics for determining thermonuclear rates for nucleosynthesis and in fission and fusion reactor design [2] as well as in structure models. The purpose of this work is to extract for ^{40}Ca specific spin- and parity-dependent level densities from the high energy-resolution isoscalar giant quadrupole resonance (ISGQR) data in the excitation energy region between 10 and 20 MeV. The method involves the use of fluctuation analysis, incorporating the autocorrelation function. This type of analysis has been successfully applied recently to giant resonance data [3, 4] whereby spin- and parity-resolved level densities were extracted in the region of the Gamow-Teller (GT) and magnetic quadrupole resonances, respectively.

High energy-resolution experiments on the ISGQR have been performed [5] for the low-mass region $12 \leq A \leq 40$ using a 200 MeV proton beam produced by the Separated Sector Cyclotron (SSC) of the iThemba Laboratory for Accelerator Based Sciences (iThemba LABS), South Africa. The proton beams were dispersion-matched to the K600 magnetic spectrometer and inelastically scattered on a ^{40}Ca target with areal density of 3.0 mg/cm^2. Dispersion matching techniques were used in order to exploit the high energy-resolution capability of the spectrometer. Energy resolutions of $\Delta E \simeq 40$ keV (FWHM) were achieved which revealed pronounced fine structure of the ISGQR. The scattering angle was selected to be at the maximum of the cross section for $\Delta L = 2$ transitions into the ISGQR. In particular, the resulting measured spectra allowed for the extraction of 2^+ level densities for ^{40}Ca in the region of the ISGQR.

In this study, the level densities of 2^+ states in ^{40}Ca were extracted by means of a self-consistent procedure based on a fluctuation analysis [6] in the excitation energy interval between 10 and 20 MeV on a background-subtracted spectrum. This method employs the autocorrelation function [7] in order to obtain a measure of the cross-section fluctuations with respect to a stationary mean value. The background contributions due to the experimental instrumentation, multipole admixtures and quasi-free scattering were treated explicitly in the excitation energy region of interest. Quasi-free scattering can contribute significantly to the background in the excitation energy region just above the particle thresholds at forward scattering angles where the ISGQR is measured. Here, the excitation energies of ^{40}Ca(p,p′) in the region of the ISGQR are a few MeV above the proton and neutron thresholds. As such, the contributions from (p,2p) and (p,pn) reactions to the measured ^{40}Ca(p,p′) data were calculated within the Distorted Wave Impulse Approximation (DWIA) [8]. In addition, the shape of the non-resonant background was also determined in a nearly model independent way [3, 4, 9] using Discrete Wavelet Transform (DWT) [10] analysis techniques employing a Biorthogonal wavelet of higher vanishing moment. The fluctuation analysis provides a method to determine the level densities of excited nuclei in the region of overlapping levels. This method is applicable to the energy range where $\langle\Gamma\rangle/\langle D\rangle \ll 1$ with $\langle D\rangle < \Delta E$, with $\langle\Gamma\rangle$ representing the mean level width, $\langle D\rangle$ is the mean level spacing and ΔE is the experimental energy resolution. Wigner and Porter-Thomas distributions have been applied in the estimation of mean widths and resonance spacings as explained in the calculation of nuclear reaction data [1]. The resulting spectrum is shown in Fig. 1a, b and c. The stationary spectrum (Fig. 1c) is a direct measure of the local intensity fluctuations. The experimental autocorrelation function can be approximated by an analytical expression [6] which depends only on the experimental energy resolution ΔE and can be simplified as

$$C(\epsilon) - 1 = \frac{\alpha\langle D\rangle}{2\Delta E\sqrt{\pi}} f(\epsilon, \Delta E), \tag{1}$$

where the function $f(\epsilon, \Delta E)$ is normalised such that $f(\epsilon = 0) = 1$ and ϵ is the energy increment at excitation energy E_x. The value $C(\epsilon = 0) - 1$ is known as the variance. From Eq. (1), the value of the variance of the stationary spectrum is proportional to the mean level spacing $\langle D\rangle$. Thus, $\langle D\rangle$ can be extracted directly once the value of the normalised variance α is known. Figure 1d shows results of autocorrelation functions for both the theoretical model and the experimental data. Finally, the level density in each excitation energy interval of 2 MeV can be calculated as the reciprocal of the mean level spacing.

Basic research on nuclear level densities has been concerned with various phenomenological and microscopic models, their validation by comparison with measurements and the extraction of level density parameters from experimental data. For example, the Back-Shifted Fermi Gas (BSFG) model [11] provides a fit on stable nuclei across the nuclear chart by including parameters for an improvement of the description in local mass areas. The BSFG model of Ref. [12] performed a global fit with parameters dependent on experimental masses only. Hartree-Fock Bogoliubov (HFB) plus combinatorial [14] is an example of a microscopic model which includes rotational and vibrational degrees-of-freedom together with a combinatorial model for the occupation of single-particle states. There exists also the Hartree-Fock-Bardeen-Cooper-Schriffer (HFBCS) [15] Model. As can be seen in Fig. 2, the phenomenological models of BSFG give a good estimate of the extracted 2^+ level densities in the region of ISGQR while the microscopic calculations HFB and HF-BCS over estimates the observed level densities by a factor of two. The uncertainties of the extracted level densities were obtained by varying the input parameters of the autocorrelation function and the fluctuation analysis as well as the background subtraction method and repeating the analysis. These lead to about 17% error in the extracted level densities. In conclusion, both microscopic and phenomenological approaches are good for the level density calculations in the considered excitation energy region.

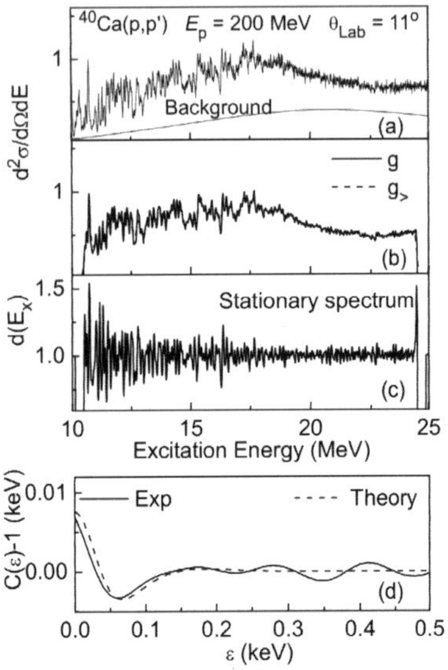

Figure 1. From top to bottom: experimental ^{40}Ca(p,p′) spectrum including background, smoothed spectra $g(E_x)$ and $g_>(E_x)$, stationary spectrum $d(E_x)$ obtained by dividing the two smoothed spectra, autocorrelation functions from experiment (full line) and theoretical (dashed line) model.

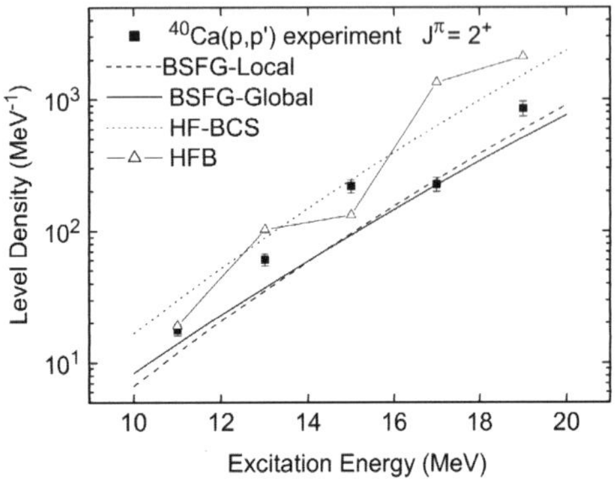

Figure 2. Extracted level density of 2^+ states in ^{40}Ca(p,p′) (Filled square) compared to local BSFG [11] (dashed line), global BSFG [12] (full line), HF-BCS [15] (dotted line) and HFB [14] (open triangle) theoretical model calculations.

References

[1] Belgya T, Bersillon O, Capote Noy R, Fukahori T, Zhigang Ge, Goriely S, Herman M, Ignatyuk A V, Kailas S, Koning A J, Oblozinsky P, Plujko V and Young P G, 2006 *Handbook for calculations of nuclear reaction data, (RIPL-2)* Reference Input Parameter Library-2, IAEA-TECDOC-1506.
[2] Bhat M R, 1983 *IAEA Advisory Group Meeting on Basic and Applied problems of Nuclear Level Densities*, Brookhaven National laboratory, Long Island, New York
[3] Kalmykov Y *et al.*, 2006 *Phys. Rev. Lett.* **96** 012502
[4] Kalmykov Y, Özen C, Langanke K, Martinez-Pinedo G, von Neumann-Cosel P and Richter A 2007 *Phys. Rev. Lett.* **99** 202502
[5] Usman I, Ph.D. thesis 2009 University of the Witwatersrand
[6] B. Jonson, E. Hagberg, P. G. Hansen, P. Hornsöj and P. Tidemand-Petersson 1976 *Proc. 3rd Int. Conf. on Nuclei far from Stability*, Cargese, CERN **76 - 13** 277
[7] Hansen P G 1979 *Ann. Rev. Nucl. Part. Sci.* **29** 69
[8] Chant N S 1998 THREEDEE, University of Maryland (unpublished)
[9] Shevchenko A, Cooper G R J, Carter J, Fearick R W, Förtsch S V, Kalmykov Y, von Neumann-Cosel P, Ponomarev V, Richter A, Usman I and Wambach J 2008 *Phys. Rev.* C **77** 024302
[10] Mallat S 1998 *A Wavelet Tour of Signal Processing* (Academic Press, San Diego)
[11] Rauscher T, Thielemann F K and Kratz K L 1997 *P*hys. Rev. C **58** 1613
[12] von Egidy T and Bucurescu D 2009 *P*hys. Rev. C **80** 054310
[13] Hilaire S and Goriely S 2006 *N*ucl. Phys. A **779** 63
[14] Goriely S, Hillaire S and Koning A J 2008 *Phys. Rev.* C **78** 064307
[15] Demetriou P and Goriely S 2001 *N*ucl. Phys. A **695** 95

Nuclear Physics in Astrophysics V IOP Publishing
Journal of Physics: Conference Series **337** (2012) 012035 doi:10.1088/1742-6596/337/1/012035

Work on a large-area MRPC-based time-of-flight detector for high energy neutrons

D. Yakorev[1] and D. Bemmerer[1] for the R^{3}B collaboration

E-mail: d.yakorev@hzdr.de

E-mail: d.bemmerer@hzdr.de
[1]Helmholtz-Zentrum Dresden-Rossendorf (HZDR), Dresden, Germany

Abstract. The NeuLAND detector at the R^{3}B experiment of the future FAIR facility in Darmstadt aims to detect fast neutrons (0.2-1.0 GeV) with high time and spatial resolutions (σ_t<100 ps, $\sigma_{x,y,z}$<1 cm).

Here, development work on a possible NeuLAND solution based on the multigap resistive plate chamber (MRPC) technology is presented. In this hypothesis, the final detector consists of 50 consecutive MRPC stacks. Each stack contains a 4 mm thick anode made of iron converter material, with an additional 4 mm of converter material between two stacks. The secondary charged particles stemming from hadronic interactions of the high energetic neutrons in the converter will be detected in the MRPC's. A number of prototypes have been developed and built. They have been tested with a beam of single electrons, with picosecond time resolution at the superconducting linac ELBE (Dresden, Germany).

The present prototypes show good time resolution and efficiency. However, much better multihit capability is expected for a fully active cube of 2.5 × 2.5 × 3 m scintillator material, so this is the solution adopted for the NeuLAND detector in the meantime.

The New Large Area Neutron Detector (NeuLAND) is the successor of the Large Area Neutron Detector (LAND) at GSI [1]. NeuLAND should have three times better time and spatial resolution and greatly improved multineutron detection capability with respect to LAND. It is part of the future R^{3}B experiment at FAIR [2]. As first part of the R&D activity, the best parameters for the detector had been studied with 40 × 20 cm large prototypes [4]. Here, some preliminary findings from the development and study of 200 × 50 cm large prototypes are reported.

The center of the detector consists of 19 stainless steel anode strips (0.4 cm × 2.5 cm × 200 cm), which also serve as converters, converting the primary neutron to a secondary charged particle. On either side of the anode there is a standard MRPC glass-gap structure with two gas gaps, in order to detect the secondary charged particle. Semiconductive mylar sheets provide high voltage and ground to the glass. Further out on each side in the signal cathode made by copper strips for both side. The covers of the box are made from 2 mm stainless steel for conversion. For mechanical reasons, two such structures are built in one frame with a common gas volume. There is a the 2.5 mm gas gap between signal cathode and converter plate box that is needed to increase cathode impedance. Rubber lines kept the gas gap at its nominal thickness (figure 1).

For testing the response of the prototypes when irradiated with minimum ionizing particles, the electron beam of the 40 MeV Electron Linac with high Brilliance and low Emittance (ELBE)

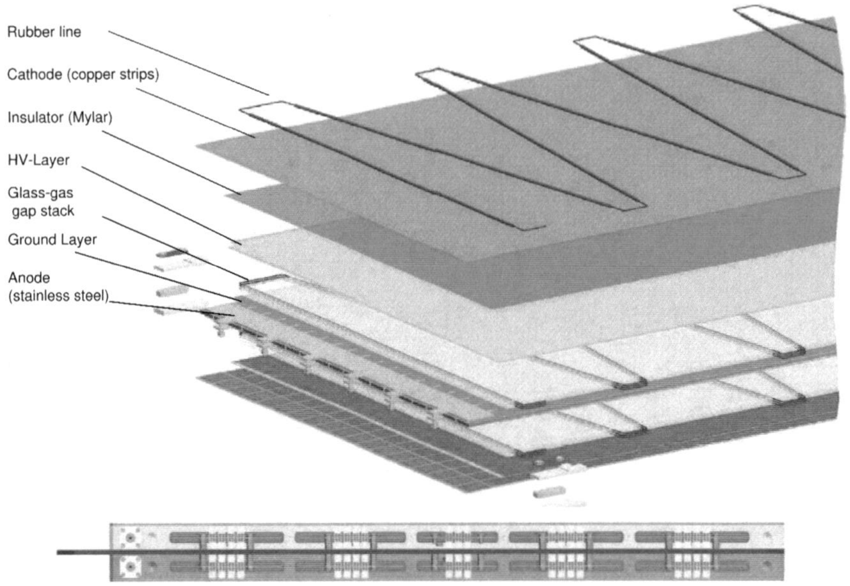

Figure 1. Top: schematic view of one prototype. Bottom: side view of a full experimental module, which contains two detector units.

facility[1] at Helmholtz-Zentrum Dresden-Rossendorf, Dresden/Germany, has been used. The tests were performed with a low-intensity electron beam of 31 MeV kinetic energy, right above the energy for the minimum of ionization. As a time reference, the radio frequency (RF) signal of the electron source was used. A new mode of operation of the ELBE accelerator was used, called one electron mode. In this mode, the gate voltage of the electron gun is reduced much below usual operating in which most accelerated bunches are empty and the rest has only one electron per bunch [3]. Two scintillators were used for trigger in coincidence with the RF signal. The flux of single electrons was 5-20 Hz/cm^2 for the present measurements, much higher than can be obtained with cosmic rays.

Hereafter, efficiency, resolution and cross-talks are defined as in Ref. [4]. The number of gaps was chosen to be 2x2, because small prototype developments showed that this is enough to reach 90% efficiency, while still avoiding the spark regime. The strip impedance is about 13 Ω, defined by the geometry of detector and the limitation on the adopted strip width of 25 mm. In order to reach 50 Ω impedance, the strips should have a width of just 3-4 mm, which is unrealistic as it would require too many electronics channels. Because of this impedance mismatch there is an unavoidable reflection at 2 times the distance from the place of interaction to the far end of the strip, i.e. 2 + 1 m for the case when the beam is in the center and 2 x 7.5 cm for the case when the beam is on the left side. On the figure 2 there are 2D plots (time over charge) for both cases. The amplitude difference for signal between the left (plot a and c) and right side (plot b and d) is just a difference in electronic amplification of that particular channel, it has no physical meaning. In principle, the system of strips is a complicate transmission lines system, and it shows the same behavior, i.e. the degradation of the signal with passed distance (plot c and d).

Due to the large width of 50 cm and the relatively thin housing of just 2 mm steel, the box bulged outward by 1.5 cm in the center. This can be a problem for mechanical stability of inner structure. Therefore, the gas volume was connected to a pump generating about 100 Pa

[1] Internet address `http://www.hzdr.de/elbe`

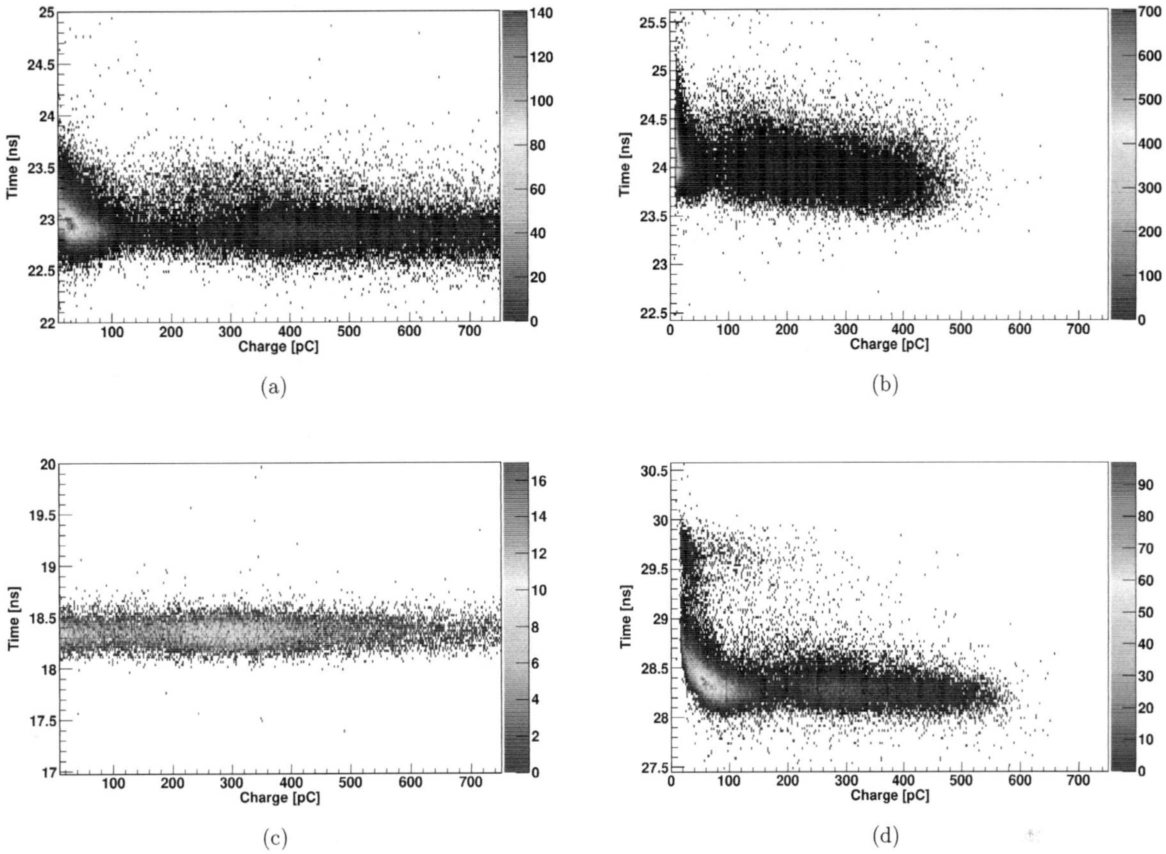

Figure 2. Time over charge signal dependence. Top plots: the beam at the center of strip (a - left end, b - right end) Bottom plots: the beam at the left edge of strip (c - left end, d - right end).

of underpressure inside the chamber, compensating the deformation.

The first beam-tests of full size prototypes (200 × 50 cm) show promising results for efficiency (>90%) and time resolution ($\sigma < 100\,\text{ps}$) for the whole area of the detector. In 2012, a test with neutrons from the breakup of the deuteron at GSI (Darmstadt, Germany) will be done including mainly scintillator-based NeuLAND solutions, but for comparison also this prototype.

Acknowledgments
This work was supported in part by BMBF (06DR9038I) and by GSI FuE (DR-ZUBE).

References
[1] Blaich Th *et al* 1992 A large area detector for high-energy neutrons *Nucl.Instr.Meth A* **314** 136
[2] Aumann T *et al* 2007 Prospects of nuclear structure at the future GSI accelerators *Prog. Part. Nucl. Phys.* **59** 3
[3] Naumann L *et al* 2008 Anordnung und Verfahren zur Erzeugung einzelner relativistischer Elektronen *Patent pending: DE10 2008 054 676.3*
[4] Yakorev D *et al* 2011 Prototyping and tests for an MRPC-based time-of-flight detector for 1 GeV neutrons *Nucl.Instr.Meth A* **654** 79

Nuclear Physics in Astrophysics V IOP Publishing
Journal of Physics: Conference Series **337** (2012) 012036 doi:10.1088/1742-6596/337/1/012036

Anatomy of an Inversion - Argon Isotopes

L Zamick[1], Y Y Sharon[1] and S J Q Robinson[2]

[1] Department of Physics and Astronony, Rutgers University, Piscataway, New Jersey, 08854 USA

[2] Department of Physics, Millsaps College, 1701 N State Street Jackson MS 39157 USA

E-mail: lzamick@physics.rutgers.edu

Abstract.
Two different interactions give similar results for excitation energies and g factors of 2_1^+ states for most even-even argon isotopes except ^{46}Ar. This is explained in terms of an inversion in J=$\frac{1}{2}^+$ and J=$\frac{3}{2}^+$ levels in ^{47}K which is successfully obtained by one of the interactions but not the other. This example shows the possible dangers of nuclear astrophysics extrapolations.

1. Even-Even Argon Isotopes-Inversion

We here discuss some works which were previously published [1],but here emphasize certain points relevant to Nuclear Physics in Astrophysics. In that work we examined the properties of the even A isotopes of Argon between A=38-46. There we found that with two different interactions WBT[2] and SPDF-U[3, 4] we get very similar results for the properties of all even-even Ar isotopes except A=46. We show these results below in Table 1. We also include for the sake of completeness results from new calculations on ^{36}Ar.

Table 1. Excitation energies (in MeV) and g factors of the first 2^+ state for the even mass Argon isotopes.

Mass	36	38	40	42	44	46
E(2_1^+)						
WBT	1.927	2.010	1.424	1.292	1.172	1.143
SPDF-U	1.909	2.022	1.281	1.154	1.087	1.592
Experiment	1.970	2.167	1.464	1.208	1.144	1.550
g(2_1^+)						
WBT	0.487	0.308	-0.197	-0.095	-0.002	+0.100
SPDF-U	0.490	0.319	-0.228	-0.084	-0.040	+0.513
Expt	0.52(18)	0.24(12)	-0.02(2)[5, 6]			

The change of sign in the g factor is due to shell structure.

To understand the big difference in the excitation energy and g factor for A=46 we should look at the odd K isotopes. Here we examine the J=$\frac{3}{2}^+$/J=$\frac{1}{2}^+$ energy splitting. At A=47 there is an experimental inversion (J=$\frac{1}{2}^+$ higher than J=$\frac{3}{2}^+$) that is correctly obtained by SPDF-U but not by WBT. This odd-even proton inversion leads to important changes in even-even properties.

Note that only when one is very close to the inversion does one get significant differences-^{46}Ar is strongly affected but not ^{44}Ar or the lighter isotopes. This shows possible dangers of nuclear astrophysics extrapolations. At the very least it underlies the importance of finding and being aware of these regions of inversion on the nuclear periodic table.

Table 2. J=$\frac{3}{2}^+$ J=$\frac{1}{2}^+$ proton splitting (MeV) in the K isotopes.

	Expt	WBT	SPDF-U
41	0.980	1.106	0.854
43	0.561	1.109	0.672
45	0.475	0.871	0.345
47	-0.360	0.507	-0.320
49	-0.092	0.729	0.087

With the WBT interaction we get good agreement for even-even Argon isotopes A=36,38,40,42,44 but the extrapolation to A=46 could yield wrong results.

It would be great of one could measure the g factor g(2_1^+) for A=46 and compare with the two interactions. We can understand however why are the computed g factors of J=2_1^+ states in ^{46}Ar so different.

There are two competing proton configurations for the 2^+ state,

1. $d_{3/2}$ $d_{3/2}$ This has a very small g factor +0.083 (Schmidt g factor)
2. $s_{1/2}$ $d_{3/2}$ this has a very large g factor g= $[\mu_{(s_{1/2})} + \mu_{(d_{3/2})}]/2 = 1.46$

If there is an inversion ($s_{1/2}$ higher than $d_{3/2}$) configuration 2 is favored over configuration 1. Interaction SPDF-U correctly gives the inversion and hence a larger g factor (+0.513) than WBT (+0.100).

2. N=Z Nuclei-Comparison with Sulfer.

In ref [7] measured g factors of 2+ states of N=Z nuclei ^{32}S and ^{36}Ar are given. The results expt.[theory] are respectively 0.45(7) [0.501] and 0.52(18) [0.488]. These are isoscalar g factors. It has been noted by many that that these g factors always come out to be very close to 0.5. This is consistent with an L=2 S=0 J=2+ state or with the collective result g=Z/A. There is a remarkable insensitivity to the details of the nuclear interaction that is used.

As shown in [7] the singly magic nucleus ^{36}S has a very large g factor 1.3 (5) [1.16]. This was explained by using the configuration 2 in the previous section, $s_{1/2}$ $d_{3/2}$. On the other hand the semi-magic ^{38}Ar has a small positive g factor because the dominant proton configuration is 1 from above $d_{3/2}$ $d_{3/2}$. In references [8, 9] it is noted that ^{40}S has a very small g factor, very far away from the collective value Z/A. However the authors say that other aspects of ^{40}S are consistent with it being deformed, especially a lower lying 2^+ state compared with neighbors and a calculated negative static quadrupole moment of the 2^+ state indicating a prolate deformation. . However in [5] the other authors conclude that ^{40}Ar displays near spherical behaviour. This in itself is not a contradiction, but what is surprising is that the measured B(E2) (see [7]) for the near spherical nucleus is larger than or at best equal that of the deformed one. 374(13) vs 334(36) e^2fm^4. The calculated values of Q in [9] with SPDF (SDF) in units of efm^2 are -19.3(-9.8) for ^{40}S (prolate) and +5.2(+13.5) for ^{40}Ar (oblate).

3. The B(E2) problem in ^{46}Ar

Whereas many of the calculated properties of ^{46}Ar for the 2 different interactions- WBT and SPDF-U are quite different (due to the inversion) the B(E2)'s from ground to the $J = 2_1^+$ state are nearly equal -about 535 e^2fm^4 [1]. In that paper we quoted and compared our results to an experimental value due to Scheid et al.[10] which was much smaller 196 e^2fm^4. Since in the single j shell model ^{46}Ar has a closed $f_{7/2}$ neutron shell our large calculated result appears to be counterintuitive. However the large calculated result has been obtained by others including Scheid et al in the same paper where the small value was measured [10]. Most recently however a large measured value for this B(E2), consistent with our calculation [1] , was published by Mengoni et al.[11] While this is quite gratifying for us it would be nice to have a confirming measurement. The Mengoni result for ^{46}Ar 570 e^2fm^4 is larger than the measured value for ^{44}Ar 334 e^2fm^4. However from isotope shift measurements it is known that the charge radius of ^{46}Ar is smaller than that of ^{44}Ar, see the paper of Blaum et al. [12]. Usually the radius increases with deformation. This is not a definitive argument against a larger B(E) for ^{46}Ar but all of these things will have to be sorted out.

Besides the full space (0 $\hbar\omega$) calculations we also performed, for comparison, sinlge j shell calculations. For these we found some amusing results. The spectra, g facors and B(E2)'s in ths small model space were the same for ^{40}Ar and ^{44}Ar. This could be explained by the fact that the two protons are at midshell and we have two $f_{7/2}$ neutrons in ^{40}Ar and two $f_{7/2}$ neutron holes in ^{44}Ar. Deviaiotns from this symmetry are seen in in Table 1.

Acknowledgments

One of us (L.Z.) acknowledges support as a Morris Belkin visiting professor at the Weizmann Institute,where much of this work was done.

References

[1] S.J.Q. Robinson ,Y.Y. Sharon and L Zamick,Phys. Rev. C **79**, 054322 (2009).
[2] E. K. Warburton and B.A. Brown, Phys. Rev. C **46**, 923 (1992).
[3] F. Nowacki and A. Poves, Phys. Rev. C **79**, 014310 (2009).
[4] S. Nummela et al., Phys. Rev. C **63**, 044316 (2001).
[5] E.A. Stefanova et al. Phys. Rev. C**72**, 014309 (2005).
[6] K.-H. Speidel et al. Phys. Rev. C **78**, 017304 (2008).
[7] K.-H. Speidel et al. Phy. Lett. B **632**, 207 (2006).
[8] A.D. Davis et al. Phys. Rev. Lett. **96**, 112503 (2006).
[9] A.E. Stuchbery et al. Phys. Rev. C **74**, 054307 (2006) .
[10] K. Scheidt et al. Phys. Rev. Lett. **77**, 3967 (1996)..
[11] D. Mengoni et al. Phys. Rev. C **82**, 04308 (2011).
[12] K. Blaum et al. ,Nucl. Phys.A **799**, 30 (2008).

Nuclear Physics in Astrophysics V IOP Publishing
Journal of Physics: Conference Series **337** (2012) 012037 doi:10.1088/1742-6596/337/1/012037

Variation of fundamental constants and the triple-alpha reaction in Population III stars and BBN

Alain Coc

Centre de Spectrométrie Nucléaire et de Spectrométrie de Masse (CSNSM), CNRS/IN2P3, Université Paris Sud, UMR 8609, Bâtiment 104, F–91405 Orsay Campus, France

E-mail: Alain.Coc@csnsm.in2p3.fr

Abstract. The effect of variations of the fundamental constants on the thermonuclear rate of the triple alpha reaction, ^{4}He$(\alpha\alpha,\gamma)^{12}$C, that bridges the gap between ^{4}He and ^{12}C is investigated. We have followed the evolution of 15 and 60 $M_\odot$ zero metallicity stellar models, up to the end of core helium burning. The calculated oxygen and carbon abundances resulting from helium burning can then be used to constrain the variation of the fundamental constants. To investigate the effect of an enhanced triple alpha reaction rate in Big-Bang Nucleosynthesis, we first evaluated Standard Big-Bang Nucleosynthesis CNO production with a network of more than 400 reactions using the TALYS code to calculate missing rates.

1. Introduction

The equivalence principle is a cornerstone of metric theories of gravitation and in particular of General Relativity. This principle, including the universality of free fall and the local position and Lorentz invariances, postulates that the local laws of physics, and in particular the values of the dimensionless constants, such as the fine structure constant (α_{em}), must remain fixed. It follows that by testing the constancy of fundamental constants one actually performs a test of General Relativity, that can be extended on astrophysical and cosmological scales (for a review, see [1]). The ^{4}He$(\alpha\alpha,\gamma)^{12}$C, or triple-$\alpha$-reaction, at the origin of the production of ^{12}C, sets constraints on the values of the fine structure and strong coupling constants. Its resonant reaction rate is very sensitive to the nuclear interaction : a 1% variation of the Nucleon-Nucleon interaction ($\delta_{NN} = 10^{-2}$) would induce a ~20 orders of magnitude change in the rate at 0.1 GK. First, we have considered the very first generation of stars which are thought to have been formed a few 10^8 years after the Big–Bang, at redshifts of $z \sim 10-15$, and with *zero initial metallicity*. For the time being, there are no direct observations of those (Population III) stars but one may expect that their chemical imprints could be observed indirectly in the most metal-poor halo (Population II) stars. Second, Big–Bang nucleosynthesis (BBN) can provide complementary constraints on the variation of constants at a much earlier epoch ($z \sim 10^8$). To extend our previous BBN study[2] that included variations of the weak and n(p,γ)D rates to a possibly inflated 3α-reaction rate, we first need to calculate the Standard BBN, ^{12}C (and CNO) production, bypassing the A=8 gap. The direct detection of primordial CNO isotopes seems highly unlikely with the present observational techniques but it is important for other applications. Hydrogen burning in low mass Pop. III stars proceeds through the slow pp chains until enough carbon is produced, through the triple-alpha reaction, to activate the CNO cycle.

The minimum value of the initial CNO mass fraction that would affect Pop. III stellar evolution was estimated to be 10^{-10} [3] or even as low as 10^{-12} for the less massive ones [4]. This is only two orders of magnitude above the Standard Big-Bang Nucleosynthesis CNO yields obtained using current nuclear reaction rate *estimates*. Indeed, the main difficulty in BBN calculations up to CNO is the extensive network (>400 reactions) needed, including n, p, α, but also d, t and ^{3}He, induced reactions on both stable and radioactive targets. Most of their cross sections cannot be extracted from experimental data only, hence, previous studies (e.g. Ref. [5]) have apparently extensively used the old prescriptions of [6] to estimate those rates. A detailed analysis of all reaction rates and associated uncertainties would be desirable but is impractical for such a large network. So, in this study we have used at first more reliable rate estimates provided by the TALYS [7] code, next we performed a sensitivity study and, finally improve the rate estimates of the most important reactions by dedicated evaluations.

2. The triple–alpha reaction

The ^{4}He$(\alpha\alpha,\gamma)^{12}$C reaction is very sensitive to the position of the "Hoyle state" [8] as the corresponding resonance width is very small (a few ~eV) as compared with the competing reaction ^{12}C$(\alpha,\gamma)^{16}$O, dominated by broad (~100 keV) resonances and subthreshold levels. Hence, in this work, we only considered variations of the ^{4}He$(\alpha\alpha,\gamma)^{12}$C reaction rate. Assuming *i)* thermal equilibrium between the ^{4}He and ^{8}Be nuclei, so that their abundances are related by the Saha equation and *ii)* the sharp resonance approximation for the alpha capture on ^{8}Be, the ^{4}He$(\alpha\alpha,\gamma)^{12}$C rate, can be expressed [9] as:

$$N_{\rm A}^2\langle\sigma v\rangle^{\alpha\alpha\alpha} = 3^{3/2}6N_{\rm A}^2\left(\frac{2\pi}{M_\alpha k_{\rm B}T}\right)^3\hbar^5\omega\gamma\exp\left(\frac{-Q_{\alpha\alpha\alpha}}{k_{\rm B}T}\right) \tag{1}$$

with $\omega\gamma \approx \Gamma_\gamma(^{12}\mathrm{C})$, the radiative partial width of the "Hoyle level" and $Q_{\alpha\alpha\alpha}$ =380 keV its energy relative to the triple alpha threshold. The variation of $Q_{\alpha\alpha\alpha}$ with the nucleon nucleon interaction dominates the variation of the reaction rate in Eq. (1). However, a more accurate calculation [10] requires a numerical integration, as done by [11], to take into account *i)* the two step process, two alpha particle fusion into the ^{8}Be ground state at a resonance energy of $E_R(^8\mathrm{Be})$ followed by another alpha capture to the Hoyle state at a resonance energy of $E_R(^{12}\mathrm{C})$ $[Q_{\alpha\alpha\alpha} \equiv E_R(^8\mathrm{Be})+E_R(^{12}\mathrm{C})]$ and *ii)* the energy dependent finite widths of those two resonances. In order to analyze their variation with the nuclear interaction, we have used a microscopic cluster model (see [12] and references therein). The nucleon-nucleon interaction $V(\mathbf{r})$ depends on the relative coordinate $\mathbf{r}$, and is written as

$$V(\mathbf{r}) = V_C(\mathbf{r}) + (1+\delta_{NN})V_N(\mathbf{r}). \tag{2}$$

where $V_C(\mathbf{r})$ is the Coulomb force, $V_N(\mathbf{r})$ the nuclear interaction [13], and δ_{NN} the parameter that characterizes the change in the nucleon-nucleon interaction. (The variation of the Coulomb interaction is assumed to be negligible compared to the nuclear interaction). We use a cluster approximation in which the wave functions of the ^{8}Be and ^{12}C nuclei are approximated by a cluster of respectively two and three α particle wave functions. This approach has been shown to be well adapted to cluster states, and in particular to the ^{8}Be and ^{12}C levels of interest [14]. We obtained: $\Delta B_D/B_D = 5.7701 \times \delta_{NN}$ for the binding energy of deuterium, $E_R(^8\mathrm{Be}) = (0.09208 - 12.208 \times \delta_{NN})$ MeV, and $E_R(^{12}\mathrm{C}) = (0.2877 - 20.412 \times \delta_{NN})$ MeV, for the resonance energies. The strong energy dependence of the particle widths $\Gamma_\alpha(E)$ follows the Coulomb penetrability factor [9] while the radiative width $\Gamma_\gamma(E)$ scales as $E^{2\lambda+1}$ where λ is the multipolarity (here 2 for $E2$) of the electromagnetic transition. Taking into account the variation of $E_R(^8\mathrm{Be})$ and $E_R(^{12}\mathrm{C})$ and of the partial widths, as a function of δ_{NN}, after numerical integration, we obtain the variation of the rate spanning many orders of magnitude for $|\delta_{NN}| \lesssim 1\%$.

3. Stellar evolution

We considered Population III stars with typical masses, 15 and 60 $M_\odot$ and zero metallicity and used the Geneva code, assuming no rotation (see [4] and references therein). The computations were stopped at the end of the core helium burning (CHeB). Note that the ^{4}He$(\alpha\alpha,\gamma)^{12}$C reaction is also important during the hydrogen burning phase at zero metallicity. It is clearly visible in the HR diagram [15] that in the absence of initial CNO, stars keep contracting while core hydrogen burning proceeds trough the slower pp-chain until sufficient ^{12}C has been produced by the triple alpha reaction. Figure 1 shows the central abundances at the end of core He burning for the 15 and 60 $M_\odot$ models. From these characteristics, we can distinguish four different cases: **1)** During He burning, ^{12}C is produced, until the central temperature is high enough for the ^{12}C$(\alpha,\gamma)^{16}$O reaction to become efficient: in the last part of the CHeB, the ^{12}C is processed into ^{16}O. The star ends the CHeB phase with a core composed of a mixture of ^{12}C and ^{16}O. **2)** If the 3α rate is weaker, the ^{12}C is produced at a slower pace, and T_c is high from the beginning of the CHeB phase, so the ^{12}C$(\alpha,\gamma)^{16}$O reaction becomes efficient very early: as soon as some ^{12}C is produced, it is immediately transformed into ^{16}O. The star ends the CHeB phase with a core composed mainly of ^{16}O, with very little^{12}C and an increasing fraction of ^{24}Mg for decreasing δ_{NN}. **3)** For still weaker 3α rates, the central temperature during CHeB is such that the ^{16}O$(\alpha,\gamma)^{20}$Ne$(\alpha,\gamma)^{24}$Mg chain becomes efficient, reducing the final ^{16}O abundance. The star ends the CHeB phase with a core purely composed of ^{24}Mg. **4)** If the 3α is strong (ΔB$_\mathrm{D}$/B$_\mathrm{D}$ > 0), the ^{12}C is very rapidly produced, but the T_c is so low that the ^{12}C$(\alpha,\gamma)^{16}$O reaction can hardly enter into play: the ^{12}C is not transformed into ^{16}O. The star ends the CHeB phase with a core almost purely composed of ^{12}C.

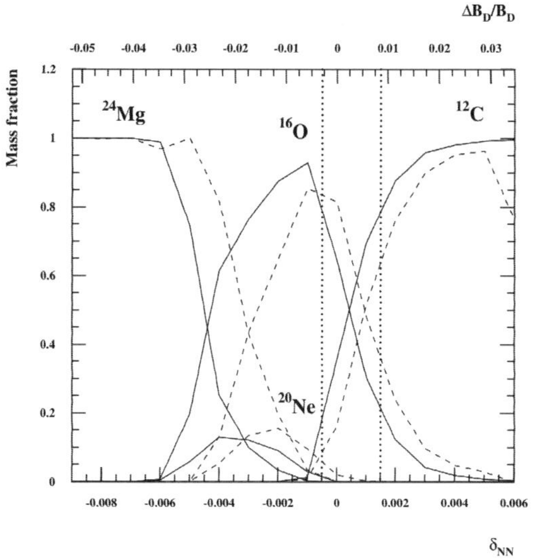

Figure 1. Central abundances at the end of CHe burning as a function of δ_{NN} for 15 (solid lines) and 60 (dashed lines) $M_\odot$ stars. The dotted line corresponds to the $-0.0005 < \delta_{NN} < 0.0015$ adopted interval for which both C and O are produced.

4. Limits on the variations of the fundamental constants in Pop III

The results of the 15 and 60 $M_\odot$ models [15] shows that the variation of the N–N interaction should be in the range $-0.0005 < \delta_{NN} < 0.0015$ (dotted lines in Fig. 1) to insure conservatively that the C/O ratio be of the order of unity. The parameter δ_{NN} is not directly related to a set of fundamental constants such as the gauge and Yukawa couplings. In order to make this connection, we use our computation of the deuterium binding energy B_D to obtain $-0.003 < \Delta B_D/B_D < +0.009$. Using a potential model, the dependence of B_D on the the nucleon, σ-meson and ω-meson mass has been estimated [16] so that it can be related to the

u, d and s quark masses and thus to the corresponding Yukawa couplings, the Higgs vacuum expectation value (vev) v and the QCD scale Λ. For instance, as explained in [2] these parameters can be replaced by (h, v, α_{em}) if grand unification is taken into account and if one assumes that all the Yukawa couplings vary identically. It can then be reduced to e.g. (h, α_{em}) in models in which the weak scale is determined by dimensional transmutation and even to α_{em} $[\Delta h/h = 1/2\Delta\alpha_{em}/\alpha_{em}]$, in certain string models where the couplings are related by a dilaton [17], and finally to:

$$\left|\frac{\Delta B_D}{B_D}\right| = [6.5(1+S) - 18R]\left|\frac{\Delta\alpha_{em}}{\alpha_{em}}\right| \tag{3}$$

Here R is a, GUT model dependent, parameter that links QCD and fine structure constant variations, $\Delta\Lambda/\Lambda = R\Delta\alpha_{em}/\alpha_{em}$, and is in the range $-20 \lesssim R \lesssim +40$ [18] while S reflects the sensitivity of the Higgs vev to the parameters of the supersymmetric model and is in the range $80 \lesssim S \lesssim 500$ [19]. The factor in Eq. (3) is hence very model dependent and can take values between 0.1 and 1000. For typical values of R=36 and S=240, the limits given by BBN (i.e. at a redshift of $z \sim 10^8$) on the variation of the fine structure constant are given by $-3.2 \times 10^{-5} < \Delta\alpha_{em}/\alpha_{em} < 4.2 \times 10^{-5}$[2]. Within the same model and parameter values, the constraints obtained above would lead to $-3. \times 10^{-6} < \Delta\alpha_{em}/\alpha_{em} < 10^{-5}$ at z = 10–15.

5. CNO production in Standard Big-Bang Nucleosynthesis

In this study, we have included 59 nuclides from the neutron to ^{23}Na, linked by 391 reactions involving n, p, d, t and ^{3}He induced reactions and 33 beta decay processes. Reaction rates were taken primarily from [11, 20, 21] and other evaluations when available. The complete list of reactions with associated references to the origin of the rates can be found in Ref. [22]. In comparison with previous works, the present study includes two specific features, namely the introduction of new evaluations of experimental reaction rates and the extensive use of the TALYS reaction code [7] to evaluate the unknown reaction rates. This code is based on the Hauser-Feshbach statistical model which is, *a priori*, not well suited for the description of the reaction involving light nuclei. However, it was found to provide rather fair estimates of those rates that can be used as a first guess for a sensitivity analysis. Indeed, comparison with experimentally determined reaction rates[11, 21] shows differences rarely exceeding three orders of magnitude at BBN relevant temperatures. The results of the calculation, for the stable CNO isotopes, are displayed in Fig. 2: the *total* CNO mass fraction is $\approx 0.5 \times 10^{-14}$. This is in overall agreement with Ref. [5] (with some differences in the isotopic ratios e.g. ^{14}C) and two orders of magnitude below the abundance needed to influence Pop. III stellar evolution. Our results on ^{4}He, D, ^{3}He and ^{7}Li are within 2% of the central value of our latest Monte–Carlo study[23] of BBN nuclear uncertainties, partly due to WMAP most recent η value [24].

However, as many reaction rates are uncertain due to the lack of experimental data, we performed a sensitivity study for each of 271 reactions whose rates are obtained either from TALYS or whose uncertainties have not been estimated. Based on our comparison between TALYS calculated rates on the one hand and experimental rates on the other hand (see Fig. 3 and Ref.[22]) we consider three orders of magnitude variations around the standard reaction rates. Hence to estimate the impact of reaction rates on Standard Big-Bang Nucleosynthesis for each reaction, we performed six calculations after changing its rate by factors of 0.001, 0.01, 0.1, 10., 100. and 1000. and calculate the relative change in isotopic abundances. In many cases, even if the rate uncertainties have not been explicitly calculated it is clear that, e.g. because of experimental works, that these uncertainties are much lower than the three orders of magnitudes assumed here. However, to keep the procedure simple, we applied the same procedure. Reactions whose rate variation induce a relative change in at least one isotope yield larger than 20% were flagged. (Mass fractions of isotopes with A$\geq$12 are added up together.) We found that for all but a few reactions, a three order of magnitude change in their rate

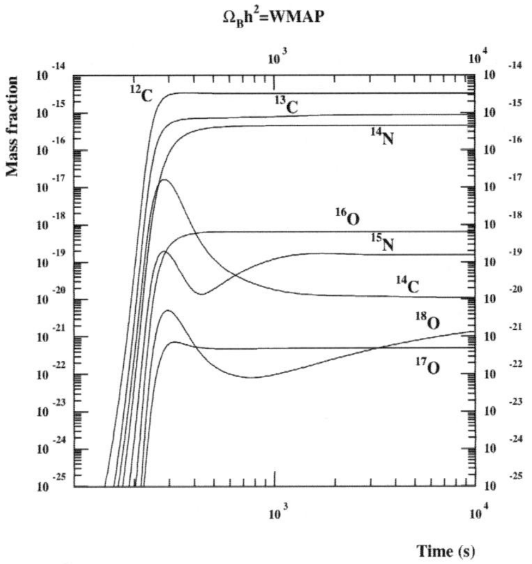

Figure 2. CNO stable isotopes production, in mass fraction, in Standard BBN. (^{18}O increases due to the decay of ^{18}F, not shown.)

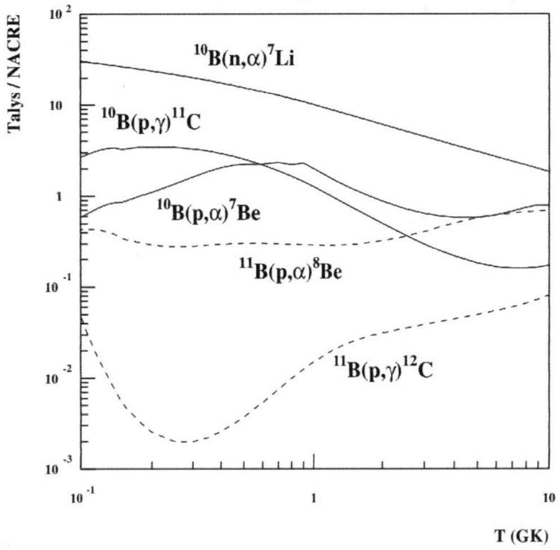

Figure 3. Comparison between TALYS theoretical rates and experimental rates from the NACRE compilation for light targets.

had no significant impact on BBN. We recall that we excluded the known important reactions for which uncertainties are small and whose impact, e.g. on ^{4}He, D, ^{3}He and ^{7}Li, have been extensively studied. Following this study, very few TALYS calculated reaction rates were found to be crucial for CNO production in BBN. Examples are the ^{11}B(d,p)^{12}B and ^{11}B(d,n)^{12}C whose rate variations induced a factor of up to 10 and 300 respectively in CNO production. A detailed analysis of the available experimental data supplemented by theoretical calculations enabled us to reduce the rate uncertainties of the flagged reactions. As a result, changes in reaction rates induced effects nearly compensating each other, conforting the value of $\approx 0.5 \times 10^{-14}$ CNO mass fraction. We are now in position to compare this value with the CNO production in Standard Big-Bang Nucleosynthesis by an inflated ^{4}He$(\alpha\alpha,\gamma)^{12}$C rate.

6. CNO production in BBN with an inflated ^{4}He$(\alpha\alpha,\gamma)^{12}$C rate

To disentangle standard CNO production as described in the previous section with the one proceeding through the triple–alpha reaction, we reduced the network to the 15 reactions involved in A< 8 nucleosynthesis plus ^{4}He$(\alpha\alpha,\gamma)^{12}$C. As we are investigating a possible enhancement of CNO production we only considered positive δ_{NN} values that lead to a higher triple–alpha reaction rate. We limited ourselves to $\delta_{NN} \leq 0.007$ as above, ^{8}Be would be bound requiring a different treatment for the calculation of the rate. This is not a limitation as the CNO yield displays a maximum at $\delta_{NN} \approx 0.006$ which is ($\approx 10^{-20}$ in mass fraction) six orders of magnitude below the standard production. This maximum is due to the sharp resonances (both ^{8}Be g.s. and "Hoyle state") mechanism dominating the cross section becomes highly inefficient when their positions are far from the Gamow window. We can hence conclude that CNO production in BBN cannot be increase through a modification of the triple–alpha reaction.

Acknowledgments
I am indebted to all my collaborators on this topic: Pierre Descouvemont, Sylvia Ekström, Stéphane Goriely, Georges Meynet, Keith Olive, Jean-Philippe Uzan, Elisabeth Vangioni and

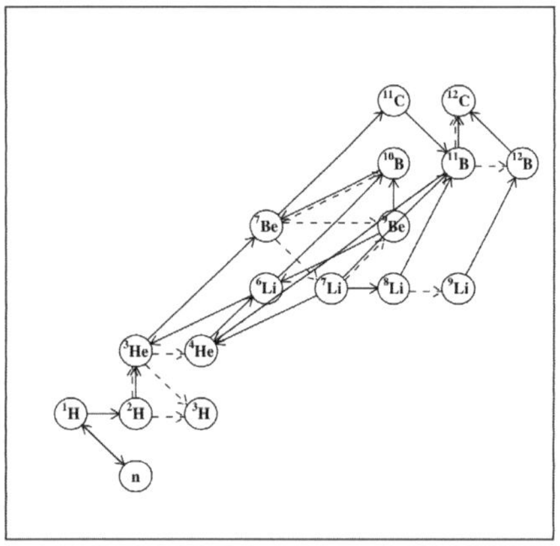

Figure 4. Most important reactions for the production of ^{4}He, D, ^{3}He and ^{7}Li, ^{9}Be, ^{10}B, ^{11}B and CNO isotopes at $\eta = \eta_{WMAP}$.

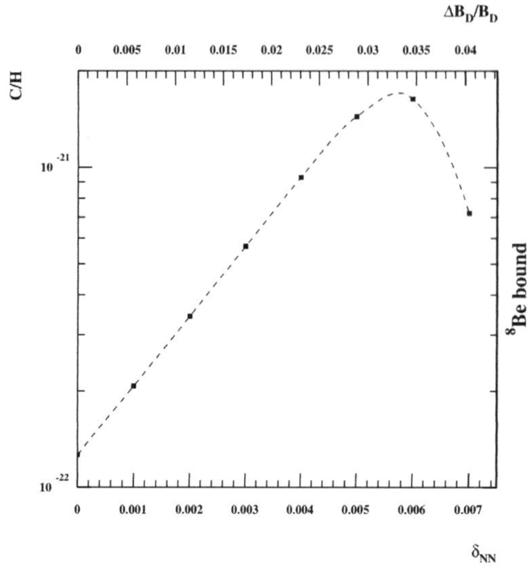

Figure 5. ^{12}C production, in number of atoms relative to H, by the triple alpha reaction as a function of the NN-interaction.

Yi Xu .

References

[1] Uzan J-P 2003 *Rev. Mod. Phys.* **75** 403
[2] Coc A, Nunes N J, Olive K A, Uzan J-P and E. Vangioni 2007 *Phys. Rev.* **D76** 023511
[3] Cassisi S and Castellani V 1993 *Astrophys. J. Supp.* **88** 509
[4] Ekström S, Meynet G, Chiappini C, Hirschi R and Maeder A 2008 *Astron. Astrophys.* **489** 685
[5] Iocco F, Mangano G, Miele G, Pisanti O and Serpico P D 2007 *Phys. Rev.* **D75** 7304
[6] Wagoner R V, Fowler W A and Hoyle F 1967 *Astrophys. J.* **148** 3
[7] Goriely S, Hilaire S, Koning A J 2008 *Astron. Astrophys.* **487** 767
[8] Hoyle F, Dunbar D N F, Wenzel W A, Whaling W 1953 *Phys. Rev.* **92** 1095
[9] liadis C, Nuclear Physics of Stars, Wiley-VCH, Weinheim 2007 pp 439–442
[10] Nomoto N, Thielemann F K and Miyaji S 1985 *Astron. Astrophys.* **149** 239
[11] Angulo C, Arnould M, Rayet M, Descouvemont P, Baye D, et al. (NACRE) 1999 *Nucl. Phys.* **A656** 3 and http://pntpm.ulb.ac.be/nacre.htm.
[12] Korennov S and Descouvemont P 2004 *Nucl. Phys.* **A740** 249
[13] Thompson D, LeMere M and Tang Y 1977 *Nucl. Phys.* **A286** 53
[14] Suzuki Y, Matsumura H, Orabi M et al. 2008 *Phys. Lett.* **B659** 160
[15] Ekström S, Coc A, Descouvemont P, Meynet G, Olive K A, Uzan J-P and E. Vangioni 2010 *Astron. Astrophys.* **514** A62
[16] Dmitriev V F, Flambaum V V and Webb J K 2004 *Phys. Rev.* **D69**, 063506
[17] Campbell B A and Olive K A 1995 *Phys. Lett.* **B345** 429
[18] Dine M, Nir Y, Raz G and Volansky T 2003 *Phys. Rev.* **D67** 015009
[19] Ellis J, Olive K A and Santoso Y 2002 *New J. Phys.* **4** 32
[20] Descouvemont P, Adahchour A, Angulo C, Coc A and Vangioni–Flam E 2004 *Atomic Data and Nuclear Data Tables* **88** 203
[21] Iliadis C, Longland R, Champagne A E, Newton J R, Ugalde C, Coc A and Fitzgerald R 2010 *Nucl. Phys.* **A841** 31
[22] Coc A, Goriely S, Yi Xu, Saimpert M and Vangioni E, submitted to *Astrophys. J.*
[23] Coc A, and Vangioni E 2010, *Journal of Physics Conference Series* **202** 012001
[24] Komatsu, E., et al. 2011, *Astrophys. J. Supp.*, 192, 18

Journal of Physics: Conference Series **337** (2012) 012038 doi:10.1088/1742-6596/337/1/012038

Recent advances in the modelling of classical novae and type I X-ray bursts

J José[1,3], J Casanova[1], A Parikh[1], and E García-Berro[2]

[1]Dept. Física i Enginyeria Nuclear, EUETIB, Univ. Politècnica de Catalunya, C. Comte d'Urgell 187, 08036 Barcelona, Spain, & Institut d'Estudis Espacials de Catalunya, Ed. Nexus-201, C. Gran Capità 2-4, 08034 Barcelona, Spain
[2]Dept. Física Aplicada,Univ. Politècnica de Catalunya, C. Esteve Terrades 5, Castelldefels, Spain, & Institut d'Estudis Espacials de Catalunya, Ed. Nexus-201, C. Gran Capità 2-4, 08034 Barcelona, Spain

E-mail: jordi.jose@upc.edu

Abstract. Classical nova outbursts and type I X-ray bursts are thermonuclear stellar explosions driven by charged-particle reactions. Extensive numerical simulations of nova explosions have shown that the accreted envelopes attain peak temperatures between 0.1 and 0.4 GK, for about several hundred seconds, and therefore, their ejecta is expected to show signatures of significant nuclear activity. Indeed, it has been claimed that novae play some role in the enrichment of the interstellar medium through a number of intermediate-mass elements. This includes ^{17}O, ^{15}N, and ^{13}C, systematically overproduced in huge amounts with respect to solar abundances, with a lower contribution to a number of species with A<40, such as ^{7}Li, ^{19}F, or ^{26}Al. In this review, we present new 1-D hydrodynamic models of classical nova outbursts, from the onset of accretion up to the explosion and ejection phases. Special emphasis is put on their gross observational properties (including constraints from meteoritic presolar grains and potential gamma-ray signatures) and on their associated nucleosynthesis. Multidimensional models of mixing at the core-envelope interface during outbursts will also be presented. The impact of nuclear uncertainties on the final yields will be also outlined. Detailed analysis of the relevant reactions along the main nuclear path for type I X-ray bursts has only been scarcely addressed, mainly in the context of parameterized one-zone models. Here, we present a detailed study of the nucleosynthesis and nuclear processes powering type I X-ray bursts. The reported bursts have been computed by means of a spherically symmetric (1D), Lagrangian, hydrodynamic code, linked to a nuclear reaction network that contains 325 isotopes (from ^{1}H to ^{107}Te), and 1392 nuclear processes. These evolutionary sequences, followed from the onset of accretion up to the explosion and expansion stages, have been performed for two different metallicities to explore the dependence between the extension of the main nuclear flow and the initial metal content. We carefully analyze the physical parameters that determine the light curve (including recurrence times, ratios between persistent and burst luminosities, or the extent of the envelope expansion). Results are in qualitative agreement with the observed properties of some well-studied bursting sources.

---

[3] Corresponding author.

1. Classical novae

Classical novae are stellar explosions that have captivated the interest of astronomers for more than two millennia. They are characterized by a sudden rise in optical brightness (from 8 to 18 magnitudes in 1 or 2 days), with peak luminosities reaching $10^4 - 10^5$ $L_{\odot}$. In the event, $10^{-4} - 10^{-5}$ $M_{\odot}$ of nuclear–processed material are ejected into the interstellar medium, at typical velocities > 10^3 km s^{-1}. Nova explosions occur in cataclysmic binary systems, consisting of a compact, white dwarf star (usually, CO- or ONe-rich) and a low mass companion (typically, a K or M dwarf of solar composition, although there is evidence pointing towards more evolved companions in some cases). The system is very close (with orbital periods < 10 – 12 hr), allowing mass transfer episodes caused by Roche Lobe overflow of the main sequence star. Since this flow of hydrogen-rich material carries angular momentum, it forms an accretion disk that surrounds the white dwarf star. Ultimately, a fraction of this material spirals in and piles up on the white dwarf surface (at a rate ~ 10^{-9} -10^{-10} $M_{\odot}$ yr^{-1}), building up an envelope in semi-degenerate conditions until a thermonuclear runaway (hereafter, TNR) ensues [1-3].

The build up of the envelope is dominated by the operation of both the proton-proton chains as well as the cold CNO cycle, mainly through $^{12}C(p, \gamma)^{13}N(\beta^+)^{13}C(p, \gamma)^{14}N$. Moreover, the dominant nuclear reaction flow, even during the explosive stages, proceeds close to the valley of stability and is dominated by (p,γ), (p,α), and β^+-decays. It is worth noting that neutron and alpha-capture reactions are completely negligible in the physical conditions that characterize a classical nova outburst. Models of nova nucleosynthesis point towards an endpoint around Ca, in agreement with observations of nova shells, with significant overproduction of ^{13}C, ^{15}N and ^{17}O (figure 1), the likely fingerprints of a large number of nova explosions in the overall Galactic history. Explosions hosting CO white dwarfs are also characterized by large overabundances of ^7Li while those occurring on ONe white dwarfs produce a wealth of intermediate-mass elements. Indeed, explosions on massive ONe white dwarfs would overproduce Si-Cl nuclei (figure 1, right panel).

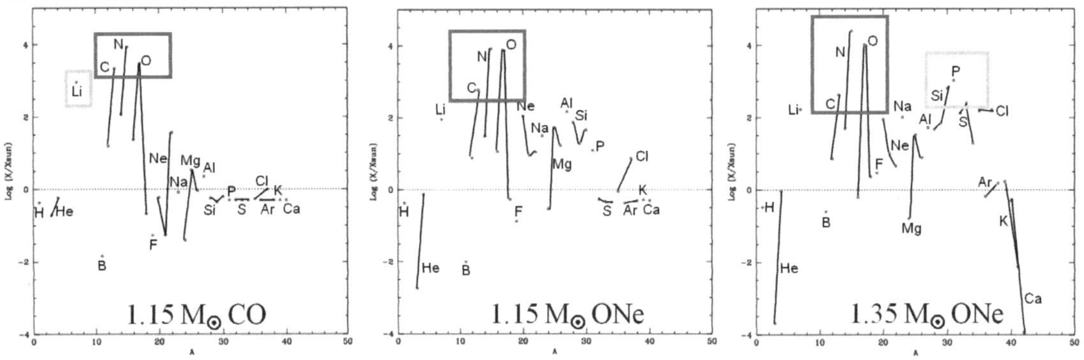

Figure 1. Overproduction factors, relative to solar abundances, obtained in three classical nova models involving 1.15 $M_{\odot}$ CO, 1.15 $M_{\odot}$ ONe and 1.35 $M_{\odot}$ ONe white dwarfs. Calculations have been performed with the 1-D *Shiva* code [4], using updated nuclear reaction rates [5].

Crude comparisons have shown, in general, good agreement between the abundance patterns inferred from observations and those derived from numerical simulations, once the stream of matter transferred from the companion is *conveniently* pre-enriched in metals, in an attempt to mimic the mixing process that should naturally occur at the core-envelope interface. Unfortunately, there is a lack of consensus concerning the nature of such mixing process, required to account for the non-solar patterns inferred in the ejecta, with metallicities around 0.3 (CO) and 0.5 (ONe), but as high as 0.86 in some extreme cases. Recent theoretical efforts have focused on multidimensional simulations in an attempt to circumvent the mixing length theory of convection usually adopted in 1-D simulations. So far, two independent 2-D studies [6,7], based upon the same initial 1-D model, led to nearly opposite conclusions about the strength of the runaway and its capability to power a fast nova. The origin of these differences was carefully analyzed [8], suggesting that the early stages of the explosion, prior to the onset of the TNR – when the evolution is almost quasi-static – are extremely sensitive to the outer boundary conditions. To disentangle the existing controversy, 2-D simulations have been recently performed with the hydrodynamic code FLASH [9, 10]: results show that a shear flow at the core-envelope interface drives mixing through Kelvin-Helmholtz instabilities. The conservation of vorticity imposed by the adopted 2-D geometry induces large convective eddies close to the core-envelope interface, with a size comparable to the height of the envelope, that mix CO-rich material from the outermost layers of the underlying white dwarf into the accreted envelope (figure 2). The metallicity enrichment achieved in the envelope, Z ~0.30, is in agreement with the values inferred in the ejecta of CO novae. 3-D simulations aimed at describing more realistically how mixing at the core-envelope interface and convection develop are currently being computed by the Barcelona group.

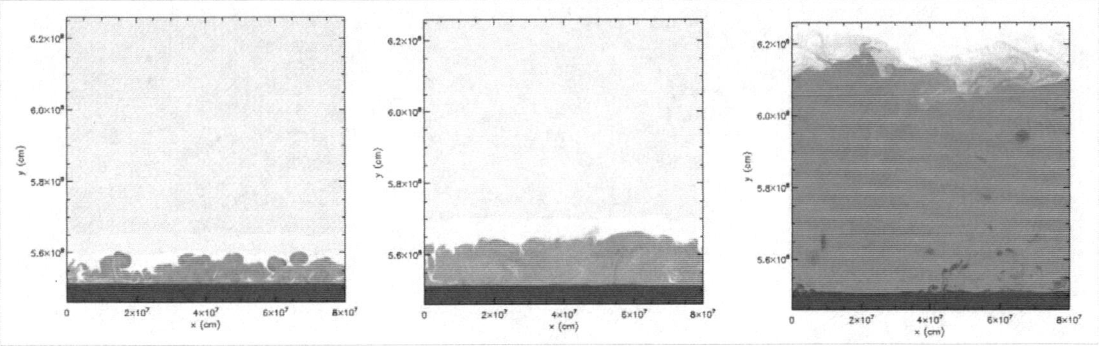

Figure 2. Snapshots of the development of Kelvin-Helmholtz instabilities shown in terms of the ^{12}C mass fraction (in logarithmic scale; darker regions correspond to higher CO mass fractions). Times correspond to 235 s (left panel), 279 s (central panel) and 498 s (right panel) since T at the core-envelope interface reaches ~10^8 K. The mean CNO mass fraction at the end of the simulations reaches ~0.30, by mass. Adapted from [9].

Unfortunately, spectroscopic determinations rely only on atomic abundances. Better nucleosynthesis constraints could be achieved through the pioneering detection of specific γ-ray signatures (such as the lines at 1275 keV –^{22}Na–, 478 keV –^{7}Li–, 511 keV or the continuum down to 20-30 keV –both from electron-positron annihilation powered mainly by ^{13}N and ^{18}F [11]) or from laboratory analyses of presolar grains of a putative nova origin [12,13].
Finally, it is worth mentioning that most of the reactions important for novae (i.e., proton-induced reactions) have been (or will be soon) measured directly in the laboratory, or rely on experimental information. Usual reaction rate uncertainties, typically less than 30% are sufficiently accurate for quantitative nova model predictions [1]. ^{18}F(p,α), ^{25}Al(p,γ), and ^{30}P(p,γ) are probably the most outstanding examples of reactions whose uncertainties in nova conditions must yet be improved. In that sense, nova nucleosynthesis predictions are today limited by uncertainties in the hydrodynamic description of those events rather than by nuclear physics uncertainties.

2. Type I X-ray bursts

With a neutron star as the underlying compact object hosting the explosion, temperatures and densities in the accreted envelope reach during type I X-ray bursts $T_{peak} > 10^9$ K, and ρ ~10^6 g cm^{-3}. Hence, nucleosynthesis studies require the use of hundreds of isotopes, up to the SnSbTe mass region [14] (or beyond [15]), linked through thousands of nuclear interactions.

The main nuclear reaction flow is driven by the *rp-process* (rapid proton-captures and β^+-decays), the 3α-reaction, and the *αp-process* (a sequence of (α,p) and (p,γ) reactions), and is expected to proceed far away from the valley of stability, merging with the proton drip-line beyond A = 38 [13-18]. The existence of a nucleosynthetic endpoint in XRB is still a matter of debate, since recent studies [19,20] suggest that the SnSbTe-mass region is maybe harder to reach.

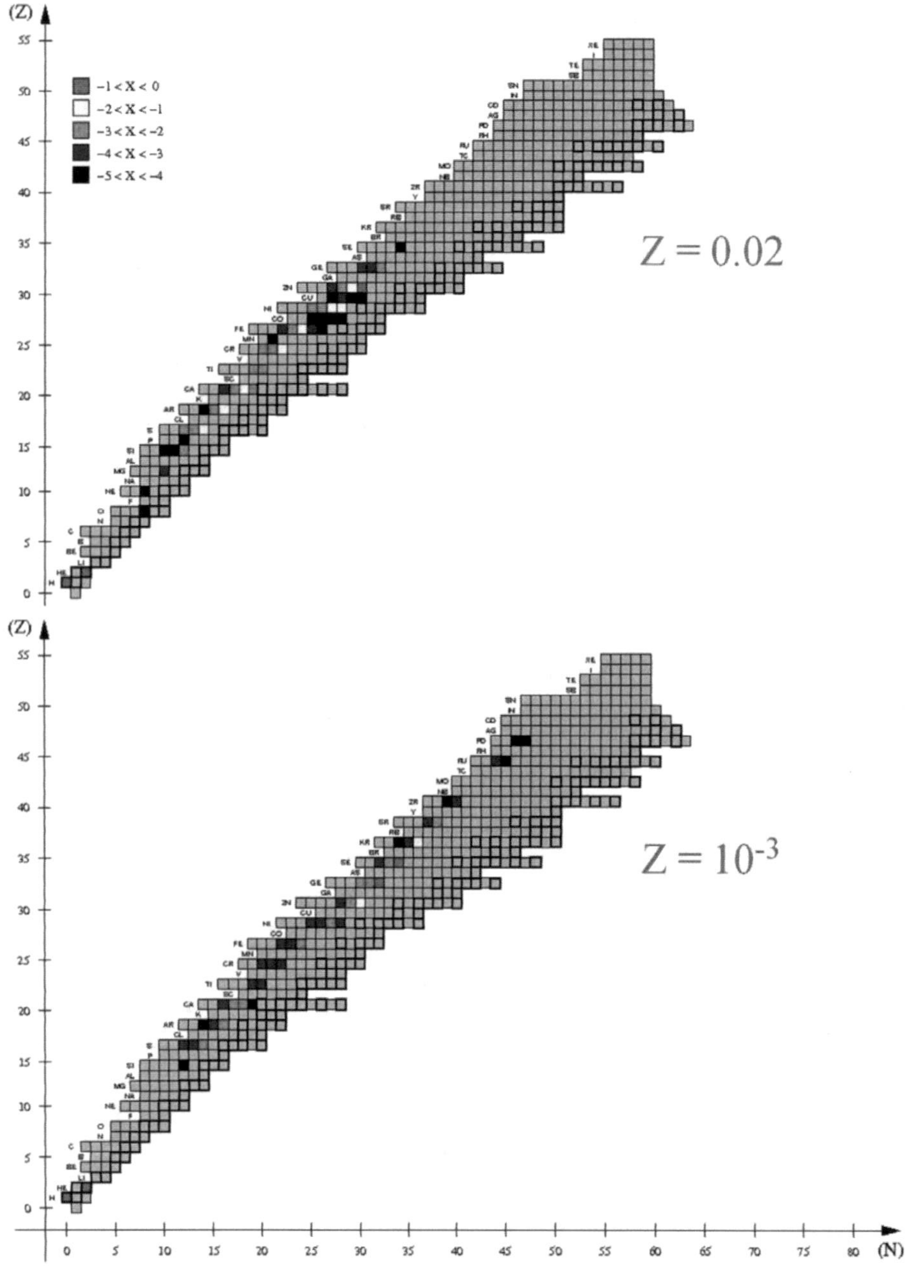

Figure 3. Main nuclear activity at peak temperature during the first burst, shown in terms of the most abundant species at the innermost envelope shell, for two different XRB models. Color legends indicate different ranges of mass fractions (in logarithmic scale). Calculations have been performed with the 1-D hydrodynamic code SHIVA, and involve 1.4 $M_\odot$ neutron stars, accreting mass at rate 1.8 $\times 10^{-9}$ $M_\odot$ yr^{-1}. The composition of the accreted material is assumed to be solar-like (Z = 0.02; upper panel), or metal-deficient (Z =0.001 = $Z_\odot$/20; lower panel). Adapted from [18].

Most of the reaction rates used in XRB nucleosynthesis calculations lack experimental information and often rely on theoretical Hauser-Feschbach estimates. Recent studies [21, 22] have led to the identification of the key reactions whose uncertainty most deeply influences the predicted XRB nucleosynthesis: this is a very limited subset (~40 reactions), with the most influential ones being ^{65}As(p, γ) and ^{61}Ga(p, γ). Mass measurements around waiting point nuclei are also required to improve our current nucleosynthesis predictions [22]. This includes ^{62}Ge, ^{66}Se, ^{70}Kr, ^{84}Nb, ^{85}Mo, ^{86,87}Tc, ^{96}Ag, ^{97}Cd, and ^{103}Sn (since ^{65}As, ^{69}Br and ^{106}Sb have been recently measured [23, 24]). Better precision for the experimentally-known masses of ^{71}Br, ^{83}Nb, and ^{86}Mo may be also required as these are known to only ±568, 315, and 438 keV, respectively. Finally, mass measurements of ^{66}Se ($t_{1/2}$ = 33 ms) as well as spectroscopy of ^{66}Se are also needed to improve our knowledge of the important ^{65}As(p, γ) reaction rate in XRBs.
Several attempts to link the extension of the nuclear path in XRBs with the metallicity of the accreted material have been published in recent years [16, 18]. In particular, we will summarize here the main conclusions reached from two sets of hydrodynamic models of type I XRBs, containing 324 species and 1392 nuclear processes [18]. In the first one, a 1.4 $M_\odot$ neutron star accretes material of solar composition from its stellar companion at a rate 1.75×10^{-9} $M_\odot$ yr^{-1}. Recurrence times between bursts of ~5 hours, peak temperatures of about $(1.1- 1.3)\times10^9$ K, and ratios between persistent and burst luminosities, α, in the range ~ 30−40 (with L_{peak} ~ $(4 - 8)\times10^{38}$ erg s^{-1}), have been obtained for the 4 bursts computed in this model. These values are qualitatively in agreement with those determined in the XRB sources [25] GS 1826-24 [τ_{rec}= 5.74±0.13 hr, α=41.7±1.6], 4U 1323-62 [τ_{rec}=5.3 hr, α=38±4], or 4U 1608-52 [τ_{rec}=4.14–7.5 hr, α=41–54]. Morever, the moderate peak temperatures achieved in this model restrict most of the nuclear activity to mass ~60, and hence, no large concentrations in the SnSbTe-mass region are found Figure 3, upper panel), in agreement with results reported previously [16,17]. This model also yields a very small post-burst abundance of ^{12}C, below the threshold amount required to power *superbursts* ($X[^{12}C]_{min} > 0.1$; see [26]).
A second set of bursts have been computed assuming now that the stream of material transferred from the companion onto the neutron star is metal-deficient, with $Z = Z_\odot/20 = 0.001$. Longer recurrence times of ~9 hours, peak temperatures of about $(1.3-1.4)\times10^9$ K, and ratios between persistent and burst luminosities of α ~ 20−30 (with L_{peak} ~10^{38} erg s^{-1}) have been obtained in the 5 bursts computed in this model, values that are also in agreement with those measured in the XRB sources [25] 1A 1905+00 [τ_{rec}=8.9 hr], 4U 1254-69 [τ_{rec}=9.2 hr], or XTE J1710-281 [τ_{rec}=8.9 hr, α=22–100]. These results reveal a clear dependence of burst properties on the metallicity of the accreted material: the smaller the metal content, the larger the recurrence time (and the smaller the α). In turn, explosions in metal-deficient envelopes are characterized by lower peak luminosities and longer decline times, in agreement with the pattern described in [16]. This, together with the longer exposure times to high temperatures (driven by a slower decline phase), cause an extension of the main nuclear path towards the SnSbTe-mass region (figure 3, lower panel).
A debated aspect of the nucleosynthesis predicted during X-ray bursts is their role as nuclear factories of the Galactic light p-nuclei ^{92,94}Mo and ^{96,98}Ru [14]. Two main aspects have to be considered here: first, the synthesis of those nuclei during XRBs is questionable since detailed hydrodynamic simulations show that the abundances of many species synthesized during the bursts decrease remarkably towards the outer envelope layers, because of limited convective transport. This shows the limitations posed by one-zone nucleosynthesis calculations (like those used in [14]), in which the chemical species synthesized in the innermost layers are, by default, assumed to represent the whole (chemically homogeneous) envelope. Indeed, the mass fractions of these p-nuclei, drop by more than an order of magnitude in the outer envelope layers (as compared with the values achieved at the innermost shells). The resulting overproduction factors are several orders of magnitude smaller than those required to account for the origin of these problematic nuclei [19], in sharp contrast with the results obtained in one-zone calculations [14]. And second, ejection from a neutron star is energetically unlikely: notice, for instance, that matter accreted onto a neutron star of mass M and radius R releases GMm_p/R ~ 200 MeV $nucleon^{-1}$, whereas only a few MeV $nucleon^{-1}$ are released from

thermonuclear fusion. However, it has been suggested that radiation-driven winds during photospheric radius expansion *may lead* to the ejection of a tiny fraction of the envelope. Unfortunately, this idea has not been properly addressed through detailed calculations.

[1] José J, Hernanz M and Iliadis C 2006 *Nucl. Phys. A* **777** 550
[2] José J and Shore S N 2008 *Classical Novae* ed M F Bode and A Evans (Cambridge: Cambridge Univ. Press) p 121
[3] Starrfield S, Iliadis C and Hix W R 2008 *Classical Novae* ed M F Bode and A Evans (Cambridge: Cambridge Univ. Press) p 77
[4] José J and Hernanz M 1998 *Astrophys. J.* **494** 680
[5] Iliadis C, Longland R, Champagne A E, Coc A and Fitzgerald R 2010 *Nucl. Phys. A* **841** 31
[6] Glasner S A, Livne E and Truran J W 1997 *Astrophys. J.* **475** 754
[7] Kercek A, Hillebrandt W and Truran J W 1998 *Astron. Astrophys.* **337** 379
[8] Glasner S A, Livne E and Truran J W 2005 *Astrophys. J.* **625** 347
[9] Casanova J, José J, García-Berro E, Calder A and Shore S N 2010 *Astron. Astrophys.* **513** L5
[10] Casanova J, José J, García-Berro E, Calder A and Shore S N 2011 *Astron. Astrophys.* **527** A5
[11] Hernanz M 2008 *Classical Novae* ed M F Bode and A Evans (Cambridge: Cambridge Univ. Press) p 252
[12] José J, Hernanz M, Amari S, Lodders K and Zinner E 2004 *Astrophys. J.* **612** 414
[13] José J and Hernanz M 2007 *Meteoritics Planetary Sci.* **42** 1135
[14] Schatz H, Aprahamian A, Barnard V, Bildsten L, Cumming A, Ouellette M, Rauscher T, Thielemann F-K and Wiescher M 2001 *Phys. Rev. Letters* **86** 3471
[15] Koike O, Hashimoto M, Kuromizu R and Fujimoto S 2004 *Astrophys. J.* **603** 242
[16] Woosley S E, Heger A, Cumming A, Hoffman R D, Pruet J, Rauscher T, Fisker J L, Schatz H, Brown B A and Wiescher M 2004 *Astrophys. J. Suppl.* **151** 75
[17] Fisker J L, Schatz H and Thielemann F-K 2008 *Astrophys. J. Suppl.* **174** 261
[18] José J, Moreno F, Parikh A and Iliadis C 2010 *Astrophys. J. Suppl.* **189** 204
[19] Bazin D, Montes F, Becerril A, Lorusso G, Amthor A, Baumann T, Crawford H, Gade A, Ginter T, Guess C J, Hausmann M, Hitt G W, Mantica P, Matos M, Meharchand R, Minamisono K, Perdikakis G, Pereira J, Pinter J, Portillo M, Schatz H, Smith K, Stoker J, Stolz A and Zegers R G T 2008 *Phys. Rev. Letters* **101** 252501
[20] Elomaa V-V, Vorobjev G K, Kankainen A, Batist L, Eliseev S, Eronen T, Hakala J, Jokinen A, Moore I D, Novikov Yu N, Penttilä H, Popov A, Rahaman S, Rissanen J, Saastamoinen A, Schatz H, Seliverstov D M, Weber C and Äystö J 2009 *Phys. Rev. Letters* **102** 252501
[21] Parikh A, José J, Moreno F and Iliadis C 2008 *Astrophys. J. Suppl.* **178** 110
[22] Parikh A, José J, Iliadis C, Moreno F and Rauscher T 2009 *Phys. Rev. C* **79** 045802
[23] Tu X L, Xu H S, Wang M, Zhang Y H, Litvinov Yu A, Sun Y, Schatz H, Zhou X H, Yuan Y J, Xia J W, Audi G, Blaum K, Du C M, Geng P, Hu Z G, Huang W X, Jin S L, Liu L X, Liu Y, Ma X, Mao R S, Mei B, Shuai P, Sun Z Y, Suzuki H, Tang S W, Wang J S, Wang S T, Xiao G Q, Xu X, Yamaguchi T, Yamaguchi Y,Yan X L, Yang J C, Ye R P, Zang Y D, Zhao H W, Zhao T C, Zhang X Y and Zhan W L 2011 *Phys. Rev. Letters* **106** 112501
[24] Rogers A M, Lynch W G, Famiano M A, Wallace M S, Amorini F, Bazin D, Charity R J, Delaunay F, de Souza R T, Elson J, Gade A, Galaviz D, Hudan S, Lee J, Lobostov S, Lukyanov S, Matos M, Mocko M, Tsang M B, Shapira D, Sobotka L G and Verde G 2010 *INPC 2010 conference proceedings* [arXiv:1009.2950]
[25] Galloway D K, Muno M P, Hartman J M, Savov P, Psaltis D and Chakrabarty D 2008 *Astrophys. J. Suppl.* **179** 360
[26] Cumming A and Bildsten L 2001 *Astrophys. J.* **559** L127

Nuclear Physics in Astrophysics V IOP Publishing
Journal of Physics: Conference Series **337** (2012) 012039 doi:10.1088/1742-6596/337/1/012039

R-process nucleosynthesis during the decompression of neutron star crust material

S. Goriely[1], A. Bauswein[2], H.-T. Janka[2]

[1] Institut d'Astronomie et d'Astrophysique, ULB, CP.226, 1050 Bruxelles, Belgium
[2] Max-Planck-Institut für Astrophysik, Postfach 1317, 85741 Garching, Germany

E-mail: sgoriely@astro.ulb.ac.be

Abstract. About half of the nuclei heavier than iron observed in nature are produced by the so-called rapid neutron capture process, or r-process, of nucleosynthesis. The identification of the astrophysics site and the specific conditions in which the r-process takes place remains, however, one of the still-unsolved mysteries of modern astrophysics. The present contribution describes the r-process nucleosynthesis taking place during the decompression of neutron star crust material after its ejection from neutron star merger events and presents first consistent results based on relativistic hydrodynamics simulations. It includes the ejection not only of the inner crust but also of the outer crust, although essentially inner crust material can be expected to contribute significantly to the galactic enrichment. The nuclear mechanisms and the sensitivity of the final abundance calculation with respect to some nuclear physics inputs are described.

1. Introduction

The r-process, or the rapid neutron-capture process, of stellar nucleosynthesis is invoked to explain the production of the stable (and some long-lived radioactive) neutron-rich nuclides that are heavier than iron and observed in stars of various metallicities, as well as in the solar system (for a review, see [1]). In recent years nuclear astrophysicists have developed more and more sophisticated r-process models, trying to explain the solar system composition in a satisfactory way by adding new astrophysical or nuclear physics ingredients. The r-process remains the most complex nucleosynthetic process for modelling from the astrophysics as well as nuclear-physics points of view. The site(s) of the r-process has (have) not been identified yet, since all the proposed scenarios face serious problems. Complex—and often exotic—sites have been considered in the hope of identifying astrophysical conditions in which the production of neutrons is high enough to give rise to a successful r-process.

Progress in the modelling of type-II supernovae and γ-ray bursts has raised a lot of excitement about the so-called neutrino-driven wind environment. However, until now a successful r-process has not been obtained *ab initio* without tuning the relevant parameters (neutron excess, entropy, expansion timescale) in a way that is not supported by the most sophisticated existing models. Although these scenarios remain promising, especially in view of their potential to significantly contribute to the galactic enrichment [2], they remain handicapped by large uncertainties associated mainly with the still incompletely understood mechanism that is responsible for the supernova explosion and the persistent difficulties obtaining suitable r-process conditions in self-consistent dynamical explosion and neutron-star cooling models [3]. In addition, predictions of the detailed composition of the ejected matter remain difficult owing to the remarkable sensitivity of r-process nucleosynthesis to uncertainties of the ejecta properties.

Early in the development of the theory of nucleosynthesis, an alternative to the r-process in high-temperature supernova environments was proposed [4]. It relies on the tendency of gravitationally bound, cool or cold matter in beta-equilibrium at high densities (typically $\rho > 10^{10}$ gcm^{-3}) to be composed of nuclei lying on the neutron-rich side of the valley of nuclear stability as a result of endothermic free-electron captures. The astrophysical relevance of this fact in accounting for the production of the r-nuclides has long been questioned. It remained largely unexplored until the study of the decompression of cold neutronized matter resulting from tidal effects of a black hole on a neutron star (NS) companion [5, 6]. Recently, special attention has been paid to NS mergers following the confirmation by hydrodynamic simulations that a non-negligible amount of matter can be ejected [7, 8, 9] and the confirmation that such material should be enriched in r-process nuclei [10, 11]. The ejection of initially cold, decompressed NS matter might also happen in other astrophysical scenarios like giant flares in soft-gamma repeaters, the explosion of a NS eroded below its minimum mass, or the equatorial shedding of material from very rapidly rotating supramassive or ultramassive NSs (see [1] for more details).

Recent three-dimensional simulations of NS–NS mergers have been performed using a general relativistic Smoothed Particle Hydrodynamics scheme [9, 12] with the Shen et al. equation of state [13]. Section 2 presents first nucleosynthesis results on the composition of the inner crust material ejected from a $1.35M_\odot$–$1.35M_\odot$NS merger event. Although the outer crust is far less massive than the inner crust, the ejection of the inner crust cannot take place without at the same time leading to the ejection of at least some outer crust material. Section 3 presents the composition of the outer crust material in nuclear statistical equilibrium (NSE) at temperatures around $T = 8\ 10^9$ K, before and after its ejection to the interstellar medium.

2. Composition of the matter ejected from NS mergers

In the hydrodynamic simulation of a double $1.35M_\odot$NS binary, about $3\ 10^{-3}M_\odot$of the system is found to become gravitationally unbound. The ejected "particles" (i.e., mass elements) originate essentially from the inner crust and more precisely from two different locations, i.e from layers at densities below and above roughly $0.3 \times \rho_S$ (where ρ_S is the nuclear saturation density), as shown in the density histogram of Fig. 1. The former low-density set of surface particles is ejected through tidal forces without being much affected by the coalescence. In addition to these particles, when the NSs come into contact for the first time, some particles deeper in the inner crust are shocked and ejected perpendicularly to the orbital plane. These particles have a higher initial density ($\rho \gtrsim 0.4\ \rho_S$) than the crust particles ejected by tidal forces. Both these types of particles have a low initial electron fraction ranging between 0.015 and 0.050 (see Fig. 1 where about 45% of the mass is characterized by $Y_e = 0.015$). Note that for the first 27 ms, the density history is consistently followed by the numerical simulation. After that time, the determined ejecta escape the system and are assumed to expand with a constant velocity. In this case, the clump size increases linearly with time t and consequently the particle density becomes proportional to $1/t^3$.

As far as the temperature history is concerned, most of the particles are heated during the ejection process to temperatures above 1 MeV. When heated to such temperatures, the composition of a particle is determined by NSE at given density and Y_e. At the time the drip density is reached, most of the ejected matter has cooled below 1 MeV (see Fig. 1, right panel) and the NSE has frozen out. As soon as the particle temperature has dropped below 10^{10} K, the particle composition, initially at NSE, is followed by a full network calculation and the temperature evolution is newly determined on the basis of the laws of thermodynamics allowing for possible nuclear heating through β-decays, fission and α-decay processes, as described in [6].

The reaction network includes all 5000 species from protons up to Z=110 lying between the

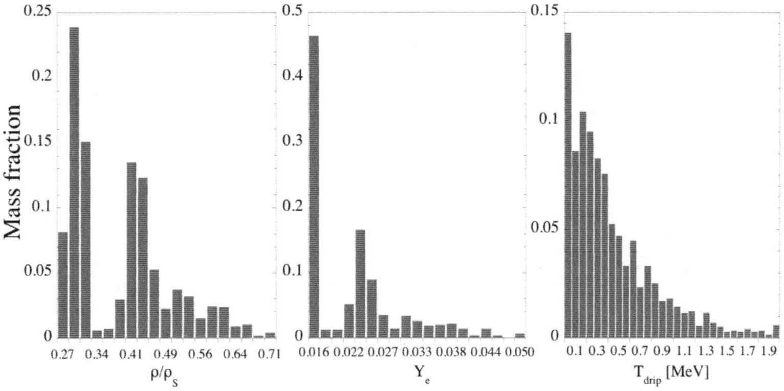

Figure 1. Left: histogram representing the mass distribution of the ejected particles in different bins of initial density ρ (relative to the saturation density $\rho_S \simeq 2.6\ 10^{14}\text{g/cm}^3$). Middle: same as the left panel for the electron fraction Y_e; Right: Same as the left panel for the temperature reached at the drip density $\rho_{drip} \simeq 4.2\ 10^{11}\text{g/cm}^3$

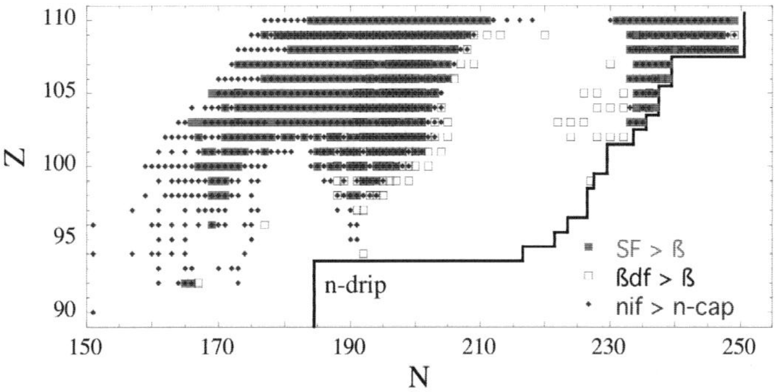

Figure 2. Representation in the (N,Z) plane of the dominant fission regions. Nuclei for which spontaneous fission is estimated to be faster than β-decays are shown by full squares, those for which β-delayed fission is faster than the β-decays by open squares and those for which the neutron-induced fission is faster than the radiative neutron capture at $T = 10^9$ K by diamonds.

valley of β-stability and the neutron-drip line. All fusion reactions on light elements that may play a role when the NSE freezes out are included in addition to the radiative neutron captures and photodisintegrations. The reaction rates on the light species are taken from the NETGEN library, which includes all the latest compilations and evaluation of experimentally determined reaction rates [14]. Experimentally unknown reactions are estimated with the TALYS code [15] on the basis of the HFB-21 nuclear mass model [16]. On top of these reactions, fission and β-decay processes are also included, i.e neutron-induced fission, spontaneous fission, β-delayed fission, photofission, as well as β-delayed neutron emissions. The β-decay processes have been taken from the updated version of the Gross Theory [17] based on the HFB-14 Q-values, while all fission processes have been estimated on the basis of HFB-14 fission path and the full calculation of the corresponding barrier penetration [18]. The main fission region is illustrated in Fig. 2. The fission fragment distribution is taken from [19] and the fragment mass- and charge-asymmetry is derived from the HFB-14 prediction of the left-right asymmetry at the outer saddle point.

In this specific r-process scenario, the neutron density is initially so high ($N_n \simeq 10^{33-35}\text{cm}^{-3}$)

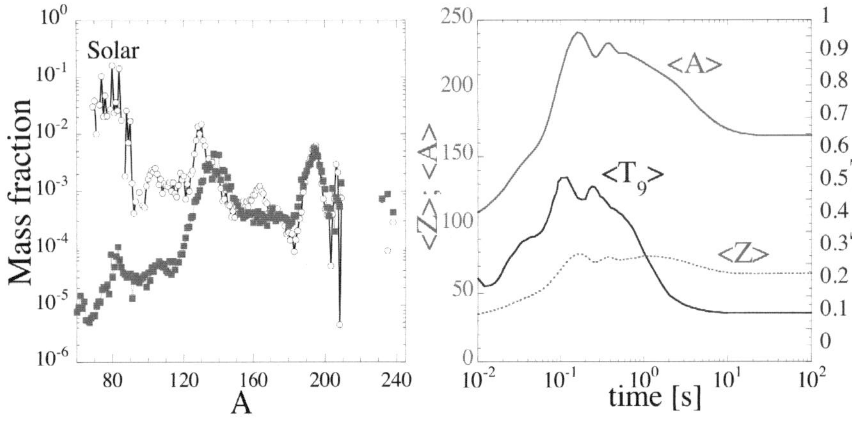

Figure 3. Left: Final abundance distribution of the matter ejected from the 1.35–1.35$M_{\odot}$NS merger as a function of the atomic mass. The distribution is compared with the solar r-abundance distribution (dotted circles). Right: time evolution of the charge number $\langle Z\rangle$, mass number $\langle A\rangle$ and temperature $\langle T\rangle$ mass-averaged over all the ejected particles.

that the nuclear flow follows for the first hundreds of ms a path licking the neutron drip line. For $Z \geq 103$, fission becomes efficient (Fig. 2) and recycling takes place two to three times before the neutrons are totally exhausted, as shown in Fig. 3 by the time evolution of the charge $\langle Z\rangle$ and mass number $\langle A\rangle$ mass-averaged over all the ejected particles. After several hundred ms, the density has dropped by a few orders of magnitude and the neutron density experiences a dramatic fall-off when neutrons get exhausted by captures. During this period of time, the nuclear flow around the $N = 126$ region follows the isotonic chain. When the neutron density reaches some $N_n = 10^{20}$ cm^{-3}, the timescale of neutron capture for the most abundant $N = 126$ nuclei becomes larger than a few seconds, and the nuclear flow is dominated by β-decays back to the stability line (as well as fission and α-decay for the heaviest species). The average temperature remains rather low during the decompression, around 0.3–0.5 GK (Fig. 3), so that photoreactions do not play a major role.

The final mass-averaged composition of the ejected particles is shown in Fig. 3. The $A = 195$ abundance peak related to the $N = 126$ shell closure is produced in solar distribution and found to be almost insensitive to all input parameters, such as the initial abundances, the expansion timescales or the adopted nuclear models. In contrast, the second peak around $A = 140$ originates exclusively from the fission recycling, which is found to take place in the $A \simeq 280$–290 region at the time all neutrons have been captured. These nuclei are found to fission symmetrically, as seen in Fig. 3 by the $A \simeq 140$ peak corresponding to the mass-symmetric fragment distribution. It should be emphasized that significant uncertainties still affect the prediction of fission probabilities and fragment distributions, so that the exact strength and location of the $A \simeq 140$ fission peak depend on the adopted nuclear model. Improvement of the fission predictions represents a major challenge in this specific scenario.

3. Composition of the matter ejected from the outer crust

Although the outer crust (characterized by densities $\rho \leq \rho_{drip} \simeq 4.2\ 10^{11}$g/cm^3) is far less massive than the inner crust, the ejection of the inner crust cannot take place without at the same time leading to the ejection of at least some outer crust material. The whole outer crust typically amounts to 10^{-5} to 10^{-4} $M_{\odot}$, depending on the NS mass and radius [20], so that its contribution even for complete ejection remains negligible compared to the 10^{-3} to 10^{-2} $M_{\odot}$of the inner crust material ejected during a NS–NS merger event. Still, we have estimated the composition of the outer crust material initially heated to NSE at a temperature of about 10^{10} K before and after the decompression that follows its ejection. Note that the amount of unbound outer crust material and the dynamics of its ejection cannot be reliably calculated from the hydrodynamic model. Details about the adopted decompression model can be found in [21].

Before decompression and at temperatures corresponding to 8–10 $\times$ 10^9 K, the Coulomb effect due to the high densities in the crust leads to an overall composition of the outer crust in neutron-rich nuclei with a mass distribution close to the solar system r-abundance distribution, as shown in Fig. 4 (Left panel). Such distributions differ, however, from the solar one due to a shift in the second peak to lower mass numbers. During the decompression, the free neutrons (initially liberated by the high temperatures) are re-captured leading to a final distribution of stable neutron-rich nuclei with a mass distribution of $80 \leq A \leq 140$ nuclei in excellent agreement with the solar distribution, provided the outer crust is initially at a temperature around 8×10^9 K and all layers of the outer crust are ejected. The decompression of the outer NS crust provides suitable conditions for a robust r-processing of the light species, i.e. r-nuclei with $A \leq 140$. The final composition should carry the imprint of the temperature at which the NSE is frozen prior to the ejection. Such temperatures typically around 8–9 $\times$ 10^9 K correspond to those at which NSE can be dynamically achieved in cooling events [21].

Uncertainties in the nuclear physics input can influence the calculation of the final abundances. In particular, the exact strength, location, and width of the $N = 50$ and $N = 82$ r-process peaks are sensitive to the mass model adopted, as shown in Fig. 4 (right panel), where three different mass models are considered, namely HFB-19, HFB-21 [16] and D1M [22]. The production of the $90 \lesssim A \lesssim 120$ r-nuclei can also be significantly affected. The importance of nuclear masses in this specific r-process nucleosynthesis scenario is essentially linked to the initial NSE conditions, which define not only the most abundant species initially present in the outer crust, but also the amount of free neutrons available for the neutron-capture process during the decompression. In contrast, the impact of uncertainties affecting the β-decay rates is rather limited, though β-delayed neutron emission plays some role in smoothing the final abundance pattern. More details can be found in [21].

4. Conclusions

Most of the problems in understanding the origin of r-process elements are related to our ignorance of the astrophysical site that is capable of providing the required large neutron flux. In this respect, explaining the production of r-process elements is essentially an astrophysics issue that will require improved hydrodynamic models to shed better light on possible scenarios. Different sites have been proposed, with neutrino-driven outflows during supernova or γ-ray burst explosions being the currently most popular ones, but facing severe difficulties concerning exact conditions and robustness. Here we have shown that decompressed NS matter is an extremely promising source. In particular, it provides suitable conditions for a robust r-processing. The resulting r-abundance distribution is very similar to the solar one, at least for $A > 140$ nuclei from the inner NS crust. Nuclei with $A < 140$ with solar distribution could originate from the outer crust, but the amount of such matter is very small. The underlying nuclear mechanisms differ significantly from those at action in in supernova scenarios, in particular fission plays a major role in recycling heavy material. The similarity between the predicted and solar abundance patterns as well as the robustness of the prediction against variations of input parameters make this site one of the most promising that deserves further exploration concerning various aspects such as nucleosynthesis, hydrodynamics, and galactic chemical evolution. From the nuclear point of view, this site also implies a new challenge, since it involves the production of neutron-drip nuclei, for which β-decay, neutron capture and fission rates need to be determined.

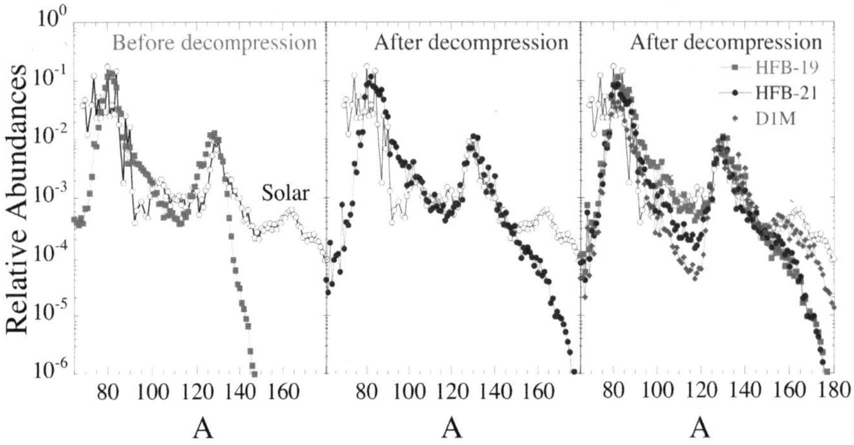

Figure 4. Left: Abundance distribution of the whole outer crust material when initially heated to NSE at a temperature of $T = 8 \times 10^9$ K. The calculation was performed with the HFB-19 masses [16] when experimental masses are not available. The distribution is compared with the solar r-abundance distribution (open circles). Middle: Same as the left panel, after ejection of the outer crust. Right: Same as the middle panel for three different mass models (and corresponding reaction rates) [16, 22]. Experimental masses are used whenever available.

Acknowledgments
Support by DFG through grants SFB/TR7, SFB/TR27, and EXC 153, and computer time at LRZ Munich and RZG Garching are acknowledged. S.G. acknowledges the financial support of the "Actions de recherche concertées (ARC)" from the "Communauté française de Belgique".

References

[1] Arnould M, Goriely S and Takahashi K 2007 *Phys. Rep.* **450** 97
[2] Argast D, Samland M, Thielemann F-K and Qian Y 2004 *Astron. Astrophys.* **416** 997
[3] Hüdepohl L, Müller B, Janka H-T, Marek A and Raffelt G G 2010 *Phys. Rev. Lett.* **104** 251101
[4] Tsuruta S and Cameron A G W 1965 *Can. J. Phys.* **43** 2056
[5] Lattimer J M, Mackie F, Ravenhall D G and Schramm D N 1977 *Astrophys. J* **213** 225
[6] Meyer B S 1989 *Astrophys. J.* **343** 254
[7] Janka H-T, Eberl T, Ruffert M and Fryer C L 1999 *Astrophys. J.* **527** L39
[8] Rosswog S, Speith R and Wynn G A 2004 *Mon. Not. R. Astron. Soc.* **351** 1121
[9] Oechslin R, Janka H-T and Marek A 2007 *Astron. Astrophys.* **467** 395
[10] Goriely S, Demetriou P, Janka H-T and Pearson J M 2005 *Nucl. Phys.* A **758** 587c
[11] Freiburghaus C, Rosswog S, Thielemann F-K 1999 *Astrophys. J* **525** L121
[12] Bauswein A, Janka H-T and Oechslin R 2010 *Phys. Rev.* D **82** 084043
[13] Shen H, Toki H, Oyamatsu K and Sumiyoshi K 1998 *Nucl. Phys.* A **637** 435
[14] Xu Y, Takahashi K, Goriely S and Arnould M 2011 *Frontiers in Nuclear Structure, Astrophysics and Reactions* (eds. P. Demetriou et al., AIP Conference proc.) in press (see also *http://www.astro.ulb.ac.be/Netgen*)
[15] Goriely S, Hilaire S and Koning A J 2008 *Astron. Astrophys.* **487** 76
[16] Goriely S, Chamel N and Pearson J M 2010 *Phys. Rev.* C **82** 035804
[17] Tachibana T, Yamada M and Yoshida Y 1990 *Prog. Theor. Phys.* **84** 641
[18] Goriely S, Hilaire S, Koning A J, Sin M and Capote R 2009 *Phys. Rev.* C **79** 024612
[19] Kodoma T and Takahashi K 1975 *Nucl. Phys A* **239** 489
[20] Pearson J M, Chamel N and Goriely S 2011 *Phys. Rev.* C, in press
[21] Goriely S, Chamel N, Janka H-T and Pearson J M 2011 *Astron. Astrophys.* in press
[22] Goriely S, Hilaire S, Girod M and Péru S 2009 *Phys. Rev. Lett.* **102** 242501

^{56}Ni, Explosive Nucleosynthesis, and SNe Ia Diversity

James W Truran[1,2], Ami S Glasner[3], and Yeunjin Kim[1]

[1]Department of Astronomy and Astrophysics, University of Chicago, 5640 South Ellis Avenue, Chicago, IL 60637, USA

[2]Physics Division, Argonne National Laboratory, Argonne, IL 60439, USA

[3]Racah Institute of Physics, The Hebrew University, Jerusalem 91904, ISRAEL

E-mail: truran@nova.uchicago.edu

Abstract. The origin of the iron-group elements titanium to zinc in nature is understood to occur under explosive burning conditions in both Type Ia (thermonuclear) and Type II (core collapse) supernovae. In these dynamic environments, the most abundant product is found to be ^{56}Ni ((τ = 8.5 days) that decays through ^{56}Co (τ = 111.5 days) to ^{56}Fe. For the case of SNe Ia, the peak luminosities are proportional to the mass ejected in the form of ^{56}Ni. It follows that the diversity of SNe Ia reflected in the range of peak luminosity provides a direct measure of the mass of ^{56}Ni ejected. In this contribution, we identify and briefly discuss the factors that can influence the ^{56}Ni mass and use both observations and theory to quantify their impact. We address specifically the variations in different stellar populations and possible distinctions with respect to SNe Ia progenitors.

1. Introduction

Dating back to the early work by Hoyle and Fowler [1], the 'standard' model for a Type Ia supernova has been assumed to involve the explosion of a carbon-oxygen (CO) white dwarf, with iron peak nuclei being a major product. The history of our understanding of supernova iron production began with Hoyle's [2] identification of the nucleosynthesis mechanism with a "nuclear statistical equilibrium" process occurring at high temperatures and densities in stellar cores. The classic papers by Burbidge et al. [3] and Cameron [4] both expanded upon this picture and sought to identify the physical conditions (T, ρ, Y_e) that provided the best fit to Solar System abundances for an iron peak dominated by ^{56}Fe. These efforts encountered difficulties. Subsequently Cameron [5] noted the critical dependence of explosive nucleosynthesis of iron-peak nuclei on neutron enrichment and Truran, Arnett and Cameron [6] demonstrated that representative supernova conditions yield ^{56}Ni as the principal product *in situ*. The first paper to incorporate the effects of ^{56}Ni decay in a study of SNe Ia light curves was that of Colgate and McKee [7].

Why ^{56}Ni? The physics that dictates the *in situ* formation of a ^{56}Ni dominated iron abundance peak under explosive supernova conditions is straightforward. For both Type Ia and Type II supernovae, the pre-explosion compositions involve primarily nuclei of Z=N, viz: ^{12}C, ^{16}O, and ^{28}Si. Explosive burning at temperatures T > 4 x 10^9 K typically occurs on timescales of fractions of seconds, on which timescales weak interaction processes effecting 'neutronization' of the matter proceed too slowly to convert any significant fraction of protons to neutrons. It follows that the main products of explosive burning *in situ* are proton-rich nuclei of Z = N, viz: ^{44}Ti, ^{48}Cr, ^{52}Fe, ^{56}Ni, ^{60}Zn, and ^{64}Ge. An intriguing outgrowth of these same nucleosynthesis conditions is written in the isotopic compositions of iron peak elements. Specifically, in the relevant supernova environments ^{50}Cr is formed directly as ^{50}Cr, while ^{52}Cr and ^{53}Cr are decay products of ^{52}Fe and ^{53}Fe, ^{54}Fe is synthesized as ^{54}Fe, while ^{56}Fe and ^{57}Fe are decay products of ^{56}Ni and ^{57}Ni, and ^{58}Ni is synthesized as ^{58}Ni, while ^{60}Ni, ^{61}Ni, and ^{62}Ni are decay products of ^{60}Zn, ^{61}Zn, and ^{62}Zn – but the resulting isotopic patterns for these respective elements are consistent with the isotopic composition of Solar System/Cosmic matter. This characteristic of the isotopic patterns of iron-peak nuclei formed in explosive nucleosynthesis is apparent in the figure below that identifies both the proton-rich nuclei along the alpha line and the stable isotopes in the iron peak regime to which they decay. This behavior strongly confirms that the iron-peak elements are indeed formed under neutron-poor ($Y_e \approx 0.5$) explosive burning conditions, for which ^{56}Ni is the dominant product.

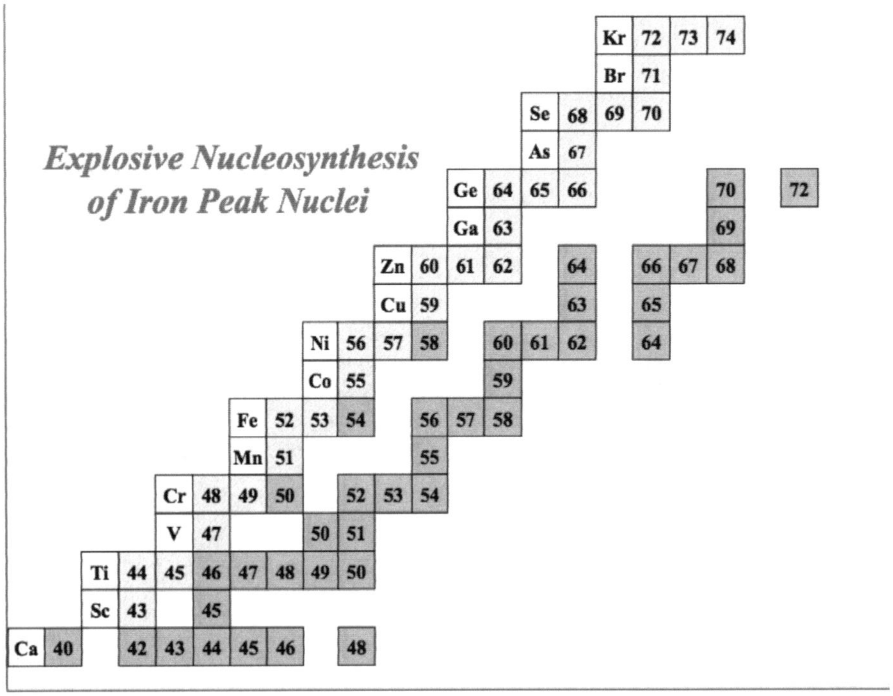

Figure 1: The nuclear chart in the iron peak regime from calcium to krypton. Stable isotopes are shown in blue. Unstable isotopes in proximity to the Z=N line that constitute significant radiogenic decay parents of abundant stable isotopes of iron-peak nuclei are shown in yellow. This illustrates the character of the synthesis of iron-peak nuclei occurring under proton-poor burning conditions in explosive supernova environments.

In this paper we are particularly concerned with Type Ia (thermonuclear) supernovae. These brightest of supernovae can be used effectively as distance indicators. Renewed interest in these events was aroused by the fact that their use over the past decade has provided compelling evidence for an accelerating cosmic expansion (Reiss et al. [8], Perlmutter at al. [9]). In this context, the Phillips [10] relation allows one to compensate for the observed variations (e.g. "diversity") in peak supernova luminosity (brighter SNe Ia decline more slowly) to provide a "standard candle." The critical factor here again is the synthesis of iron (^{56}Fe) as ^{56}Ni, with the peak luminosity being proportional to the mass of ^{56}Ni ejected (Arnett [11]).

2. The Observed Diversity of Type Ia Supernovae

One of the most striking characteristics of the observed diversity of Type Ia supernovae is the broad range in their luminosities at maximum. This diversity in peak luminosity and its correlation with the properties/characteristics of the astronomical environments in which SNe Ia occur has now been well documented observationally. We now fully appreciate that, while Type Ia supernovae are the only type of supernova event seen to occur in early type (elliptical) galaxies, they are also observed to occur in younger stellar populations. The view that SNe Ia can have short-lived progenitors (~ 10^8 years) was first emphasized many years ago by Oemler and Tinsley [12], based upon their recognition that the SNe Ia rate per unit mass is very high in irregular galaxies and roughly proportional to their present star formation rates in spiral galaxies. More recently, Scannapieco and Bildsten [13] have essentially expanded upon this in their discussion of a two component - prompt and extended – Type Ia supernova rate.

The observational support for both short lived and long lived progenitor models for SNe Ia has been noted and explored by a number of sources [14,15,16,17]. The observed trends relevant to our discussions in this paper can be briefly summarized as follows: (1) A rather broad range of peak Type Ia supernova luminosities is observed quite independent of galaxy type (environment), that is both for young stellar populations (e.g. spiral and irregular galaxies) and for older stellar populations (e.g. elliptical or S0 galaxies). For spiral galaxies, the spread can be greater than a factor ~ 5 (stretch factors ~ 0.7 to 1.15), while for elliptical galaxies the spread is reduced by perhaps 20-30 percent (stretch factors ~ 0.7 to 1.050 [15]. (2) The mean luminosities of SNe Ia observed in spiral galaxies are clearly higher than those of elliptical galaxies [15,16]. A very significant factor here is the absence of the brightest SNe Ia in elliptical and S0 galaxies. While this would appear to be a population based correlation having to do with e.g. the metallicity or the age of the underlying stellar component, the cause of this is not unambiguous. It is the factors that can impact the diversity – as reflected in the variations in peak luminosities – which we wish now to identify and discuss.

3. Creating Diversity

A critical clue to understanding the observed diversity of Type Ia is the recognition that their peak luminosities are proportional to the mass of ^{56}Ni ejected [11]. For ^{56}Ni to be a dominant feature of supernova ejecta itself imposes constraints on supernova models: (1) the peak temperature must be sufficient (typically $T > 4\text{-}5\times10^9$ K) to insure that nuclear transformations proceed rapidly enough to burn the available carbon and oxygen fuels to iron-peak nuclei on a hydrodynamic timescale; (2) the densities must be high enough to insure that these high temperatures can be achieved as a consequence of the advance of the burning front through the outer regions of the star (typically, for densities below ~ 10^6 g cm^{-3}, the nuclear burning products can involve rather intermediate mass nuclei in the range silicon to calcium or titanium); and (3) the degree of neutron enrichment of the matter (Y_e) must lie in the range compatible with the synthesis of ^{56}Ni and not more neutron-rich elements like ^{50}Ti, ^{54}Cr, and ^{58}Fe, or the isotopes ^{54}Fe and ^{58}Ni, or even the direct production of ^{56}Fe itself (this constrains the densities in the critical regions to values below ~ 3 x 10^9 g cm^{-1}, so that weak interactions will not neutronize too large a fraction of the core mass.) Concerns with the possible overproduction of such species have eased because of the lower rates of electron captures predicted by the calculations of Langanke and Martinez-Pinedo [30].

Journal of Physics: Conference Series **337** (2012) 012040 doi:10.1088/1742-6596/337/1/012040

Factors that can be understood to be capable of influencing the light curves then include anything that can impact ^{56}Ni production. Two important considerations are the following:

(1) The primordial composition of the white dwarf star can introduce a level of neutronization (in the form of ^{22}Ne) over the entire inner regions of the white dwarf that will reduce the ^{56}Ni abundance at the expense of forming greater concentrations of ^{54}Fe and ^{58}Ni (Timmes et al. [18]). This neutron enrichment arises from first the conversion of CNO isotopes to ^{14}N during the hydrogen burning phase and second the conversion of this to ^{22}Ne by the reactions

$$^{14}N(\alpha,\gamma)^{18}F(e^{+}\nu)^{18}O(\alpha,\gamma)\ ^{22}Ne$$

For matter of solar composition, this yields a ^{22}Ne concentration of about 2.5 % by mass.

(2) The density at which nuclear burning proceeds in the outermost regions also constrains ^{56}Ni production. For densities $< 10^{6}$ g cm^{-1}, the peak temperature is not sufficient to burn to nickel, and the main burning products will typically be intermediate mass nuclei (~silicon-to calcium) at the expense of ^{56}Ni, and dim Type Ia supernovae will result.

The manner in which these factors can influence peak supernova brightness is evident from some features of Figure 2, where the abundance pattern is that predicted for the "carbon deflagration" nucleosynthesis calculations of Thielemann et al. [19]. For this model, the velocities achieved in the outermost layers containing intermediate mass nuclei reached ~ 15,000 km s^{-1}. Note the high degree of neutron enrichment of the innermost layers (~ 0.2 solar masses) by means of electron captures, the broad region of mass (~0.8 solar masses) dominated by ^{56}Ni, and the ~ 0.4 solar masses of intermediate mass nuclei in the outermost regions.

4, Implied Constraints on the Diversity of Interesting Type Ia Supernova Models:

The history of nucleosynthesis models has been strongly guided and constrained by observations. The early 'carbon detonation' model of Arnett [20] was successful in predicting an iron peak *in situ* that was dominated by ^{56}Ni and for which many isotopes in this region were synthesized in approximately solar proportons [21], but failed in that virtually all the mass was ejected as ^{56}Ni. This was in conflict with observations by Branch [22] of the presence of "a mixture of intermediate mass elements from carbon to calcium" in the ejecta.

A significant and needed characteristic of 'deflagration' models is the (pre-)expansion of the outermost regions of the white dwarf prior to the arrival of the burning front. It is this feature that leads to the synthesis of intermediate mass nuclei, in that the temperatures achieved in the burning regions at lower densities are insufficient to drive burning to iron. The high velocities of the early ejecta confirm, however, that the velocities in the outer regions must approach sound speed. This has led to considerations of models in which the flame begins as a deflagration and undergoes a "deflagration to detonation transition" or DDT [23, 24, 25].

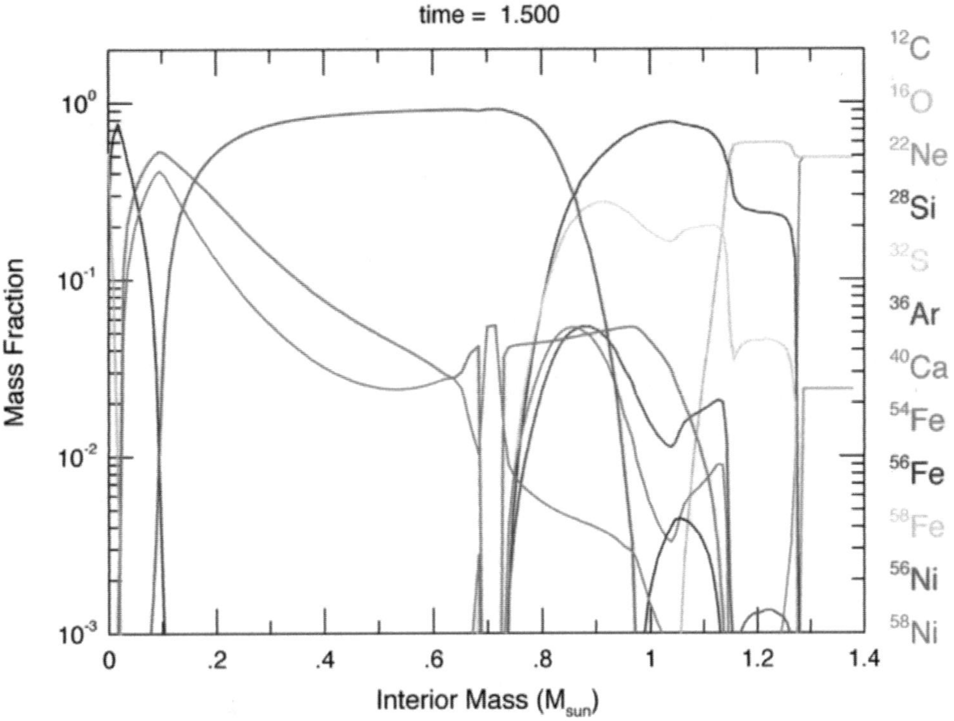

Figure 2: The composition of the ejected matter predicted for the carbon-deflagration model of Thielemann, Nomoto, and Yokoi [19]. The velocity history of the ejecta began with low velocities that increased systematically as the burning front continued outward through the star, such that the velocities of the outermost ejecta reached values ~ 15,000 km s^{-1}, consistent with observations. (Figure courtesy F.-X. Timmes)

Another interesting and studied model for Type Ia supernovae is the 'gravitationally confined detonation' or GCD model [26, 27]. In this mechanism, ignition occurs at one or several off-center points, resulting in a burning bubble of hot gas that is propelled rapidly to the surface by buoyancy, breaks through the surface of the star, and collides at a point opposite the breakout on the stellar surface. Detonation conditions are then achieved as a result of the inwardly directed jet that is produced by the compression of unburnt surface matter when the surface flow collides with itself. A diversity reflected here in the range of peak brightness can naturally arise as a consequence of the number of ignition points and the positioning of these ignition points. Pre-expansion here again will yield lower density conditions in the outermost regions, constraining nickel production and producing concentrations of intermediate mass nuclei.

All of the models we have discussed above are 'single degenerate' models that are generally assumed to involve a Chandrasekhar mass carbon-oxygen white dwarf in a close binary system. (This includes as well models based upon pulsationally delayed detonation.) The challenge in all cases is to identify the nature of the progenitor system within which the white dwarf has grown to the critical mass, presumably by accretion from its binary companon. No definitive model for such growth has yet to have been identified. For this reason, there is renewed interest in "sub-Chandrasekhar" mass white dwarf models that involve either off-center ignition of helium at sub-Chandrasekhar masses or perhaps rather white dwarf-white dwarf collosions/mergers in double degenerate systems. As examples, we note the recent studies by Fink et al. [28] and Sim et al. [29] of detonations/double-detonation in sub-Chandrasekhar CO white dwarfs.

In conclusion, we emphasize again the important feature of SNe Ia outbursts – that it is the energy release in the decay of ^{56}Ni that is reflected in the observed peak luminosities of these events - and that this provides an extremely important tool to explore and to constrain theoretical models of Type Ia supernova events.

Acknowledgments:
This work is supported in part at the University of Chicago by the National Science Foundation under Grant PHY 02-16783 for the Frontier Center "Joint Institute for Nuclear Astrophysics" (JINA), and in part at the Argonne National Laboratory by the U.S. Department of Energy, Office of Nuclear Physics, under contract DE-AC02-06CH11357.

References
[1] Hoyle, F. and Fowler, W. A. 1960, ApJ 132, 565-590
[2] Hoyle, F. 1946, MNRAS 106, 343-383
[3] Burbidge, E. M., Burbidge, G. R., Fowler, W. A., and Hoyle, F. 1957, RMP 29, 547-650
[4] Cameron, A. G. W. 1957, Chalk River Reports, CRL-41 and PASP 69, 201-241
[5] Cameron, A. G. W. 1963, Yale University Lectures in Nuclear Astrophysics
[6] Truran, J. W., Arnett, W. D., and Cameron, A. G. W. 1967, Can. J.Phys. 45, 2315-2332
[7] Colgate, S. A. and McKee, C. 1969, ApJ 157, 623-643
[8] Riess et al. 1998, AJ 116, 1009-1038
[9] Perlmutter, S. 1999, ApJ 517, 565-586
[10] Phillips, M. M. 1993, ApJ 413, L105
[11] Arnett, W. A. 1992, ApJ 253, 785-797
[12] Oemler, A. Jr. and Tinsley, B. M. 1979, AJ 84, 985-992
[13] Scannapieco, E. and Bildsten, L. 2005, ApJ 629, L85-L88
[14] Hamuy, M. et al. 2000, AJ 120, 1479-1486
[15] Sullivan, M. et al. 2006, ApJ 648, 868-883
[16] Wang, B. et al. 2008, Chin. J. Astron. Astrophys.8, 71-80
[17] Sullivan, M. et al. 2010, MNRAS 406, 782-802
[18] Timmes, F, X., Brown, E. F., and Truran, J. W. 2003, ApJ 590, L83-L86
[19] Thielemann, F.-K., Nomoto, K., and Yokoi, K. 1986, A&A 158, 17-33
[20] Arnett, W. D. 1969, Ap&SS 5, 180-212
[21] Arnett, W. D., Truran, J. W., and Woosley, S. A. 1971, ApJ 165, 87-103
[22] Branch, D. 1987, Physica Scripta 35, 787-791
[23] Hillebrandt, W. and Niemeyer, J. C. 2000, ARA&A 38, 191-225
[24] Khokhlov, A. M., Oran, E.S., and Wheeler, J. C. 1997, ApJ 478, 678-688
[25] Schmidt, W. et al. 2010, APj 710, 1683-1693
[26] Jordan, G. C. et al. 2008, ApJ 681, 1448-145726
[27] Jordan, G. C. et al. 2009, ASP Conference Series 406, 92-97
[28] Fink, M. et al. 2010, A&A 514, 53-62
[29] Sims, S. A. 2010, ApJ 714, 52-57
[30] Langanke, K,-H. and Martinez-Pinedo, G. 2000, Nucl. Phys. A 673, 481

Nuclear Physics in Astrophysics V IOP Publishing
Journal of Physics: Conference Series **337** (2012) 012041 doi:10.1088/1742-6596/337/1/012041

"Heavy" elements produced in neutrino-driven winds

Almudena Arcones

Department of Physics, University of Basel, Klingelbergstraße 82, 4056, Basel, Switzerland

E-mail: a.arcones@unibas.ch

Abstract. We present nucleosynthesis studies based on trajectories of hydrodynamical simulations for core-collapse supernovae and their subsequent neutrino-driven winds. Based on recent hydrodynamical simulations, heavy r-process elements ($Z > 56$) cannot be synthesized in the neutrino-driven winds because the entropy is too low and ejected matter is proton-rich. We have shown that the lighter heavy elements (e.g., Sr, Y, Zr) are produced in neutron- and proton-rich winds and could explain the abundance observed in some very old halo stars.

1. Introduction

The astrophysical site where half of the heavy elements are produced by the r-process (rapid neutron capture compared to beta decay) remains unknown. The necessary neutron-rich conditions, i.e., 100 free neutrons per seed nuclei, point to violent events like core-collapse supernova and neutron star mergers (see [1] for a review). Core-collapse supernovae and the subsequent neutrino-driven winds have attracted vast attention as candidates for the production of r-process elements because they occur early and frequently enough to account for the abundances observed in old halo stars and in the solar system [2]. The necessary conditions to produce heavy elements ($A > 130$) are identified [3] (high entropies, low electron fractions, and short expansion timescales), however these are not found in the most recent long-time supernova simulations [4–7].

Most of the recent progress in understanding the origin of elements commonly associated with the r-process is due to observations of ultra metal-poor (UMP) stars (see [8] for recent review). The elemental abundances observed in the atmosphere of these very old stars come from a few nucleosynthesis events. These stars generally present a robust pattern for "heavy" elements $56 < Z < 83$, in agreement with the expected contribution of the r-process to the solar system, but show some scatter for "light" elements $Z < 47$ [8]. This suggests that at least two types of events contribute to the r-process abundances (see e.g., [9]). The process leading to elements with $A < 130$ has been called in the literature the weak r-process [10], charged-particle reaction (CPR) process [11–13], and Light Elemental Primary Process (LEPP) [14, 15]. We refer to this as LEPP because such name does not make any reference to the specific nuclear reactions or astrophysical environment. The term LEPP was first introduced in Ref. [14] which used a galactic chemical evolution model to search for possible astrophysical environments producing the elements such as Sr, Y, and Zr. Montes et al. in Ref. [15] suggested that the LEPP observed in UMP stars show a hint of robustness of the process and could have contributed to the solar system abundances.

2. Neutrino-driven wind and nucleosynthesis

When a supernova explodes, matter surrounding the proto-neutron star is heated by neutrinos and expands very fast reaching sometimes even supersonic velocity [16]. This neutrino-driven wind moves through the early supernova ejecta and eventually collides with it. The interaction of the wind with the slow-moving ejecta results in a wind termination shock or reverse shock where kinetic energy is transformed into internal energy [4]. Therefore, the expansion velocity drops and the temperature (and thus the entropy) increases after the reverse shock. Recently, we have extended our spherically symmetric study [4] to two dimensional simulations [7]. In these multidimensional explosions we have showed that the neutrino-driven wind remains spherically symmetric due to the isotropic neutrino emission from a neutron star that stays spherical in the absence of rotation. Although the wind develops identically in all directions, our results demonstrate that the anisotropic distribution of early supernova ejecta has a great impact on the position of the wind termination shock and the long-time evolution (see Fig. 1). This suggests that hydrodynamic instabilities after core bounce and the resulting asymmetries have important effects on the conditions relevant to the synthesis of heavy nuclei.

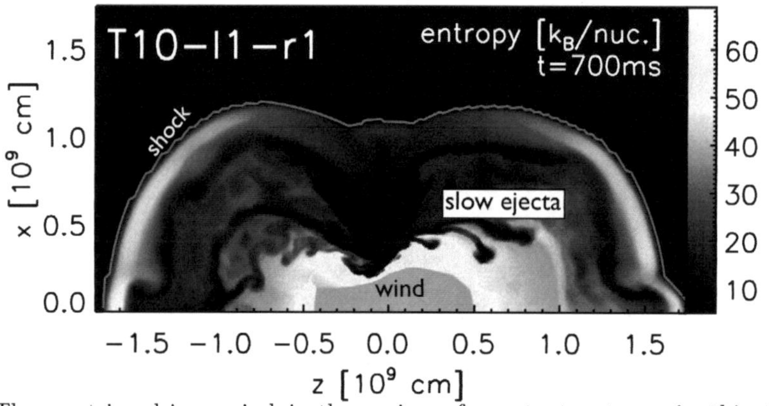

Figure 1. The neutrino-driven wind is the region of constant entropy in this two-dimensional simulation [7]. Notice the anisotropic distribution of the slow, early supernova ejecta.

In most recent hydrodynamic simulations with detailed neutrino transport, the neutrino-driven wind has relative low entropy and is proton-rich [5, 6]. The electron fraction is extremely sensitive to details of the neutrino interactions and transport around the neutrinosphere where neutrinos decouple from matter (see Fig. 2). The evolution of this region depends on the nuclear equation of state and on neutrino interactions, which are both key inputs for supernova simulations, but still very uncertain. Therefore, the electron fraction is only known approximately and its variation can lead to three different nucleosynthesis regimes in the neutrino-driven wind [17] assuming the same expansion: 1) for low electron fraction ($Y_e < 0.4$), the r-process can produce heavy nuclei up to uranium (also entropy $S > 100\,k_{\rm B}$/nuc is required); 2) in slightly neutron-rich winds ($Y_e \sim 0.49$ and $S \sim 50 - 100\,k_{\rm B}$/nuc) the weak r-process produces elements up to $A = 90$ corresponding to neutron magic number $N = 50$; 3) proton-rich winds are the site for the νp-process [18] that can explain the origin of light p-nuclei.

At present it is unclear if neutrino-driven winds are extreme enough for a successful r-process, but it is certain that core-collapse supernovae are fascinating hosts where various nucleosynthesis processes produce neutron-rich and neutron-deficient nuclei. As suggested in Ref. [13], neutrino-driven winds can also be the site where lighter heavy elements (e.g., Sr, Y, Zr) are produced by the weak r-process or by the νp-process. These can account for the suggested

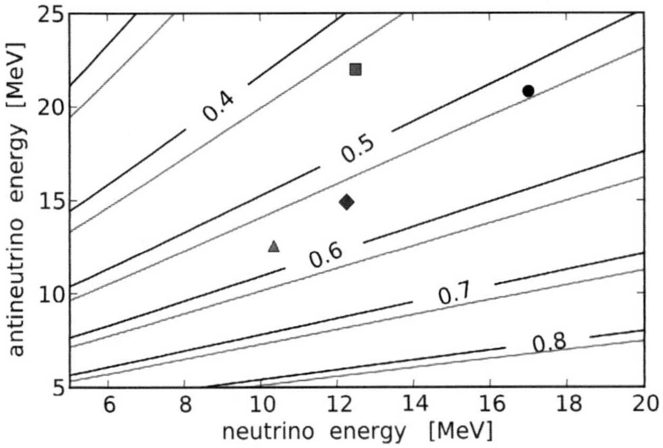

Figure 2. Contours represent the electron fraction based on the approximation of Ref. [19]. The points indicate approximately the electron neutrino and antineutrino energies for different supernova models: the green square from Ref. [20], the black circle from model M15-l1-r6 of Ref. [4], the red triangle is from a 10 $M_\odot$ progenitor of Ref. [5], and the blue diamond from Ref. [6], all at 10 s after bounce. For more details see Ref. [17].

LEPP contribution [14], responsible for the abundances of these elements at low metallicities in very old stars where the s-process is not active yet.

3. The origin of LEPP nuclei in supernovae

We have shown that the lighter heavy elements [17] can be synthesized in neutrino-driven winds as suggested in Ref. [13]. We have performed the first comparison between the LEPP abundances observed in some UMP stars and nucleosynthesis calculations based on long-time hydrodynamic simulations of core-collapse supernovae. Our results show that neutrino-driven winds can explain the observed LEPP pattern in proton- and neutron-rich conditions.

The exact calculation of the electron fraction remains a very challenging open problem [6]. As shown in Fig. 2, the antineutrino energy has decreased as the neutrino reactions and transport have been improved leading to proton-rich winds in the most recent simulations as shown by the Y_e contours. This motivated our exploration of the impact of the electron fraction on the production of the LEPP elements. Figure 3 illustrates that the LEPP elements can be obtained for different proton- and neutron-rich conditions.

Left top panel in Fig. 3 shows that the LEPP pattern (dots obtained from observation as in Ref. [21]) is reproduced in proton-rich winds. Moreover, we found that this abundance pattern is quite robust under variations of the evolution of Y_e. However, elements heavier than iron-group nuclei can be produced only when the neutrino fluxes are high enough to allow a successful νp-process [18, 22, 23]. The left bottom panel in Fig. 3 shows the production factor for various isotopes (see [17] for detailed discussion). There is not strong overproduction (indicated by the dotted line) and the elemental abundances nicely reproduce the observed LEPP pattern in UMP stars. Almost only neutron-deficient isotopes are produced suggesting that proton-rich winds contribute to synthesize light p-nuclei [24].

When the electron fraction is assumed to evolve towards neutron-rich conditions, the LEPP pattern can be also reproduced but it is very sensible under variations of Y_e (right column in Fig. 3). This scenario may contribute to the LEPP elements found in the solar system abundances (marked with circles in the production factor figures) because neutron-rich isotopes are produced. However, we find an overproduction around $A \sim 90$ that was already pointed out in previous nucleosynthesis studies based on supernova simulations (see e.g.,[25]). This overproduction problem and the fact that most recent supernova simulations [5, 6] favor proton-rich winds could suggest that neutron-rich winds are rare events.

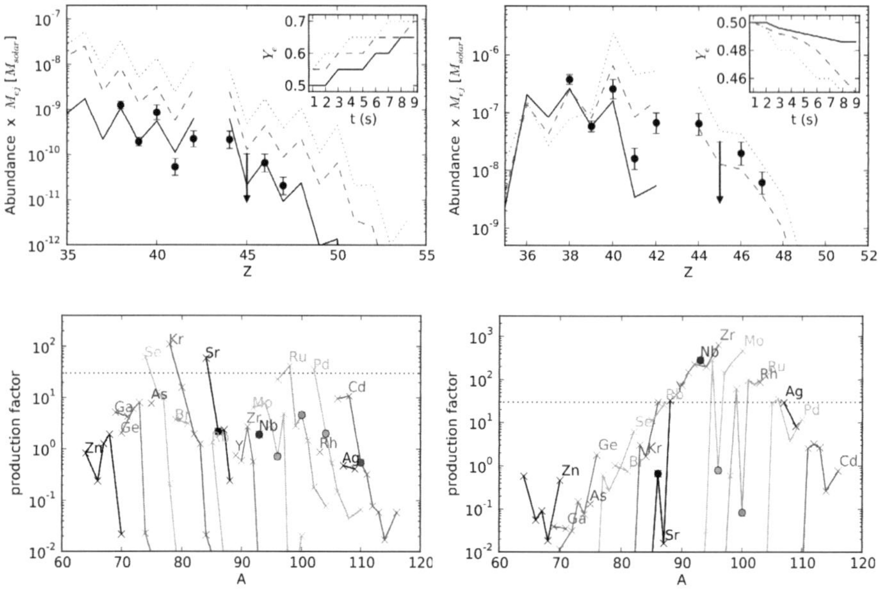

Figure 3. Elemental abundances are shown in the upper panels for different electron fraction evolutions and compared to observations from UMP stars (dots). The production factors correspond to the electron fraction evolution given by the green dashed lines.

4. Conclusions

Recent long-time supernova simulations do not produce r-process elements because the wind entropy is too low and the electron fraction high, even staying proton rich during several seconds [6]. However, the LEPP elements (e.g., Sr, Y, Zr) can be produced in proton- and neutron-rich neutrino-driven winds following the abundance distribution of some UMP star.

Observation of isotopic abundances in UMP stars are very promising to constraint the neutron richness of the neutrino-driven wind and thus the evolution of the electron fraction and the neutrino properties in supernovae.

References

[1] M. Arnould, S. Goriely, and K. Takahashi, *Phys. Repts.* **450**, 97 (2007).
[2] D. Argast, M. Samland, F.-K. Thielemann, and Y.-Z. Qian, *Astrophys. J.* **416**, 997 (2004).
[3] B.S. Meyer et al. *Astrophys. J.* **399**, 656 (1992).
[4] A. Arcones, H.-Th. Janka, and L. Scheck *Astron. & Astrophys.*, **467**, 1227 (2007).
[5] T. Fischer et al., *Astron. & Astrophys.* **517**, A80 (2010).
[6] L. Hüdepohl et al., *Phys. Rev. Lett.* **104**, 251101 (2010).
[7] A. Arcones and H.-Th. Janka, *Astron. & Astrophys.* **526**, A160 (2011).
[8] C. Sneden, J.J. Cowan and R. Gallino, *Annu. Rev. Astro. Astrophys.* **46**, 241 (2008).
[9] Y.-Z. Qian and G.J. Wasserburg, *Astrophys. J.* **559**, 925 (2001).
[10] G.J. Wasserburg, M. Busso, and R. Gallino, *Astrophys. J.* **466**, L109 (1996).
[11] S.E. Woosley and R.D. Hoffman, *Astrophys. J.* **395**, 202 (1992).
[12] C. Freiburghaus et al., *Astrophys. J.* **516**, 381 (1999).
[13] Y.-Z. Qian and G.J. Wasserburg, *Phys. Rep.* **442**, 237 (2007).
[14] C. Travaglio et al., *Astrophys. J.* **601**, 864 (2004).
[15] F. Montes et al., *Astrophys. J.* **671**, 1685 (2007).
[16] T.A. Thompson, A. Burrows, and B.S. Meyer, *Astrophys. J.* **562**, 887 (2001).
[17] A. Arcones and F. Montes, *Astrophys. J.* **731**, 5 (2011).
[18] C. Fröhlich et al., *Phys. Rev. Lett.* **96**, 142502 (2006).
[19] Y.Z. Qian and S.E. Woosley, *Astrophys. J.* **471**, 331 (1996).

[20] S.E. Woosley et al, *Astrophys. J.* **433**, 229 (1994).
[21] Y.Z. Qian and G.J. Wasserburg, *Astrophys. J.* **687**, 272 (2008).
[22] J. Pruet et al., *Astrophys. J.* **644**, 1028 (2006).
[23] S. Wanajo, *Astrophys. J.* **647**, 1323 (2006).
[24] S. Wanajo, H.-Th. Janka, and S. Kubono, *Astrophys. J.* **729**, 46 (2011).
[25] R.D. Hoffman, S.E. Woosley, G.M. Fuller, and B.S. Meyer *Astrophys. J.* **460**, 478 (1996).

Disappointing model for ultrahigh-energy cosmic rays

R Aloisio[1], V Berezinsky[1,2], and A Gazizov[2,3]

[1] INFN, National Gran Sasso Laboratory, I-67010 Assergi (AQ), Italy
[2] Gran Sasso Astroparticle Center, I-67010 Assergi (AQ), Italy
[3] Institute of Physics of NASB, 68 Independence Avenue, BY-22072 Minsk, Belarus

E-mail: askhat.gazizov@lngs.infn.it

Abstract. Data of Pierre Auger Observatory show a proton-dominated chemical composition of ultrahigh-energy cosmic rays spectrum at $(1-3)$ EeV and a steadily heavier composition with energy increasing. In order to explain this feature we assume that $(1-3)$ EeV protons are extragalactic and derive their maximum acceleration energy, $E_p^{\max} \simeq 4$ EeV, compatible with both the spectrum and the composition. We also assume the rigidity-dependent acceleration mechanism of heavier nuclei, $E_A^{\max} = Z \times E_p^{\max}$. The proposed model has rather disappointing consequences: i) no pion photo-production on CMB photons in extragalactic space and hence ii) no high-energy cosmogenic neutrino fluxes; iii) no GZK-cutoff in the spectrum; iv) no correlation with nearby sources due to nuclei deflection in the galactic magnetic fields up to highest energies.

Spectra and chemical compositions of ultrahigh-energy ($E \gtrsim 1$ EeV) cosmic rays (UHECR) measured by two largest detectors, High Resolution Fly's Eye (HiRes) [1] and Pierre Auger Observatory (PAO) [2], are significantly different. The HiRes data show pure *proton* composition [3, 4], confirming such signatures of their propagation through CMBR as the GZK cutoff [5, 6] and the pair-production dip [7–11]. The PAO data, on the contrary, strongly favor the nuclei

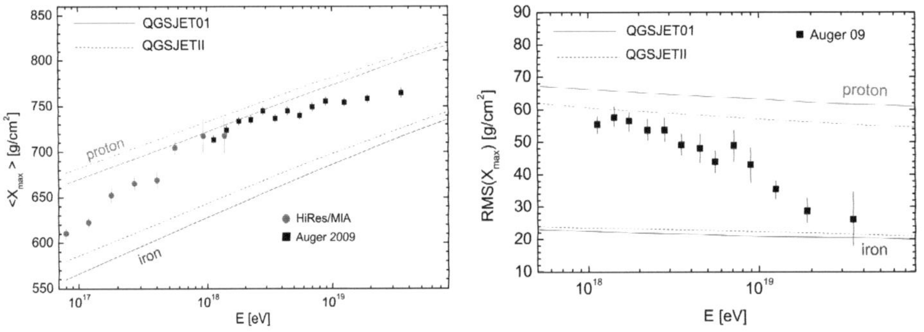

Figure 1. PAO data [12–14] on $X_{\max}(E)$ (left panel) and on RMS($X_{\max}$) (right panel). Lines for protons and Iron are according to QGSJET model [15].

composition getting progressively heavier at $E \simeq (4-40)$ EeV. This feature, in terms of energy dependence of EAS development maximum in atmosphere, $X_{\max}(E)$, and r.m.s. of this observable, RMS($X_{\max}$), is clearly seen in Fig. 1. The data also suggest that the nucleus charge number Z changes smoothly in sources. We do not attempt to determine which one of the two

Published under licence by IOP Publishing Ltd

Journal of Physics: Conference Series **337** (2012) 012042 doi:10.1088/1742-6596/337/1/012042

experiments is more likely to be correct. Instead, we focus solely on PAO data and examine the implications should the reported CR spectrum and composition be born out by future verifications. We show that the simple, but disappointing for future experiments, model [16] can naturally explain both energy spectrum and mass composition observed by the PAO.

The basic assumption of the model is the *proton composition* of UHECR spectrum at $E \simeq (1-3)$ EeV, the feature supported both by PAO and HiRes. Two more assumptions are that these protons are *extragalactic* and that acceleration of primary nuclei in sources is *rigidity-dependent*, i.e. that $E_{\rm max}^{\rm acc} = Z \times E_0$, where E_0 is a universal energy to be determined from data; Z is a nucleus charge number.

In order to determine the maximum acceleration energy of protons, $E_p^{\rm max} = E_0$, let us calculate the extragalactic diffuse proton flux, assuming the power-law generation spectrum $Q_g(E) \propto E^{-\gamma_g}$ with $E_{\rm max} = E_0$, and normalize it by the PAO flux at $(1-3)$ EeV. Varying γ_g in the range $2.0 - 2.8$, the maximum value of E_0 allowed by the PAO mass composition (see Fig. 1) and energy spectrum (see Fig. 2) may be obtained.

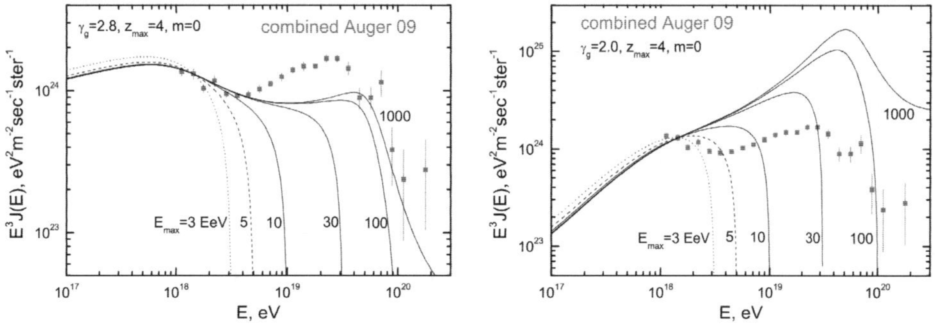

Figure 2. Calculated proton spectra compared to the combined PAO spectrum for different $E_p^{\rm max}$. Extreme cases $\gamma_g = 2.8$ and 2.0 are shown in the left and right panels, respectively.

In our calculations a homogeneous distribution of sources with no cosmological evolution ($m = 0$) was assumed; the highest redshift of sources $z_{\rm max} = 4$. As a criterion of contradiction an excess of calculated proton flux at $E \sim (4-5)$ EeV was chosen. The contradiction has different character for different values of γ_g.

For steep source generation functions with $\gamma_g \simeq 2.6 - 2.7$ the shape and flux of the PAO spectrum may be described by $E_p^{\rm max} \sim 10^{20} - 10^{21}$ eV; the contradiction occurs only in data on mass composition. The extreme case, given by $\gamma_g = 2.8$, is displayed in the left panel of Fig. 2.

For flat generation spectra (see the extreme case of $\gamma_g = 2.0$ in the right panel of Fig. 2) the contradiction is very pronounced. For $E_p^{\rm max} = 5$ EeV the calculated proton flux exceeds the observed one even at $E \approx 2$ EeV.

It is clear that with some redundancy $E_p^{\rm max} \simeq (4-6)$ EeV for all $2.0 \lesssim \gamma_g \lesssim 2.8$.

An influence of possible intergalactic magnetic fields on proton spectrum calculated in a diffusive model is shown in the left panel of Fig. 3. Here $\gamma_g = 2.3$, which might be the case for acceleration by relativistic shocks. The Kolmogorov diffusion in turbulent magnetic field with basic scales $(B_c, l_c) = (1$ nG, 1 Mpc) was assumed (see [17, 18]) and distances between sources were $d \simeq 40$ Mpc. The analysis of proton maximum energy of acceleration gives again $E_0 = E_p^{\rm max} = 4$ EeV, in a rough agreement with the analysis made for homogeneous distribution of sources. The account for diffusion brings to the flattening of the proton spectrum at $E \lesssim 1$ EeV, seen in Fig. 3 as a 'diffusive cutoff', which provides a transition from the steep galactic spectrum, most probably composed of Iron, to the flat spectrum of extragalactic protons.

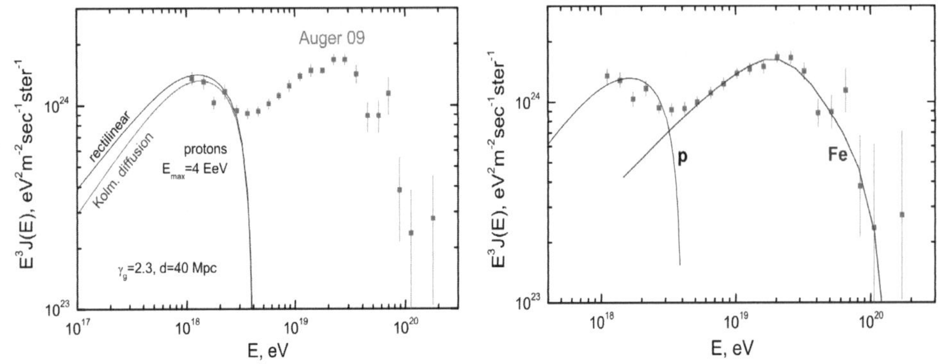

Figure 3. *Left panel:* Comparison of calculated proton spectra with the combined PAO spectrum for $\gamma_g = 2.3$ and diffusive proton propagation. The cutoff at $E_p^{\max} = 4$ EeV is needed to avoid the contradiction with data at $E > 3$ EeV. *Right panel:* The energy spectrum in two-component model with protons and Iron nuclei with $\gamma_g = 2.0$ and $E_{\max} = 4 \times Z$ EeV. The Iron nuclei spectrum is calculated for homogeneous distribution of the sources.

The basic feature of the PAO mass composition, the progressively heavier composition with energy increasing, is guaranteed in our model by the rigidity-dependent maximum energy of acceleration: at energy higher than $Z \times E_p^{\max}$ nuclei with charge $Z' < Z$ disappear, while heavier nuclei with larger Z survive. Starting from $E_p^{\max} \sim (4-6)$ EeV, the higher energies are accessible only for nuclei with progressively larger values of Z.

Let us now consider a two-component model, with only protons and Iron nuclei being produced in sources with generation index $\gamma_g = 2.0$ and the maximum acceleration energy $E_{\max} = 4 \times Z$ EeV, shown in the right panel of Fig. 3. The primary Iron nuclei spectrum is calculated as in [19, 20] for homogeneous distribution of sources. One may notice that the calculated spectrum of Iron describes well the cutoff in the PAO spectrum. This steepening is caused by the photo-disintegration of Iron nuclei.

To agree with the mass composition of PAO, the Iron spectrum in Fig. 3 must have a low-energy cutoff at $E \lesssim (20-30)$ EeV. Most naturally it is produced as a 'diffusive cutoff' which appears in models with lattice-located sources due to *magnetic horizon* [21]. Such cutoffs are shown in Fig. 4 for three different sets of parameters B_c, l_c, d. The beginning of this cutoff E_c for Iron nuclei is $Z = 26$ times higher than for protons, i.e. $E_c \approx 2.6 \times 10^{19}$ eV, which has a reasonable physical meaning. The gap between 2 EeV and 26 EeV is expected to be filled by intermediate nuclei. To provide a smooth RMS($X_{\max}$) curve seen in Fig. 1, there are many free parameters, e.g. arbitrary fractions of nuclei accelerated in distant sources.

The predictions of our model are very disappointing for the future detectors. Really, the maximum acceleration energy $E_{\max} \sim (100-200)$ EeV for Iron nuclei implies the energy per nucleon $E_p < E_{\max}/A \sim (2-4)$ EeV, well below the GZK cutoff for epochs with $z \lesssim 15$. Therefore, practically no cosmogenic neutrinos can be produced in collisions of protons and nuclei with CMB photons. Correlation with UHECR sources also is absent due to deflection of nuclei in the galactic magnetic fields. The lack of correlation in the model is strengthened by the dependence of the maximum energy on Z.

The signatures of the 'disappointing model' for the PAO detector are the mass-energy relation, already seen in the elongation curve $X_{\max}(E)$, and transition from galactic to extragalactic cosmic rays below the characteristic energy $E_c \sim 1$ EeV.

There are some uncertainties in the model presented above. The most important one relates to estimates of $E_p^{\max}$. It is determined by the lowest energy where PAO data become inconsistent

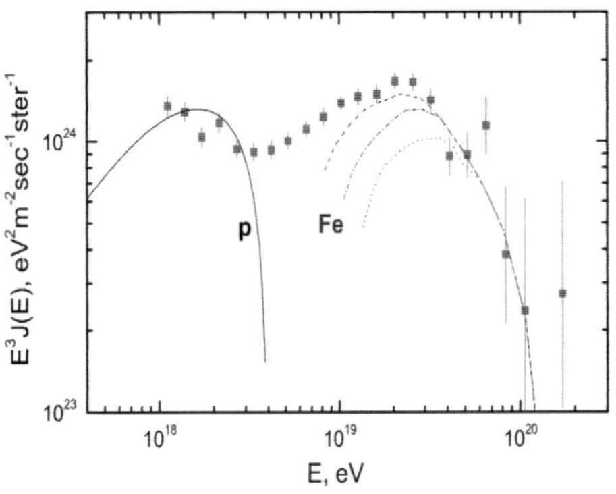

Figure 4. As in the right panel of Fig. 3, but with the ''diffusion cutoff' introduced for three different sets of parameters B_c, l_c, d. The gap between 3 EeV and $E_{\rm cut}$ (beginning of 'diffusive cutoff') is expected to be filled by intermediate nuclei with $2 \leq Z \leq 25$.

with the proton composition (the 6th low-energy bin of the PAO data in Fig. 1). If this energy increases, $E_p^{\max}$ increases, too. The model collapses when the allowed $E_p^{\max}$ reaches e.g. $(50 - 100)$ EeV.

Another case is given by the mass composition beinglight nuclei starting right from 1 EeV [18]. The cosmological evolution of sources are not included in our calculations; since this effect slightly decreases $E_p^{\max}$, it is not needed to be taken into account. In principle, it is also possible that the EeV protons detected by PAO are secondary ones, i.e. those produced in photo-dissociation of primary nuclei in collisions with CMBR and extragalactic IR/UV photons. However, in fact, as it was demonstrated in [20, 22], the flux of secondary protons in the EeV range is always smaller than the sum of primary and secondary nuclei fluxes.

Acknowledgments

The work of A. Gazizov was supported by a contract with Gran Sasso Center for Astroparticle Physics (CFA) funded by European Union and Regione Abruzzo under the contract P.O. FSE Abruzzo 2007-2013, Ob. CRO.

References

[1] Abbasi R U *et al.* (HiRes) 2008 *Phys. Rev. Lett.* **100** 101101 (astro-ph/0703099)
[2] Abraham J *et al.* (Pierre Auger) 2008 *Phys. Rev. Lett.* **101** 061101 (arXiv:0806.4302)
[3] Sokolsky P 2009 *SOCoR 2009, http://web.phys.ntnu.no/~mika/programme.html*, Department of Physics, NTNU Trondheim, Norway
[4] Sokolsky P (HiRes) 2010 (arXiv:1010.2690)
[5] Greisen K 1966 *Phys. Rev. Lett.* **16** 748–750
[6] Zatsepin G T and Kuzmin V A 1966 *JETP Lett.* **4** 78–80
[7] Berezinsky V S and Grigor'eva S I 1988 *Astron. Astrophys.* **199** 1–12
[8] Berezinsky V, Gazizov A Z and Grigorieva S I 2006 *Phys. Rev.* **D74** 043005 (hep-ph/0204357)
[9] Aloisio R *et al.* 2007 *Astropart. Phys.* **27** 76–91 (astro-ph/0608219)
[10] Berezinsky V, Gazizov A Z and Grigorieva S I 2002 (astro-ph/0210095)
[11] Berezinsky V, Gazizov A Z and Grigorieva S I 2005 *Phys. Lett.* **B612** 147–153 (astro-ph/0502550)
[12] Unger M 2009 *SOCoR 2009, http://web.phys.ntnu.no/~mika/programme.html*, Department of Physics, NTNU Trondheim, Norway
[13] Abraham J *et al.* (PAO Collaboration) 2010 *Phys. Rev. Lett.* **104** 091101 (arXiv:1002.0699)
[14] Bellido J A 2009 *Proceedings of the 31st ICRC, ŁODŹ 2009* (University of Łodź with Andrzej Soltan Institute for Nuclear Studies)
[15] Ostapchenko S 2006 *Phys. Rev.* **D74** 014026 (hep-ph/0505259)
[16] Aloisio R, Berezinsky V and Gazizov A 2011 *Astropart.Phys.* **34** 620–626 (arXiv:0907.5194)
[17] Berezinsky V and Gazizov A Z 2006 *Astrophys. J.* **643** 8–13 (astro-ph/0512090)
[18] Aloisio R, Berezinsky V and Gazizov A 2009 *Astrophys. J.* **693** 1275–1282 (arXiv:0805.1867)
[19] Aloisio R, Berezinsky V and Grigorieva S 2008 (arXiv:0802.4452)
[20] Aloisio R, Berezinsky V and Grigorieva S 2010 (arXiv:1006.2484)
[21] Aloisio R and Berezinsky V 2004 *Astrophys. J.* **612** 900–913 (astro-ph/0403095)
[22] Allard D, Busca N G, Decerprit G, Olinto A V and Parizot E 2008 *JCAP* **0810** 033 (arXiv:0805.4779)

Nuclear Physics in Astrophysics V IOP Publishing
Journal of Physics: Conference Series **337** (2012) 012043 doi:10.1088/1742-6596/337/1/012043

Quark-mass variation effect on big bang nucleosynthesis

J C Berengut[1], V V Flambaum[1] and V F Dmitriev[2,3,4]

1 School of Physics, University of New South Wales, Sydney 2052, Australia,

2 Budker Institute of Nuclear Physics SB RAS, 630090, Novosibirsk-90, Russia,

3 Novosibirsk State University, 630090, Novosibirsk-90, Russia

E-mail: dmitriev@inp.nsk.su

Abstract. We calculate the effect of variation in the light-current quark mass, m_q, on standard big bang nucleosynthesis. A change in m_q at during the era of nucleosynthesis affects nuclear reaction rates, and hence primordial abundances, via changes the binding energies of light nuclei. It is found that a relative variation of $\delta m_q/m_q = 0.016 \pm 0.005$ provides better agreement between observed primordial abundances and those predicted by theory. This is largely due to resolution of the existing discrepancies for ^{7}Li. However this method ignores possible changes in the position of resonances in nuclear reactions. The predicted ^{7}Li abundance has a strong dependence on the cross-section of the resonant reactions ^{3}He (d, p) ^{4}He and t (d, n) ^{4}He. We show that changes in m_q at the time of BBN could shift the position of these resonances away from the Gamow window and lead to an increased production of ^{7}Li, exacerbating the lithium problem.

1. Introduction

Measurements of the primordial baryon-to-photon ratio η from the cosmic microwave background from WMAP [1], coupled with precise measurements of the neutron half-life [2], have made big bang nucleosynthesis (BBN) an essentially parameter-free theory [2, 3, 4]. In this paradigm excellent agreement has been obtained between predicted and observed abundances of deuterium and ^{4}He (see, e.g. the Particle Data Group review [2] and references therin). However there is some disagreement for ^{7}Li, the only other element for which the abundance has been measured to an accuracy at which fruitful comparison with theory can be made. While the "lithium problem" has been known for some time, it has been exacerbated by recent measurements of the ^{3}He$(\alpha, \gamma)^7$Be reaction [5]. Standard BBN theory with η provided by WMAP5 overproduces ^{7}Li by a factor of 2.4 – 4.3 (around 4 – 5σ) [4].

One possible solution to the lithium problem is that the physical constants of the early Universe may have been slightly different. Recently, Ref. [6] examined the response of BBN to variation of several physical parameters, including binding energies, in a linear approximation. These were coupled with calculated dependences of binding energies on m_q in [7], which found that the ^{7}Li

[4] Corresponding author.

abundance discrepancy could be resolved by a variation in light-quark mass of $\delta m_q/m_q = 0.013\pm0.002$. Crucially, the ^{4}He and d abundances were found to be relatively insensitive to m_q and so the existing agreement between theory and observation in these elements was maintained. In this work we re-examine the dependence of light element production on variation of the dimensionless parameter $X_q = m_q/\Lambda_{QCD}$ where m_q is the light-quark mass and Λ_{QCD} is the pole in the running strong coupling constant. We follow [7] and assume that Λ_{QCD} is constant, calculating the dependence on the small parameter m_q. This is not an approximation. Rather it only means that we measure all dimensions (m_q, cross sections, etc) in units of Λ_{QCD}. Therefore $\delta m_q/m_q$ should be understood as $\delta X_q/X_q$. We take into account several effects that were not previously considered, most importantly the nonlinear dependence on m_q and variation of resonance positions.

2. Variation of binding energies

The energy released in each reaction, Q, is determined by the nuclear binding energies. As noted in [6], the Q-values affect the forward (exothermic) reaction rates via phase space and radiative emission factors. For radiative capture reactions at low energy E the Q-dependence is

$$\sigma(E) \sim E_\gamma^3 \sim (Q+E)^3. \tag{1}$$

For low-energy reactions with two nucleons in the exit channel the dependence is proportional to the outgoing channel velocity, $v \sim (Q+E)^{1/2}$. When the outgoing particles are charged, the Gamow factor of the exit channel can also contribute:

$$\sigma(E) \sim (Q+E)^{1/2} \exp(-\sqrt{E_g/(Q+E)}). \tag{2}$$

The Gamow factor appears because of the Coulomb barrier to the reaction; $E_g = 2\pi^2 Z_1^2 Z_2^2 \alpha^2 \mu c^2$ where α is the fine-structure constant, Z_1 and Z_2 are the charge numbersof the products, and μ is the reduced mass of the products. At BBN temperatures we can usually assume that $E << Q$. Expanding in Q,

$$\sigma = \sigma_0[1 + 1/2(1+\sqrt{E_g/Q})\delta Q/Q + ...], \tag{3}$$

and we see that the Gamow term in (2) is generally small (it was neglected in [6]). However it can be important for some reactions, for example in ^{7}Be (n, p) ^{7}Li, $\sqrt{E_g/Q} = 2.17$, i.e. it triples the effect of δQ on the reaction rate.

An exception to the rule (1) is found in the reaction p (n, γ) d, an important reaction because d is a precursor to all further nucleosynthesis. This reaction is sensitive not only to Q but also to the position of the virtual level with energy $\varepsilon = 0.07$ MeV. The sensitivity of this reaction to Q was calculated in [8]

$$<\sigma v> \sim [1 + (5/2 + \sqrt{Q/\varepsilon})\delta Q/Q]. \tag{4}$$

Note that [6,7] did not take variation of the virtual level into account.

We denote the sensitivity of nuclear binding energies to the light-current quark mass m_q by

$$K = (\delta E/E)/(\delta m_q/m_q). \tag{5}$$

Values of K for several light nuclei were presented in Refs. [7, 9]. We use the "best values" from these papers, given by the AV18+UIX nuclear Hamiltonians, with hadron mass variations calculated in terms of the m_q using the Dyson-Schwinger equation calculation of [10]. From these one calculates the m_q-dependence of the Q values, and therefore the reaction rates, and therefore the primordial abundances of light elements in BBN.

Comparing the observed and predicted abundances we obtain for ^{4}He, d, and ^{7}Li respectively, $\delta m_q/m_q = -0.002 \pm 0.037$, 0.012 ± 0.011, and 0.018 ± 0.006. The three data sets are therefore consistent,with weighted mean

$$\delta m_q/m_q = 0.016 \pm 0.005. \tag{6}$$

The ^{4}He abundance has a low sensitivity to m_q. It is worth pointing out that the more tightly constrained fraction, $Y_p = 0.2477 \pm 0.0029$ [11] instead of coservative fraction $Y_p = 0.249 \pm 0.009$ [12]

used in our calculation is also consistent with the variation (6). Taking into account the nonlinear dependence of BBN abundances on m_q is important, particularly for ^{7}Li. In fact, if we assume a linear response, as was done in [6, 7], we instead obtain $\delta m_q/m_q = 0.014 \pm 0.002$.

3. Resonances

Of the most important reactions in BBN, the mirror reactions ^{3}He (d, p) ^{4}He (reaction 1), and t (d, n) ^{4}He (reaction 2) are the only reactions where the cross-section is dominated by a fairly narrow resonance. Therefore, one can hope for sensitivity of primordial abundances to the position of these resonances. Both of these reactions have the cross-sections with the general form

$$\sigma(E) = \exp(-\sqrt{E_g / E}) / E \frac{P(E)}{(E - E_r)^2 + \Gamma_r^2 / 4}, \tag{7}$$

where E_g is the Gamow energy of the reactants, E_r and Γ_r are resonance parameters, and $P(E)$ is a polynomial chosen to fit the measured reaction cross-section. In this work we use the cross-section fits of Ref. [13], which give $E^{(1)}_r = 0.183$ MeV, $\Gamma^{(1)}_r = 0.256$ MeV and $E^{(2)}_r = 0.0482$ MeV, $\Gamma^{(2)}_r = 0.0806$ MeV for reactions 1 and 2, respectively.

Consider modification of the resonance positions, $E_r \rightarrow E_r + \delta E_r$, due to a variation of the fundamental constant m_q. Reaction 1 will be affected in the following way. The resonance is an excited state of ^{5}Li; that is, a compound nucleus with three protons and two neutrons: we call this state ^{5}Li*. Similarly there is a state ^{5}He* for reaction 2. Then

$$E_r^{(1)} = E_{^5Li*} - E_{^3He} - E_d, \quad E_r^{(2)} = E_{^5He*} - E_t - E_d, \tag{8}$$

and so $E_{5Li*} = -9.76$ MeV and $E_{5He*} = -10.66$ MeV. The change in the resonance position due to a variation in m_q is therefore

$$\delta E_r^{(1)} = \delta E_{^5Li*} - \delta E_{^3He} - \delta E_d = (K_{^5Li*} E_{^5Li*} - K_{^3He} E_{^3He} - K_d E_d)\delta m_q / m_q \tag{9}$$

with the K defined by (5).

Changes to the cross-section of reaction 1 affects the primordial abundances of ^{3}He and ^{7}Be, while changes in reaction 2 affect abundances of t and ^{7}Li. Since t and 3He are not well constrained observationally, we choose to focus on ^{7}Li. In figure 1 we present ^{7}Li abundance against variation of light quark mass $\delta m_q/m_q$ at $\eta = 6.23\text{x}10^{-10}$, the WMAP5 value. For such a value of η, the majority of ^{7}Li is created as ^{7}Be (which β-captures to ^{7}Li) via the reaction ^{3}He (^{4}He, γ) ^{7}Be.

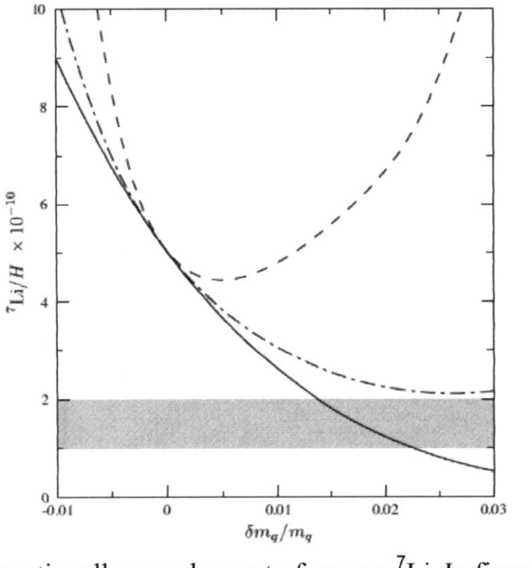

Figure 1: Calculated ^{7}Li abundance vs. relative change in light quark mass m_q/Λ_{QCD}. Solid line: no shifts in resonance positions included; dashed line: resonance shifts according to assumption that resonant state varies as much as the ground state (equations 12); dotdashed line: an averaged value of resonance-position sensitivity used. The shaded area shows the 1σ range of abundances.

We need to find $K_{^5Li*}$ (and similarly $K_{^5He*}$). One assumption is that the mass-energy of the resonance varies with the mass-energy in the incoming channel [6]; in this case the resonance does not shift. This assumption corresponds to $K_{^5Li*} = -1.54$ and $K_{^5He*} = -1.44$. It corresponds to the solid line in figure 1. A more reasonable guess is to assume that the variation of the resonant state ^{5}Li* will be approximately the same as that of the ground state ^{5}Li. This can be seen by considering the resonance and the ground state configurations as residing in the same potential. The sensitivity of the ground state ^{5}He to m_q has been calculated $K_{^5He} = -1.24$ [9]; $K_{^5Li}$ was not calculated explicitly, but its value will be very close to that of ^{5}He. Our assumption of equal variation of the ground and excited state then gives

$$K_{^5Li*} = -3.35\,, \qquad K_{^5He*} = -3.19 \tag{10}$$

This assumption corresponds to the dashed line in figure 1. The equal-variation assumption in the previous paragraph represents an upper limit on the relationship between the ground and excited state. In reality the potential-dependence of the states may be different, in which case the shift of the ^{5}Li (or ^{5}He) resonance may be smaller than the shift of the ground state. On the other hand a minimum value of K for the resonance states is that of the ground state, $K_{^5Li*} = K_{^5Li} = -1.24$. A reasonable, conservative, estimate is to take the average of these extremal values: $K_{^5Li*} = -2.29$ and $K_{^5He*} = -2.21$; this is the dot-dashed line in figure 1. Ultimately however, we require a nuclear calculation of sensitivity, of the kind presented in Refs. [6, 7]. The first calculation has been done in [14]. The result $K_{^5Li*} = -2.48$ and $K_{^5He*} = -2.16$ is remarkably close to our estimated values. The effect of $\delta E^{(1)}{}_r$ on BBN can be understood in the following way. When the cross-section is convolved with a Maxwellian distribution, the exponential term gives rise to the "Gamow window" at energy $E_0/E_g = (kT/2E_g)^{2/3}$. This reaction is most active at $kT \approx 0.07$ MeV, at which time the Gamow window is at $E_0 = 0.180$ MeV. This is remarkably close to the resonance energy for this reaction $E_r = 0.183$ MeV. Therefore movement of the resonance position in either direction will reduce the cross-section for this reaction at the relevant temperatures. In turn this reduces the amount of ^{3}He that is destroyed via reaction 1, leaving more to react with ^{4}He to produce ^{7}Be. On the other hand the effect of this reaction on d and ^{4}He abundances is minimal. The effect of $\delta E^{(2)}{}_r$ is very similar: it reduces the amount of t destroyed in reaction 2, leaving more tritium to react with ^{4}He to produce ^{7}Li directly. Despite this production channel being suppressed at high η, the effect of $\delta E^{(2)}{}_r$ is still important for ^{7}Li production because the relative effect of the variation is larger $\delta E^{(2)}{}_r/\Gamma^{(2)}{}_r > \delta E^{(1)}{}_r/\Gamma^{(1)}{}_r$. The trends seen in figure 1 are the same even at low η since both reaction pathways behave in much the same way to variation in m_q.

From figure 1 we see that taking shifts in the resonance positions into account can destroy the agreement between theory and observation previously obtained by varying m_q. In the case where the shifts in the ground and resonant states vary by the same amount (dashed line), the ^{7}Li discrepancy actually gets worse with variation in light quark mass. On the other hand the milder "averaged K" response (dot-dashed line) still significantly challenges the conclusions of Section II.

References

[1] Dunkley J et al. 2009 Astrophys. J. Suppl. Ser. **180** 306
[2] Amsler C et al. (Particle Data Group) 2008) Phys. Lett. B **667** 1
[3] Fields B D and Olive K A 2006 Nucl. Phys. A **777** 208
[4] Cyburt R H, Fields B D and Olive K A (2008) JCAP **11** 012.
[5] Cyburt R H and Davids B 2008 Phys. Rev. C **78** 064614
[6] Dent T, Stern S and Wetterich C 2007 Phys. Rev. D **76** 063513
[7] Flambaum V V and Wiringa R B 2007 Phys. Rev. C **76** 054002
[8] Dmitriev V F, Flambaum V V and Webb J K 2004 Phys. Rev. D **69** 063506
[9] Flambaum V V and Wiringa R B 2009 Phys. Rev. C **79** 034302
[10] Flambaum V V, Höll A, Jaikumar P, Roberts C D and Wright S V 2006 Few-Body Sys. **38** 31
[11] Peimbert M, Luridiana V and Peimbert A 2007 Astrophys. J. **666** 636
[12] Olive K A and Skillman E D 2004 Astrophys. J. **617** 29
[13] Cyburt R H 2004 Phys. Rev. D **70** 023505
[14] Wiringa R B private communication.

Energy-broadened proton beam for production of quasi-stellar neutrons from the ^{7}Li(p,n)^{7}Be reaction

G Feinberg[1,2], A Shor[1], D Berkovits[1], Y Eisen[1], M Friedman[2], G Giorginis[3], T Hirsh[1], A Krása[3], M Paul[2], A Plompen[3], E Tsuk[2]

[1]Nuclear Physics and Engineering Division, Soreq NRC, Yavne 81800, Israel
[2]Racah Institute of Physics, The Hebrew University of Jerusalem, Jerusalem 91904, Israel
[3]EC-JRC-IRMM, Retieseweg 111, 2440 Geel, Belgium

E-mail: gitai@phys.huji.ac.il

Abstract. Production of quasi-stellar neutrons by the ^{7}Li(p,n)^{7}Be reaction has been used for measuring s-process cross sections and efforts to upgrade the proton beam intensity with RF linear accelerators are presently ongoing. We investigated the effect of an energy-broadened proton beam, as is expected for a RF linear accelerator, on the produced 25-keV quasi-Maxwellian neutron spectrum and compared it to that of a Van de Graaff accelerator with well-defined proton energy. Neutron spectrum measurements from 0° to 80° were carried out with a pulsed proton beam at the IRMM Van de Graaff accelerator using time-of-flight techniques, both with a narrow energy spread ($\sigma \approx 1.5$ keV) proton beam and with an energy-broadened beam ($\sigma \approx 20$ keV) obtained by straggling through a Au-foil degrader. In the latter case, the neutron spectrum is closer to the Maxwellian flux distribution in the high neutron energy region.

1. Introduction

The kinematics of the ^{7}Li(p,n) reaction for 1912 keV protons, ~30 keV above threshold, in a thick lithium target is known to generate forward-collimated neutrons having an energy spectrum which closely simulates the energy dependence of the neutron flux in a stellar medium at a temperature of ~$3 \cdot 10^{8}$ K (kT≈25 keV) [1]. The neutrons produced by the near-threshold ^{7}Li(p,n) reaction have been widely used to activate stable (and more recently unstable) nuclides involved in the s-process, providing a direct measurement of the Maxwellian-averaged cross section (MACS) for neutron capture at kT≈25 keV [2]. There exists currently a strong interest in expanding such activation measurements to unstable nuclides at branching points of the s-process and off the valley of stability or to cases experimentally difficult to measure. Higher proton beam intensities (in the milliampere range) are then required, which are incompatible both with traditional Van de Graaff accelerators and with the poor thermal properties of conventional solid lithium targets. We are presently pursuing the construction of a setup based on the high-intensity proton beam of the superconducting RF linear accelerator at the Soreq Applied Research Accelerator Facility (SARAF) [3] and a high-power Liquid-Lithium Target (LiLiT) [4] to be used in such experiments. However the properties of a high-intensity ion beam from a RF linear accelerator are different from those of Van de Graaff accelerators in two main aspects: a significantly larger energy spread of the beam and a somewhat larger transverse

Published under licence by IOP Publishing Ltd

distribution, limited by space-charge effects. These will in turn affect the shape of the neutron energy and angular distribution and the yield of neutrons emitted in the ^{7}Li(*p,n*) reaction and will require to establish a new standard for neutron monitoring. Using the IRMM Van de Graaff accelerator and a solid LiF target, we investigate experimentally for the first time the effects of a broadened-energy proton beam on the neutron energy distribution and yield. The broadened energy was obtained by straggling of the incident protons with a thin gold foil placed in front of the LiF target. Neutrons were detected by ^{6}Li-glass detectors and their energy was measured by time-of-flight (TOF), using a 625-kHz pulsed proton beam. Angular distributions of the neutrons were measured and summed to obtain the integral energy distribution. The neutron energy distribution in the broad-energy case is modified by a cutoff at higher energies and is closer in this region to the $E\cdot\exp(-E/kT)$ dependence of the Maxwellian stellar flux.

2. Experimental procedure

The experiment was performed in the low-scatter target hall of the Van de Graaff laboratory at the IRMM in Geel. Neutron spectra were measured by TOF techniques using ^{6}Li-glass detectors for both a narrow-energy spread proton beam and for a broad-energy spread proton beam. All measurements were performed while minimizing the materials that may cause neutron scatter. In front of the target was a collimator that limited the beam diameter to 6 mm and an electron suppressor enabling charge integration of the proton beam at the target. Typical diameter of the beam on target is estimated as 2-3 mm. The targets were made of LiF evaporated on a 1 mm thick Cu disk. The LiF layer thickness was determined to be thick enough to reduce the proton energy significantly below the reaction threshold. To enable TOF measurements, the accelerator was operated in pulsed mode using a frequency of 625 kHz. The chopped and bunched proton beam had a pulse width of 1.5-2 ns FWHM. Cooling of the target during irradiation with chopped beams with average current of 100 nA was done by air flow on the target copper backing.

In order to mimic the energy spread expected for a proton beam from a RF linear accelerator, we placed a gold foil degrader directly upstream of the target. SRIM [5] calculations showed that 2096 keV protons traversing through a 2.06 μm gold foil will emerge with energy of 1912 keV and an energy-spread standard deviation of ~20 keV, similar to the energy spread expected for the SARAF high intensity beam at target position. The degrader was glued onto a ring holder which positioned it 1 mm upstream of the LiF layer to minimize beam growth in transversal dimensions caused by the scattering in the gold foil.

For the TOF measurements, ^{6}Li-glass detectors (Scionix Ltd.) were used. It was deemed necessary to use relatively thick detectors (1" in thickness) to keep beam time within practical limits. The ^{6}Li-glass detector could be placed on a goniometer at various angular positions with 5° spacing and at a fixed distance of 51 cm from the LiF target. Accuracy of better than ±0.3° was obtained after proper alignment of the goniometer on the target hall floor. For all TOF measurements, another ^{6}Li detector used as monitor was placed at fixed position – at a distance of 85 cm and at angle of 12° from the goniometer symmetry axis and 10° upward out of plane.

The signal from the photomultiplier of each detector was split to a fast timing circuit and to a pulse height circuit. The fast timing circuit was comprised of a timing filter amplifier followed by a constant-fraction discriminator (CFD). The pulse height circuit consisted of a preamplifier and a spectroscopy amplifier. The timing signal from each detector provided the TOF start signal in an allocated time-to-amplitude converter (TAC) channel. A beam pick up monitor placed upstream of the target provided a signal that was processed through a timing filter amplifier and a CFD and provided the common TOF stop. The total TOF resolution for the combined system, as measured for the prompt-gamma peak, was ~4 ns for most of the runs. Data for each detector were acquired independently, where the timing and pulse height information for each detector were acquired in coincidence.

Journal of Physics: Conference Series **337** (2012) 012044 doi:10.1088/1742-6596/337/1/012044

3. TOF spectra

Figure 1 displays a raw TOF spectrum, the prompt gamma peak (off-scale) and neutron spectra at longer flight times (left of the gamma peak) are observed. The insert highlighting the gamma peak exhibits a Gaussian TOF distribution with FWHM≈4 ns. The large amount of prompt gamma rays are primarily from ^{7}Li($p,p'\gamma$), ^{7}Li(p,γ), ^{19}F($p,\alpha\gamma$) and ^{19}F(p,γ) reactions.

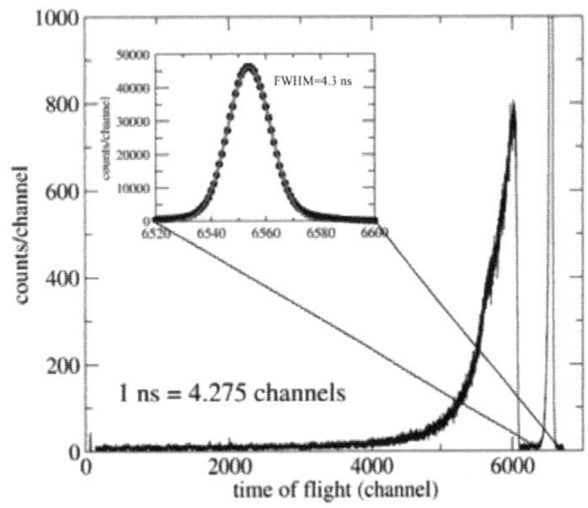

Figure 1. Raw data - TOF spectrum showing gamma peak (off-scale) and neutrons. The inset shows the gamma peak with FWHM≈4 ns.

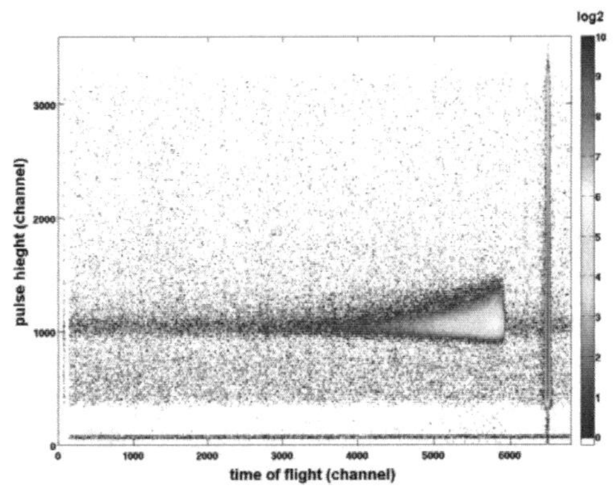

Figure 2. A 2D display of pulse height vs. TOF. Neutron-induced events appear in a horizontal band.

Figure 2 shows a 2D scatter plot of pulse height vs. TOF for the ^{6}Li-glass detector. The pulse height spectrum exhibits a neutron peak with a FWHM of ~15%. Figures 1 and 2 also contain scattered events, due to stray uncorrelated gamma rays and neutrons.

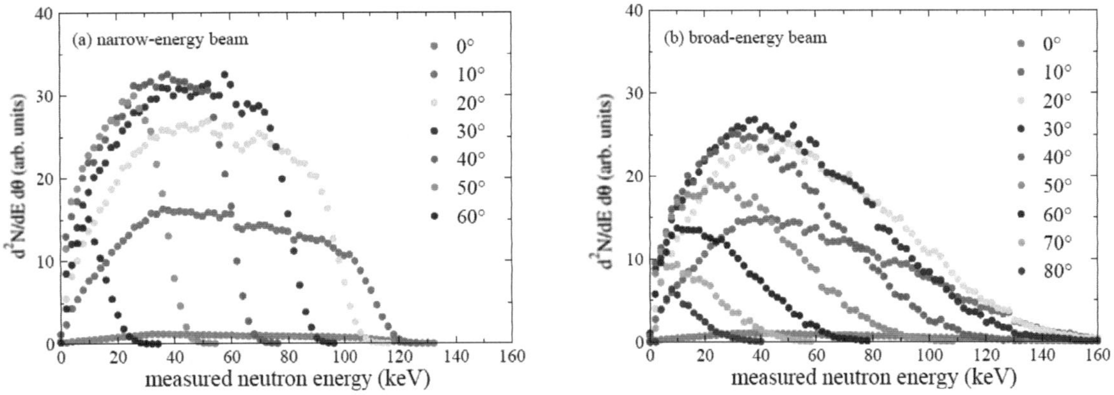

Figure 3. Measured neutron spectra at various angles for the ^{7}Li(p,n)^{7}Be reaction obtained with narrow-energy (a) and a broad-energy (b) proton beam.

The neutron angular spectra were measured for both the narrow-energy and broad-energy proton beam. Spectra were taken at all angles from 0° to 65° at 5° intervals for the narrow-energy beam, and from 0° to 80° at 5° intervals for the broad-energy beam. The stationary ^{6}Li-glass detector was used to monitor beam stability and for normalization. Conversion of TOF to neutron energy was through an iterative method which takes into consideration the finite thickness of the detector [6]. Figure 3 shows the neutron energy spectra at angles of 0°, 10°, ... , 60° for the data taken with the narrow-energy proton beam and 0°, 10°, … , 80° for the broad-energy proton beam (the double differential spectrum is integrated over 2π radians in azimuth). The individual neutron spectra taken with the broad-energy proton beam exhibit pronounced spectral broadening compared to the spectra taken with the narrow-energy proton beam. It should be noted that the measured neutron energies extend slightly beyond the kinematical endpoint due to detector thickness effects.

Figure 4 shows the combined neutron spectrum for the narrow-energy beam, and also the combined neutron spectrum for the broad-energy beam. These spectra were obtained by summing the spectra from 0°, 5°, … , 80° with appropriate normalization. The horizontal error bars reflect the uncertainty in energy due to the ^{6}Li-glass detector thickness. Although the individual spectra at each angle look markedly different for the narrow-energy and the broad-energy proton beams, the combined neutron spectra are remarkably similar, except in the high energy tail above 80 keV. Also shown in figure 4 is a fit to the distribution:

$$f(E) = a \cdot E \times e^{-E/kT}$$

corresponding to a Maxwellian flux at temperature T. For all the data from 10 to 80 keV, a best fit is obtained with kT=28 keV. The integrated neutron energy spectrum corresponding to the broad-energy proton beam shows considerably better agreement with the Maxwellian flux distribution at the higher neutron energies.

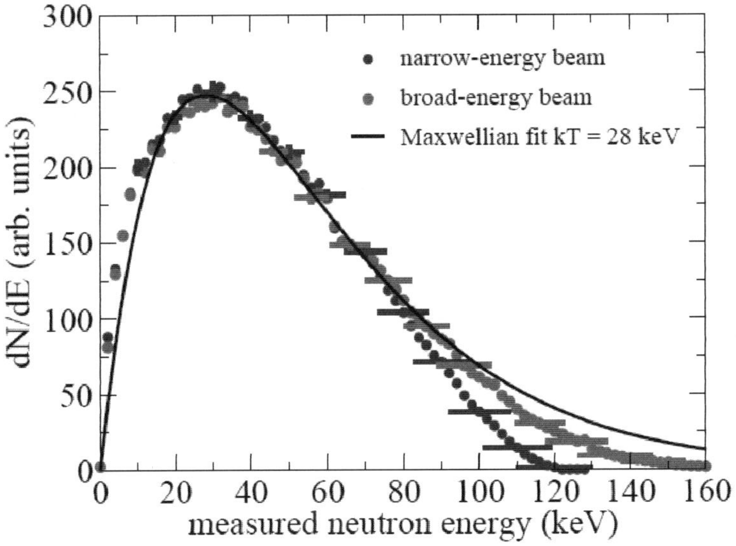

Figure 4. Measured neutron energy spectra for the ^{7}Li(p,n)^{7}Be reaction. Total angle integrated spectrum obtained for a narrow energy proton beam, $\sigma \approx 1.5$ keV (blue) and for a broad energy proton beam, $\sigma \approx 20$ keV (green). Both neutron spectra include the same normalization coefficient. Fit to a semi-Maxwellian yields kT=28 keV.

In the same experiment, measurements of the average cross section of the ^{197}Au(n,γ)^{198}Au reaction over the neutron energy distribution for both the narrow- and broad-energy spread beams were performed. The results are presently being analyzed.

This work is supported in part by the Pazi Foundation and by the German-Israeli Foundation (GIF).

References

[1] Ratynski W and Käppeler F 1988 *Phys. Rev.* C **37** 595
[2] Heil M, Käppeler F, Uberseder E, Gallino R and Pignatari M 2008 *Phys. Rev.* C **77** 015808
[3] Weissman L et al. 2010 *Proc. of linac 2010*, Tsukuba, WE102, pp 679-683
[4] Feinberg G et al. 2009 *Nucl. Phys.* A **827** 590c–592c
[5] Ziegler J and Biersack J SRIM 2008, www.srim.org
[6] Feinberg G et al, in preparation

^{197}Au(n,γ) - towards a new standard for energies relevant to stellar nucleosynthesis

C. Lederer[1], N. Colonna[2], I. Dillmann[3], C. Domingo-Pardo[4], U. Giesen[5], F. Gunsing[6], M. Heil[7], F. Käppeler[8], C. Massimi[9], A. Mengoni[10], M. Mosconi[5], R. Nolte[5], R. Reifarth[7,11], S. Schmidt[11], A. Wallner[1], and the n_TOF Collaboration (www.cern.ch/ntof)

[1]University of Vienna, Faculty of Physics, Waehringerstr. 17, 1090 Vienna, Austria.
[2]INFN - Sezione di Bari, V. Orabona 4, 70126 Bari, Italy.
[3]Physik Department E12 and Excellence Cluster Universe, Am Coulombwall 6, 85748 Garching, Germany.
[4]Instituto de Fisica Corpuscular, C/ Catedratico Jose Beltran, 2, 46980 Paterna, Spain.
[5]Physikalisch-Technische Bundesanstalt (PTB), Bundesallee 100, 38116 Braunschweig, Germany.
[6]CEA-Saclay, Irfu, 91191 Gif-sur-Yvette, France.
[7]GSI Helmholtzzentrum für Schwerionenforschung GmbH, Planckstr. 1, 4291 Darmstadt, Germany.
[8]Karlsruhe Institute of Technology (KIT), Campus Nord, Institut für Kernphysik, P.O. Box 3640, 76021 Karlsruhe, Germany.
[9]INFN - Sezione di Bologna and University of Bologna, V. Irnerio 46, 40126 Bologna, Italy.
[10]IAEA - Nuclear Data Section, Wagramer Str. 5, 1400 Vienna, Austria.
[11]Goethe University Frankfurt, Max-von-Laue-Str. 1, 60438 Frankfurt, Germany.

E-mail: `claudia.lederer@univie.ac.at`

Abstract. Two experiments were performed to resolve a discrepancy of the ^{197}Au(n,γ) cross section between the IAEA standard evaluation and the reference cross section for astrophysical experiments in the lower keV energy region. We measured the ^{197}Au(n,γ) reaction by means of the time-of-flight technique with n_TOF at CERN and extracted the cross section in the unresolved resonance region between 5-400 keV with an overall uncertainty from 3.9-6.7% for a resolution of 20 bins per energy decade. Additionally, we remeasured the neutron spectrum of the ^{7}Li$(p,n)^{7}$Be reaction with E_p=1912 keV, which was used as a neutron source for determining the astrophysical reference cross section. The n_TOF data are already fully analyzed and published. The results are in good agreement with the ENDF standard evaluation, but uncertainties don't allow to draw definite conclusions. The analysis of the neutron spectrum of ^{7}Li$(p,n)^{7}$Be is underway, preliminary results will be reported here.

1. Introduction

The ^{197}Au(n,γ) cross section is an accepted standard for thermal neutron energies and for neutron energies from 200 keV to 2.8 MeV [1]. The standard evaluation, based on several experimental datasets also covers the lower keV region starting from 2.5 keV, where the ^{197}Au(n,γ) is not considered as standard due to discrepancies of 6-8% relative to the Au cross section used as a reference in astrophysical experiments. This latter cross section is based on an activation measurement by Ratynski and Käppeler [2] in combination with a time-of-flight

measurement by Macklin *et al.* [3, 4].
The activation study was performed using the ^{7}Li$(p,n)^7$Be reaction as a neutron source with a proton energy, $E_p = 1912$ keV. Under these conditions, the emitted neutron spectrum resembles closely the spectrum required for measuring the Maxwellian Averaged Cross Section (MACS, energy dependent cross section averaged over the stellar neutron spectrum) at $kT \sim 25$ keV, thus simulating stellar conditions. Since the proton energy is only slightly above the reaction threshold of 1881 keV, neutrons are emitted in a forward peaked cone with 120 degrees opening angle. The Au sample was chosen to cover the entire neutron beam, allowing to directly determine the neutron flux by measuring the activity of the product ^{7}Be. The energy dependence of the cross section by Macklin *et al.* was used to correct for differences between the laboratory neutron distribution and a true Maxwellian spectrum. Ratynski and Käppeler found a MACS at $kT = 30$ keV of (582 ± 9) mb, in good agreement with the results obtained via the time-of-flight technique by Macklin *et al.*
In the following we report on recent experiments aimed at resolving the discrepancies mentioned above. One of them is the measurement of the ^{197}Au(n,γ) cross section at the time-of-flight facility n_TOF at CERN in the keV region [5]. As a second approach, we remeasured the neutron spectrum at Physikalisch-Technische Bundesanstalt Braunschweig (PTB) produced under the same conditions as the Ratynski and Käppeler experiment [6].

2. Measurement of ^{197}Au(n,γ) at n_TOF

The neutron time-of-flight facility n_TOF installed at CERN produces a highly intense, pulsed neutron beam by spallation reactions of 20 GeV/c protons provided by the CERN PS impinging on a massive, water moderated and cooled lead target. The short proton pulse width of 7 ns and the large neutron flight path of 185 m guarantee a high energy resolution. More information can be found elsewhere [7]. Neutron captures on ^{197}Au were studied for neutron energies from 1 eV-400 keV using a pair of C_6D_6 liquid scintillation detectors, detecting the prompt γ-emission following a neutron capture event. For energies up to 5 keV a resonance analysis was performed by Massimi *et al* [8]. Here we will concentrate on the results of the unresolved resonance region covering neutron energies from 5 to 400 keV [5].
Using the C_6D_6 detection system, the efficiency to detect a capture event, which depends on the details in the deexcitation path of the nucleus can be taken into account using the pulse-height-weighting-technique [9]. After applying this correction, the capture yield in the unresolved resonance region is obtained as

$$Y = f_{corr} \times \frac{C_W}{\Phi \times E_C}$$

where C_W denotes the weighted, background subtracted count-rate, Φ the neutron flux, E_C the excitation energy and f_{corr} takes into account corrections for the sample size. A careful analysis of the background components was performed by measurements and verified by Monte Carlo simulations including an extensive description of the experimental setup. The neutron flux was measured using a ^{235}U loaded parallel plate fission chamber from PTB [10] and continuously monitored by placing a Mylar foil deposited with Li in the beam and detecting the reaction products of ^{6}Li$(n,\alpha)^3$H with surrounding Si detectors. Furthermore, the yield was corrected for the fact that the sample is smaller than the neutron beam by normalizing the reaction yield to the 4.9 eV saturated resonance in Au. Multiple scattering and self shielding effects were calculated using the code SESH [11].
The neutron capture cross section could be extracted with a total uncertainty of 3.9-6.7% for a neutron energy resolution of 20 bins per decade. The results show good agreement with the ENDF standard evaluation as shown in Fig. 1. We also calculated the MACS from $kT = 5$ keV to $kT = 100$ keV. The MACS at 30 keV is in very good agreement with the MACS of the ENDF

standard evaluation. Our result is also in fair agreement with the value obtained by Ratynski and Käppeler (see Table 1).

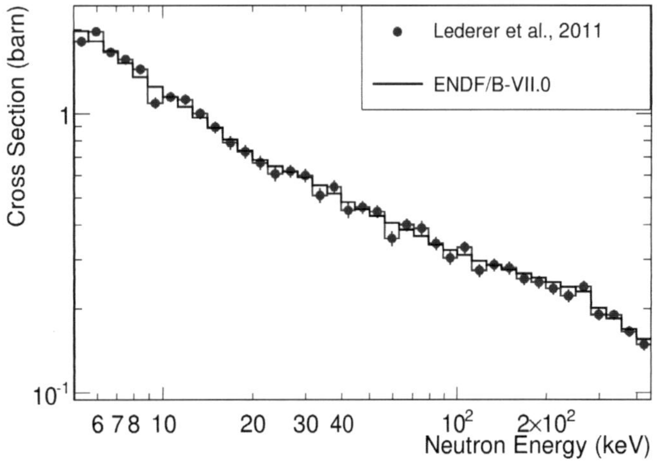

Figure 1. The ^{197}Au(n,γ) cross section measured at n_TOF [5] compared to the ENDF/B-VII standard evaluation [12]

Table 1. Maxwellian Averaged Cross Section at $kT = 30$ keV

Author	$<\sigma>_{30keV}$ (mb)
ENDF/B-VII [12, 13]	614.1
Ratynski and Käppeler, 1988 [2]	582 ± 9
Lederer *et al.*, 2011 [5]	611 ± 22

3. Measurement of the ^{7}Li$(p,n)^7$Be neutron spectrum at PTB

The neutron spectrum of the reaction ^{7}Li$(p,n)^7$Be was measured at PTB Braunschweig via the time-of-flight technique. A pulsed proton beam, 3 ns in width and (1912 ± 1) keV in energy was provided by the Van de Graaff Accelerator of the PTB Ion Accelerator Facility PIAF. The metallic Li target, evaporated onto a Ta backing was 565 μg/cm^2 in thickness, sufficient to slow down all protons below the reaction threshold. The time-of-flight spectra were recorded in steps of 5 degrees with a moveable Li-glass detector at angular positions from 0-65 degrees with respect to the proton beam. Two runs were performed using different flight paths of 35 cm and 70 cm. The shorter flight path provided a more detailed investigation of the low energy end of the spectra because of the reduction of wrap-around effects. A long counter at an angular position of 16 degrees relative to the proton beam and a distance of approximately 6 m from the target served as fluence monitor. The stability of the setup was checked by regular measurements at a reference position of 70 cm flight path and 0 degree angle.
The background and dead-time corrected time-of-flight spectra were transformed to an energy grid and normalized for the neutron fluence provided by the long counter. We applied corrections for the detection efficiency performing Monte Carlo simulations, taking into account the details of the experimental setup. The individual angular spectra were scaled for their solid angle and

summed. Figure 2 shows the summed spectrum at a flight path of 70 cm compared to the measurement by Ratynksi and Käppeler. The agreement is fairly well, our measurement shows deviations specifically in the high neutron energy range and around 20 keV. However, the effect on the averaged Au cross section due to these differences is negligibly small.

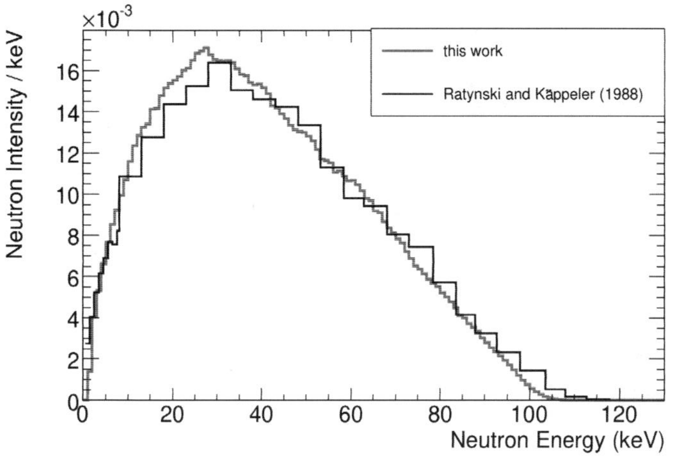

Figure 2. Angle integrated neutron spectrum measured at PTB compared to the results of Ratynski and Käppeler [2]. Both spectra are normalized to an area of 1.

4. Summary

We performed two different experiments to resolve the discrepancy of the ^{197}Au(n,γ) cross section between the ENDF standard evaluation and the reference cross section used for astrophysical experiments below 200 keV. Our results for the ^{197}Au(n,γ) cross section measured at n_TOF are in good agreement with the ENDF/B-VII standard evaluation, and, within the quoted uncertainty with the activation study by Ratynski and Käppeler and the time-of-flight measurement performed by Macklin *et al.* Preliminary results on the neutron spectrum of the ^{7}Li$(p,n)^7$Be reaction show only small deviations from the Ratynski and Käppeler measurement and thus this aspect cannot be regarded as a reason for a discrepancy of $6-8\%$ to the ENDF/B-VII standard evaluation. Currently, further activities are being pursued like a time-of-flight measurement of the Au cross section at the Geel-Electron-Linear-Accelerator at IRMM [14] and a study describing an independent definition of the ^{7}Li$(p,n)^7$Be spectrum and an activation measurement of the Au cross section in that spectrum [15]. Additional systematic studies concerning the ^{7}Li$(p,n)^7$Be spectrum and spectrum averaged cross sections are planned in framework of EUFRAT2010 [16].

Acknowledgments

This work was partly supported by the Austrian Science Fund (FWF), projects P20434 and I428, and by the European Commission within the Sixth Framework Programme through I3-EFNUDAT (EURATOM contract no. 036434).

[1] A.D. Carlson *et al.*, Nucl. Data Sheets, **110**, 3215 (2009)
[2] W. Ratynski and F. Käppeler, Phys. Rev. C **37**, 595 (1988).
[3] R.L. Macklin, J. Halperin, and R.R. Winters, Phys. Rev. C., **11**, 1270 (1975).
[4] R. Macklin, private communication to Mughabghab, S.F. (1982), see also www.nndc.bnl.gov/nndc/ EXFOR 12720.002 (unpublished).
[5] C. Lederer, and the n_TOF collaboration, Phys. Rev. C, **83** 034608 (2011).
[6] C. Lederer *et al.*, in preparation.
[7] U. Abbondanno *et al.*, Technical report CERN-SL-2002-053 ECT, *CERN n_TOF Facility: Performance Report*, CERN, Geneva (2003).

Journal of Physics: Conference Series **337** (2012) 012045 doi:10.1088/1742-6596/337/1/012045

[8] C. Massimi, C. Domingo Pardo, and the n_TOF collaboration, Phys. Rev. C, **81** 044616 (2010).
[9] R.L. Macklin and R.H. Gibbons, Phys. Rev. **159** 1007 (1967).
[10] C. Borcea *et al.*, Nucl. Instrum. Methods Phys. Res. A **513**, 524 (2003).
[11] F.H. Fröhner, SESH computer code, GA-8380, Gulf General Atomic, (1968).
[12] M.B. Chadwick *et al.*, *ENDF/B-VII.0: Next generation evaluated nuclear data library for nuclear science and technology*, Nucl. Data Sheets **107**, 2931 (2006).
[13] I. Dillmann, M. Heil, F. Käppeler, R. Plag, T. Rauscher, and F.-K. Thielemann, *KADoNiS - The Karlsruhe Astrophysical Database of Nucleosynthesis in Stars*, AIP Conf. Proc. **819** , 123; online at http://www.kadonis.org.
[14] C. Lampoudis, *et al.*, IRMM, Private Communication (2011).
[15] G. Feinberg, *et al.*, *Quasi-stellar neutrons from the* $^7Li(p,n)$ *reaction with an energy-broadened proton beam*, publication in preparation.
[16] F. Käppeler, *et al.*, *Neutron capture cross section of* $^{197}Au(n,\gamma)$ *at the VdG using a* $^7Li(p,n)$ *neutron beam*, EUFRAT proposal, 2010.

Nuclear Physics in Astrophysics V IOP Publishing
Journal of Physics: Conference Series **337** (2012) 012046 doi:10.1088/1742-6596/337/1/012046

Exotic Nuclei and Astrophysics

Yu.Penionzhkevich
Joint Institute for Nuclear Research, Dubna, Russia
E-mail: pyuer@lnr.jinr.ru

Abstract. In recent years, nuclear-physics investigations into the laws of the microscopic world have contributed to extend significantly our knowledge of phenomena occurring in the macroscopic world (Universe) and made a formidable contribution to the development of astrophysical and cosmological theories. First of all, this concerns the expanding-universe model, the evolution of stars, and the abundances of elements, as well as the properties of various stars and cosmic objects, including "cold" and neutron stars, black holes, and pulsars. Without claiming to give a full account of all cosmological problems, we will dwell upon those of them that, in my opinion, have much in common with nuclear-matter properties manifesting themselves in nuclear interactions.

1. NUCLEOSYNTHESIS

In the course of evolution, the Universe is permanently enriched in ever heavier chemical elements. A some what different mechanism governs the production of nuclei heavier than iron—this is the mechanism of sequential reactions of radioactive neutron capture by elements of the iron group. The fact that the curve of element abundances features double peaks in the curve of abundances of elements [1] suggests the occurrence of two different neutron-capture processes—these are so-called *r* and *s* processes, which correspond to different neutron densities. An enhanced abundance of the $N =$ 50, 82, and 126 isotopes confirms the existence of *r* and *s* processes. It was shown experimentally that the abundance of elements is in inverse proportion to the total neutron-capture cross sections. For nuclei characterized by magic numbers, these cross sections are several orders of magnitude smaller than the respective cross sections for other neighboring nuclei. From the point of view of nuclear physics, this result is a manifestation of magic numbers. For astrophysics, this proves the existence of the *s* process. Also, the existence of the $Z = 114$ shell and, accordingly, the improvement of stability of superheavy nuclei in the vicinity of the doubly magic nucleus $^{298}114$ (114 protons and 184 neutrons) were also predicted. Calculations on the basis of the shell model revealed the possible existence of this superheavy nucleus with a half-life of up to about 10^8 yr [2].

2. NUCLEAR-PHYSICS EXPERIMENTS AND ASTROPHYSICS

From the foregoing, it follows that energy deposition processes in the formation and explosion of stars and nucleosynthesis—they refer to the most fundamental problems in astrophysics—are directly related to particle physics and to nuclear-physics investigations of effective reaction cross sections, masses of nuclei, and their lifetimes with respect to various modes of their decay. Nuclear reactions involving the capture of protons, neutrons, or alpha particles by various nuclei, including unstable ones, play an important role in nucleosynthesis processes. It is difficult to determine experimentally the rate of such reactions. The cross section for such processes also depends strongly on the temperature of the object. In the non explosive evolution of a star, its temperature is relatively low, in which case the effective reaction cross section ranges between a few pb and a few nb units (10^{-36}–10^{-33} cm^2). In the case of the explosion of stars, the temperature is very high (about 10^9 K), in which case the effective reaction cross section reaches a few mb units (10^{-27} cm^2). In order to simulate such processes under laboratory conditions, we must have in our disposal a broad range of radioactive nuclei, which play a dominant role in the explosion of stars. In recent years, such a possibility appeared owing to the advent of accelerators for beams of radioactive nuclei. With the aid of such beams, one studies features of nuclear reactions, employing targets from hydrogen and helium. Another, indirect, means for determining reaction rates and for studying nucleosynthesis consists in obtaining, by nuclear-physics methods, spectroscopic data on nuclei far from the drip line. These data are used directly to describe *s* and *r* processes. High precision knowledge of the properties of decays of nuclei at the so-called turning points of the *r* process is of paramount importance here. In [3], the scheme of fast neutron capture in the region of S, Cl, and Ar nuclei was obtained on the basis of data from the joint Dubna (Russia)–GANIL (France)–Mainz (Germany) experiment in heavy-ion-accelerator beams. The half-lives ($T_{1/2}$) and neutron-decay probabilities (P_n) were measured in those experiments to a high precision for the neutron-rich isotopes of S, Cl, and Ar nuclei. The maximum time of irradiation with neutrons, the neutron flux, and the corresponding temperature of the star emitting neutrons on the time scale were calculated in accordance with this scheme (Fig.1).

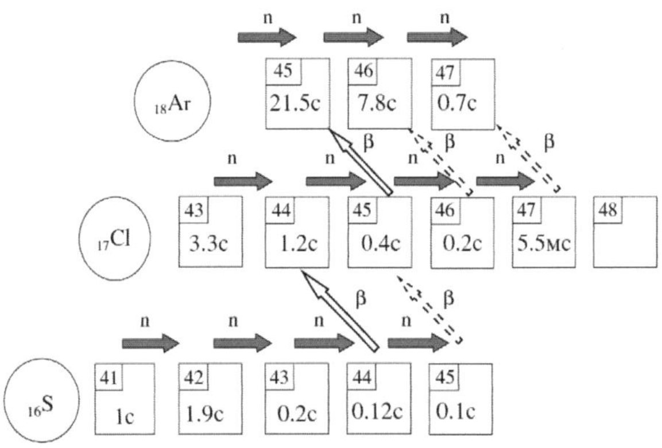

Figure.1 Neutron capture path in the $_{16}S$ to $_{18}Ar$ chains for stellar temperature of $T=8.10^8$ K and neutron density 5.10^{-5} mol cm^{-3} with our experimental data [3]. At both N=28 “turning point” isotopes, ^{44}S and ^{45}Cl, β- decay back to stability starts to dominate over further neutron capture. Hence, the possible A=46, 47 progenitors of $^{48}Ca/^{46}Ca$ ratios can be obtained, as required to explain the observed abundances in EK-1-4-1 meteorite.

The discovery and investigation of new forms of radioactive decay (beta-delayed fission, protonic decay, shape isomerism) are among the last successes of nuclear physics. The inclusion of these decay forms has a substantial effect on the calculation of the nuclide-production chain and is an important problem that relates nuclear physics and astrophysics. Reactions involving loosely bound nuclei and proceeding at energies close to the Coulomb barrier are of particular interest for astrophysics. These reactions have many special features, which have been discovered in recent years by using beams of radioactive nuclei. An enhancement of interaction cross sections in the subbarrier energy region is one of such features (Fig.2)

Journal of Physics: Conference Series **337** (2012) 012046 doi:10.1088/1742-6596/337/1/012046

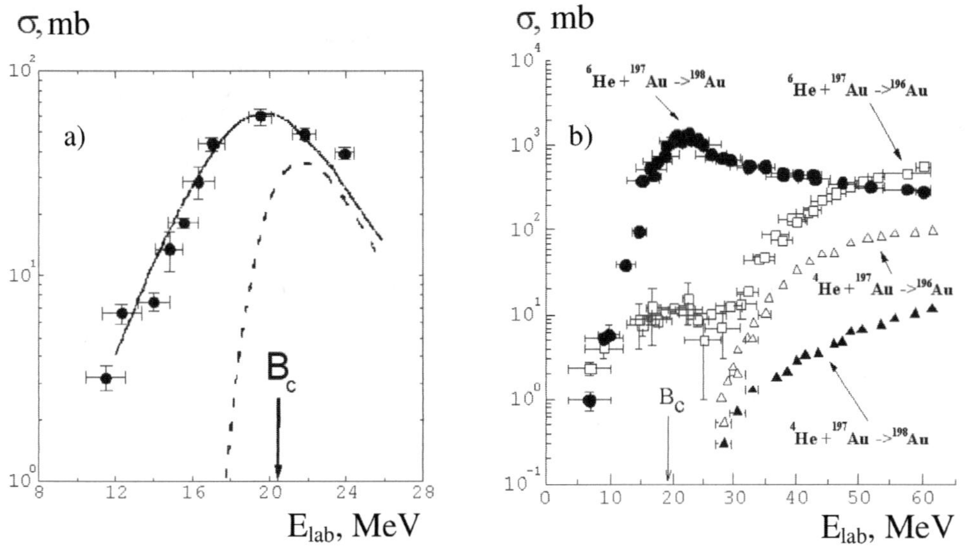

Figure.2 Excitation functions for the reactions (close boxes) ^{206}Pb(^{6}He, $2n$)^{210}Po (a). The dashed curves represent the results of the statistical-model calculations. Solid curve stands of the results of the calculation within the sequential-fusion model [4]. Excitation functions for ^{197}Au + ^{6}He reactions (b) leading to the production of the isotopes (open boxes) ^{196}Au and (closed cycles) ^{198}Au, open and closed triangulars are corresponding to the ^{4}He induced reactions. Here, B_c is the Coulomb barrier for ^{6}He + ^{197}Au reactions. For details, see Refs. 6 and 7.

This effect manifests itself most strongly for cluster nuclei (^{6,9}Li) and for nuclei involving a neutron halo (^{6}He) [5,6]. Deep-inelastic transfer reactions, breakup reactions, and complete-fusion reactions are dominant interaction channels for such nuclei. However, the fusion process has a more complicated character in the case of the interaction between loosely bound nuclei, because, with a high probability, such nuclei may undergo breakup followed by the capture of a residual nucleus (incomplete fusion). This complicates substantially the description of interaction in such systems and leads to new, unexpected, effects at energies in the vicinity of the Coulomb barrier, such as deep-subbarrier fusion and reactions of cluster transfer from loosely bound nuclei, which, as a rule, have a cluster structure. By way of example, we indicate that, in the reaction of neutron transfer in case of interaction of ^{6}He nuclei, the cross section reaches a value of several barns and has a maximum in the vicinity of the Coulomb barrier. A large cross section for the transfer of one neutron and its smooth decrease toward the region of low energies (to 5 MeV) may be indicative of the mechanism of interaction between a quasifree neutron of the ^{6}He nucleus with the target nucleus. A similar example is well known for *(d, p)* reactions, here, in the deep subbarrier region, there is a significant increase in the cross section (so-called Oppenheimer–Phillips resonance) because of the polarization of a loosely bound deuteron. In our case, this effect may be more pronounced because of a lower binding energy for the ^{6}He nucleus than for the deuteron and because of stronger Coulomb forces of repulsion of alpha particles in the ^{6}He nucleus in relation to the proton in the deuteron. It is of importance to obtain deeper insight into the mechanism of such reactions, especially with allowance for astrophysical aspects of such processes involving light nuclei at low energies. An interesting result was obtained recently on the basis of an analysis of excitation functions for deep inelastic transfer reactions on ^{6}Li nuclei [7]. This analysis revealed that the capture of a deuteron from ^{6}Li by the target nucleus is a dominant mechanism for this reaction. Thus, the excitation of the projectile nucleus in the field of the target nucleus and its polarization simplify the mechanism of deuteron capture, the Coulomb barrier for it being lower than in the case of alpha-particle capture. This mechanism is peculiar to the majority of loosely bound nuclei that have a cluster structure. The above results are of paramount importance for solving astrophysics problems- in particular, for obtaining deeper insight into the mechanism of production of light elements in the Universe. In nucleosynthesis, a large cross section for the interaction of cluster loosely bound nuclei (^{6}He, ^{9}Li, ^{7}Be, etc.) may change the chain of beta decays leading to the production of various elements and, hence, the whole scenario of nucleosynthesis [8]. New-generation radioactive-beam accelerating facilities also play an important role in the development of astrophysics studies. Only a few examples illustrating connections between nuclear physics and macroscopic physics have been given above. Despite the smallness of the number of particles involved in the formation of nuclei (not more 300), they constitute a unique system for simulating problems of the macroscopic world. Under laboratory conditions, nuclei that have unusual properties, so-called exotic nuclei (those that have a high temperature or a high angular momentum, those that consist of neutrons almost exclusively, and super deformed nuclei), which can in principle be produced in cosmic processes, are obtainable by means of nucleus–nucleus collisions implemented at present-day accelerators of heavy ions at energies of several hundred MeV units per nucleon. Fig. 3 shows the temperature of a nucleus as a function of its excitation energy, which is proportional to the projectile energy. These data were obtained in three experiments: in Dubna (Russia) [9], at low energies, at CERN (Switzerland) at excitation energies in the range 2–8 MeV, and in Darmstadt (Germany) at energies in the range 8–15 MeV. One can see that, as the excitation energy increases in the range 2–8MeV, the nuclear temperature does not increase, but, at low energies, the temperature increases sharply, in just the same way as it does at very high energies. This sensational result first demonstrates that one may reach, in nuclear reactions, temperatures that exists only at the instant of star formation and second confirms the occurrence of phase transitions in hot objects, which was assumed in considering the expanding universe and the explosion of stars.

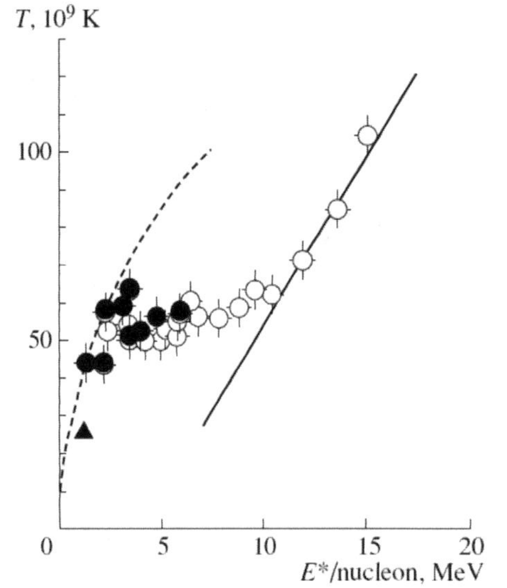

Figure 3. Experimental dependences [(closed triangles) JINR (Dubna), (open circles) GSI (Darmstadt, Germany), and (closed circles) CERN (Switzerland), see for details Ref. 9] of the temperature T of a nucleus on its excitation energy. The solid and dashed curves represent the dependences in question for a perfect gas and a liquid, respectively.

In the programs for research centers involved in the construction of new-generation accelerator facilities, including accelerators of intense beams of radioactive nuclei, the production of new exotic nuclei and investigation of their properties with an eye to simulating processes proceeding in the Universe are among the main tasks.

3. ACKNOWLEDGMENTS
This work was supported by the Russian Foundation for Basic Research (projects no. 07-02-00271-a, 09- 02- 91056).

4. REFERENCES

[1] D.G.Baranov et al. , Izv. Akad. Nauk., Ser. Fiz., 69, 832, 2005
[2] Yu.Ts.Oganessian, J.Mod.Phys. e**16**, 949, 2007
[3] O.Sorlin et al., Phys.Rev. C **47**, 2941, 1993.
[4] V.I.Zagrebaev, Phys.Rev.C **67**, 061601 (R), 2003
[5] M.Dasgupta et al., Phys.Rev. C **70**, 024606, 2004
[6] Yu.E.Penionzhkevich et al., Eur.Phys.J. A **31**, 185, 2007
[7] Yu.Penionzhkevich et al. J.Phys.G **36**, 025104, 2009
[8] V.Zagrebaev et al., Phys.Rev. C **75**, 035809, 2007
[9] Yu.Penionzhkevich, Pys. Atomic Nucl.,v.**73**, n.8, 1460, 2010

Current Status of the ^{22}Ne+α s-Process Neutron Source

Richard Longland[1] **and Christian Iliadis**[2]

[1]Departament de Física i Enginyeria Nuclear, EUETIB, Universitat Politècnica de Catalunya, 08036 Barcelona, Spain

[2]Department of Physics and Astronomy, University of North Carolina at Chapel Hill, Chapel Hill, NC 27599, USA

E-mail: richard.longland@upc.edu

Abstract. The reaction rates of the s-process neutron producing ^{22}Ne(α,n)^{25}Mg reaction and its competing ^{22}Ne(α,γ)^{26}Mg reaction are needed to accurately predict nucleosynthesis in massive stars and Asymptotic Giant Branch stars. Here, we present a re-evaluation of the reaction rates by incorporating recent data and by using a Monte Carlo uncertainty propagation method. The effect of those results on nucleosynthesis in massive stars is studied. We show that our new rates lead to similar final abundances, but with significantly reduced uncertainties in comparison to literature rates.

1. Introduction

Neutron capture reactions are the main producer of nuclei heavier than iron in our solar system; slow neutron captures (the s-process) being responsible for a large fraction of them. The ^{22}Ne(α,n)^{25}Mg reaction is a key source of neutrons in both the main and weak components of the s-process. For the main component, where the ^{13}C(α,n)^{16}O reaction produces the majority of neutrons, the ^{22}Ne(α,n)^{25}Mg reaction affects mainly branchings in the s-process path. For the weak component, the ^{22}Ne(α,n)^{25}Mg reaction is the main source of neutrons. Knowing the ^{22}Ne(α,n)^{25}Mg reaction rates in these environments *as well as* those of its competing ^{22}Ne(α,γ)^{26}Mg reaction are therefore essential to fully understand s-process nucleosynthesis.

Recent experimental results motivate a complete re-evaluation of the available nuclear physics data for the ^{22}Ne+α reactions. Here we present the results of that evaluation using a Monte Carlo method for propagation of uncertainties. The impact of these results on nucleosynthesis in massive stars will also be discussed.

2. ^{22}Ne+α Reaction Rates

Rates for the ^{22}Ne+α reactions based on experimental results were published in Refs. [1] and [2] for the ^{22}Ne(α,γ)^{26}Mg and ^{22}Ne(α,n)^{25}Mg reactions respectively. Since those publications, a number of improvements have been made in the available experimental data concerning properties of states in the ^{26}Mg compound nucleus. Most notable are the ^{25}Mg(n,γ)^{26}Mg analysis by Ref. [3] and the ^{26}Mg(γ,γ')^{26}Mg results published in Refs. [4, 5]. The former experiment used an R-matrix fit to assign quantum numbers and spins to important states, as well as determining partial widths that allow us to reliably integrate wide resonances numerically in reaction rate calculations. The ^{26}Mg(γ,γ')^{26}Mg experiment resolved definite spin-parities of several states above and below the neutron threshold. Most significantly, they found that a state corresponding to a α-particle resonance energy of $E_r^{\rm lab}$ = 630 keV

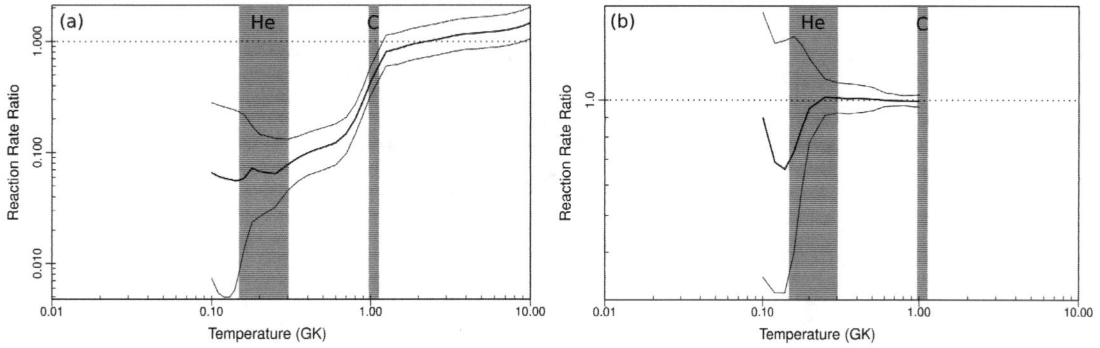

Figure 1. Comparison of our new rates for (a) the ^{22}Ne(α,γ)^{26}Mg reaction in comparison to those presented in Ref. [1] and (b) the ^{22}Ne(α,n)^{25}Mg reaction in comparison to those presented in Ref. [2]. The thick solid line represents our recommended rate while the thinner lines represent our 68% uncertainties. The red regions indicate the temperatures most important for helium ('He') and carbon ('C') burning.

has an unnatural parity of $J^{\pi} = 1^{-}$. It therefore cannot contribute to the reaction rate, contrary to previous assumptions.

The calculation of reaction rates is well understood, and is presented in detail elsewhere (see, for example, Ref. [6]). However, it is important to also calculate statistically meaningful uncertainties on the rates for them to be useful in determining the reliability of nucleosynthesis predictions. To calculate the rates for the ^{22}Ne+α reactions, we have employed a Monte Carlo method, which takes into account the probability density functions of every nuclear physics input [7].

The reaction rates are compared with the literature results of Refs. [1] and [2] in Fig. 1. Although our ^{22}Ne(α,n)^{25}Mg rates do not differ significantly from those of Ref. [2] (they are slightly lower because the E_r^{lab} = 630 keV resonance is now removed), the ^{22}Ne(α,γ)^{26}Mg rates are significantly lower, which should lead to a new increase in the neutron flux. The reason for this discrepancy is mostly because the contribution of wide resonances did not include experimental values for the resonance widths in the literature rates. The measured widths from Refs. [2] and [3] are found to be significantly smaller than those assumed in Ref. [1] and hence the rate at low temperatures is dramatically reduced.

3. Impact on Nucleosynthesis

The effect of our new rates for the ^{22}Ne+α reactions have been calculated for nucleosynthesis during convective core helium burning in massive stars. We have utilised a single zone temperature-density profile for the core helium burning stage of $25M_{\odot}$ star from Ref. [8]. The impact on nucleosynthesis yields is shown in Fig. 2.

The most significant change is in the production of ^{26}Mg, whose abundance at the end of core helium burning is about 5 times less than with the previous rates. The p-process nuclides: ^{74}Se, ^{78}Kr, and ^{84}Sr are also present in lower quantities by greater destruction through neutron capture. The uncertainties in p-nuclide abundances is, however, still quite large. With our new rates, the increased neutron flux does not extend the reach of the weak s-process component, but rather increases production of nuclei situated around the iron peak by a small amount. One reason for this insensitivity of nucleosynthesis to neutron flux could be that the neutron poison, ^{14}N(n,γ)^{15}N, dilutes the effect. In addition to modified yields, the uncertainties in nucleosynthesis have been significantly reduced by roughly an order of magnitude with the new reaction rates.

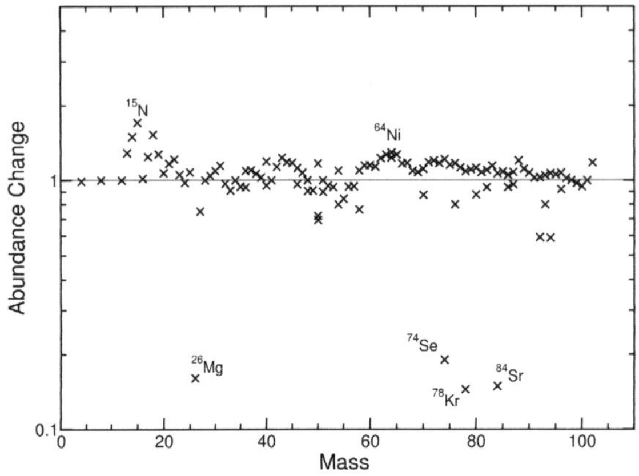

Figure 2. Change in isotopic abundances produced at the end of core helium burning in comparison to those obtained using the literature rates. Points above unity represent an increase in production when using the new rates.

4. Conclusions

The ^{22}Ne+α reactions are an important source of neutrons for the s-process in massive stars and AGB stars. Uncertainties in their rates can therefore give rise to large uncertainties in s-process nucleosynthesis of nuclei heavier than iron. Here, we have presented the reaction rates that arise from a full re-evaluation of the rates using up-to-date nuclear data and a Monte Carlo uncertainty propagation method.

We have shown that our new rates affect the production of nuclei around the iron peak, but most significantly affect the production of ^{26}Mg, which is reduced by a factor of five at the end of core helium burning for our massive star model. The uncertainties in nucleosynthesis are also reduced significantly with our new rates. However, we must emphasise that further studies should be performed to resolve ambiguities in the nuclear physics data, particularly for low energy resonances below α-particle bombarding energies of $E_r^{\rm lab}$ = 1000 keV.

References

[1] Angulo C, Arnould M, Rayet M, Descouvemont P, Baye D, Leclercq-Willain C, Coc A, Barhoumi S, Aguer P, Rolfs C, Kunz R, Hammer J W, Mayer A, Paradellis T, Kossionides S, Chronidou C, Spyrou K, Degl'Innocenti S, Fiorentini G, Ricci B, Zavatarelli S, Providencia C, Wolters H, Soares J, Grama C, Rahighi J, Shotter A and Lamehi Rachti M 1999 *Nucl. Phys. A* **656** 3–183

[2] Jaeger M, Kunz R, Mayer A, Hammer J W, Staudt G, Kratz K L and Pfeiffer B 2001 *Phys. Rev. Lett.* **87** 202501

[3] Köhler P E 2002 *Phys. Rev. C* **66** 055805

[4] Longland R, Iliadis C, Rusev G, Tonchev A P, DeBoer R J, Görres J and Wiescher M 2009 Phys. Rev. C **80** 055803

[5] Deboer R J, Wiescher M, Görres J, Longland R, Iliadis C, Rusev G and Tonchev A P 2010 *Phys. Rev. C* **82** 025802

[6] Iliadis C 2007 *Nuclear physics of stars* (Wiley-VCH)

[7] Longland R, Iliadis C, Champagne A E, Newton J R, Ugalde C, Coc A and Fitzgerald R 2010 Nucl. Phys. A **841** 1–30 (*Preprint* 1004.4136)

[8] The L, El Eid M F and Meyer B S 2000 *Astro. Phys. J* **533** 998–1015

Nuclear Physics in Astrophysics V IOP Publishing
Journal of Physics: Conference Series **337** (2012) 012048 doi:10.1088/1742-6596/337/1/012048

Cross-Section Measurements of the ^{86}Kr(γ, n) Reaction to Probe the *s*-Process Branching at ^{85}Kr

R Raut[1,5], A Banu[2], C Iliadis[3,5], J H Kelley[4,5], G Rusev[1,5], R Schwengner[6], A P Tonchev[1,5], W Tornow[1,5]

[1] Duke University, Durham, North Carolina 27708-0308, USA
[2] James Madison University, Harrisonburg, Virginia 22807, USA
[3] University of North Carolina at Chapel Hill, Chapel Hill, North Carolina 27695-8202, USA
[4] North Carolina State University, Raleigh, North Carolina 27695-8202, USA
[5] Triangle Universities Nuclear Laboratory, Durham, North Carolina 27708-0308, USA
[6] Institut für Strahlenphysik, Forschungzentrum, Dresden-Rossendorf, D-01314 Dresden, Germany

E-mail: `raut@tunl.duke.edu`

Abstract. For the first time, cross-section measurements were carried out for the ^{86}Kr(γ, n) reaction in order to probe the *s*-process branching point nucleus ^{85}Kr. The branching point nuclei in the *s*-process path are of importance in testing and constraining the nucleosynthesis models. Cross-section measurement for the photoneutron reaction on the neighbouring stable isotope is carried out to deduce the aforesaid neutron capture cross-section, as has been adopted in the present case of ^{85}Kr branching point. The cross-section results from the ^{86}Kr(γ, n) reaction were compared with the statistical model calculations using TALYS code.

1. Introduction

The *s*-process branching point nuclei, with the branching ratio depending on the stellar conditions (temperature, neutron density) are excellent tests to the nucleosynthesis models [1]. The particular branching at ^{85}Kr is significant since it is independent of temperature and depends only on the neutron density [2]. Knowledge of this parameter is crucial, in particular, for constraining models of the Asymtotic Giant Branch (AGB) stars. However, extraction of the neutron density from isotopic abundance ratios (for ^{82}Kr and ^{86}Kr) that has been measured in solar system materials or in presolar grains of primitive meteorites demands accurate knowledge of the ^{85}Kr(n,γ)^{86}Kr cross section. Unfortunately, the nucleus ^{85}Kr is unstable ($T_{1/2}$ = 10.76 years) and the ^{85}Kr(n,γ) reaction has not been measured yet. Consequently, the neutron capture cross section of ^{85}Kr is listed with a factor of 2 uncertainty even in the latest Kadonis V0.3 [3] evaluation of the (n,γ) rates. It has been shown by M. Lugaro [4], for example, that the current models of the AGB stars can easily reproduce the whole range of the ^{86}Kr/^{82}Kr ratios observed in the SiC presolar grains simply by varying the ^{85}Kr(n,γ) cross section by a factor of 2.
Owing to the short β-decay half-life of the branching point nucleus, it is not possible to carry out neutron capture experiments, apart from the general inconvenience of working with radioactive targets. Theoretical estimates based on statistical model calculations vary considerably, based on the choice of parametrizations of level density, gamma ray strength functions and neutron-nucleus optical potential. Neuton capture cross sections on branching point nuclei, nevertheless,

Figure 1. The ^{86}Kr gas target and the ^{3}He proportional counter, used in the present measurements

can be derived from probing the (γ,n) photodisintegration cross section of the neighbouring stable nucleus. The impetus is to find a parameter set that reproduces the (γ,n) cross sections and to apply the same set for the reverse (n,γ) reaction to predict the capture cross section on the neighbouring branching point nucleus. Such prediction is expected to be more accurate compared to choice of parameters based on global or local systematics used to derive the parameters from the neighbouring nuclei [1].

2. Experiment and Data Analysis

The ^{86}Kr(γ, n) cross-section measurements were carried out at the High Intensity Gamma-Ray Source (HIγS) facility located on the campus of Duke University. The facility has a 0.24-1.2 GeV storage ring where quasi-monoenergetic γ-beam is produced by the intracavity Compton back-scattering of the free electron laser photons from the electron bunches. At present HIγS delivers the most intense monoenergetic γ-beams, with an average flux of $\sim 10^{7}$-10^{8} γ/s in the energy range from 1 to 20 MeV and typical energy spread (FWHM) of 1-3%. The latter depends on the collimator diameter used for particular measurements [6]. The first measurements were carried out at beam energies E_γ = 9.970±0.238, 10.115±0.275, 10.170±0.234, 10.290±0.270, 10.360±0.277, 10.490±0.258, and 11.005±0.250 MeV.

The target was 1012 mg of Kr gas enriched to 99.4% in ^{86}Kr, contained in a stainless steel cell as shown in Fig. 1. An empty cell of identical material and dimensions was used to subtract the background contribution. The emitted neutrons from the (γ, n) reaction was detected using a 4π assembly of ^{3}He proportional counters, fabricated into a single unit, also shown in Fig. 1. The assembly consists of 18 individual counters, arranged in two concentric rings. Each counter was filled with ^{3}He gas at 6 atm pressure. The space around the counters was stuffed with polyethelene that themalizes the emitted neutrons before they are captured (detected) in the ^{3}He counters. The counter produces TTL outputs corresponding to the detected events, that were recorded on a scaler in a VM-USB based data acquisition system. The detector has an axial cylidrical hollow wherein the target was placed for the present measurements. The efficiency characteristics of the detector were studied previously [5] and were used in the present work. The γ-flux incident on the target was measured using a plastic scintillator that was cross-calibrated at three different γ-energies, between 10.0 and 11.0 MeV using the Au-foil activation [6].

At each beam energy, consecutive measurements were carried out using the gas cell and the empty cell and the net number of neutrons from the ^{86}Kr(γ, n) reaction was extracted. The photoneutron cross-section was determined from the formula,

$$\sigma(E_\gamma) = \frac{N_n}{N_\gamma N_t \epsilon_n g} \tag{1}$$

where N_n is the net number of neutrons, N_γ is the number of incident γ-rays, N_t is the number of target atoms, ϵ_n is the neutron detection efficiency and g is the fraction of the incident γ-rays with energy higher than the neutron threshold (S_n=9.86 MeV).

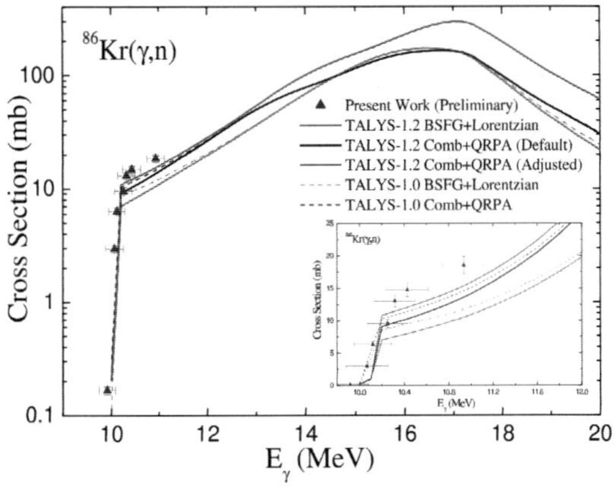

Figure 2. Experimental cross-sections for ^{86}Kr(γ, n) reaction plotted alongwith the calculated exciation functions using the TALYS code. The different parameter sets used in the calculations are explained in the text

3. Results and Discussions

The experimental ^{86}Kr(γ, n) cross-section from the present measurement is plotted in Fig. 2, along with the statisical model calculations using the TALYS [7] code. Two sets of parameters, differeing in the choice of the Nuclear Level Density (NLD) and Gamma Strength Function (GSF) models, were used to fit the experimental results. One set of parameters used the macroscopic models, generalized energy dependent Lorentzian model for GSF and back-shifted Fermi gas (BSFG) model for NLD. The other set comprised of microscopic models, Skryme-Hartree-Fock-Bogoliubov plus quasiparticle Random Phase Approximation (QRPA) for GSF and Hartree-Fock-Bogoliubov plus a combinatorial method, adopted for NLD. Apart from the calculations with the default values, the microscopic GSF was adjusted by scaling factors in TALYS-1.2, for better agreement with the measured values. It was generally observed that the microscopic models led to better agreement with the experimental data, as illustrated in Fig. 2.

This was the first measurement of the low energy tail of the Giant Dipole Resonance (GDR) of the ^{86}Kr(γ, n) reaction. The GDR for ^{86}Kr has not been experimentally measured, till day. The GDR parameters used in the aforesaid statistical model calculations were from the systematics. This could have resulted in the observed differences between the experimental results and the model predictions. Measurements are in progress at higher energies to map the entire GDR distribution. The experimentally derived GDR can help constrain the statistical model predictions and result in a credible parameter set that can be subsequently applied to extract the neutron capture cross section on ^{85}Kr branching point nucleus. It is noteworthy to mention that at incident photon energies just above the neutron separation energy, the GDR seldom decays into the high spin ground state ($J^{\pi}=9/2^{+}$) of ^{85}Kr, since such a decay would require f-wave neutrons that are hindered by small penetrability. More significant role is expected by the low spin isomer ($J^{\pi}=1/2^{-}$) of ^{85}Kr. Measurements are underway to investigate the exclusive isomeric cross section of ^{86}Kr$(\gamma, n)^{85m}$Kr reaction using activation technique.

This work is supported by the US DOE grants DE-FG52-09NA29448, DE-PS52-08NA28920 and DE-FG02-97ER41033.

References

[1] F. Kappeler *Prog. Part. Nucl. Phys.* **43**, 419(1999)
[2] K. Takahashi *et al. At. Nucl. Data Tables* **36**, 375(1987)
[3] Karlsruhe Astrophysical Database of Nucleosynthesis in Stars, *www.kadonis.org*
[4] Maria Lugaro *Stardust from Meteorites*, World Scientific, 2005
[5] C.W. Arnold *et al.*, *Nucl. Instr. Meth. Phys. Res. A* doi:10.1016/j.nima.2011.01.177(2011)
[6] A.P. Tonchev *et al. Phys. Rev. C* **82**, 054620(2010)
[7] A.J. Koning, S. Hillaire, M.C. Duijvestijn *AIP Conference Proceedings* **769**, 1154(2005)

Nuclear Physics in Astrophysics V IOP Publishing
Journal of Physics: Conference Series **337** (2012) 012049 doi:10.1088/1742-6596/337/1/012049

The Last Neutrino Mixing angle θ_{13}

A.B. Balantekin

Physics Department, University of Wisconsin, Madison, WI 53706 USA

E-mail: baha@physics.wisc.edu

Abstract. Among the still unmeasured neutrino properties, the third neutrino mixing angle, θ_{13}, is likely to be the one we will next find out. In this contribution, first a brief summary of the limits and the preliminary measurements of this angle is given. Second a critical assessment of a widely used formula connecting two- and three-flavor evolution is provided.

1. Introduction

Solar, accelerator, and reactor experiments performed during the last two decades firmly established that neutrinos mix. In general, for three flavors, the neutrino mixing matrix contains three mixing angles, one CP-violating phase and two Majorana phases. The Majorana phases are accessible only in double beta decay experiments. Two of the mixing angles are well measured. From solar and reactor experiments we know the value of one of them (usually denoted θ_{12}). In the atmospheric neutrino experiments we measured the other one (usually denoted θ_{23}). Contrary to the original (and naive) expectations, both angles are large. In fact one gets $\theta_{23} \sim \pi/4$. To have the complete information on neutrino mixing, we still need to know three more pieces of information: i) the value of the remaining mixing angle, θ_{13}, ii) the mass hierarchy; and iii) the value of the CP-violating phase. Of these three, the remaining mixing angle is likely to be the next piece to be measured.

One popular parametrization of the neutrino mixing matrix is

$$\mathbf{T}_{23}\mathbf{T}_{13}\mathbf{T}_{12} = \begin{pmatrix} 1 & 0 & 0 \\ 0 & C_{23} & S_{23} \\ 0 & -S_{23} & C_{23} \end{pmatrix} \begin{pmatrix} C_{13} & 0 & S_{13}e^{-i\delta_{CP}} \\ 0 & 1 & 0 \\ -S_{13}e^{i\delta_{CP}} & 0 & C_{13} \end{pmatrix} \begin{pmatrix} C_{12} & S_{12} & 0 \\ -S_{12} & C_{12} & 0 \\ 0 & 0 & 1 \end{pmatrix} \quad (1)$$

where $C_{ij} = \cos\theta_{ij}$, $S_{ij} = \sin\theta_{ij}$, and δ_{CP} is the CP-violating phase. Note that although the CP-violating phase multiplies S_{13} in Eq. (1), there is nothing special about this choice; one could have placed δ elsewhere in the mixing matrix. In fact the combination

$$P(\bar{\nu}_\mu \to \bar{\nu}_e) - P(\nu_\mu \to \nu_e), \quad (2)$$

which quantifies the CP-violation in neutrino oscillations, is proportional to the sines of *all* the mixing angles. However, since we know the other mixing angles are *non-zero*, observation of CP-violation in neutrino oscillations hinges on a non-zero value of θ_{13}.

The current limit on θ_{13} quoted by the Particle Data Group is $\sin^2(2\theta_{13}) < 0.15$ at 90% C.L. [1]. There are strong hints for a non-zero value of θ_{13}. Already several years ago, it was pointed out that the slight discrepancy between solar neutrino experiments and the KamLAND reactor neutrino experiment could be explained by a small, but non-zero value of θ_{13} [2]. Indeed global fits indicate a slight preference for a non-zero value of θ_{13} [3]. (A recent brief summary of the θ_{13} phenomenology is given in [4]). Recent low-threshold analysis of the Sudbury Neutrino Observatory yields $\theta_{13} = 7.2^{+2.0}_{-2.8}$ degrees, albeit with non-Gaussian errors [5]. A global analysis by the KamLAND collaboration indicates $\sin^2\theta_{13} = 0.009^{+0.013}_{-0.007}$ at the 79% C.L. [6]. Currently three reactor neutrino experiments (Double Chooz [7], Daya Bay [8], and RENO [9]) as well as the T2K accelerator experiment [10] are working on a measurement of this angle. So far the best hint for a non-zero value of θ_{13} came from the T2K experiment which announced a 2.5σ result: $0.03 < \sin^2 2\theta_{13} < 0.28$ for the normal and $0.04 < \sin^2 2\theta_{13} < 0.34$ for the inverted hierarchies, both for $\delta_{CP} = 0$ [11].

2. Reduction of three-flavor evolution

To extract θ_{13} from solar neutrino data many times the formula

$$P_{3\times3}(\nu_e \to \nu_e) = \cos^4\theta_{13}\, P_{2\times2}(\nu_e \to \nu_e \text{ with } N_e \cos^2\theta_{13}) + \sin^4\theta_{13} \tag{3}$$

is used. Since many fits to solar neutrino data that give a non-zero value of of θ_{13} use Eq. (3), here we wish to give an assessment of the accuracy of this formula. To do so, following Ref. [12], we introduce the combinations

$$\tilde{\Psi}_\mu = \cos\theta_{23}\Psi_\mu - \sin\theta_{23}\Psi_\tau, \tag{4}$$

$$\tilde{\Psi}_\tau = \sin\theta_{23}\Psi_\mu + \cos\theta_{23}\Psi_\tau, \tag{5}$$

which leads to the MSW evolution equation

$$i\frac{\partial}{\partial t}\begin{pmatrix}\Psi_e\\ \tilde{\Psi}_\mu\\ \tilde{\Psi}_\tau\end{pmatrix} = \tilde{\mathbf{H}}\begin{pmatrix}\Psi_e\\ \tilde{\Psi}_\mu\\ \tilde{\Psi}_\tau\end{pmatrix} \tag{6}$$

where

$$\tilde{\mathbf{H}} = \mathbf{T}_{13}\mathbf{T}_{12}\begin{pmatrix}E_1 & 0 & 0\\ 0 & E_2 & 0\\ 0 & 0 & E_3\end{pmatrix}\mathbf{T}_{12}^\dagger\mathbf{T}_{13}^\dagger + \begin{pmatrix}V_{e\mu} & 0 & 0\\ 0 & 0 & 0\\ 0 & 0 & 0\end{pmatrix}. \tag{7}$$

In writing Eq. (7), since it is proportional to the identity, the neutral current contribution is dropped and

$$V_{e\mu}(x) = \sqrt{2}G_F N_e(x). \tag{8}$$

To write the evolution operator associated with Eq. (6),

$$i\frac{\partial}{\partial t}\tilde{\mathbf{U}} = \tilde{\mathbf{H}}\tilde{\mathbf{U}}, \tag{9}$$

following the procedure of Ref. [13] we introduce the evolution operator $\mathbf{U}$:

$$\tilde{\mathbf{U}} = \mathbf{T}_{13}\mathbf{U}\mathbf{T}_{13}^\dagger \tag{10}$$

which satisfies the equation

$$i\frac{\partial}{\partial t}\mathbf{U} = \mathcal{H}\mathbf{U} \tag{11}$$

with

$$\mathbf{T}_{13}^{\dagger}\tilde{\mathbf{H}}\mathbf{T}_{13} = \mathcal{H} = \begin{pmatrix} \frac{1}{2}\tilde{V} - \Delta_{21}\cos 2\theta_{12} & \frac{1}{2}\Delta_{21}\sin 2\theta_{12} & \frac{1}{2}V_{\mu e}\sin 2\theta_{13} \\ \frac{1}{2}\Delta_{21}\sin 2\theta_{12} & -\frac{1}{2}\tilde{V} + \Delta_{21}\cos 2\theta_{12} & 0 \\ \frac{1}{2}V_{\mu e}\sin 2\theta_{13} & 0 & \frac{1}{2}(\Delta_{31}+\Delta_{32}) + V_{\mu e} - \frac{3}{2}\tilde{V} \end{pmatrix}, \tag{12}$$

where we took the CP-violating phase to be zero and introduced the modified matter potential

$$\tilde{V} = V_{\mu e}\cos^2\theta_{13} \tag{13}$$

as well as the quantity

$$\Delta_{ij} = \frac{m_i^2 - m_j^2}{2E} = \frac{\delta m_{ij}^2}{2E}. \tag{14}$$

Writing the Hamiltonian in Eq. (12) as

$$\mathcal{H} = \mathcal{H}_0 + \mathcal{H}_1 \tag{15}$$

with

$$\mathcal{H}_0 = \begin{pmatrix} \frac{1}{2}\tilde{V} - \Delta_{21}\cos 2\theta_{12} & \frac{1}{2}\Delta_{21}\sin 2\theta_{12} & 0 \\ \frac{1}{2}\Delta_{21}\sin 2\theta_{12} & -\frac{1}{2}\tilde{V} + \Delta_{21}\cos 2\theta_{12} & 0 \\ 0 & 0 & \frac{1}{2}(\Delta_{31}+\Delta_{32}) + V_{\mu e} - \frac{3}{2}\tilde{V} \end{pmatrix}, \tag{16}$$

and

$$\mathcal{H}_1 = \frac{1}{2}\sin 2\theta_{13} V_{\mu e}\begin{pmatrix} 0 & 0 & 1 \\ 0 & 0 & 0 \\ 1 & 0 & 0 \end{pmatrix}, \tag{17}$$

we see that Eq. (11) can be solved perturbatively, assuming θ_{13} is small. (In seeking such solutions we will take all the evolution operators at the neutrino production (t=0) to be the identity operator). The contributions which are lowest order in θ_{13} will come from from the quantity $\tilde{V}$ and from the transformation in Eq. (10). Eq.(16) implies that if θ_{13} were zero, $\tilde{\Psi}_\tau$ would decouple from the other flavors. Writing $\mathbf{U} = \mathbf{U}_0\mathbf{U}_1$, the solution of

$$i\frac{\partial}{\partial t}\mathbf{U}_0 = \mathcal{H}_0\mathbf{U}_0 \tag{18}$$

is

$$\mathbf{U}_0 = \begin{pmatrix} \mathbf{S} & 0 \\ 0 & \beta \end{pmatrix}, \tag{19}$$

where $\mathbf{S}$ is the evolution operator for the two-flavor problem with the modified matter potential $\tilde{V}$ and

$$\beta = exp\left(-i\frac{1}{2}(\Delta_{31}+\Delta_{32})t - i\int_0^t dt'\left(V_{\mu e}(t') - \frac{3}{2}\tilde{V}(t')\right)\right). \tag{20}$$

The quantity $\mathbf{U}_1$ is the solution of

$$i\frac{\partial}{\partial t}\mathbf{U}_1 = (\mathbf{U}_0^{\dagger}\mathcal{H}_0\mathbf{U}_0)\mathbf{U}_1 \tag{21}$$

where

$$\mathbf{U}_0^{\dagger}\mathcal{H}_0\mathbf{U}_0 = \begin{pmatrix} 0 & J \\ J^{\dagger} & 0 \end{pmatrix} \tag{22}$$

with

$$J = \frac{1}{2}\sin 2\theta_{13} V_{\mu e} \mathbf{S}^{\dagger} \begin{pmatrix} 1 \\ 0 \end{pmatrix} \beta. \tag{23}$$

A perturbative solution of Eq. (21) is

$$\mathbf{U}_1(T) = \begin{pmatrix} 1 - \int_0^T dt \int_0^t dt' J(t) J^{\dagger}(t') & -i \int_0^T dt J(t) \\ -i \int_0^T dt J^{\dagger}(t) & 1 - \int_0^T dt \int_0^t dt' J^{\dagger}(t) J(t') \end{pmatrix} + \mathcal{O}(\sin^3 2\theta_{13}). \tag{24}$$

Consider the integral appearing in Eq. (24):

$$\begin{aligned} I(T) &= \int_0^T dt J(t) \\ &= \frac{1}{2}\sin 2\theta_{13} \int_0^T dt e^{-i(\Delta_{31}+\Delta_{32})t/2} \exp\left(-i\int_0^t dt' \left(V_{\mu e}(t') - \frac{3}{2}\tilde{V}(t')\right)\right) V_{\mu e} \mathbf{S}^{\dagger} \begin{pmatrix} 1 \\ 0 \end{pmatrix}. \end{aligned} \tag{25}$$

In this integral the first exponential $e^{-i(\Delta_{31}+\Delta_{32})t/2}$ is a rapidly oscillating function in the Sun, averaging to zero inside the integral. (Note that $\Delta_{32} R_{\odot} \sim 10^5$ for a 10 MeV neutrino). Indeed, Riemann-Lebesgue lemma ensures that

$$\int_a^b dt f(t) e^{ixg(t)} \rightarrow 0$$

as $x \rightarrow \infty$, provided that

$$\int_a^b dt |f(t)|$$

exists. The leading contribution to the integral in Eq. (25) is [14]

$$\int_a^b dt f(t) e^{ixg(t)} \sim \left. \frac{f(t)}{ix(dg/dt)} e^{ixg(t)} \right|_{t=a}^{b} \tag{26}$$

as $x \rightarrow \infty$. (For a proof of this statement with integration by parts see [13]). Using Eq. (26) to calculate the integral in Eq. (25) yields

$$I(R_{\odot}) = -i \sin 2\theta_{13} \left(\frac{1}{\Delta_{32} + \Delta_{31}} V_{\mu e}(r=0) \right) \begin{pmatrix} 1 \\ 0 \end{pmatrix}. \tag{27}$$

For a 10 MeV neutrino with $\delta m^2_{32} \sim 2 \times 10^{-3}$ eV^2, assuming an electron density of 100 N_A / cm^3 in the neutrino production region, the quantity inside the parentheses in Eq. (27)

$$\alpha \equiv \frac{1}{\Delta_{32} + \Delta_{31}} V_{\mu e}(r=0) \sim 3 \times 10^{-2}. \tag{28}$$

If one takes $\alpha = 0$, one finds, e.g. Ψ_e on the surface of the Sun to be

$$\Psi_e = \cos^2\theta_{13} \psi_1 + \sin^2\theta_{13} \beta, \tag{29}$$

where ψ_1 is the electron neutrino wave function obtained in two-flavor evolution with the modified matter potential $\tilde{V}$. Upon averaging over the Earth-Sun distance (see e.g. [15]) one gets Eq. (3). Including the first order correction in $\mathbf{U}_1$, which is proportional t α of Eq. (28), gives

$$P_{3\times 3}(\nu_e \rightarrow \nu_e) = \cos^4\theta_{13}(1 - 4\sin^2\theta_{13}\alpha) P_{2\times 2}(\nu_e \rightarrow \nu_e \,\text{with}\, N_e \cos^2\theta_{13}) + \sin^4\theta_{13}(1 + 4\cos^2\theta_{13}\alpha). \tag{30}$$

A different derivation of this formula was given in Ref. [16]. Even for the largest value of θ_{13} quoted by T2K experiment, this correction to the survival probability is less than one percent for a 10 MeV neutrino.

3. CP-violating phase

It is rather straightforward to show that the CP-violating phase factorizes out in the neutrino evolution Hamiltonian of Eq. (6):

$$\tilde{\mathbf{H}}(\delta_{CP}) = \mathbf{S}\tilde{\mathbf{H}}(\delta_{CP} = 0)\mathbf{S}^{\dagger} \tag{31}$$

with

$$\mathbf{S} = \begin{pmatrix} 1 & 0 & 0 \\ 0 & 1 & 0 \\ 0 & 0 & e^{i\delta_{CP}} \end{pmatrix}. \tag{32}$$

Eq. (31) implies that the neutrino evolution operator can be similarly factorized. This factorization gives us interesting sum rules: Electron neutrino survival probability, $P(\nu_e \to \nu_e)$ is independent of the value of the CP-violating phase, δ; or equivalently, in the absence of sterile neutrino mixing, the combination $P(\nu_\mu \to \nu_e) + P(\nu_\tau \to \nu_e)$ at a fixed energy is independent of the value of the CP-violating phase [17]. However the probability $P(\nu_\mu \to \nu_e)$ depends on the δ_{CP}. This is why the disappearance experiments such as those that use reactor antineutrinos can quote a probability independent of δ_{CP}, whereas the appearance experiments, such as those that use accelerator neutrinos, cannot disentangle θ_{13} from δ_{CP}. It is possible to derive similar sum rules for other amplitudes [18]. A discussion of the breakdown of this formula when sterile neutrinos are present is given in Ref. [19].

4. Concluding Remarks

We provided a brief summary of the limits and preliminary measurements of the still poorly known neutrino mixing angle, θ_{13}. Many analyses of the solar neutrino data utilize a formula connecting two- and three-flavor evolution. We gave a critical assessment of the validity of this formula.

If the angle θ_{13} is rather large, there may be significant astrophysical implications. A brief review of the impact of the value of the mixing angle θ_{13} on various astrophysical phenomena was given in [20].

Acknowledgments

I thank A. Palazzo for discussions and correcting a typo in the earlier version of the manuscript. This work was supported in part by the U.S. National Science Foundation Grant No. PHY-0855082 and in part by the University of Wisconsin Research Committee with funds granted by the Wisconsin Alumni Research Foundation.

References

[1] Nakamura K *et al.* [Particle Data Group] 2010 *J. Phys. G* **37** 075021.
[2] Balantekin A B and Yilmaz D 2008 *J. Phys. G* **35** 075007 (Preprint arXiv:0804.3345 [hep-ph])
[3] Fogli G L, Lisi E, Marrone A, Palazzo A and Rotunno A M 2008 *Phys. Rev. Lett.* **101** 141801 (Preprint arXiv:0806.2649 [hep-ph]); Schwetz T, Tortola M A and Valle J W F 2008 2008 *New J. Phys.* **10** 113011 (Preprint arXiv:0808.2016 [hep-ph]); Gonzalez-Garcia M C, Maltoni M and Salvado J 2010 *JHEP* **1004** 056 (Preprint arXiv:1001.4524 [hep-ph]).
[4] Mezzetto M and Schwetz T 2010 *J. Phys. G* **37** 103001 (Preprint arXiv:1003.5800 [hep-ph]).
[5] Aharmim B *et al.* [SNO Collaboration] 2010 *Phys. Rev. C* **81** 055504 (Preprint arXiv:0910.2984 [nucl-ex]).
[6] Gando A *et al.* [The KamLAND Collaboration] 2011 *Phys. Rev. D* **83** 052002 (Preprint arXiv:1009.4771 [hep-ex]).
[7] Ardellier F *et al.* [Double Chooz Collaboration], (Preprint arXiv:hep-ex/0606025).
[8] Guo X*et al.* [Daya Bay Collaboration], (Preprint arXiv:hep-ex/0701029).
[9] Ahn J K *et al.* [RENO Collaboration], (Preprint arXiv:1003.1391 [hep-ex]).
[10] Abe K *et al.* [T2K Collaboration], (Preprint arXiv:1106.1238 [hep-ex]).
[11] Abe K *et al.* [T2K Collaboration], (Preprint arXiv:1106.2822 [hep-ex]).

[12] Balantekin A B and Fuller G M 1999 *Phys. Lett. B* **471**, 195 (Preprint arXiv:hep-ph/9908465).
[13] Balantekin A B and Yuksel H 2003 *J. Phys. G* **29** 665 (Preprint arXiv:hep-ph/0301072]).
[14] Bender C M and Orszag SA 1978 *Advanced Mathematical Methods for Scientists and Engineers* (New York: McGraw-Hill) p 278
[15] Balantekin A B 1999 *Phys. Rept.* **315** 123 (Preprint arXiv:hep-ph/9808281).
[16] Fogli G L, Lisi E and Palazzo A 2002 *Phys. Rev. D* **65**, 073019 [arXiv:hep-ph/0105080].
[17] Balantekin A B, Gava J and Volpe C 2008 *Phys. Lett. B* **662** 396 (Preprint arXiv:0710.3112 [astro-ph]).
[18] Kneller J P and McLaughlin G C 2009 *Phys. Rev. D* **80** 053002 (Preprint arXiv:0904.3823 [hep-ph]).
[19] Palazzo A 2011 *Phys. Rev. D* **83** 113013 [arXiv:1105.1705 [hep-ph]].
[20] Balantekin A B 2010 AIP Conf. Proc. **1269**, 195 (arXiv:1006.2836 [nucl-th]).

Journal of Physics: Conference Series **337** (2012) 012050 doi:10.1088/1742-6596/337/1/012050

Neutrinos Associated with Cosmic Rays

Francis Halzen
Department of Physics, University of Wisconsin, Madison, WI 53706, USA
and
DESY, Zeuthen, Germany

ABSTRACT

Construction of the first kilometer-scale neutrino observatory has been completed; IceCube has been fully commissioned and has been taking data since May 2011. Its present performance exceeds expectations in both the neutrino collection area (by a factor $2 \sim 3$ depending on energy) and angular resolution. It continues to improve with ongoing refinements in calibration, software tools and our understanding of the optics of the natural ice. IceCube was designed more than a decade ago with the goal of observing the sources of both Galactic and extragalactic cosmic rays with good statistical significance after 5 years. Because the origin of cosmic rays is still unresolved, the exercise is inevitably performed on models. We here revisit three illustrative examples chosen because they are predictive, although with relatively large errors associated with the astrophysics of the sources: Galactic supernova remnants, gamma-ray bursts and GZK neutrinos produced in interactions of cosmic rays with the microwave background. We conclude that the IceCube design, as well as the prospect for observing neutrinos from cosmic-ray sources, have survived the test of time.

1. The First Kilometer-Scale Neutrino Detector: IceCube

A series of first-generation experiments[1)] have demonstrated that high-energy neutrinos with $\sim$ 10 GeV energy and above can be detected by observing Cherenkov radiation from secondary particles produced in neutrino interactions inside large volumes of highly transparent ice or water instrumented with a lattice of photomultiplier tubes. Construction of the first second-generation detector, IceCube, at the geographic South Pole has been completed in December 2010[2)]; see Fig.1.

IceCube consists of 80 strings, each instrumented with 60 10-inch photomultipliers spaced by 17 m over a total length of 1 kilometer. The deepest module is located at a depth of 2.450 km so that the instrument is shielded from the large background of cosmic rays at the surface by approximately 1.5 km of ice. Strings are arranged at apexes of equilateral triangles that are 125 m on a side. The instrumented detector volume is a cubic kilometer of dark, highly transparent and sterile Antarctic ice. Radioactive background is dominated by the instrumentation deployed into this natural ice.

Each optical sensor consists of a glass sphere containing the photomultiplier and the electronics board that digitizes the signals locally using an on-board computer. The digitized signals are given a global time stamp with residuals accurate to less than 3 ns and are subsequently transmitted to the surface. Processors at the surface continuously collect these time-stamped signals from the optical modules; each functions independently. The digital messages are sent to a string processor and a global event trigger. They are subsequently sorted into the Cherenkov patterns emitted by secondary muon tracks, or electron and tau showers, that reveal the direction of the parent neutrino[3)].

Based on data taken during construction with 40 of the 59 strings, the anticipated effective area of the completed IceCube detector is shown in Fig.2. Notice the factor 2 to 3 increase in effective area over what had been anticipated[4)]. The neutrino collecting area will continue to increase with improved calibration and development of optimized software tools for the 86-string detector operating stably in its final configuration. Already reaching an angular resolution of better than 0.5 degree for high energies, reconstruction is also superior to what was anticipated.

Despite its discovery potential touching a wide range of scientific issues from the search of dark matter to the physics of neutrinos themselves, construction of IceCube has been largely motivated

Journal of Physics: Conference Series **337** (2012) 012050 doi:10.1088/1742-6596/337/1/012050

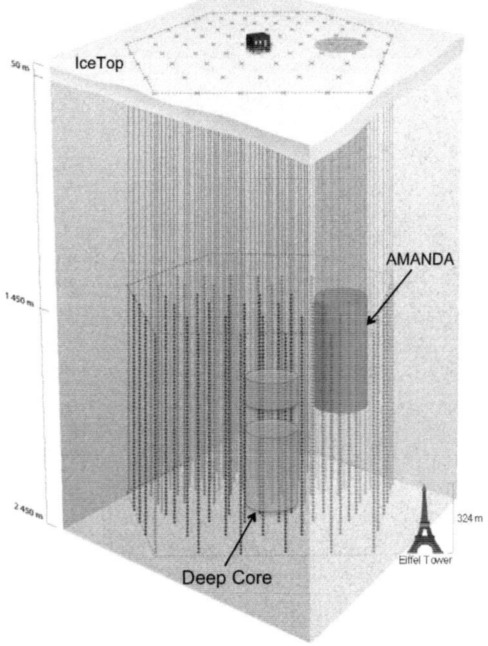

Figure 1: The IceCube detector, consisting of IceCube and IceTop and the low-energy sub-detector DeepCore. Also shown is the first-generation AMANDA detector.

by the possibility of opening a new window on the Universe using neutrinos as cosmic messengers. Specifically, we will revisit IceCube's prospects for detecting cosmic neutrinos associated with cosmic rays and for revealing their sources at a time when we are commemorating the 100th anniversary of their discovery by Victor Hess in 1912.

Cosmic accelerators produce particles with energies in excess of 10^8 TeV; we still do not know where or how[6)]. The flux of cosmic rays observed at Earth is shown in Fig.3. The energy spectrum follows a sequence of three power laws. The first two are separated by a feature dubbed the "knee" at an energy* of approximately 3 PeV. There is evidence that cosmic rays up to this energy are Galactic in origin. Any association with our Galaxy disappears in the vicinity of a second feature in the spectrum referred to as the "ankle"; see Fig.3. Above the ankle, the gyroradius of a proton in the Galactic magnetic field exceeds the size of the Galaxy, and we are witnessing the onset of an extragalactic component in the spectrum that extends to energies beyond 100 EeV. Direct support for this assumption now comes from two experiments [7)] that have observed the telltale structure in the cosmic-ray spectrum resulting from the absorption of the particle flux by the microwave background, the so-called Greissen-Zatsepin-Kuzmin (GZK) cutoff. Neutrinos are produced in GZK interactions; it was already recognized in the 1970s that their observation required kilometer-scale neutrino detectors. The origin of the cosmic-ray flux in the intermediate region covering PeV-to-EeV energies remains a mystery, although it is routinely assumed that it results from some high-energy extension of the reach of Galactic accelerators.

Acceleration of protons (or nuclei) to TeV energy and above requires massive bulk flows of relativistic charged particles. These are likely to originate from exceptional gravitational forces in the vicinity of black holes or neutron stars. The gravity of the collapsed objects powers large currents of charged particles that are the origin of high magnetic fields. These create the opportunity for

*We will use energy units TeV, PeV and EeV, increasing by factors of 1000 from GeV energy.

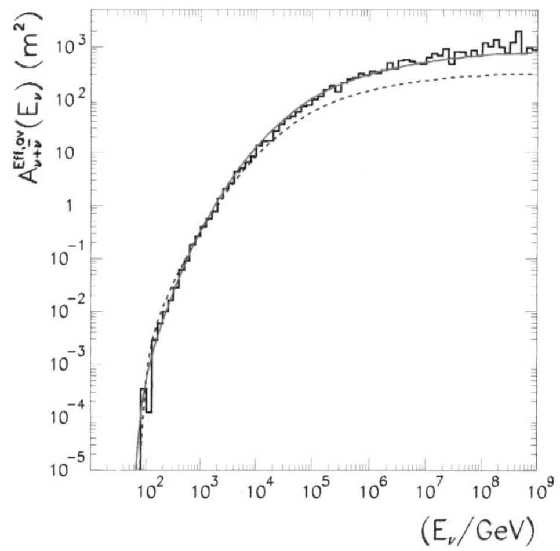

Figure 2: The neutrino effective area for point sources of cosmic neutrinos (averaged over the Northern Hemisphere) from IceCube simulation (black histogram) is compared to the convolution of the approximate muon effective area from reference[5)] (solid red line) that we will use in the various estimates of event rates throughout this paper. The neutrino area exceeds the design area (shown as the dashed blue line) [4)] at high energy.

particle acceleration by shocks. It is a fact that electrons are accelerated to high energy near black holes; astronomers detect them indirectly by their synchrotron radiation. Some must accelerate protons because we observe them as cosmic rays.

How many gamma rays and neutrinos are produced in association with the cosmic-ray beam? Generically, a cosmic-ray source should also be a beam dump. Cosmic rays accelerated in regions of high magnetic fields near black holes inevitably interact with radiation surrounding them, e.g., UV photons in active galaxies or MeV photons in gamma-ray-burst fireballs. In these interactions, neutral and charged pion secondaries are produced by the processes

$$p + \gamma \to \Delta^+ \to \pi^0 + p \text{ and } p + \gamma \to \Delta^+ \to \pi^+ + n.$$

While secondary protons may remain trapped in the high magnetic fields, neutrons and the decay products of neutral and charged pions escape. The energy escaping the source is therefore distributed among cosmic rays, gamma rays and neutrinos produced by the decay of neutrons, neutral pions and charged pions, respectively. In the case of Galactic supernova shocks, cosmic rays mostly interact with the hydrogen in the Galactic disk producing equal numbers of pions of all three charges in hadronic collisions $p + p \to n\,[\pi^0 + \pi^+ + \pi^-] + X$; n is the pion multiplicity. The flux should be enhanced in interaction of the cosmic rays with high-density molecular clouds that are ubiquitous in the star-forming regions where supernovae are more likely to explode.

Kilometer-scale neutrino detectors have the sensitivity to reveal generic cosmic-ray sources with an energy density in neutrinos comparable to their energy density in cosmic rays[8)] and pionic TeV gamma rays[9)].

2. Sources of Galactic Cosmic Rays

Supernova remnants were proposed as possible sources of Galactic cosmic rays as early as 1934 by Baade and Zwicky[10)]; their proposal is still a matter of debate after more than 70 years[11)]. Galactic

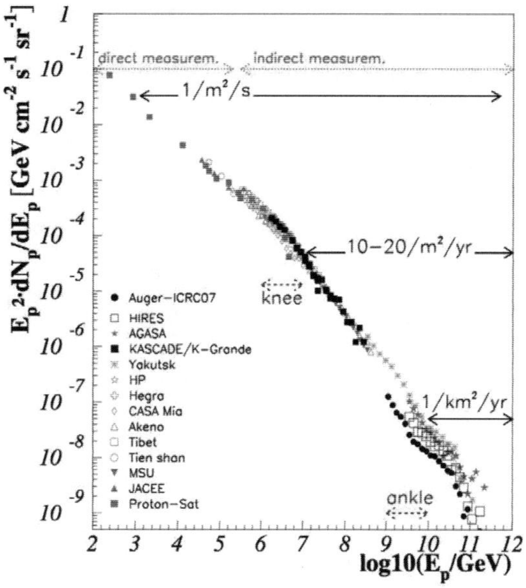

Figure 3: At the energies of interest here, the cosmic-ray spectrum follows a sequence of 3 power laws. The first 2 are separated by the "knee", the 2nd and 3rd by the "ankle". Cosmic rays beyond the ankle are a new population of particles produced in extragalactic sources.

cosmic rays reach energies of at least several PeV, the "knee" in the spectrum. Their interactions with Galactic hydrogen in the vicinity of the accelerator should generate gamma rays from the decay of secondary pions that reach energies of hundreds of TeV. Such sources should be identifiable by a relatively flat energy spectrum that extends to hundreds of TeV without attenuation, because the cosmic rays themselves reach at least several PeV near the knee; they have been dubbed PeVatrons. Straightforward energetics arguments are sufficient to conclude that present air Cherenkov telescopes should have the sensitivity necessary to detect TeV photons from PeVatrons[5,12)].

They may have been revealed by the highest-energy all-sky survey in ~ 10 TeV gamma rays with the Milagro detector[13)]. A subset of sources located within nearby star-forming regions in Cygnus and in the vicinity of Galactic latitude $l = 40$ degrees are identified; some cannot be readily associated with known supernova remnants or with non-thermal sources observed at other wavelengths. Subsequently, directional air Cherenkov telescopes were pointed at three of the sources, revealing them as PeVatron candidates with an approximate E^{-2} energy spectrum that extends to tens of TeV without evidence for a cutoff [14,15)], in contrast with the best studied supernova remnants RX J1713-3946 and RX J0852.0-4622 (Vela Junior).

Some Milagro sources may actually be molecular clouds illuminated by the cosmic-ray beam accelerated in young remnants located within ~ 100 pc. One expects indeed that multi-PeV cosmic rays are accelerated only over a short time period, when the remnant transitions from free expansion to the beginning of the Sedov phase and the shock velocity is high. The high-energy particles can produce photons and neutrinos over much longer periods when they diffuse through the interstellar medium to interact with nearby molecular clouds[16)]. An association of molecular clouds and supernova remnants is expected, of course, in star-forming regions.

Despite the rapid development of both ground-based and satellite-borne instruments with improved sensitivity, it has been impossible to conclusively pinpoint supernova remnants as the sources of cosmic-ray acceleration by identifying accompanying gamma rays of pion origin. In fact, recent data from Fermi LAT have challenged the hadronic interpretation of the GeV-TeV radiation from one of the best-studied candidates RX J1713-3946[17)]. In contrast, detecting the accompanying neutrinos provides incontrovertible evidence for cosmic-ray acceleration in cosmic- ray sources. Particle physics dictates the relation between pionic gamma rays and neutrinos and basically predicts the production of a $\nu_\mu + \bar{\nu}_\mu$ pair for every two gamma rays seen by Milagro. This calculation can be performed in a more sophisticated way with approximately the same outcome. We conclude that, within uncertainties in the source parameters, confirmation that Milagro mapped sources of Galactic cosmic rays should emerge after operating the complete IceCube detector for several years; see Fig.4.

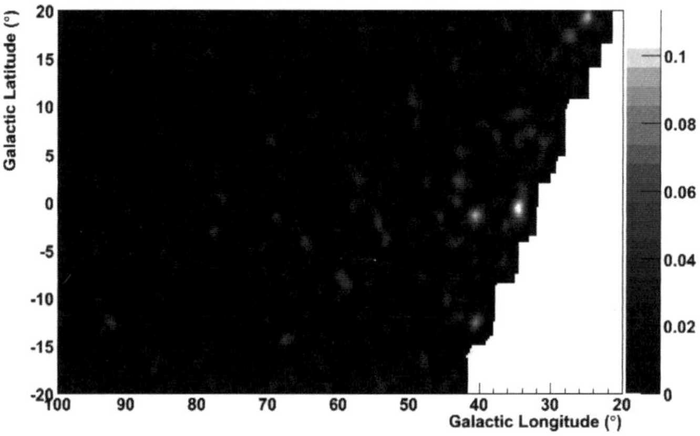

Figure 4: Simulated sky map of IceCube in Galactic coordinates after 5 years of operation of the completed detector. Two Milagro sources are visible "by eye" with 4 events for MGRO J1852+01 and 3 events for MGRO J1908+06 with energy in excess of 40 TeV. These, as well as the background events, have been randomly distributed according to the resolution of the detector and the size of the sources.

The quantitative statistics can be summarized as follows. For average values of the parameters describing the flux, we find that the completed IceCube detector could confirm sources in the Milagro sky map as sites of cosmic-ray acceleration at the 3σ level in less than one year and at the 5σ level in three years[5)]. We here assume that the source extends to 300 TeV, or 10% of the energy of the cosmic rays near the knee in the spectrum. These results agree with previous estimates[18)]. There are intrinsic ambiguities in this estimate of an astrophysical nature that may reduce or extend the time required for a 5σ observation[5)]. In the absence of observation of TeV-energy supernova neutrinos by IceCube in a period of 10 years, the nature of sources that produce cosmic rays near the knee of the spectrum will remain unresolved.

3. Sources of the Extragalactic Cosmic Rays

Although there is no direct evidence that supernovae accelerate cosmic rays, the idea is generally accepted because of energetics: three supernovae per century converting a reasonable fraction of a solar mass into particle acceleration can accommodate the steady flux of cosmic rays in the Galaxy. Originally, energetics also drove speculations on the origin of extragalactic cosmic rays.

By integrating the cosmic-ray spectrum in Fig.3 above the ankle, we find that the energy density of the Universe in extragalactic cosmic rays is $\sim 3\times10^{-19}\,\mathrm{erg\,cm}^{-3}$ [8)]. The power required for a population of sources to generate this energy density over the Hubble time of 10^{10} years is $\sim 3\times10^{37}\,\mathrm{erg\,s}^{-1}$ per $(\mathrm{Mpc})^3$. (In the astroparticle community, this flux is also known as $5\times10^{44}\,\mathrm{TeV\,Mpc}^{-3}\,\mathrm{yr}^{-1}$). A gamma-ray-burst (GRB) fireball converts a fraction of a solar mass into the acceleration of electrons, seen as synchrotron photons. The energy in extragalactic cosmic rays can be accommodated with the reasonable assumption that shocks in the expanding GRB fireball convert roughly equal energy into the acceleration of electrons and cosmic rays[19)]. It so happens that $\sim 2\times10^{52}$ erg per GRB will yield the observed energy density in cosmic rays after 10^{10} years, given that the rate is of order 300 per Gpc^3 per year. Hundreds of bursts per year over Hubble time produce the observed cosmic-ray density, just like three supernovae per century accommodate the steady flux in the Galaxy.

Problem solved? Not really: it turns out that the same result can be achieved assuming that active galactic nuclei (AGN) convert $\sim 2 \times 10^{44}$ erg s^{-1} per AGN into particle acceleration. As is the case for GRB, this is an amount that matches their output in electromagnetic radiation. Whether GRB or AGN, the observation that these sources radiate similar energies in photons and cosmic rays is consistent with the beam-dump scenario previously discussed. In the interaction of cosmic rays with radiation and gases near the black hole, roughly equal energy goes into the secondary neutrons and neutral pions whose energy ends up in cosmic rays and gamma rays, respectively.

Unlike what is the case for Galactic cosmic rays, there is no straightforward γ-ray path to the neutrino flux expected from extragalactic cosmic-ray accelerators. Neutrino fluxes from AGN are difficult to estimate. For GRB, the situation is qualitatively better, because neutrinos of PeV energy should be produced when protons and photons coexist in the GRB fireball[21]. As previously discussed, the model is credible because the observed cosmic-ray flux can be accommodated with the assumption that roughly equal energy is shared by electrons, observed as synchrotron photons, and protons. The GRB neutrino flux is related to the cosmic ray flux by

$$\begin{aligned}\frac{dN_\nu}{dE_\nu} &= \left[1-\left(1-e^{-n_{int}}\right)\right]\frac{1}{3}x_\nu\frac{dN_p}{dE_p}\left(\frac{E_p}{x_\nu}\right) f_{GZK}\\ &\simeq n_{int}\ x_\nu\ \frac{dN_p}{dE_p}\left(\frac{E_p}{x_\nu}\right),\end{aligned} \tag{1}$$

where $x_\nu \simeq 0.05$ is the average relative energy of the neutrino and the parent proton and n_{int} ($\simeq 1$) is the average number of interactions of the proton with fireball photons before it becomes optically transparent and the photons escape. Neutrinos reach us from sources distributed over all redshifts, while cosmic rays do so only from local sources inside the so-called GZK radius of less than 100 Mpc. The evolution of the sources will boost the neutrino flux by a factor $f_{GZK} \simeq 3$ that depends on the redshift distribution of GRB.

The critical quantity normalizing the GRB neutrino flux is n_{int}; its calculation is relatively straightforward. The phenomenology that successfully accommodates the astronomical observations is that of the creation of a hot fireball of electrons, photons and protons that is initially opaque to radiation. The hot plasma therefore expands by radiation pressure, and particles are accelerated to a Lorentz factor Γ that grows until the plasma becomes optically thin and produces the GRB display. From this point on, the fireball coasts with a Lorentz factor that is constant and depends on its baryonic load. The baryonic component carries the bulk of the fireball's kinetic energy. The energetics and rapid time structure of the burst can be successfully explained by shocks (shells) of width ΔR generated in the expanding fireball. The rapid temporal variation of the gamma-ray burst, t_v, is of the order of milliseconds, and can be interpreted as the collision of internal shocks with a varying baryonic load leading to differences in the bulk Lorentz factor. Electrons accelerated by first-order Fermi acceleration radiate synchrotron gamma rays in the strong internal magnetic field, and thus produce the spikes observed in the burst spectra.

The number of interactions is determined by the optical depth of the fireball shells to pγ interactions

$$n'_{int} = \frac{\Delta R'}{\lambda_{p\gamma}} = (\Gamma c t_v)\left(n'_\gamma \sigma_{p\gamma}\right). \tag{2}$$

The primes refer to the burst rest frame; unprimed quantities are in the observer frame. The density of fireball photons depends on the total energy in the burst $E_{GRB} \simeq 2 \times 10^{52}$ erg, the characteristic photon energy of $E_\gamma \simeq 1\,MeV$ and the volume V' of the shell

$$n'_\gamma = \frac{E_{GRB}/E_\gamma}{V'}, \tag{3}$$

with

$$V' = 4\pi R'^2 \Delta R' = 4\pi \left(\Gamma^2 c t_v\right)^2 (\Gamma c t_v)\,. \tag{4}$$

The only subtlety here is the Γ^2 dependence of the shell radius R'; for a simple derivation see ref. 21.

For typical choices of the parameters $\Gamma \sim 300$ and $t_v \sim 10^{-2}s$, about 100 events per year are predicted in IceCube, a flux that is already challenged[24] by the limit on a diffuse flux of cosmic neutrinos obtained with one-half of IceCube in one year[25]. Facing this negative conclusion, Ahlers *et al.* have investigated the dependence of the predicted neutrino flux on the cosmological evolution of the sources as well as on the parameters describing the fireball, most notably E_{GRB}, Γ and t_v. Although these are constrained by the electromagnetic observation and by the the requirement that the fireball must accommodate the observed cosmic-ray spectrum, the predictions can be stretched to the point that it will take 3 years of data with the now-completed instrument to conclusively rule out the GRB origin of the extragalactic cosmic rays; see Fig.5. Alternatively, detection of their neutrino emission may be imminent.

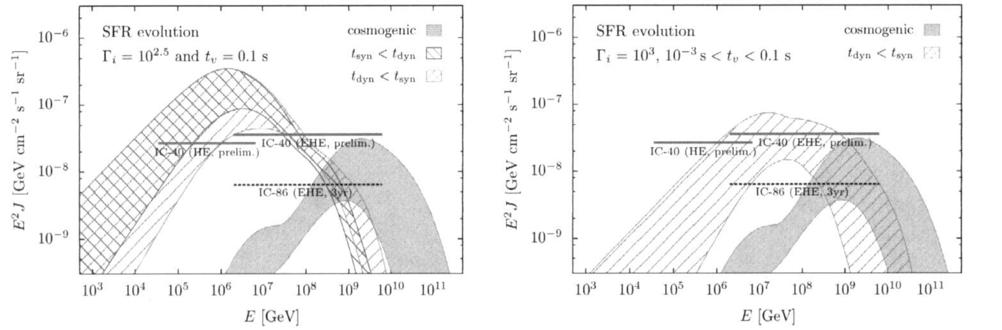

Figure 5: GRB neutrino spectra (the prompt spectrum emitted by the sources and neutrino spectrum generated in GZK interactions are shown separately), assuming the luminosity range $0.1 < (\epsilon_B/\epsilon_e)\, L_{\gamma,52} < 10$ and star-forming redshift evolution of the sources. Here $\epsilon_{e,B}$ are the fractional energies in the fireball carried by the electrons and the magnetic field; the two are equal in the case of equipartition. $L_{\gamma,52}$ is the photon energy in units of 10^{52} erg. We show the prompt spectra separately for models where the fireball's dynamical timescale t_{dyn} is smaller(larger) than the synchrotron loss time scale t_{syn} (green right-hatched and blue left-hatched respectively). Here the dynamical time scale is just the variability scale $t_{dyn} = t_v$ and $t'_{dyn} = t_v\Gamma$. The IceCube limits[25] on the total neutrino flux from the analysis of high-energy and ultrahigh-energy muon neutrinos with the 40 string sub-array assume 1:1:1 flavor composition after oscillation. We also show the sensitivity of the full IceCube detector (IC-86) to muon neutrinos after 3 years of observation. The gray solid area shows the range of GZK neutrinos expected at the 99% C.L.

Although it is well known that GRBs satisfy the necessary conditions for accelerating protons to UHE, it is problematic how these protons may eventually be ejected as cosmic rays: protons are magnetically connected to the expanding fireball, and its adiabatic cooling will reduce the maximum proton energy significantly. However, this does not apply to neutrons that are produced in pγ-interactions of accelerated protons with fireball photons. Cosmic-ray protons could thus be identified as neutrons that can escape from the magnetic environment and decay back to protons at a safe distance. As was already discussed, a smoking-gun test of this scenario is the production of PeV neutrinos from decay of the charged pions inevitably produced along with the neutrons. Identifying the observed cosmic rays with secondary neutrons rather than fireball protons significantly raises the contributions of individual GRB to the energy budget in the Universe of $\sim 3 \times 10^{-19}\,\mathrm{erg\ cm^{-3}}$.

Is the GRB origin of sources of the highest-energy cosmic rays challenged? Recall that calculation of the GRB neutrino flux is normalized to the observed total energy in extragalactic cosmic rays of $\sim 3\times10^{-19}\,\mathrm{erg\ cm^{-3}}$, a value that is highly uncertain because it critically depends on the assumption that no cosmic rays above the ankle are Galactic in origin. Although direct fits to the spectrum support this assumption[24)], by shifting this transition to higher energies one can reduce the energy budget by as much as an order of magnitude. The lower value of 0.5×10^{44} TeV Mpc^{-3} yr^{-1} can be accommodated with a more modest fraction of $\sim 2 \times 10^{51}$ erg (or $\sim 1\,\%$ of a solar mass) going into particle acceleration in individual bursts. We will revisit this issue in the context of GZK neutrinos.

While this temporarily remedies the direct conflict with the present diffuse limit, IceCube has the alternative possibility to perform a direct search for neutrinos in spatial and time coincidence with GRB observed by the Swift and Fermi satellites. In this essentially background-free search, 14 events are expected when IceCube operated with 40 and 59 strings during 2 years of construction, even for the lowest value of the cosmic-ray energy budget of 0.5×10^{44} TeV Mpc^{-3} yr^{-1}. Two different and independent searches failed to observe this flux at the 90% confidence level[26)]. Also, the reliability of calculating the neutrino flux is an issue here and, as already discussed, IceCube has the potential to confirm or rule out GRB as the sources of the highest-energy cosmic rays within 3 years of operation.

If, on the contrary, AGN were indeed the sources, the proximity of the Fanaroff-Riley I active galaxies Cen A and M87 singles them out as potential accelerators[27,28)]. As we did for the Milagro sources, we have attempted to translate their TeV gamma rays into a neutrino flux, although interpreting TeV gamma-ray observations is in this case challenging. The high-energy emission of AGN is indeed extremely variable, and it is difficult to compare multi-wavelength data taken at different times. A best guess of the gamma-ray flux yields a neutrino flux from a single source such as Cen A that is small, typically less than one event per year, even if all gammas in the TeV range are assumed to be of pionic origin. The diffuse flux from all FRI yields a more comfortable event rate of between 19 and 0.5 neutrinos per year, assuming a spectral index between 2 and 3. A detailed discussion of these estimates has been presented elsewhere[29)].

4. Neutrinos from GZK Interactions

Whatever the sources of the extragalactic cosmic rays may be, a cosmogenic flux of neutrinos originates from the interactions of cosmic rays with the cosmic microwave background (CMB). Produced within a GZK radius from a source located at a cosmological distance, GZK neutrinos point back to it with good precision. The calculation of the GZK neutrino flux is relatively straightforward, and its magnitude is very much determined by their total energy density in the universe; this brings back the crossover from the Galactic to the extragalactic component as a critical parameter. Recent calculations[31)] are shown in Fig.6. It is also important to realize that, among the $p\,\gamma$ final state products produced via the decay of pions, the neutrinos are accompanied by electrons, positrons and γ-rays that quickly cascade on the CMB and intergalactic magnetic fields to lower energies. An electromagnetic cascade develops with a maximum in the GeV-TeV energy region. Here the total energy in the electromagnetic cascade is constrained by recent Fermi-LAT measurements of the diffuse extragalactic γ-ray background[30)].

The increased performance of IceCube at EeV energy has opened the possibility for IceCube to detect GZK neutrinos. We anticipate 2.3 events in 3 years of running the completed IceCube detector, assuming the best fit in Fig.6, and 4.8 events for the highest flux consistent with the Fermi constraint.

Throughout the discussion, we have assumed that the highest-energy cosmic rays are protons. Experiments disagree on the composition of particles around 10^{20} eV. Little is known about the chemical composition from just below to beyond the GZK cutoff, where the most significant contribution to cosmogenic neutrinos is expected. In any case, uncertainties in extrapolation of the proton-air interaction cross-section, elasticity and multiplicity of secondaries from accelerator measurements to the high energies characteristic for air showers are large enough to undermine any definite conclusion on the chemical composition[32)]. Therefore, the conflicting claims by these experiments most likely illustrate that the particle physics is not sufficiently known to derive a definite

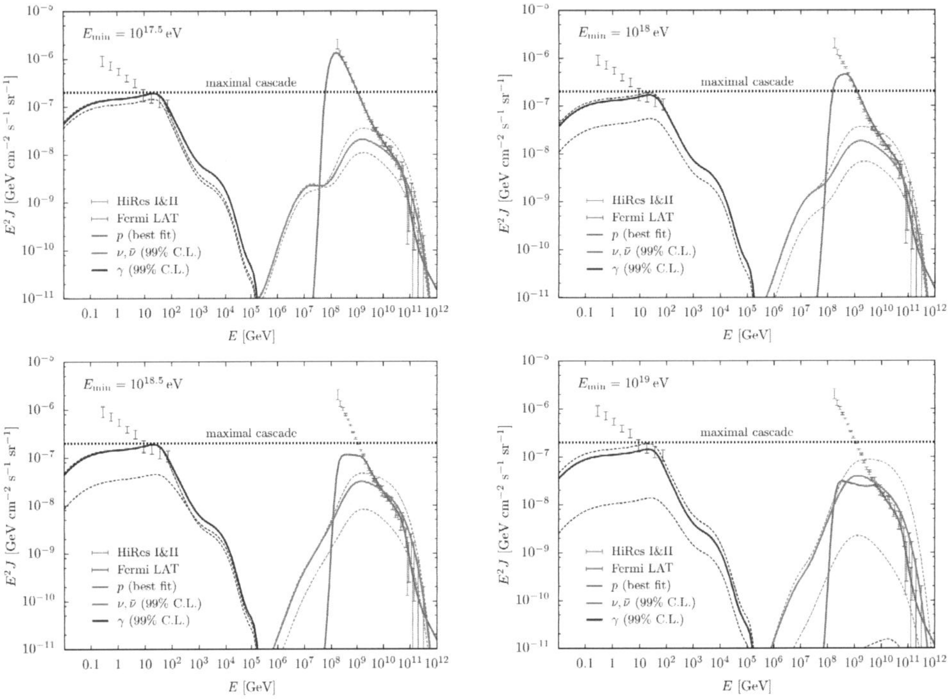

Figure 6: Comparison of proton, neutrino and gamma ray fluxes produced in interactions on the CMB by cosmic-ray protons fitted to HiRes data. We repeat the calculation for 4 values of the crossover energy marking the transition to the extragalactic cosmic ray flux. We show the best fit values (solid lines) as well as neutrino and gamma-ray fluxes within the 99% C.L. with minimal and maximal energy density (dashed lines). The γ-ray fluxes are marginally consistent at the 99% C.L. with the highest-energy measurements by Fermi-LAT. The contribution around 100 GeV is somewhat uncertain, due to uncertainties in the cosmic infrared background.

result. Dedicated experiments at the LHC may remedy this situation.

5. Conclusion: Stay Tuned

In summary, IceCube was designed for a statistically significant detection of cosmic neutrinos accompanying cosmic rays in 5 years. In this talk, I attempted to make the case that we are indeed closing in on supernova remnants, GRB (if they are the sources of cosmic rays) and GZK neutrinos. One should not forget, however, that the most exciting IceCube science may come from the detection of dark matter, the observation of a Galactic supernova explosion, or from the particle physics of the neutrinos themselves.

6. Acknowledgements

I would like to thank my IceCube collaborators as well as Markus Ahlers, Concha Gonzalez-Garcia, Alexander Kappes, Aongus O'Murchadha and Nathan Whitehorn for valuable discussions. This research was supported in part by the U.S. National Science Foundation under Grants No. OPP-0236449 and PHY-0354776; by the U.S. Department of Energy under Grant No. DE-FG02-95ER40896; by the University of Wisconsin Research Committee with funds granted by the Wisconsin Alumni

Research Foundation, and by the Alexander von Humboldt Foundation in Germany.

7. References

1) C. Spiering, astro-ph/08114747.
2) F. Halzen, S. R. Klein, Rev. Sci. Instrum. **81**, 081101 (2010). [arXiv:1007.1247 [astro-ph.HE]]
3) F. Halzen, Eur. Phys. J. C **46** 669 (2006); astro-ph/0602132.
4) IceCube collaboration, Astropart. Phys. **20**, 507 (2004); astro-ph/0305196.
5) M. C. Gonzalez-Garcia, F. Halzen, S. Mohapatra, Astropart. Phys. **31**, 437-444 (2009). [arXiv:0902.1176 [astro-ph.HE]].
6) For a recent review of the cosmic-ray problem, *see* P. Sommers and S. Westerhoff, astro-ph/08021267; A. M. Hillas, astro-ph/0607109; V. Berezinsky, astro-ph/08013028.
7) HiRes collaboration, Phys. Rev. Lett. **100** 101101 (2008), astro-ph/0703099; and Pierre Auger collaboration, Phys. Rev. Lett. **101** 061101 (2008), astro-ph/08064302.
8) T. K. Gaisser, OECD Megascience Forum, Taormina, Italy (1997), astro-ph/9707283; the discussion was recently revisited in Ahlers, *et al.*, Phys. Rev. D **72** 023001 (2005), astro-ph/0503229.
9) J. Alvarez-Muniz and F. Halzen, Astrophys. J. **576** L33 (2002); astro-ph/0205408.
10) W. Baade and F. Zwicky, Phys. Rev. D **46** 76 (1934).
11) Y. Butt, Nature **460**, 701 (2009). [arXiv:1009.3664 [astro-ph.HE]].
12) M. Ahlers, P. Mertsch, S. Sarkar, Phys. Rev. **D80**, 123017 (2009). [arXiv:0909.4060 [astro-ph.HE]].
13) A. A. Abdo *et al.*, Astrophys. J. **658** L33 (2007); astro-ph/0611691.
14) H.E.S.S. collaboration, 30th ICRC, Merida, Mexico (2007); astro-ph/07102418.
15) J. Albert *et al.*, astro-ph/08012391.
16) For a recent discussion, *see* S. Gabici and F. .A. Aharonian, astro-ph/07053011.
17) A.A. Abdo, Fermi LAT Collaboration, arXiv:1103.5727v1 [astro-ph.HE].
18) F. Halzen *et al.*, Phys. Rev. D **78** 063004 (2008), *see also* Nucl. Instr. and Meth. A **602** 117 (2009); astro-ph/08030314v2.
19) E. Waxman, Phys. Rev. Lett. **75** 386 (1995), astro-ph/9701231; M. Vietri, Phys. Rev. Lett. **80** 3690 (1998), astro-ph/9802241; and M. Bottcher and C. D. Dermer, astro-ph/9801027.
20) E. Waxman and J. N. Bahcall, Phys. Rev. D **59** 023002 (1999); hep-ph/9807282.
21) E. Waxman, J. N. Bahcall, Phys. Rev. Lett. **78**, 2292-2295 (1997). [astro-ph/9701231].
22) D. Guetta *et al.*, Astropart. Phys. **20** 429 (2004); astro-ph/0302524.
23) F. Halzen, D. Hooper, Rept. Prog. Phys. **65**, 1025-1078 (2002). [astro-ph/0204527].
24) M. Ahlers, M. C. Gonzalez-Garcia, F. Halzen, [arXiv:1103.3421 [astro-ph.HE]].
25) R. Abbasi *et al.* [IceCube Collaboration], [arXiv:1103.4250 [astro-ph.CO]].
26) R. Abbasi *et al.* [IceCube Collaboration], Phys. Rev. Lett. **106**, 141101 (2011). [arXiv:1101.1448 [astro-ph.HE]].
27) J. Abraham *et al.*, Science **318** 939 (2007).
28) L. A. Anchordoqui *et al.*, Phys. Lett. B **600** 202 (2004); for an earlier suggestion using beamed emission instead, *see* R. J. Protheroe *et al.*, Astropart. Phys. **19** 559 (2003), astro-ph/0210249.
29) M. C. Gonzalez-Garcia, F. Halzen, S. Mohapatra, A. O'Murchadha, "Neutrinos from cosmic ray sources," in Proceedings of the 13th International Workshop on Neutrino Telescopes, Venice, Italy (2009).
30) A. A. Abdo *et al.* [The Fermi-LAT collaboration], Phys. Rev. Lett. **104**, 101101 (2010) [arXiv:1002.3603 [astro-ph.HE]].
31) M. Ahlers, L. A. Anchordoqui, M. C. Gonzalez-Garcia, F. Halzen, S. Sarkar, Astropart. Phys. **34**, 106-115 (2010). [arXiv:1005.2620 [astro-ph.HE]]
32) R. Ulrich, R. Engel, S. Muller, F. Schussler and M. Unger, Nucl. Phys. Proc. Suppl. 196, 335 (2009) [arXiv:0906.3075 [astro-ph.HE]].

Journal of Physics: Conference Series **337** (2012) 012051 doi:10.1088/1742-6596/337/1/012051

The Fate of a WD Accreting H-Rich Material at High Rates

Irit Idan[1], Nir J. Shaviv[2], Giora Shaviv[1]

[1] Department of Physics, Israel Institute of Technology, Haifa 32000, Israel
[2] Racah Institute of Physics., The Hebrew University, Jerusalem 91904, Israel

Abstract. We study a white dwarf (WD) accreting solar-composition material at super-Eddington accretion rates. We find that after many cycles of hydrogen ignition and burning, several large Helium burning flashes expel the accreted envelope, leaving no net mass accumulation. This implies that such a system will not undergo accretion induced collapse.

1. Introduction

The standard progenitor for SN type Ia is that of a WD accreting at very high rates, having a net accumulation of mass. One can distinguish two cases, accretion of H-rich and accretion of He-rich material. Equally important is the initial mass of the accreting WD. Although a small mass WD can in principle grow in mass at high accretion rates, if a massive WD does not grow as well, no eventual explosion will ensue. For a detailed review of progenitor scenarios for SN Type Ia, see Podsiadlowski et al (2008).

Here we wish to address the specific case of high rate H-accretion. We know from the work of Prialnik and Kovetz (1995), that if one is to capture the long term secular behavior of the WD, a relatively large number of cycles is required. For this reason, we should be prepared to simulate several thousands of cycles. We begin by describing the model, continue with the results and end with a discussion about the important energy budget in the system.

2. Model Description

The initial model includes a $1M_\odot$ C/O WD. The central core temperature was 10^7K. No rotation was included. We assumed that the dissipation of the rotational energy in the boundary layer is complete and radiated away.

Special attention was paid to have a fine mass division, mainly near the surface of the WD and on the outer part of the accreted envelope. The cyclic phenomenon demands very fine division during mass loss and peak burning. Therefore, the minimal mass shell chosen was $10^{-7}M_\odot$. The program determines automatically the size of the mass shells and their number according to the specified accuracy. At the peak, over 4000 mass shells were used in the dynamic model. We included all relevant nuclear reactions up to A=40 and used the intermediate electrostatic screening correction.

Two basic models were calculated, one with a WD mass of $1M_\odot$, and one with $1.2M_\odot$ despite the recent finding by Weidemann (2000) that this is too high. In both models the accretion rate was $10^{-6}M_\odot/yr$. According to Nomoto (1982), this accretion rate is high enough to puff the accreting system into a red giant like star burning at steady state and lead into a situation in which the model evolves on a timescale determined by the nuclear burning.

A crucial issue is the treatment of the mass loss. At each time step, we check whether an optically thick wind can exist (Kovetz 1998) and assume that if such a wind can exist, it does. A solution for a thick wind is then found and used to calculate the expected mass loss rate. The mass of the outermost shell is then corrected for the mass-loss. When the ratio between the masses of the last two outermost mass shells exceeds a prescribed condition, a new mass division is established in which the ratio between two successive mass shells does not exceed a prescribed value.

The parameters are chosen in such a way that the readjustments due to the mass loss and the merger of shells does not induce unacceptable fluctuations and is effectively smooth. If the procedure is inhibited artificially, the mass which had to be removed expands to infinity, and reduces the time step to unacceptably small values once the shell reaches many solar radii. We stress that in no case did the removal of the mass found to induce additional mass removal, namely no rarefaction wave was created leading to endless mass loss. The possibility of an optically thin wind was not checked in this calculation.

Next comes the question of the initial conditions. Most calculations assume a bare WD with an unperturbed temperature and density distributions. But in reality, since many flashes can take place on the surface of the WD, mass can be ejected and heat can flow into the WD, the aforementioned initial conditions are obviously inappropriate. As we shall see, even high accretion rates lead to flashes, which are not very powerful but still cause some mass loss. As a consequence, the cumulative effect of successive flashes, whether small or large, cause a secular change in the structure of the WD and dramatically affect the long term outcome. The tacit assumption is that any initial conditions will converge to a unique dynamic behavior after a sufficiently long time.

In view of the above basic questions, we calculated a series of models in which the above limitations were lifted and pose the following questions: Does the mass of the WD increase? And if the answer is yes, how far can it increase and can it reach the Chandrasekhar mass? Is a WD accreting at a high rate a viable model for the progenitor of a SN of any kind?

3. Results

In contrast with Fujimoto (1982), we find that the hydrogen does not burn steadily, but does so in flashes instead. Hydrostatic calculations do not properly describe the problem.

We show in fig. 1 the general behavior of the total luminosity as a function of time. The hydrogen burning is unstable, and exhibits cyclic behavior with a period of 7-8 years. Over the cycle, the luminosity changes by over a factor of 10, while the changes from cycle to cycle are quite small. The flashes are characterized by a very sharp rise and a much slower decline. A refined picture is shown fig. 2, where we see that the system becomes super-Eddington for a couple of years during which the rate of decline in luminosity is relatively low, but towards about three quarters of the cycle, the rate of decline increases. The minimum state is very short.

Despite the fact that the luminosity during the flash is above the Eddington luminosity, no mass loss developed. This is because the luminosity was used to expand the thin hydrogen shell, without allowing the velocity to reach the escape speed.

The burnt hydrogen accumulates as helium. It is important to note that the burning is complete, no hydrogen is left and the composition of the envelope is that of the original heavy elements, surrounded by a thick layer of He. At this phase, the system appears as a perfect recurrent object which accretes the ashes onto the original WD. Hardly any secular changes are noticeable during the first few cycles.

The surprise appeared after 4153 cycles, once the accreted mass accumulated to $10^{-2}M_{\odot}$. The temperature inside the helium layer reached 1.2×10^{8}K (see fig. 3), while the WD core

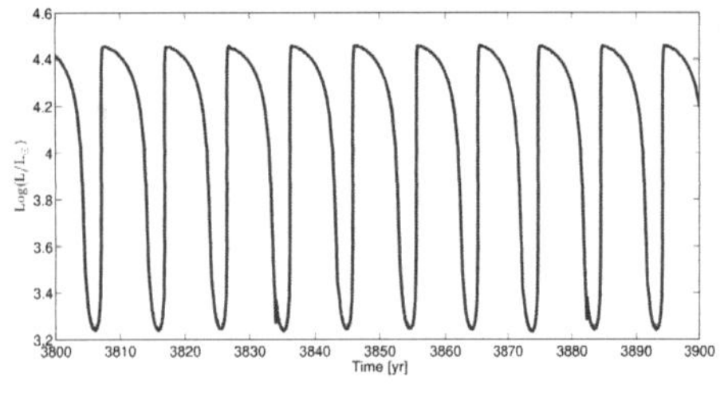

Figure 1. The secular sequence of flashes.

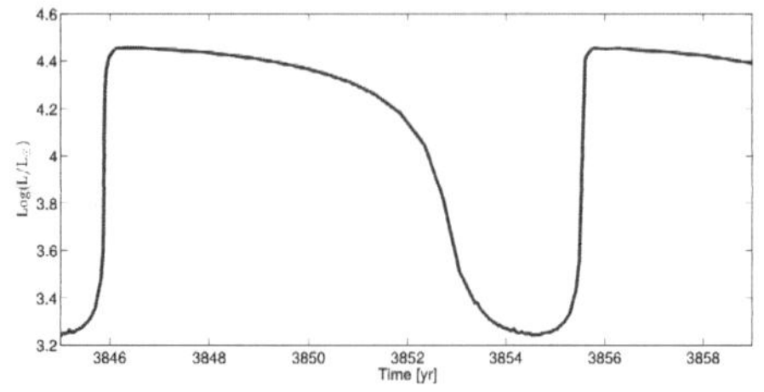

Figure 2. A flash cycle in more detail.

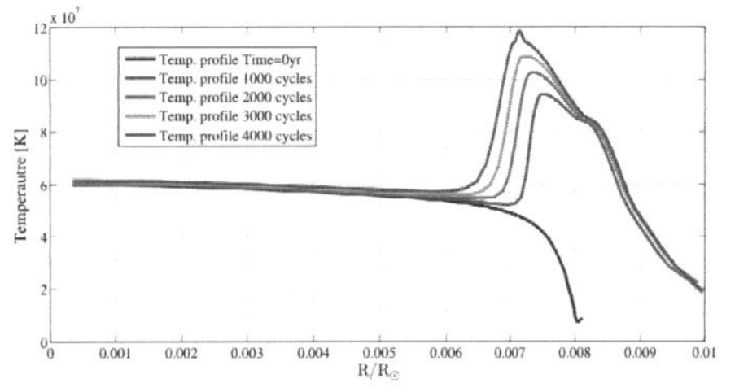

Figure 3. Temperature profiles.

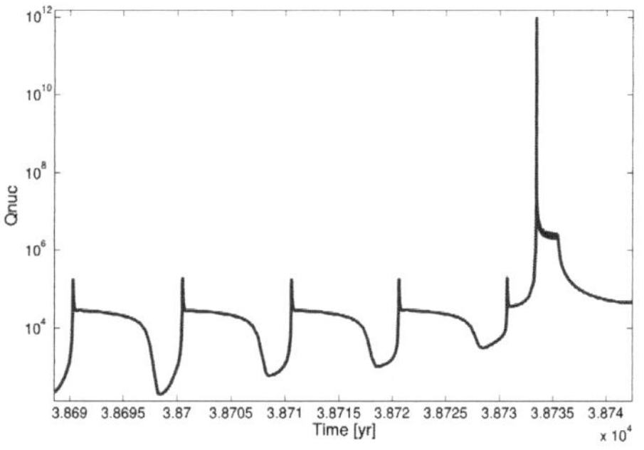

Figure 4. The approach to the giant flash which ejected 1/3 of the accreted envelope.

temperature increased during these 10,000 years of accretion by 3% (from 6×10^7K to 6.2×10^7K). At that point, helium ignited in a strong flash (see fig. 4). The peak nuclear energy generation reached $10^{12} erg/sec$ and the maximum temperature reached a peak of 7×10^8K. Note that the peak nuclear luminosity, which is contributed solely by the hydrogen burning, hardly changes in the hydrogen flashes that precede the strong helium flash, on the other hand, the minimum luminosity rises. This is due to the slow rise in helium burning.

The very strong He flash caused a temperature wave which propagated to the hydrogen burning at the top of the pure helium layer, raised it to a high temperature, accelerated the burning accordingly, and caused an ejection of about 1/3 of the helium layer accreted. This phenomenon repeated two more times with the outcome that the entire accreted helium layer was ejected.

4. Energetics: How come?

As is well known, the energy released in helium burning per unit mass is insufficient to eject it from the (original) surface of the WD. According to the simple classical estimate, the gravitational binding energy of the envelope is given by $E_{grav} = GM_{WD}m_{env}/R_{WD}$, while the available energy is $E_{nuc} = Qm_{env}$. As the envelope is very small and compact, the radius which enters the binding energy calculation is essentially the radius of the white dwarf. Under these conditions, one finds that $E_{nuc,Helium} < E_{grav}$ and no mass can be ejected through helium burning.

The present results implies that one of the assumptions in the simplest calculation breaks down. Indeed, as the mass accumulated over more than 4000 cycles, a very small but continuous and steady change builds up in the accreted envelope. As a consequence, the envelope heats to a large extent through adiabatic compression.

When the small fraction of the luminosity from the stellar boundary layer was artificially omitted, the final result hardly changed. The envelope expanded mostly through adiabatic compression and an envelope with a large radial extent was created. The WD attempted to become a red giant. Thus, most of the energy needed for the ejection of the envelope was supplied to it during the accretion process and not during the eruption. Once the envelope became bloated, the mass ejection became possible and indeed took place.

The theory of super-Eddington luminosities predicts a decrease in the effective radiation absorption coefficient (Shaviv 1998) which results from instabilities which lead to inhomogeneities. We rerun the problem with the proper correction to the effective radiative absorption coefficient (which is part of the corrections required by the theory) and found practically the same result: There is no net mass accreted by the WD in the long run. The next correction by the theory requires a change in the mass loss and we believe that its introduction will not change the outcome.

5. Conclusions

The general belief until now was that the high accretion rate prevents degeneracy, thereby giving rise to quiet burning. We find that this is not the case. The accretion process is accompanied by thermonuclear pulses having luminosity increases of more than a factor of ten.

We also find that all the accreted hydrogen burns out completely and the ejecta therefore contain no hydrogen. This implies that the non existence of hydrogen in observed ejecta does not imply that the accreted matter did not contain any hydrogen as well. Recall that the lack of hydrogen in the ejecta is considered as one of the strongest observational evidences for the idea of helium accretion.

Last, the high accretion rate does not lead in this case to a secular increase in the mass of the white dwarf. On the contrary, the WD is even eroded through the giant eruptions.

We guess that a more massive WD behaves in the same way. The cycles will be shorter but we expect the higher mass WDs to have a similar ratio between secular changes and the cycle length, and therefore give rise to essentially the same result. This point is now under investigation.

This research has been supported by the Israel Science Foundation, grant 1589/10.

References

Fujimoto, M. Y., ApJ, **257**, 767, (1982)
Kovetz, A., ApJ, **495**, 401, 1998.
Nomoto, K. ApJ, **257**, 780, 1982.
Podsiadlowski P., Mazzali P., Lesaffre P., Han Z., Förster F., New Astron. Rev., **52**, 381, (2008)
Prialnik, D. & Kovetz, A. ApJ., **445**, 789, (1995).
Shaviv, N. ApJ., **494**, 177, (1998)
Weidemann, V., A&A, **363**, 647, (2000)

On identifying the neutron star that was born in the supernova that placed ^{60}Fe onto the Earth

R Neuhäuser, N Tetzlaff, T Eisenbeiss, M M Hohle

Astrophysikalisches Institut, University Jena, Schillergäßchen 2-3, D-07745 Jena, Germany

E-mail: rne@astro.uni-jena.de

Abstract. Recently, ^{60}Fe was found in the Earth crust formed in a nearby recent supernova (SN). If the distance to the SN and mass of the progenitor of that SN was known, then one could constrain SN models. Knowing the positions, proper motions, and distances of dozens of young nearby neutron stars, we can determine their past flight paths and possible kinematic origin. Once the birth place of a neutron star in a SN is found, we would have determined the distance of the SN and the mass of the SN progenitor star.

1. Introduction: Link between 60Fe as SN debris on Earth and Neutron Stars

^{60}Fe was found in the Earth crust, which arrived on Earth $\sim$ 2 Myr ago (Knie et al. 2004). Production of ^{60}Fe in red giants cannot explain the observed peak, so that its origin can only be a nearby (40-100 pc) supernova (SN). If one could estimate the mass ejected by the SN and the mass of the progenitor star, one could constrain SN models. To estimate the amount of ^{60}Fe ejected, one needs to know the uptake factor (fraction of material deposited in the crust among the whole material arrived on Earth), the SN distance, the half-life of ^{60}Fe, and the time since the SN (i.e. the sum of the time mentioned above, $\sim$ 2 Myr, since the deposition in the crust, plus the travel time from the SN to the Earth's crust through interstellar space and Earth atmosphere). The uptake factor is probably few per cent, but hardly known with high precision; it is possible to constrain the factor, if other material would also be found in the Earth's crust from the same SN event. The travel time from the SN to the Earth crust is most likely negligible compared to the $\sim$ 2 Myr, since the SN must have been nearby, roughly 40 pc (within a factor of 2): If the SN would have been more distant than $\sim$ 100, the SN ejecta would not have been able to penetrate into the Solar System given the Solar wind ram pressure; if the SN would have been closer than $\sim$ 10 pc, a large cosmic ray flux on Earth would have caused major climatic changes and a large mass extinction of life. Hence, the distance towards the SN remains uncertain within the limits given, so that one cannot derive the amount of ^{60}Fe ejected from the amount found on Earth. Also, the mass of the SN progenitor star remains unknown.

If we can identify the neutron star (NS) born in that SN and, by tracing back the motion of that NS, also the location of that SN event, we would know the distance towards the SN, i.e. could solve the degeneracy. We may even be able to determine the mass of the progenitor star from the difference between the age of the parent association (if the NS was born in an association) and NS kinematic age (flight time), if all stars were formed at once in that association, as it is assumed by star formation theory and observed for most clusters. Such results would be very important and valuable for SN theory, ejecta models, nucleosynthesis, etc.

2. Tracing back neutron stars to their birth place in a supernova

NSs are born when massive stars ($\sim$ 8 to 30 $M_{\odot}$) explode as SNe after their short life-times of a few Myr. Hence, many of them should still be embedded in their star forming region, stellar association, or cluster, when they explode as SNe. In such SNe, NSs are formed, which receive a kick and fly away with large space motion. Given the measured 2D proper motions of NSs, one can construct the likely 3D (and 1D radial velocity) distributions (Arzoumanian et al. 2002, Hobbs et al. 2005). NSs also cool down fast after the SN, but are still hot enough for detection as faint optical or bright X-ray sources for up to 5 Myrs (see cooling curves in Gusakov et al. 2005, Popov et al. 2006). Young NSs are detectable (and have been detected) as optical and/or X-ray sources within 1 kpc (e.g. review in Haberl 2007). If the SN takes takes place in a binary system, then the companion star gets ejected and may then be observable as a run-away star (Blaauw 1961); the exact direction and velocity depends on the former binary orbit and the location of the NS and the companion during the SN. In principle, it is possible to trace back the motion of a NS to find out, whether it may have flown through a star forming region or young stellar association. If that is the case, and if the present mass function of that association is consistent with at least one very massive star, i.e. with at least one past SN, then the NS may have been born in that association. To get additional evidence for such a scenario, and also in case that a particular NS could have flown through more than one association, it would be best, if one could also identify a run-away star as former companion of the SN progenitor, also to have been at the same time at the same location as the NS and the association. Then, the place of origin and the age of the NS would be known, as well as the distance towards the SN. It is indeed possible to trace back both NSs and run-away stars to find close possible encounters in space and time, i.e. evidence for a SN event in a binary, e.g. Hoogerwerf et al. (2001), Tetzlaff et al. (2011b). Even without the identification of a former companion, the parent association of a NS can be found (Tetzlaff et al. 2009, 2010, 2011b). Hence, it might be possible, but not guaranteed, to find the NS that was formed in the SN that also placed ^{60}Fe onto Earth.

3. Tracing back neutron star RXJ0720 and a run-away star as former companion

To identify potential parent SN sites of a NS, we calculate the trajectories backwards into the past for the NS, the centre of all nearby young associations, and all known young nearby run-away stars. The sample of known young nearby NSs can be obtained from the ATNF database[1] (Manchester et al. 2005). A list of all known young nearby associations has been put together in Tetzlaff et al. (2010, 2011b). For a catalog of nearby run-away stars, see Tetzlaff et al. (2011a). For details on the procedure, see Tetzlaff et al. (2010, 2011b). For run-away stars and associations, distances and 3D space motions are known. For NSs, distances and 2D proper motions on sky are known, but the radial velocity is unknown for most NSs. Hence, we perform Monte-Carlo simulations by drawing the radial velocity from the most likely distribution.

Recently, Eisenbeiss (2011) re-determined the parallaxe of the NS RXJ0720 using archival Hubble Space Telescope data yielding $\pi = 3.6 \pm 1.6$ mas (Eisenbeiss 2011), i.e. $\sim$ 278 pc (192-500 pc); this is consistent with an earlier determination by Kaplan et al. (2007), (270-530 pc) within large error bars. We use the parallaxe and proper motion from Eisenbeiss (2011) and vary the unknown radial velocity v_r in a Monte-Carlo simulation using (a) the known distribution of NS velocities and, after we found that v_r for RXJ0720 is probably relatively small, (b) a flat distribution from -300 to $+300$ km/s).

There are nine possible parent associations, through which RXJ0720 could have flown, i.e. in which it could have been born (TWA, Tuc-Hor, β Pic-Cap, HD 141569 group, AB Dor group, Col 140, Tr-10, CarA, Argus). Then, there are also three possible former run-away companion stars (i.e. former companions of the SN progenitor star), which could have been at the same

[1] http://www.atnf.csiro.au/research/pulsar/psrcat/

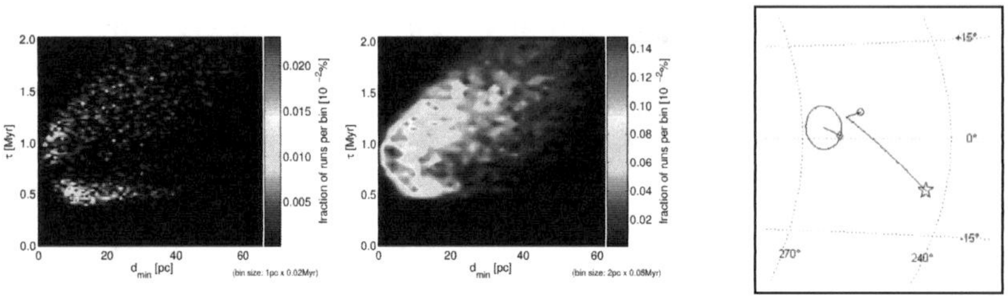

Figure 1: Contour plots for age (t in Myr since SN, left) and separation (d_{min} in pc, middle) distributions for the two cases, where NS RXJ0720 and HIP 43158 as run-away star (with radial velocity from Hobbs et al. 2005, left, and for a flat distribution from -300 to $+300$ km/s, middle) could have been at the same place (in the Tr-10 association) at the same time in the past, namely ~ 0.7 to ~ 1 Myr ago (Tetzlaff et al. 2011b). Right: Past trajectories (~ 1 Myr, full lines) for RXJ0720 (star), the run-away star HIP 43158 (small circle), and Tr 10 (diamond, radius 23 pc as large circle) projected on a Galactic coordinate system (for a particular set of input parameters). Present positions are marked with the symbols.

place at the same time as RXJ0720, for a certain NS radial velocity, namely HIP 43158, 57269, and 76304 (run-away stars from Tetzlaff et al. 2011a). If HIP 57269 or 76304 would have been the former companion, the SN would have taken place at ~ 27 or ~ 52 pc, respectively, but the current measured distance of RXJ0720 would be inconsistent with the predictions in such cases. For HIP 43158, however, the current measured distance of RXJ0720 is fully consistent with the prediction in case that it would have been the former companion of the RXJ07020 SN progenitor: The SN would have been ~ 1 Myr ago in Tr-10 at the distance of ~ 370 pc (during the SN), Fig. 1. Then, RXJ0720 should now have a radial velocity of ~ -100 km/s. The difference between the kinematic age of the NS (~ 1 Myr) and the Tr-10 association age (15–35 Myr, Tetzlaff et al. 2010 and references therein) then yield the life-time and mass of the progenitor star, namely ~ 13 to 15 $M_\odot$. Hence, RXJ0720 is not the NS born in the SN that placed the detected Fe onto Earth. See Tetzlaff et al. (2011b) for details.

3.1. Acknowledgments
We would like to thank DFG in SFB TR 7 and Carl-Zeiss-Stiftung for financial support.

4. References
[1] Arzoumanian Z, Chernoff D F, Cordes J M 2002 *ApJ* **568** 289
[2] Blaauw A 1961 Bull. Astron. Inst. Netherlands *15* 265
[3] Eisenbeiss E 2011 PhD thesis U Jena
[4] Gusakov M E, Kaminker A D, Yakovlev D G, Gnedin O Y 2005 *MNRAS* **363** 555
[5] Haberl F 2007 *ApSS* **308** 181
[6] Hobbs G, Lorimer D R, Lyne A G, Kramer M 2005 *MNRAS* **360** 974
[7] Hoogerwerf R, de Bruijne J H J, de Zeeuw P T 2001 *A&A* **365** 49
[8] Kaplan D L, van Kerkwijk M H, Anderson J 2007 *ApJ* **660** 1428
[9] Knie K, Korschinek G, Faestermann T, Dorfi E A, Rugel G, Wallner A 2004 *PRL* **93** 1103
[10] Manchester R N, Hobbs G B, Teoh A, Hobbs M 2005 *AJ* **129** 1993
[11] Popov S B, Grigorian H, Blaschke D 2006 *RPC* **74** 5803
[12] Tetzlaff N, Neuhäuser R, Hohle M M 2009 *MNRAS* **400** L99
[13] Tetzlaff N, Neuhäuser R, Hohle M M, Maciejewski G 2010 *MNRAS* **402** 2369
[14] Tetzlaff N, Neuhäuser R, Hohle M M 2011a *MNRAS* **410** 190
[15] Tetzlaff N, Eisenbeiss E, Neuhäuser R, Hohle M M 2011b *MNRAS* in press, astro-ph/1107.1673

Nuclear Physics in Astrophysics V IOP Publishing
Journal of Physics: Conference Series **337** (2012) 012053 doi:10.1088/1742-6596/337/1/012053

Direct measurement of the $^2H(\alpha,\gamma)^6Li$ cross section at energies of astrophysical interest

A Bellini[1] **for the LUNA-Collaboration**

[1] Università di Genova and INFN Sezione di Genova, Genova, Italy

E-mail: `abellini@ge.infn.it`

Abstract. The knowledge of the ^{2}H(α,γ)^{6}Li cross section is very important to understand the production of ^{6}Li during the Big-Bang Nucleosynthesis (BBN). Presently, no direct measurement of this reaction in the energy region of interest exists (the only direct measurements actually available are above 1 MeV in the center of mass reference frame and around the 711 keV resonance). The existing data in the energy region of interest for BBN (40 keV < E_{cm} < 400 keV) have been taken with indirect techniques which are affected by large uncertainties because they strongly depend on theoretical assumptions. The direct measurement is a difficult task because the expected cross section is of the order of a few pico-barn. Thanks to the natural shielding offered by the about 1400 meters of rocks, the cosmic background is reduced by a factor 10^6 at the Laboratori Nazionali del Gran Sasso (LNGS). Here, at the LUNA (Laboratory for Underground Nuclear Astrophysics) facility it is possible to explore with a direct measurement of the ^{2}H(α,γ)^{6}Li reaction cross section in the BBN energy region. This measurement can be done with a ^{4}He intense beam hitting a windowless deuterium gas target. The emitted photons are detected with a High-Purity Germanium detector (HPGe) in close geometry. In this paper the experimental setup will be described and the measurement strategy based on the first tests already performed will be discussed.

1. The ^{2}H(α,γ)^{6}Li nuclear reaction

One of the main achievement of the standard Big-Bang model of the Universe is the consistency of the observed light element abundances with the Big-Bang Nucleosynthesis predictions. The primordial nucleosynthesis ended with the creation of the lightest nuclei, ^{2}H, ^{3}He and ^{4}He. Also some ^{7}Li was produced and probably ^{6}Li too. ^{6}Li has been found in very old low metallicity halo stars, in an unexpectedly high amount (2-3 orders of magnitude higher compared to available BBN network predictions [1]). The only way to produce ^{6}Li during the BBN is the ^{2}H(α,γ)^{6}Li direct capture, with a very weak expected cross section. The global picture is shaded by the the fact that, up to now, direct measurements of this reaction at BBN energies do not exist.

A summary of the ^{2}H(α,γ)^{6}Li available S-factor is shown in Figure 1. Direct measurements cover only the energy region between 1 and 3.5 MeV and the resonance at 711 keV. At BBN energies, only indirect measurements based on Coulomb breakup exist which are affected by large uncertainties. Therefore current estimates differ by about one order of magnitude [7]. For this reason a direct measurement of the ^{2}H(α,γ)^{6}Li S-factor in the BBN energy region is mandatory and the LUNA collaboration is performing this measurement for the first time.

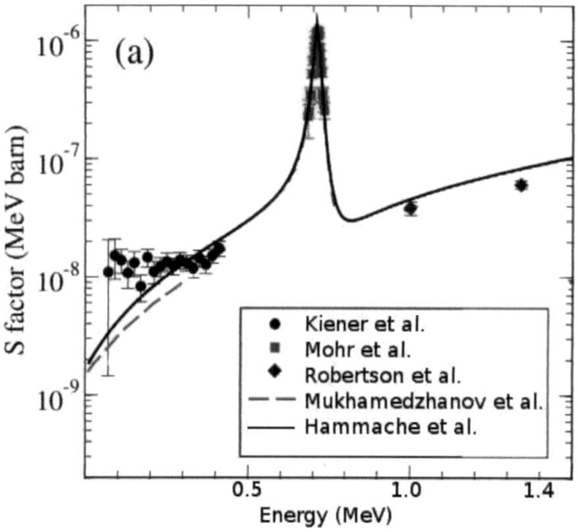

Figure 1. $^2H(\alpha,\gamma)^6Li$ S-factor as a function of the interaction energy. Direct measurements: Robertson *et al.* [2] (blue diamonds), and Mohr *et al.* [3] (red squares). Coulomb breakup experiments: Kiener *et al.* [4] (black circles), Mukhamedzhanov *et al.* [5] (red long-dashed line) and Hammache *et al.* [6] (black line).

2. The experimental apparatus

Since the expected cross section in the BBN energy region is of the order of few picobarns, the signal-to-noise ratio must be improved as much as possible. The LUNA facility is located underground at the Laboratori Nazionali del Gran Sasso (LNGS) where about 1400 m of rocks (3800 m.w.e.) reduce the muon flux of a factor 10^6 and the neutron flux of a factor 10^3 with respect to the surface. At LUNA a custom 400 keV electrostatic accelerator is used to meet requirements of nuclear astrophysics experiments [8]. The machine can provide high intensity (up to 300 μA) α-beams with an excellent energy resolution (less then 100 eV) over long run periods. The setup consists of a D_2 windowless differentially pumped gas target. A High-Purity Germanium detector (HPGe) with 135% relative efficiency is used for γ detection [9]. The detector is in close geometry, 20 mm far from the beam line. A calorimeter is used for beam intensity measurements. In order to minimize the natural background, all the setup is surrounded by a 4π lead shielding. Around the lead shielding an anti-Rn-box is built to keep a constant ^{222}Rn daughters γ counting rate. The Q-value of the reaction is 1.473 MeV and the Region-of-Interest (RoI) for the $^2H(\alpha,\gamma)^6Li$ reaction with a 400 keV α beam ($E_{cm} \approx 130$ keV) is given by $E\gamma = Q + E_{cm} + \Delta E_{Doppler} - \Delta E_{Recoil} = 1585 \div 1625$ keV.

The $^2H(\alpha,\alpha)^2H$ Rutherford scattering of the beam inside the gas target is a side effect that has to be dealt with. It induces the $^2H(^2H,n)^3He$ and $^2H(^2H,p)t$ reactions. While the $^2H(^2H,p)t$ reaction is not a problem, the neutrons produced by the $^2H(^2H,n)^3He$ reaction (average neutron energy 2.45 MeV) can induce (n,γ) and (n,n$'\gamma$) reactions on different target atoms like Pb, Ge, Cu, and Fe. These reactions generate a considerable beam-induced background in the γ-ray spectrum that has to be dealt with. In order to reduce the diffused deuteron mean free path a steel tube with squared cross section is placed on the beam axis, as shown in Figure 2.

The amount of Cu in the setup is reduced as much as possible since a (n,n$'\gamma$) peak on ^{65}Cu at 1623.4 keV is inside the ^{2}H+α RoI. It is possible to monitor the neutron produced by the $^2H(^2H,n)^3He$ reaction by measuring the proton produced by the $^2H(^2H,p)t$ reaction since both fusion reaction cross sections are very well known. For this reason a silicon detector is placed above the steel tube inside the scattering chamber. An aluminum foil is placed between the silicon detector and the deuterium target in order to stop ^{3}He, tritium ions and α scattered particles. Figure 3 shows a silicon detector measured spectrum while Figure 4 shows the simulation done with the LUNA Monte-Carlo code based on GEANT3 [10].

Moreover, with a germanium detector, the neutron-induced peaks are usually identifiable

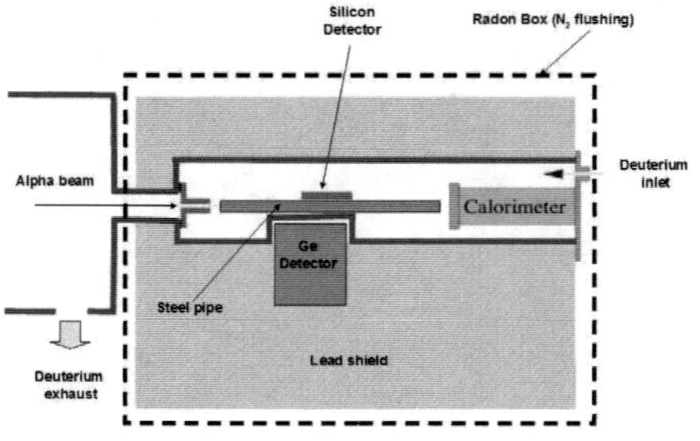

Figure 2. Scattering chamber scheme. In the scheme are well identifiable the HPGe and Si detectors, the calorimeter for beam current measurements and the steel tube used for reducing the scattered deuterium mean free path.

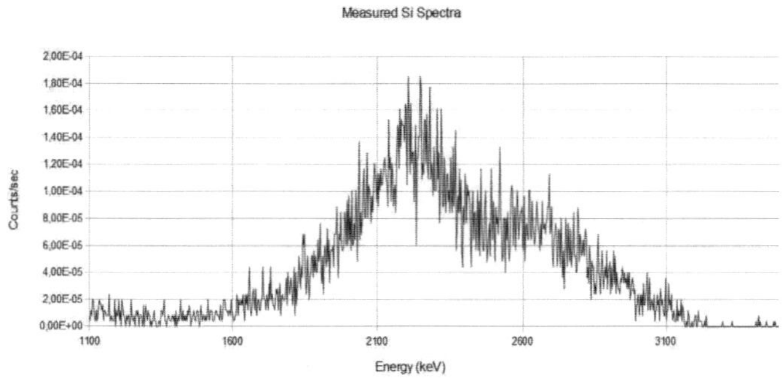

Figure 3. Silicon detector spectrum of the proton produced by the ^{2}H(^{2}H,p)t reaction measured with a 400 keV α beam, 226 μA beam current and 0.2 mbar deuterium gas target pressure.

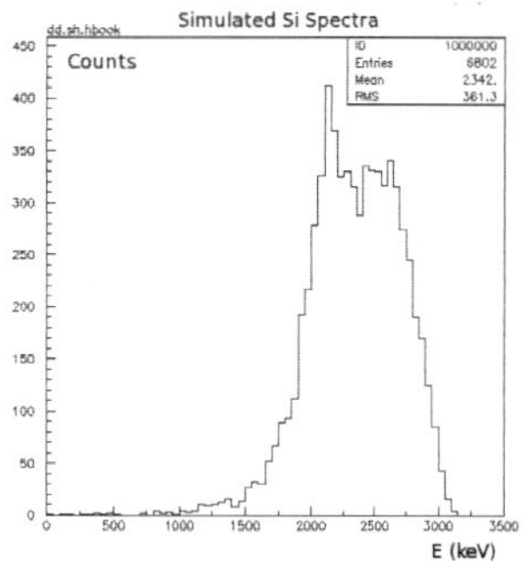

Figure 4. Silicon detector spectrum of the proton produced by the ^{2}H(^{2}H,p)t reaction simulated with the LUNA Monte-Carlo code based on GEANT3.

because their widths are normally larger then that for γ-ray-induced events. As indicated by Knoll [11], it is possible to calculate the neutron produced by measuring the γ counting rate in the neutron inelastic scattering edge on ^{72}Ge at 693.4 keV. This is a rough estimation of the neutron fluence because the calculation depends on detector and setup geometries but can be compared with the Si detector measurements.

3. Measurements and future outlook

In spite of the natural background reduction offered by LNGS and by the lead shielding, the beam-induced background make this measurement not an easy task. Figure 5 shows a HPGe ^{2}H+α spectrum measured with the best experimental conditions and compared with the natural background. The natural background γ counting rate in the ^{2}H+α RoI is $\sim 55\ \frac{counts}{day}$, while the total counting rate is $\sim 446\ \frac{counts}{day}$. According to Hammache ^{2}H+α cross section value [6] in these experimental conditions we expect to have $\sim 44\ \frac{counts}{day}$, ten times lower than the total γ counting rate. Therefore the beam-induced background is the dominant γ contribution and it should be measured with the best possible accuracy. For this reason a possible solution was studied, that is to produce a similar beam-induced background by measuring the ^{2}H(α,γ)^{6}Li with a 280 keV α beam. The ^{2}H(α,γ)^{6}Li signal is obtained by the subtraction of these two spectra. At this lower energy the ^{2}H+α RoI is 1550÷1580 keV, so there is no overlapping between the two different energy RoIs. The GEANT4 simulation are encouraging in this concern. Measurements will be done during 2011.

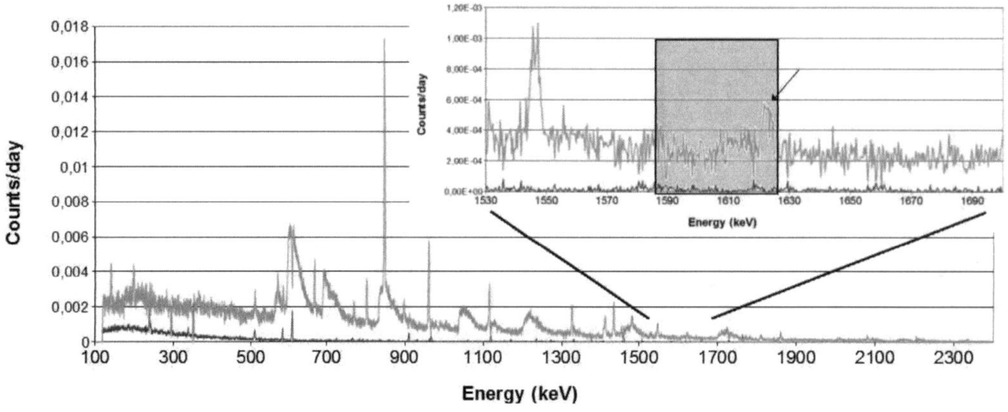

Figure 5. HPGe ^{2}H+α spectrum measured for 9 days with a 400 keV α beam, 259 μA beam current and 0.2 mbar deuterium gas target pressure (yellow) and natural background spectrum (pink). In the ^{2}H+α spectrum are well visible the inelastic scattering n-edges on Ge isotopes. The panel on the top-right corner shows a zoom of both spectra around the Region of Interest of the ^{2}H(α,γ)^{6}Li reaction, marked by the light blue box. The black arrow indicate the (n,n$'\gamma$) peak on ^{65}Cu at 1623.4 keV.

References

[1] NACRE, Nuclear Astrophysics Compilation of Reaction Rates, http://pntpm.ulb.ac.be/Nacre/nacre.htm
[2] Robertson R G H *et al.* 1981 *Phys. Rev. Lett.* **47** 1867-70
[3] Mohr P *et al.* 1994 *Phys. Rev.* C **50** 1543-49
[4] Kiener J *et al.* 1991 *Phys. Rev.* C **44** 2195-2208
[5] Mukhamedzhanov A M *et al.* 1995 *Phys. Rev* C **52** 3483-87
[6] Hammache F *et al.* 2010 *Phys. Rev.* C **82** 065803
[7] Descouvemont P and Kharbach A 1998 *Phys. Rev.* C **58** 1066-72
[8] Formicola A *et al.* 2003 *Nucl. Inst. and Meth.* A **507** 609-16
[9] Caciolli A *et al.* 2009 *Eur. Phys. J.* A **39** 179
[10] Arpesella C *et al.* 1995 *Nucl. Instrum. Meth.* A **52** 607
[11] Knoll G F 1986 *Radiation Detection and Measurements* (John Wiley and Sons Inc.)

Studies in Nuclear Astrophysics with an Optical Readout TPC (O-TPC) at HIγS

Moshe Gai
for the UConn-Yale-Duke-Weizmann-PTB-UCL Collaboration

Laboratory for Nuclear Science at Avery Point
University of Connecticut, 1084 Shennecossett Rd, Groton, CT 06340-6097, USA

E-mail: moshe.gai@yale.edu

Abstract. An Optical Readout Time Projection Chamber (O-TPC) operating with the gas mixture of CO_2(80%) + N_2(20%) at 100 torr with gamma beams from the HIγS facility of TUNL at Duke University were used to study the formation of carbon and oxygen during helium burning. Measurements were carried out with circularly polarized gamma-ray beams at energies: E_γ = 9.08, 9.38, 9.58, 9.78, 10.08, 10.38 and 10.68 MeV. We have begun the process of extracting complete angular distributions for the $^{16}O(\gamma,\alpha)^{12}C$ reaction and the $^{12}C(\gamma,3\alpha)$ reaction (eventually with energy bins of approximately 100 keV) in order to determine the values of S_{E1}, S_{E2} and the mixing phase ϕ_{12} of the $^{16}O(\gamma,\alpha)^{12}C$ reaction. The rate of carbon formation at high temperatures (T > 3 GK) was suggested to increase due to contributions from a higher lying 2^+ state. We have measured an angular distribution of (essentially) pure E2 transition at E_γ = 9.78 MeV of the $^{12}C(\gamma,3\alpha)$ reaction, providing conclusive evidence for the elusive 2^+_2 state in ^{12}C.

1. Introduction

The outcome of helium burning is the formation of the two elements: carbon and oxygen [1]. The ratio of carbon to oxygen at the end of helium burning has been identified three decades ago as one of the key open questions in Nuclear Astrophysics [1] and it is still today an open question. To solve this problem one must extract the p-wave [$S_{E1}(300)$] and d-wave [$S_{E2}(300)$] cross section factors of the $^{12}C(\alpha,\gamma)^{16}O$ reaction at the Gamow peak (300 keV) with an accuracy of approximately 10% or better. Our current knowledge of these astrophysical cross section factors is not near the required accuracy. In particular the data on the beta-delayed alpha-particle emission of ^{16}N does not determine whether the interference between the bound 1^- state at 7.117 and the broad (Γ = 420 keV) 1^- state at 9.585 MeV in ^{16}O is constructive or destructive leading to large S_{E1} = 80 keV-b or small S_{E1} = 10 keV-b, respectively, since the measured minima of χ^2 for the ^{16}N data are identical for both cases [2] as shown in Fig. 1.

Carbon is formed during stellar helium burning in the triple-alpha process, the $^8Be(\alpha,\gamma)^{12}C$ reaction, that is mostly governed by the contribution of the 0^+ Hoyle state at 7.654 MeV. At high temperatures (T > 3 GK) higher lying states in ^{12}C may contribute. Indeed a broad (Γ = 560 keV, Γ_γ = 0.2 eV) 2^+ state at 9.11 MeV in ^{12}C was included in the NACRE compilation [3] that increases the production of carbon at temperatures in excess of 1 GK by up to a factor of 15.

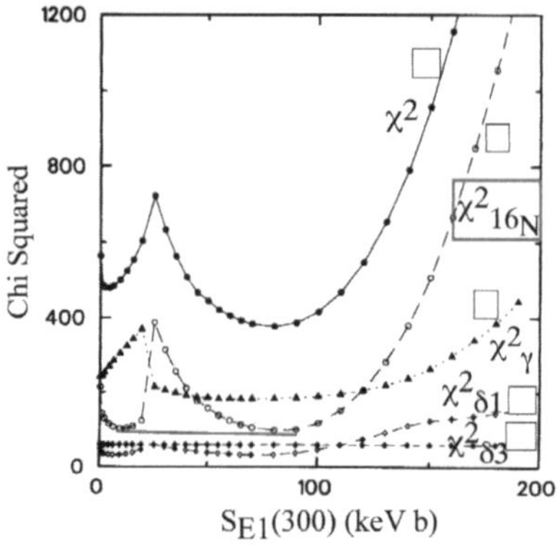

Figure 1. The χ^2 obtained from world data as shown in Ref. [2]. The χ^2 for the ^{16}N data show two minima at the same low values of χ^2 which does not allow us to distinguish between the destructive and constructive interference solutions as discussed in the text.

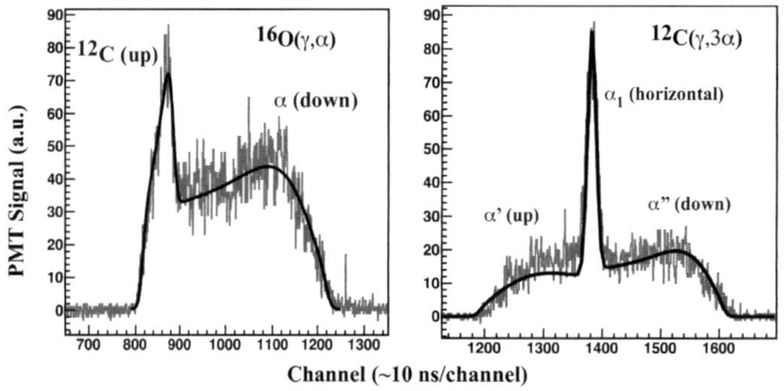

Figure 2. The measured PMT signal compared to the predicted line shape for ^{16}O (left) and ^{12}C (right) dissociation events.

2. Measurements of the photo-dissociation of ^{16}O and ^{12}C

We used our Optical-Readout Time Projection Chamber (O-TPC) [4] operating with CO_2 gas with circularly polarized gamma-ray beams extracted from the HIγS facility of TUNL at Duke University [5] to study the $^{16}O(\gamma,\alpha)$ and the $^{12}C(\gamma,3\alpha)$ reactions at E = 9.08, 9.38, 9.58, 9.78, 10.08, 10.38 and 10.68 MeV. The outgoing particle resulting from the photo-disscoiation of target nuclei are fully determined by the tracks recorded in the O-TPC in three dimensions. Thus all relevant kinematical variables are measured by the O-TPC [4]. The total energy deposited (grid charge-signal) in the O-TPC detector is determined by the Q-value for the dissociation event; $E_{total} = E_\gamma - Q$; Q = 6.227, 7.162 and 7.275 for the dissociation of ^{18}O, ^{16}O and ^{12}C, respectively. The Q-values for ^{16}O and ^{18}O dissociations are sufficiently different (935 keV) and differ considerably more than the beam width of FWHM $\approx$ 300 keV, hence these events are well separated in the total energy spectrum measured by the charge signal. However one major draw back for using CO_2 gas is that the difference of Q-values (112 keV) for the dissociation of ^{16}O and ^{12}C is considerably smaller than the beam width and comparable to the detector resolution of approximately 90 keV [4]. In addition the larger quenching factor for the low energy ^{12}C projectiles leads to a smaller grid charge-signal from the dissociation of ^{16}O with very similar energy as for the dissociation of ^{12}C. Hence the total energy deposit cannot be used to separate (and thus identify) ^{12}C and ^{16}O dissociation events.

Journal of Physics: Conference Series **337** (2012) 012054 doi:10.1088/1742-6596/337/1/012054

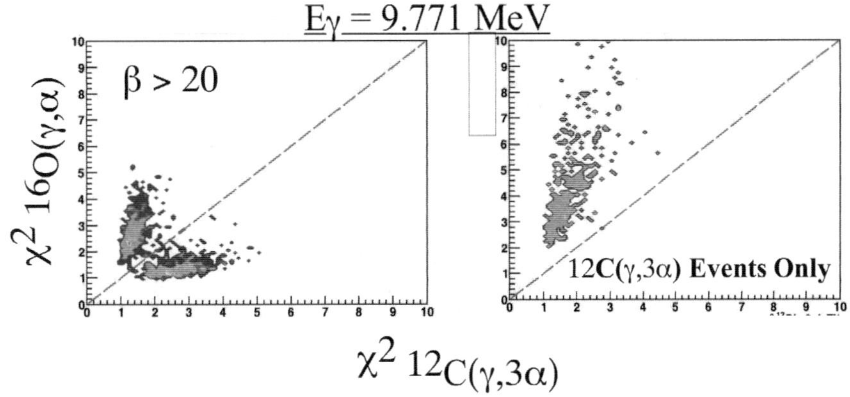

Figure 3. The χ^2 obtained by fitting the PMT signal with the line shape predicted for ^{16}O and ^{12}C events and for selected well identified ^{12}C events (right).

To identify and distinguish ^{12}C dissociation events we relied on the line shape of the PMT signal which is very well determined by the calculated dE/dX along the track. In Fig. 2 we compare the observed PMT signal to the calculated line shape for a co-linear α + ^{12}C event. In this calculation we used the drift velocity of 1.14 cm/μs (at 100 torr) measured with a well collimated ^{148}Gd source [4]. The measured line shape of the PMT signal is compared to 181 calculated functions of the out-of-plane angle (in $\beta = 1°$ increments) to determine the best fit. The calculated line shape shown in Fig. 2 has essentially only one free parameter, the out of plane angle (β). A good χ^2 is found for the ^{16}O dissociation events shown in Fig. 2. The line shape of ^{12}C dissociation events arise from a considerably more complicated dE/dx of the three body (non-colinear) decay pattern. The line shape of ^{12}C events requires 181x9 = 1629 functions that are calculated with sufficient angular bin size ($\beta = 1°$) as well as sufficient bin size (30°) for the θ and ϕ angles of the two alphas (α' and α'') from the decay of ^8Be. For (α_1) decay into the excited states, as shown in Fig. 2, we also considered the energy of the excited ^8Be. A good χ^2 is found for the ^{12}C dissociation events shown in Fig. 2.

Unfortunately the noise level in the CCD camera was too high and it did not permit line shape analysis of the pixel-content. Hence only out of plane events with out of plane angle β larger than 20° (approximately 40% of the data) could be analyzed in the current setup. A new cleaner camera has been installed that permits including all data. In addition the resolution of the optical system did not permit resolving the two outgoing alphas emitted from the decay of the ground state of ^8Be. Due to the poor resolution such decays most of the time appear co-linear in the image recorded by the CCD camera but are clearly distinguished from ^{16}O events in the PMT signal. In contrast the two outgoing alphas emitted in the decay of the first excited state of $^8Be^*(3.0)$ are well resolved as shown in Fig. 2. The very low energy of the $^8Be^*(3.0)$ yield a decay pattern which are almost as for a decay in rest.

3. A measurement of the $^{12}C(\gamma,\alpha)$ reaction.

The in plane angle (α) measured by the track registered in the CCD image and the out-of-plane angle (β) measured by the Time projection signal of the PMT allow us to deduce for each event the scattering angle (θ) and the azimuthal angle (ϕ) of the polar coordinate system used in scattering theory: $\cos\theta = \cos\beta$ x $\cos\alpha$ and $\tan\phi = \tan\beta/\sin\alpha$. The so obtained angular distribution is shown in Fig. 4 together with that predicted for a pure $0^+ \rightarrow 2^+$ E2 transition. For these data we used only in plane ($\beta < 20°$) data for which the scattering angle (θ) is determined with high accuracy.

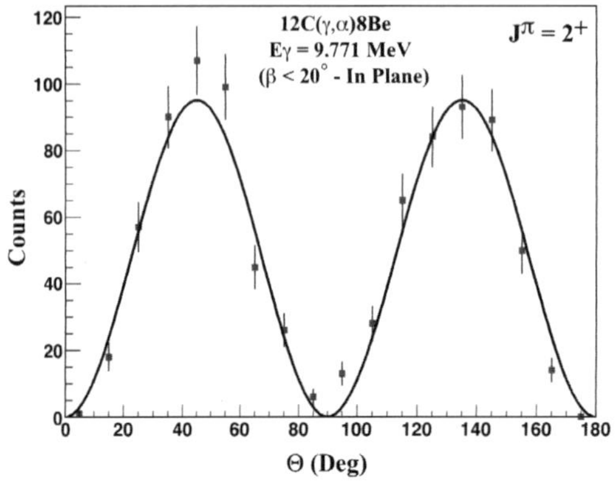

Figure 4. Measured angular distribution for in plane ($\beta < 20°$) $^{12}C(\gamma, 3\alpha)$ events compared to the prediction for a pure $0^+ \rightarrow 2^+$ E2 transition.

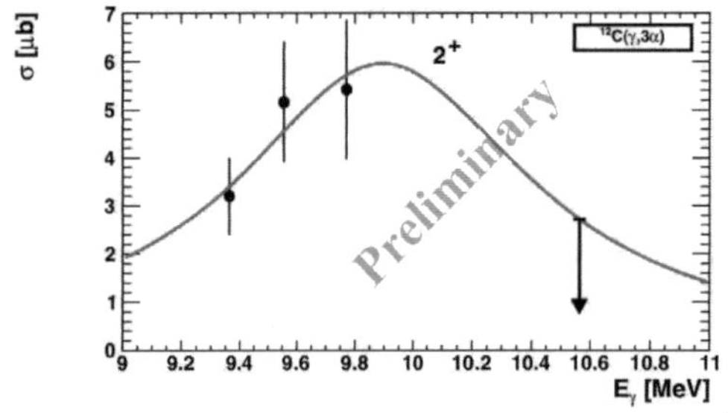

Figure 5. Preliminary measured excitation curve for the $^{12}C(\gamma, 3\alpha)$ reaction.

We also measured an excitation curve for the $^{12}C(\gamma, \alpha)$ reaction. The accuracy of these data is limited by the measured beam intensity. A considerably more accurate beam monitoring system has been developed using the known cross section of the photo-dissociation of deuterium. The result of the limited accuracy study is shown in Fig. 5.

4. Conclusions

Dissociation events from the $^{12}C(\gamma, 3\alpha)$ reaction were identified in our measurement using an O-TPC detector and clear evidence is observed for a pure E2 angular distribution most likely arising from a 2^+ state just below 10.0 MeV. These data are being remeasured with an improved setup including a CCD camera with lower background. This study is in progress.

5. References

[1] W.A. Fowler, Rev. Mod. Phys. 56(1984)149.
[2] R.E. Azuma *et al.*; Phys. Rev. C**50**(1994)1194.
[3] C. Angulo *et al.*; Nucl. Phys. **A656**(1999)3.
[4] M. Gai, M.W. Ahmed, S.C. Stave, W.R. Zimmerman, A. Breskin, B. Bromberger, R. Chechik, V. Dangendorf, Th. Delbar, R.H. France III, S.S. Henshaw, T.J. Kading, P.P. Martel, J.E.R. McDonald, P.-N. Seo, K. Tittelmeier, H.R. Weller, and A.H. Young; JINST **5**(2010)12004.
[5] H.R. Weller, M.W. Ahmed, H. Gao, W. Tornow, Y. Wu, M. Gai, R. Miskimen; Prog. Part. Nucl. Phys. **62**(2009)257.

Journal of Physics: Conference Series **337** (2012) 012055 doi:10.1088/1742-6596/337/1/012055

Microscopic calculation of the ^{3}He(α,γ)^{7}Be reaction rate using realistic interactions

Thomas Neff, Hans Feldmeier, and Karlheinz Langanke

GSI Helmholtzzentrum für Schwerionenforschung GmbH, Planckstraße 1, 64291 Darmstadt, Germany

E-mail: t.neff@gsi.de

Abstract. The cross sections for the ^{3}He(α,γ)^{7}Be and the ^{3}H(α,γ)^{7}Li radiative capture reactions are calculated in the fully microscopic Fermionic Molecular Dynamics approach using a realistic effective interaction obtained in the Unitary Correlation Operator Method. The model space is divided in an external region where bound and scattering states are described by antisymmetrized products of ^{4}He and ^{3}He/^{3}H ground states and an internal region where additional many-body wave functions obtained by variation after parity and angular momentum projection enlarge the Hilbert space. These additional configurations, representing polarized cluster configurations, are necessary for a successful description of the bound and scattering states. The calculated S-factor for the ^{3}He(α,γ)^{7}Be reaction is in good agreement with recent experimental data both in absolute normalization and energy dependence. In case of the isospin mirror reaction ^{3}H(α,γ)^{7}Li the calculated S-factor is larger than the experimental data by about 15%. Dipole matrix elements are analyzed in terms of overlap functions calculated from the A-body wave functions.

1. Introduction

The ^{3}He(α,γ)^{7}Be radiative capture reaction plays an important role in the solar proton-proton chains and determines the production of ^{7}Be and ^{8}B neutrinos [1, 2]. Together with the ^{3}H(α,γ)^{7}Li reaction it is the main source for the uncertainty in the primordial ^{7}Li production [3]. In recent years lots of experimental efforts went into remeasuring the cross reaction [4, 5, 6, 7, 8]. However it is still not possible to reach the low energies relevant for solar burning.

From the theory side this reaction has been investigated using simple potential models where ^{3}He and ^{4}He are treated as point-like particles interacting via an effective nucleus-nucleus potential, e.g., [9] or microscopic cluster models, e.g., [10, 11] where the ^{7}Be bound and scattering states are constructed from microscopic ^{3}He and ^{4}He clusters interacting via an effective nucleon-nucleon interaction. *Ab-initio* calculations using variational Monte Carlo [12] and no-core shell model wave functions [13] were used to calculate asymptotic normalization coefficients for the bound states but relied on potential models for the scattering states and the calculation of the capture cross sections. In [14] we presented the first calculation of the capture cross section using fully microscopic wave functions and a realistic effective interaction.

2. Unitary Correlation Operator Method

Starting from the realistic Argonne V18 interaction [15] we derive an effective low-momentum interaction using the unitary correlation operator method (UCOM). The basic idea of the UCOM approach is to explicitly include short-range central and tensor correlations into the many-body state by means of a unitary operator [16, 17, 18]. The correlation operators can also be mapped on the Hamiltonian to define a correlated interaction which not only contains two-body but also higher order contributions. However, the UCOM interaction is defined as the two-body part the correlated Hamiltonian. It provides the same phase shifts as the original interaction but behaves differently in many-body systems. No-core shell model calculations show that the two-body UCOM interaction gives a good description of s- and light p-shell nuclei [18].

3. Fermionic Molecular Dynamics

We employ fermionic molecular dynamics (FMD) as our many-body approach. FMD uses Slater determinants

$$|Q\rangle = \mathcal{A}\left\{|q_1\rangle \otimes \ldots \otimes |q_A\rangle\right\} \tag{1}$$

as intrinsic many-body basis states. The single-particle states are Gaussian wave-packets

$$\langle\vec{x}|q_k\rangle = \frac{(\vec{x}-\vec{b}_k)^2}{2a_k} \otimes |\chi_k^{\uparrow}, \chi_k^{\downarrow}\rangle \otimes |\xi_k\rangle\,, \tag{2}$$

where the complex parameters $\vec{b}_k$ encode the mean positions and momenta of the wave-packets and a_k their widths. The spins can assume any direction, isospin is ± 1 denoting a proton or a neutron. The wave-packet basis is very flexible — it contains harmonic oscillator shell model and Brink-type cluster states as special limiting cases. To restore the symmetries of the Hamiltonian the intrinsic basis states are projected on parity, angular momentum and total linear momentum. The total linear momentum projection decouples the internal motion of the clusters from their relative motion and the total center-of-mass motion.

For the ^{3}He(α,γ)^{7}Be reaction we divide the model space into two regions. In the external region bound and scattering states are described by ^{3}He and ^{4}He clusters in their FMD ground states. These microscopic wave functions can also be rewritten as resonating group (RGM) wave functions. Using that boundary conditions for bound and scattering states can be included by matching to Whittaker and Coulomb functions at the channel radius (a=12 fm) employing the microscopic R-matrix method developed by the Brussels group [19]. In the interaction region additional FMD many-body configurations that are obtained by variation after parity and angular momentum projection on spin-parity $1/2^+$, $3/2^+$, $5/2^+$ and $3/2^-$, $1/2^-$, $7/2^-$, $5/2^-$ are included. Furthermore a constraint on the radius of the intrinsic states is used to vary the distance between the clusters. By that polarized cluster configurations are generated.

4. Capture Cross Section

Up to energies of about 2.5 MeV only the capture from S- and D-wave scattering states into the $3/2^-$ and $1/2^-$ bound states has to be considered. The addition of polarized configurations is essential — using the frozen configurations only, ^{7}Be is bound by 200 keV. Including the polarized configurations the $3/2^-$ state is bound by 1.49 MeV and the $1/2^-$ state by 1.31 MeV with respect to the cluster threshold. The splitting between the two states is too small compared to the experimental value of 430 keV. This is related to a deficiency of the two-body interaction – additional spin-orbit strength is assumed to be coming from three-body forces. Fortunately it turns out that the capture cross section depends strongly on the centroid energy, which is reproduced nicely, but only very weakly on the splitting. The calculated charge radius of 2.67 fm agrees well with the experimental value of 2.647(17) fm [20]. This is important as the dipole matrix element and therefore for the capture cross section depends on an accurate description of the tail of the wave function.

In Fig. 1 we compare the calculated phase shifts for the S-, D- and P-waves with the phase shifts extracted from the scattering data [21, 22]. There is a noticeably change in the results

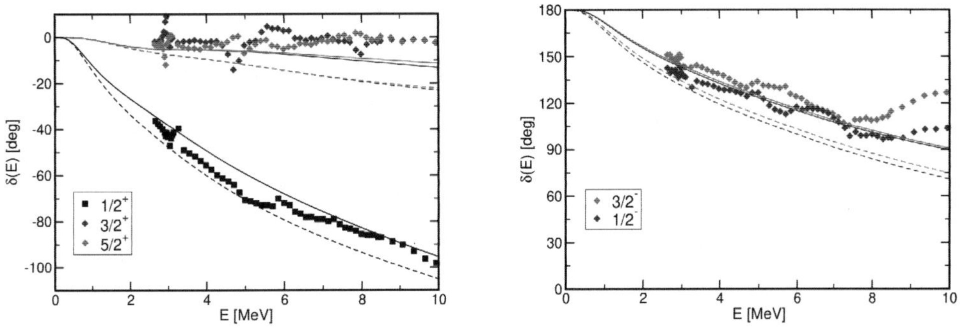

Figure 1. ^{4}He-^{3}He scattering phase shifts. Dashed lines show results using only frozen configurations, solid lines show results with full FMD model space. Left: S- and D-wave phase shifts. Right: P-wave phase shifts. Experimental results are from [21] and [22].

going from the model space containing only the frozen configurations to the full FMD model space. In case of the S- and D-wave phase shifts we find a good agreement with the experimental data. Also the P-wave phase shifts are essentially in good agreement with experiment, with the exception of the splitting.

The total cross section for ^{3}He$(\alpha,\gamma)^7$Be capture in form of the astrophysical S-factor is shown in the left part of Fig. 2. It agrees very well with the recent experimental data both in absolute normalization and in the energy dependence. On the right side of Fig. 2 we show the results for the isospin mirror reaction ^{3}H$(\alpha,\gamma)^7$Li. Whereas the energy dependence of the calculated S-factor agrees very well with the data the absolute cross section is larger then the data by Brune *et al.* by about 15%. The uncertainty of the cross section with respect to the effective interaction will be addressed in a future publication.

In the left part of Fig. 3 the S- and D-wave contributions to the total S-factor are shown. If we compare our results for example with the microscopic cluster model calculations by Kajino [11] we find that the biggest differences are found in the S-wave contribution. Our results also deviate from the empirical correlation between radius or quadrupole moment of the ^{7}Be ground state with the S-factor at zero energy that was found in the microscopic cluster model using different phenomenological interactions.

The A-body bound and scattering state wave functions can be analyzed in terms of overlap functions where the full microscopic wave function is projected on the subspace of cluster

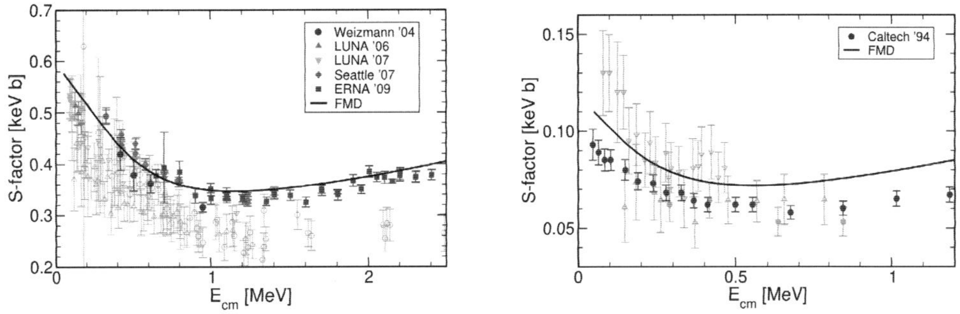

Figure 2. Left: S-factor for the ^{3}He$(\alpha,\gamma)^7$Be reaction. Recent experimental data [4, 5, 6, 7, 8] are shown as colored symbols, older data as gray symbols. Right: S-factor for the ^{3}H$(\alpha,\gamma)^7$Li reaction. Most recent data [23] is shown as colored symbols, older data as gray symbols.

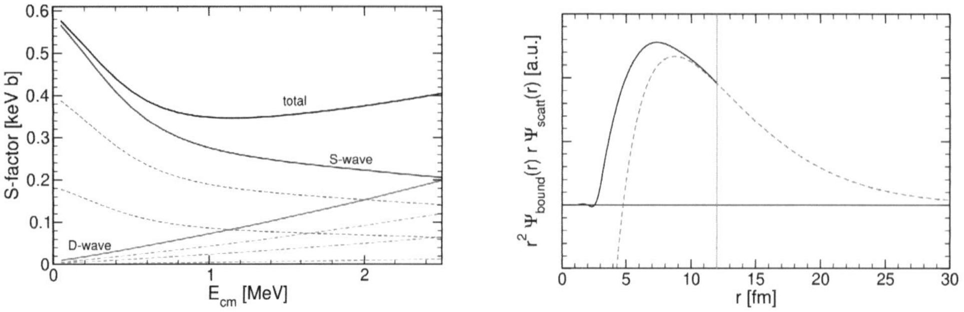

Figure 3. Left: S- and D-wave contributions to the S-factor for the $^3He(\alpha,\gamma)^7Be$ reaction. Dashed lines show contributions of individual transitions. Right: Dipole strength between the $3/2^-$ bound state and the 50 keV $1/2^+$ scattering state calculated from the overlap functions (solid line), from the Whittaker and Coulomb functions matched at the channel radius (dashed line).

configurations. If one takes antisymmetrization between the clusters into account properly (by folding with the square-root of the RGM norm kernel) these overlap functions can be interpreted as the relative wave functions of point-like ^3He and ^4He clusters. It is interesting that the overlap functions deviate from the Whittaker and Coulomb functions, that describe the asymptotic behavior of the cluster motion for bound and scattering states, up to distances of about 9 fm. These deviations from the asymptotic behavior are even more pronounced in the dipole strength. The dipole matrix element calculated from the overlap functions as shown in the right part of Fig. 3 agree with the matrix elements from the microscopic wave functions within 2%. They have sizable contributions already for distances as small as 3 fm and deviate significantly from matrix elements calculated from the asymptotic Whittaker and Coulomb functions up to 10 fm. It is therefore not really justified to treat the reaction in a pure external capture picture.

References

[1] Adelberger E G *et al.* 1998 *Rev. Mod. Phys.* **70** 1265
[2] Adelberger E G *et al.* 2011 *Rev. Mod. Phys.* **83** 195
[3] Burles S, Nollett K M, Truran J W and Turner M S 1999 *Phys. Rev. Lett.* **82** 4176
[4] Nara Singh B S, Hass M, Nir-El Y and Haquin G 2004 *Phys. Rev. Lett.* **93** 262503
[5] Bemmerer D *et al.* 2006 *Phys. Rev. Lett.* **97** 122502
[6] Confortola F *et al.* 2007 *Phys. Rev. C* **75** 065803
[7] Brown T A D *et al.* 2007 *Phys. Rev. C* **76** 055801
[8] Di Leva A *et al.* 2009 *Phys. Rev. Lett.* **102** 232502
[9] Kim B T, Izumoto T and Nagatani K 1981 *Phys. Rev. C* **23** 33
[10] Langanke K 1986 *Nucl. Phys.* **A457** 351
[11] Kajino T 1986 *Nucl. Phys.* **A460** 559
[12] Nollett K M 2001 *Phys. Rev.* **C63** 054002
[13] Navrátil P, Gueorguiev V G, Vary J P, Ormand W E and Nogga A 2007 *Phys. Rev. Lett.* **99** 042501
[14] Neff T 2011 *Phys. Rev. Lett.* **106** 042502
[15] Wiringa R B, Stoks V G J and Schiavilla R 1995 *Phys. Rev. C* **51** 38
[16] Feldmeier H, Neff T, Roth R and Schnack J 1998 *Nucl. Phys.* **A632** 61
[17] Neff T and Feldmeier H 2003 *Nucl. Phys.* **A713** 311
[18] Roth R, Neff T and Feldmeier H 2010 *Prog. Part. Nucl. Phys.* **65** 50
[19] Descouvemont P and Baye D 2010 *Rep. Prog. Phys.* **73** 036301
[20] Nörtershäuser W *et al.* 2009 *Phys. Rev. Lett.* **102** 062503
[21] Boykin W R, Baker S D and Hardy D M 1972 *Nucl. Phys.* **A195** 241
[22] Spiger R J and Tombrello T A 1967 *Phys. Rev.* **163** 964
[23] Brune C R, Kavanagh R W and Rolfs C 1994 *Phys. Rev. C* **50** 2205

An Experimentalist's Overview of Solar Neutrinos

Scott M. Oser
University of British Columbia, Department of Physics & Astronomy, Vancouver V6T 1Z1, Canada
E-mail: oser@phas.ubc.ca

Abstract.
Four decades of solar neutrino research have demonstrated that solar models do a remarkable job of predicting the neutrino fluxes from the Sun, to the extent that solar neutrinos can now serve as a calibrated neutrino source for experiments to understand neutrino oscillations and mixing. In this review article I will highlight the most significant experimental results, with emphasis on the latest model-independent measurements from the Sudbury Neutrino Observatory. The solar neutrino fluxes are seen to be generally well-determined experimentally, with no indications of time variability, while future experiments will elucidate the lower energy part of the neutrino spectrum, especially *pep* and *CNO* neutrinos.

1. The Standard Solar Model

The Sun is a prolific source of ν_e's with energies in the ~0.1-20 MeV range, produced by the fusion reaction

$$4p + 2e^- \rightarrow^4 \text{He} \ + \ 2\nu_e \ + \ 26.731 \text{ MeV}. \quad (1)$$

Solar fusion proceeds through a chain of sub-reactions called the *pp* chain, consisting of several steps [1]. Each neutrino-producing reaction in the *pp* chain produces a characteristic neutrino energy spectrum that depends only on the underlying nuclear physics, while the rates of the reactions must be calculated through detailed astrophysical models of the Sun. Experimentally the *pp*, ^{8}B, and ^{7}Be reactions are the most important neutrino-producing steps of the *pp* chain. A very small fraction of the Sun's energy is produced through the alternate *CNO* chain.

Figure 2 illustrates the solar neutrino energy spectra. At energies above 2 MeV only ^{8}B neutrinos have been directly detected by water Cherenkov experiments, while the *hep* reaction, which extends slightly higher in energy but is 1000 times smaller in flux, eludes detection still. Radiochemical experiments count the integral number of neutrinos from all reactions above their energy thresholds without resolving the separate contributions of each reaction. To date only the gallium experiments have sensitivity to *pp* neutrinos produced in the initiating reaction in the *pp* chain, with a rate calculable to a couple of percent just from the Sun's luminosity. Recently the Borexino experiment has succeeded in the first real-time measurements of ^{7}Be neutrinos, which are produced in a monoenergetic line just below 1 MeV [3].

2. Pre-SNO results

The pioneering solar neutrino experiment was Ray Davis's chlorine experiment in the Homestake mine near Lead, South Dakota [4]. This experiment measured solar neutrinos by observing

Published under licence by IOP Publishing Ltd

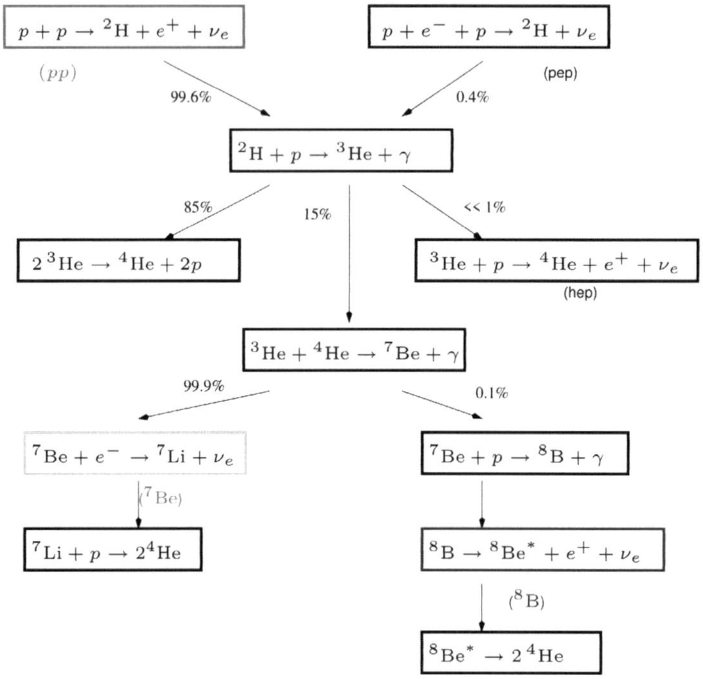

Figure 1. The *pp* reaction chain. The three most important neutrino-producing reactions are highlighted in color.

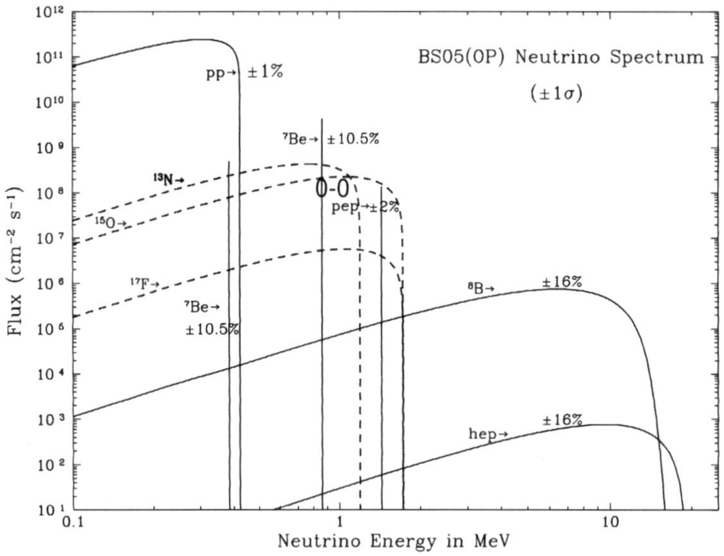

Figure 2. Energy spectra of solar neutrinos separated by reaction type, reproduced from reference [2].

the rate of Ar atom production through the reaction $\nu_e + {}^{37}\mathrm{Cl} \to {}^{37}\mathrm{Ar} + e^-$. By placing 600 tons of tetrachloroethylene deep underground (to shield it from surface radiation), and using radiochemistry techniques to periodically extract and count the number of argon atoms in the tank, Davis inferred a solar neutrino flux that was just $\sim 1/3$ of that predicted by solar model calculations [4, 2].

When scrutiny of both the Davis experiment and the solar model calculations failed to uncover any clear errors, other experiments were built to measure solar neutrinos in other ways. The Kamiokande and Super-Kamiokande water Cherenkov experiments have measured elastic scattering of electrons by ^{8}B solar neutrinos, using the directionality of the scattered electrons to confirm that the neutrinos in fact are coming from the Sun [5]. The measured elastic scattering rate is just $\sim$47% of the solar model prediction. The SAGE and GNO/GALLEX experiments have employed a different radiochemical technique to observe the $\nu_e+^{71}\text{Ge}\rightarrow^{71}\text{Ge}+e^-$ reaction, which is primarily sensitive to *pp* neutrinos, and have measured a rate that is $\sim$55% of the solar model prediction [6].

Multiple experiments using different techniques have therefore confirmed a deficit of solar ν_e's relative to the model predictions. Although interpretation of the data is complicated by the fact that each kind of experiment is sensitive to neutrinos of different energies produced by different reactions in the *pp* fusion chain, in fact there is apparently no self-consistent way to modify the solar model predictions that will bring the astrophysical predictions into agreement with the experimental results. This situation suggested that the explanation of the solar neutrino problem may not lie in novel astrophysics, but rather might indicate a problem with our understanding of neutrinos.

3. Results from the Sudbury Neutrino Observatory

The Sudbury Neutrino Observatory (SNO) has provided conclusive evidence that solar neutrinos change flavour by directly counting the rate of all active neutrino flavours, not just the ν_e rate to which the other experiments were primarily sensitive.

SNO was a water Cherenkov detector that used 1000 tonnes of D_2O as the target material [8]. Solar neutrinos interact with the heavy water by three different interactions:

$$\begin{array}{llcl} (CC) & \nu_e + d & \rightarrow & p + p + e^- \\ (NC) & \nu_x + d & \rightarrow & p + n + \nu_x \\ (ES) & \nu_x + e^- & \rightarrow & \nu_x + e^- \end{array} \tag{2}$$

Here ν_x is any active neutrino species. The reaction thresholds are such that SNO is only sensitive to ^{8}B solar neutrinos.[1] The charged current (CC) interaction measures the flux of ν_e's coming from the Sun, while the neutral current (NC) reaction measures the flux of all active flavours. The elastic scattering (ES) reaction is primarily sensitive to ν_e, but ν_μ or ν_τ also elastically scatter electrons with $\sim 1/6$th the cross section of ν_e.

SNO ran in three separate configurations. In the first neutrons were detected by observing the γ-ray emitted when they capture on deuterons. In the 2nd phase 0.2% NaCl was added to the D_2O and γ-rays from neutron capture on ^{35}Cl were included. In the final phase the salt was removed and replaced by an array of ^{3}He proportional counters. These counters detected neutrons independently of the Cherenkov signal used to measure CC and ES reaction rates.

SNO's most sensitive analysis is a low-energy threshold analysis of the Phase I and Phase II data [9]. In this analysis a simultaneous fit of signals and low-energy backgrounds is used to extract not just rates, but also to fit for a parameterized oscillation survival probability. For the favored Large Mixing Angle oscillation solution, the survival probability is well approximated by a 2nd order polynomial and a day-night asymmetry in the probability that can vary linearly with neutrino energy. This fit yields a value for the ^{8}B flux of

[1] The tiny flux of higher-energy neutrinos from the *hep* chain may be neglected here.

$5.046^{+0.159}_{-0.152}(stat)^{+0.107}_{-0.123}(syst)$ neutrinos/cm^2/s, which is in excellent agreeement with solar model predictions. Figure 3 shows the fitted day and night ν_e survival probabilities.

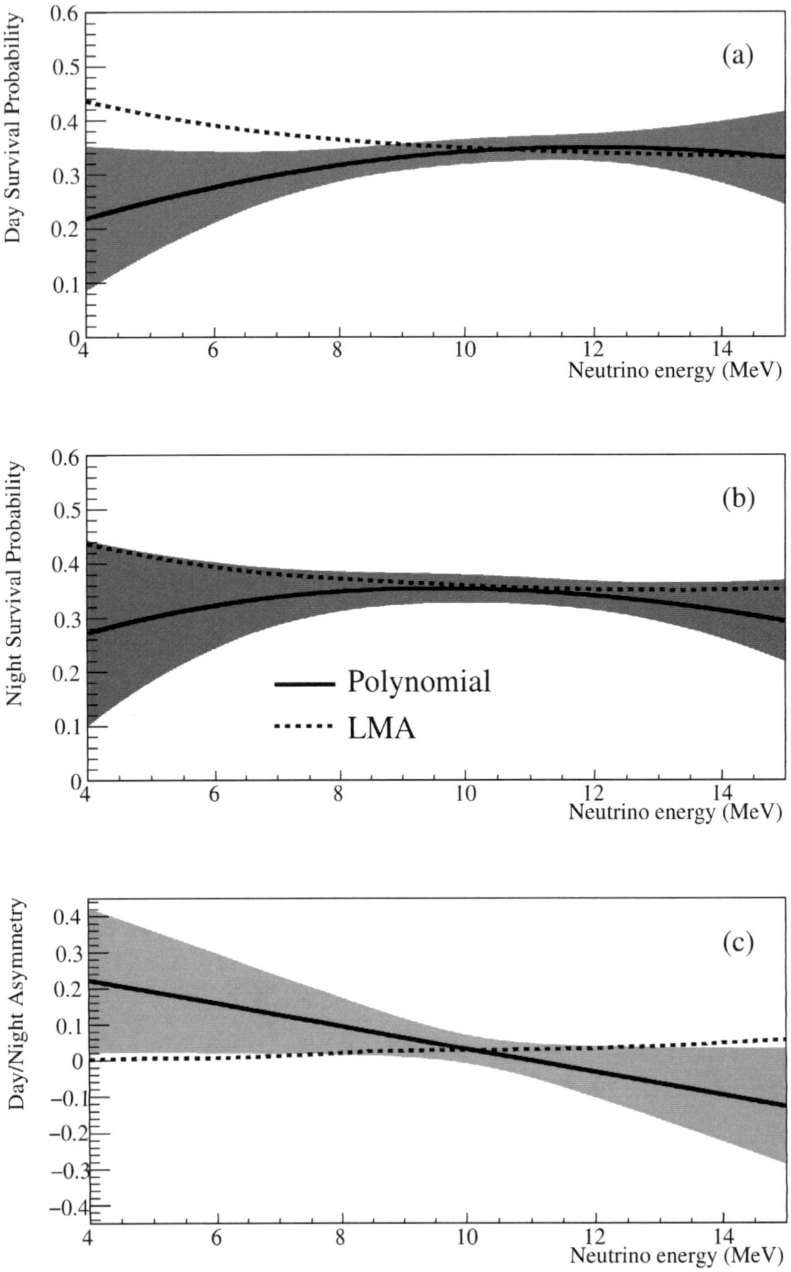

Figure 3. Fitted day (a) and night (b) oscillation probabilities, along with the day-night asymmetry formed from them (c).The colored bands show the total uncertainty as a function of energy while the dashed line show the predict of the best-fit LMA point [9].

Fits of SNO's ^{3}He proportional counter data from Phase III give a consistent value for the ^{8}B flux of $5.54^{+0.33}_{-0.31}(stat)^{+0.36}_{-0.34}(syst)$ neutrinos/cm^2/s, which although having larger uncertainties shares almost no systematics in common with other SNO analyses [10].

4. Time variability of the solar neutrino flux

Some researchers have reported periodic variations in the measured solar neutrino fluxes [11, 12, 13, 14, 15, 16, 17, 18]. Other analyses of these same data, including analyses by the experimental collaborations themselves, have failed to find such evidence [19, 20]. The reported periods have been claimed to be related to the solar rotational period. Because solar rotation should not produce variations in the solar nuclear fusion rate, non-standard neutrino properties have been proposed as an explanation. Periodicities in the solar neutrino flux, if confirmed, could provide evidence for new neutrino physics beyond the commonly accepted picture of matter-enhanced oscillation of massive neutrinos.

The SNO experiment has searched for evidence of time variability at periods ranging from 10 years down to 10 minutes [21, 22]. The sensitivity to high-frequency periodicities ($f > 1$/day) is unique to SNO, which combined real-time detection with negligible backgrounds. SNO has found no indications for any time variability of the ^{8}B flux at any timescale, including in the frequency window in which g-mode oscillations of the solar core might be expected to occur. At present there is no conclusive evidence for time variability of the solar neutrino fluxes from any experiment.

5. Real-time low energy measurements

Liquid scintillator detectors offer some unique advantages for solar neutrino research. A scintillator detector possesses a very low (sub-MeV) energy threshold, giving sensitivity to the lower energy solar neutrinos with real-time detection capability. Unfortunately scintillator detectors do not possess directional sensitivity, and only detect elastic scattering of solar neutrinos. Unlike SNO they do not have any ability to distinguish between neutrino flavors, but can access neutrinos at very low energies. They consequently face extremely stringent radiopurity requirements.

The Borexino experiment has successfully used the scintillation technique to detect the monoenergetic ^{7}Be neutrino line at 0.862 MeV. Borexino finds that the rate of ^{7}Be neutrinos is $0.52^{+0.07}_{-0.06}$ of the solar model prediction, in line with expectations from the Large Mixing Angle oscillation prediction. No day-night asymmetry is seen in the ^{7}Be rate, consistent with expectations [3].

The SNO+ experiment will reuse the SNO infrastructure, replacing the heavy water with a liquid scintillator target [23]. With its great depth SNO+ will be well-shielded against cosmogenic backgrounds such as ^{11}C. This should allow SNO+ to observe the *pep* neutrino line. The *pep* neutrinos are particularly interesting because this reaction is one of the initiating reactions of the *pp* chain, and so its rate is predicted precisely by the overall solar luminosity. Along with their favorable energy, intermediate between the MSW-dominated and vacuum-dominated regions of the oscillation, this will provide a good test of neutrino oscillation predictions and sensitivity to new physics such as non-standard neutrino interactions. SNO+ will also have sensitivity to neutrinos from the CNO cycle.

6. Conclusions

Understanding of neutrino production in the Sun has advanced to the point that almost all of the most significant reactions in the *pp* chain have been measured experimentally with varying degrees of precision. The ^{8}B flux itself is now measured to $\sim 3\%$, a constraint that forms the basis of the world's best determination of the neutrino mixing angle θ_{12}. On the whole standard solar models do a remarkable job of predicting the observed neutrino fluxes once matter-enhanced neutrino oscillations are accounted for. Future efforts will focus primarily on real-time low energy measurements of the *pep* reaction and CNO cycle.

6.1. Acknowledgments
I am grateful to the organizers of the NPA5 conference for the invitation to present this overview talk on solar neutrinos.

References
[1] John Bahcall, *Neutrino Astrophysics*, Cambridge University Press, 1989.
[2] See for example Bahcall et al., Astrophys. J. 621 (2005) L85-L88.
[3] Borexino collaboration, Phys. Rev. D 82, 033006, 201; arXiv:1104.1816; arXiv:1104.2150
[4] B. T. Cleveland et al., Astrophys. J. 496, 505 (1998).
[5] J. Hosaka et al., hep-ex/0508053, submitted to PRD; M. B. Smy et al., Phys. Rev D69 (2004) 011104; S. Fukuda et al., Phys. Lett. B539, 179 (2002).
[6] V. Gavrin, *Results from the Russian American Gallium Experiment (SAGE)*, VIIIth International Conference on Topics in Astroparticle and Underground Physics (TAUP 2003), Seattle, September 5–9, 2003; J.N. Abdurashitov *et al.*, J. Exp. Theor. Phys. **95**, 181 (2002); C. Cattadori, *Results from Radiochemical Solar Neutrino Experiments*, XXIst International Conference on Neutrino Physics and Astrophysics (Neutrino 2004), Paris, June 14–19, 2004.; E. Bellotti, *The Gallium Neutrino Observatory (GNO)*, VIIIth International Conference on Topics in Astroparticle and Underground Physics (TAUP 2003), Seattle, September 5–9, 2003; M. Altmann *et al.*, Phys. Lett. B **490**, 16 (2000); W. Hampel *et al.*, Phys. Lett. B **447**, 127 (1999).
[7] Q. R. Ahmad et al., PRL 87 (2001) 071301; Q. R. Ahmad et al., PRL 89 (2002) 011301; Q. R. Ahmad et al., PRL 87 (2001) 011302; S. N. Ahmed et al., PRL 92 (2004), 181301.
[8] SNO Collaboration, Nucl. Instr. and Meth. A449 (2000), 1972.
[9] B. Aharmin *et al.*, Phys. Rev. C 81, 055504 (53 pages), 2010.
[10] B. Aharmin et al., Phys. Rev. Lett. 101, 111301 (5 pages), 2008.
[11] P.A. Sturrock, Astrophys. J. **594**, 1102 (2003).
[12] P.A. Sturrock, Astrophys. J. **605**, 568 (2004).
[13] P.A. Sturrock *et al.*, hep-ph/0501205.
[14] P.A. Sturrock and J.D. Scargle, Astrophys. J. **550**, L101 (2001).
[15] P.A. Sturrock and D.O. Caldwell, hep-ph/0409064, submitted to Astroparticle Physics.
[16] M.A. Weber and P.A. Sturrock, Astrophys. J. **565**, 1366 (2002).
[17] A. Milsztajn, hep-ph/0301252.
[18] G. Ranucci, hep-ph/0505062.
[19] J. Yoo *et al.*, Phys. Rev. D **68**, 092002 (2003).
[20] L. Pandola, Astropart. Phys. **22**, 219 (2004).
[21] B. Aharmin *et al.*, Phys. Rev. D 72, 052010, (2005).
[22] B. Aharmin *et al.*, Ap. J. 710, 540-548 (2010).
[23] See for example "Status and Prospects for SNO+", Jose Maneira (on behalf of the SNO+ collaboration), NOW2010 conference, Otranto, Lecce, Italy, Sept 2010, to appear in Nuclear Physics B Proceedings Supplement; and Christine Kraus (on behalf of the SNO+ collaboration), Erice International School of Nuclear Physics, Erice, Sicily, Italy, Sept 2009, Prog.Part.Nucl.Phys.64:273-277, 2010

Cross section measurements of the ^{3}He(α, γ)^{7}Be reaction using DRAGON at TRIUMF

B S Nara Singh[1], S K L Sjue [2], B Davids [2], M Hass[3], P Adsley[1], L Buchmann[2], M Carmona-Gallardo[4], J Fallis [2], S P Fox[1], B Fulton[1], N Galinski [2,5], U Hager[2], D A Hutcheon[2], A M Laird[1], L Martin [2], D Ottewell[2],S Reeve[2,5],C Ruiz[2],G Ruprecht[2],A Shotter[2], O Tengblad[4]

[1]University of York, York, YO10 5DD UK
[2]TRIUMF, Vancouver, British Columbia V6T 2A3, Canada
[3]Weizmann Institute of Science, Rehovot, Israel
[4]Instituto de Estructura de la Materia, CSIC, Madrid, Spain
[5]Simon Fraser University, Burnaby, British Columbia, V5A 1S6, Canada

E-mail: narasingh.bondili@york.ac.uk

Abstract. We present our initial efforts with the DRAGON separator at TRIUMF facility towards obtaining the energy dependence of the astrophysical S-factor for ^{3}He(α, γ)^{7}Be reaction in the energy range of E_{cm}= 2 to 3 MeV that was recommended by the recent evaluations. A comparison between the existing data and our new complementary Madrid data, together with the recent theoretical calculations, is also given in the context of our ongoing work.

1. Introduction

The ^{3}He(α, γ)^{7}Be reaction was first studied half a century ago sparking interest in different fields of physics including solar neutrinos and big bang nucleosynthesis calculations [1, 2]. Since then, there have been a number of efforts both on the experimental and theoretical fronts, often resulting in disagreements in the absolute values of cross sections [2]. Obtaining accurate information on this reaction has become increasingly important in the recent years as the reaction, among the nuclear inputs, has become the second largest contributor to the uncertainty in calculating the high energy solar neutrino flux. As it is practically impossible to carry out measurements in the laboratories around the astrophysically most effective energy, 23 keV, theoretical extrapolations of the data taken at higher energies become necessary. From a recent compilation and critical analysis of the astrophysical S-factor variation with energy, it is clear that no model can give an accurate fit to all the available data and differences in the theoretical energy dependence contribute up to a 6% error in the extrapolated value of the S-factor. In particular, the models differ in energy dependence of the S-factor seen in the range of E_{cm}= 2 to 3 MeV [3, 4]. Clearly, precise data should also be obtained at medium energies. Such efforts, when combined with the measurements at lower energies, will allow an accurate model independent shape analysis of the S-factor curve and will help constraining theoretical models. Currently, data from Refs. [3] and [5] extend well above 1 MeV and disagree with each other at these medium energies. The available data on the astrophysical S-factor, S_{34}, were obtained by using different methods: the detection of prompt γ rays from ^{7}Be, of the

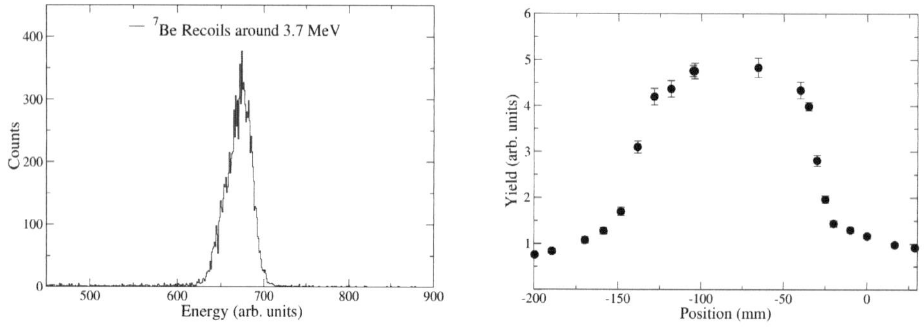

Figure 1. Left: The DSSSD spectrum showing around 11,000 ^{7}Be nuclei having recoil energies of 3.7 MeV with essentially no interference from background or the ^{4}He beam with 300 nA current at a bombarding energy of 6.5 MeV. No collimator was used between the gas target and the separator. Right: The measured target profile shows that the effective target length is 140mm; with a pressure of 8 Torr this leads to ΔE=14.8 A keV within the gas that was verified by measuring the exit energies of the beam after the gas at different pressures between 2 and 9 Torr. This is different from that of 12.3 cm for H_2 gas at 4 Torr [10]. In addition, the yield profile does not go to zero at extremes. As the gas mixture with only 3.5% of ^{3}He was used, the statistical accuracies and the quality of the γ spectra detected with a collimated BGO detector can be improved by using pure gas.

ensuing activity from ^{7}Be, and of produced ^{7}Be nuclei [2, 3, 4, 5, 6, 7, 8, 9]. In this paper, we present our initial work towards obtaining new data using the recoil detection method utilizing the DRAGON setup [10]. Our ongoing work, in relation to our new Madrid activity data and the recent calculations presented at this conference, is also discussed [11, 12].

2. Experiment and Results

As can be seen from the work by Di Leva *et al.* [3] the recoil detection method needs fairly complex setups which employ a differentially pumped windowless gas target cell, a recoil separator and a detection system at the focal plane. We utilized the two stage DRAGON recoil separator at the TRIUMF laboratory in Vancouver with ^{4}He beams of energies in the range of 3.5 MeV to 6.5 MeV and a ^{3}He windowless gas target at 8 Torr. Details on the separator can be found in Ref. [10]. Cross section measurements with an accuracy better than 10% using such a setup demand a good knowledge of, the acceptance of the separator, the charge state distributions and the background expected in the focal plane detectors for the produced ^{7}Be recoils, as well as the total numbers of beam and target particles. As we are addressing the most symmetric reaction studied ever at this separator and the recirculation of the ^{3}He gas target was required, a few modifications or developments for various components of our setup needed to be done and extensive careful tests have been performed to characterize our setup. It should be pointed out that due to the limitation in the acceptance of the separator, the reaction can best be studied in inverse kinematics, i.e. using ^{4}He as a beam and ^{3}He as a gas target.

The acceptance tests were initially carried out using an α source in conjunction with collimators that are aligned with the gas target and have geometric half-angles of 17,19 and 21 mrad. The α particles passed through the collimator and the separator before being detected in the Double Sided Si Strip Detector (DSSSD) at the focal plane typically at a rate of 1 Hz. The position information from the DSSSD and the event rates observed with collimators allowed us to infer that the separator has an acceptance cone half-angle of 21.0(6) mrad. In addition we measured the d(^{16}O,n)^{17}F reaction which has a threshold beam energy of 908.06 AkeV at which the maximum angle for the resulting ~12.0 MeV ^{17}F recoils is 0° and increases gradually with the energy. Our measurements at 971, 952, 922 AkeV utilizing the collimators mentioned before, helped us to confirm our source tests.

The existing differentially pumped gas target design was modified to adapt it for the recirculation of the ^{3}He gas avoiding any possible leaks. Since a pure gas was not available at the time of our tests, we used a gas mixture of ^{3}He (3.5%) and ^{4}He (96.5%). The He gas pressure was maintained in the target chamber at 8 Torr. The recoiling ^{7}Be nuclei were separated from the beam by DRAGON and subsequently detected by the DSSSD at the focal plane. Figure 1 (left) shows the energy spectrum of recoils detected in the DSSSD at E_{cm}=2.8 MeV. The total number of ^{4}He beam particles striking the gas target during this data collection was $\sim 2.5\times10^{16}$. Based on background measurements without beam, we expect a mean background at and below the beam energy of 38.1 events. We observed 31 events during the measurement of recoils and therefore infer a 90% confidence level lower limit on the beam suppression factor of $\sim 3\times10^{15}$. This result confirms the suitability of the DRAGON set up to study our symmetric reaction.

The target density profile was measured using the ^{3}He(^{12}C,p γ)^{14}N reaction at the E_{cm}=2.389 MeV resonance having 40 AkeV width (Γ). A ^{12}C beam at 100 pnA current and 11.985 MeV energy was delivered onto the ^{3}He target gas at a pressure of 8 Torr. Prompt-γ-rays at energies of 3.38, 3.89, 5.11 and 6.44 MeV from this reaction were detected using a BGO detector, that is a part of the DRAGON γ detection setup, which was collimated using 4 inch Pb blocks with a 5 mm gap. The detector was mounted on a movable table that could be controlled remotely. For each position we acquired data to obtain at least 1000 counts for the 6.44 MeV line in the γ spectrum. Figure 1 on the right gives the measured target profile. Although we expect the yield to fall off to zero at the extremes of the target profile, our measurements do not give such a result. Therefore, we plan for future measurements to understand this. For more details see the caption. The beam intensity was measured at regular intervals of 2 hrs. with a Faraday cup. This allows us to normalize the continuous measurement of the elastically scattered target particles detected in the two Si surface barrier detectors with precise collimators placed at 30° and 57° with respect to the beam. This provided us a way of monitoring the relative beam intensity during the course of the experiment [10]. Indeed utilizing, the target profile, the ^{7}Be observed in the DSSSD, the characteristics of the setup after the modifications mentioned above and the capture cross section from Ref. [5], we confirmed the composition of the gas mixture. Although these checks were done with uncertainties somewhat larger than 10%, this is a big step forward towards obtaining the S-factor in the medium energy range, since future tests as discussed below will lead us to better or similar accuracies compared to those from Ref. [3].

3. Discussion

In future we will perform more accurate measurements of the ^{3}He target density profile using better shielding for the BGO detectors and with an arrangement that will allow a bigger range of distances for the movement to get better definition at the falling parts of the profile. Geant simulations, using detailed geometry of the separator including any estimated misalignments of its components, particularly of the differentially pumped gas target, the capture reaction mechanisms and the resulting γ-ray angular distributions, are being developed to compare with our tests. The acceptance measurements will also be improved upon using an arrangement which will allow movements of an α source in all the three spatial directions scanning the entire gas target in a controlled manner. We also plan to carry out charge state distribution measurements using stable Be beam which will be crucial for the final analysis. As the rigidities of ^{4}He and ^{7}Be charge states are very different, online beam monitoring by collecting the unreacted beam on a Faraday cup at the far-side of the dipole magnet of the separator is also contemplated.

This ongoing TRIUMF work is complemented by our new Madrid measurements carried out using the activity method [11]. Figure 2 shows the existing data together with our new data

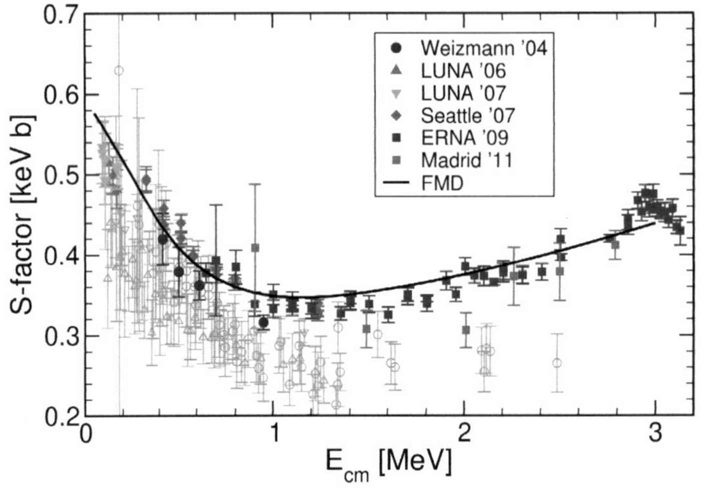

Figure 2. Existing data and new Madrid work (Madrid '11) compared with the calculations by T. Neff (FMD) [12]. Data are taken from Refs. [6] (Weizmann '04), [7, 8] (LUNA '06, LUNA '07), [9] (Seattle '07), [3] (ERNA '09). Old data (shown in grey, for details see e.g. Ref. [12]) also included data from Parker *et al.* at medium energies those do not agree with new data.

from this experiment. Although the uncertainties are large, our data clearly disagree with the energy dependence seen by Parker *et al.* [5], but agrees somewhat with that of Ref. [3]. The latter results could not be explained consistently by any of the theories when analyzed together with the other modern data [3]. This situation is somewhat changed after the work by Neff [12] that agrees with all of the modern data including that from the ERNA data and our new Madrid work. These fully microscopic calculations using a realistic effective interaction reproduce the nucleon-nucleon scattering data and do not require adjustment of any theoretical parameters including the absolute scale of the S-factor curve and gives an extrapolated value of 0.593 keVb for $S_{34}(0)$. However, we would like to stress that the current situation should be treated with a caution because these new calculations do not completely agree with the data for ^{3}H(α,γ)^{7}Li reaction. Since the data was obtained a few decades ago and has a scope for some improvement, future accurate measurements may change the scenario. Even though future calculations in different approaches will be helpful, there is still much work needed such as ours at TRIUMF on the experimental side to extract the shape of the S-factor curve in a model independent way from the data obtained in a wide energy range and using complementary techniques.

This work has been supported by the UK STFC.

References

[1] Holmgren H D and Johnston R L 1959 Phys. Rev. **113** 1556 .
[2] Adelberger E G *et al.* 1998 Rev. Mod. Phys. **70** 1265 and 2011 Rev. Mod. Phys. **83** 195
[3] Di Leva A *et al.* 2009 Phys. Rev. Lett. **102** 232502
[4] Cyburt R H and Davids B 2008 Phys. Rev. C **78** 064614
[5] Parker P D and Kavanagh R W 1963 Phys. Rev. **131** 2578
[6] Nara Singh B S *et al.* 2004 Phys. Rev. Lett. **93** 262503
[7] Bemmerer D *et al.* 2006 Phys. Rev. Lett. **97** 122502
[8] Confortola F*et al.* 2007 Phys. Rev. C **75** 065803
[9] Brown T A D *et al.* 2007 Phys. Rev. C **76** 055801
[10] Hutcheon D A *et al.* 2003 Nucl. Instrum. Methods A **498** 190
[11] Carmona-Gallardo M *et al.* 2011 at this conference.
[12] Neff T Phys. Rev. Lett. 2011 **106** 042502 (2011) and the references therein.

Decay spectroscopy for nuclear astrophysics: β- and β-delayed proton decay

L Trache[1], **A Banu, JC Hardy, VE Iacob, M McCleskey, BT Roeder, E Simmons, A Spiridon and RE Tribble** - Texas A&M University, College Station, TX, USA
A Saastamoinen, A Jokinen and J Äysto - University of Jyvaskyla, Finland
T Davinson, G Lotay and PJ Woods - University of Edinburgh, UK
E Pollacco - CEA Saclay, France

Abstract. In several radiative proton capture reactions important in novae and XRBs, the resonant parts play the capital role. We use decay spectroscopy techniques to find these resonances and study their properties. We have developed techniques to measure beta- and beta-delayed proton decay of sd-shell, proton-rich nuclei produced and separated with the MARS recoil spectrometer of Texas A&M University. The short-lived radioactive species are produced in-flight, separated, then slowed down (from about 40 MeV/u) and implanted in the middle of very thin Si detectors. This allows us to measure protons with energies as low as 200 keV from nuclei with lifetimes of 100 ms or less. At the same time we measure gamma-rays up to 8 MeV with high resolution HPGe detectors. We have studied the decay of ^{23}Al, ^{27}P, ^{31}Cl, all important for understanding explosive H-burning in novae. The technique has shown a remarkable selectivity to beta-delayed charged-particle emission and works even at radioactive beam rates of a few pps. The states populated are resonances for the radiative proton capture reactions ^{22}Na(p,γ)^{23}Mg (crucial for the depletion of ^{22}Na in novae), ^{26m}Al(p,γ)^{27}Si and ^{30}P(p,γ)^{31}S (bottleneck in novae and XRB burning), respectively. Lastly, results with a new detector that allowed us to measure down to about 80 keV proton energy are announced.

1. Introduction

The difficulties encountered in doing direct measurements in nuclear astrophysics (very low cross sections and/or involvement of unstable nuclei) lead to the use of indirect methods. Among the indirect methods, a large class is the spectroscopy of resonances. In many cases, the resonant parts are important or dominant to reactions that are important in stars. These resonances are metastable states in the compound system produced in reaction as an intermediate step. To evaluate the corresponding contributions to the reaction rates (for narrow, isolated resonances) it is sufficient to determine the location of the resonances (E_{res}) and their resonance strengths ($\omega\gamma$). These may be obtained by studying the spectroscopic properties of the corresponding metastable states, populated through a convenient method. The decay spectroscopy is one such method: instead of measuring proton capture we study its inverse, the proton decay of the same state. We study here states populated by beta decay. Of course the condition of populating exactly those states is crucial: selection rules of energy, spin and parity must be met. In this contribution we present the use of β– and β-delayed proton-decay (βp)

[1] To whom any correspondence should be addressed. E-mail: livius_trache@tamu.edu

studies to populate and characterize resonances in a few proton capture reactions on sd-shell nuclei; reactions important in H-burning in novae and X-ray bursters (XRB). A major problem is that the energies of the protons to be measured are very small, below 300-400 keV, close to the limits of our current detection techniques. We describe a method consisting in the implantation of radioactive sources in very thin Si detectors, which allows, up to a point, measuring such low energy protons from βp-decays. We did β- and βp-decay studies of ^{23}Al, ^{27}P and ^{31}Cl, aiming at determining the properties of resonances that contribute to the ^{22}Na(p,γ)^{23}Mg, ^{26m}Al(p, γ)^{26}Si and ^{30}P(p, γ)^{31}S reactions, respectively. These reactions are important in the description of thermonuclear runaways in novae and XRB [1]. Part of the results were published elsewhere [2-4], but most were not, and the full analysis is still on-going. Only the first and the third isotopes in the list are going to be briefly treated here, with emphasis on the case of ^{23}Al, which is the most complete to date.

2. Experimental methods

We have developed a simple technique to measure β- and β-delayed proton-decay of proton-rich nuclei produced and separated with the MARS recoil spectrometer at the K500 superconducting cyclotron of Texas A&M University. The short-lived radioactive species are produced in-flight with (p,2n) reactions in inverse kinematics on a cryogenic H_2 gas target, separated in MARS, then slowed down (from about 40 MeV/u) and implanted in the middle of a very thin Si detector. Then the beam is stopped and the decay is measured for times equal to about 2 half-lives of the isotope. The cycle is repeated until a convenient statistics is reached. The primary beams used were ^{24}Mg at 48 MeV/u, ^{28}Si and ^{32}S @ 40 MeV/u, respectively. The implantation of sources directly in the active part of the detector avoids the problems with detector windows or dead layers and allows us to measure protons with energies as low as 200 keV from nuclei with lifetimes of 100 ms or less. Two different double sided strip Si detectors (DSSSD), 65 or 45 μm thick (W1-65 and BB2-45, respectively), were used as proton detectors. A 1 mm thick Si detector was placed behind the proton detector to measure betas in coincidence. One or two HpGe detectors were put outside the implantation chamber, as close as possible to the implantation site, to measure gamma-rays. Protons and gamma-rays in coincidence with the beta-detector were measured simultaneously. The technique provides a valuable tool for the study of states that are resonances important in the radiative proton capture on nuclei close to the proton drip line.

3. Results

For all three cases announced, the isotopes were for the first time abundantly produced (2-4000 pps) and well separated from their isobars (purities >85% at the focal plane and close to 100% at implantation). These have allowed good β–γ measurements and to establish their decay scheme for the first time. For example, we obtained information on the location and decay of Isobaric Analog States (IAS) and possible isospin mixing. Before our measurements, most of the decay data for these isotopes were based solely on the observation of their β-delayed proton emission. The lifetimes of all 3 isotopes could also be determined with <1% uncertainties, improving considerably the previous data (uncertainties were ~25%).

Space limitations do not allow discussing here in detail the results for all cases. In brief, the most important results from the study of the decay of ^{23}Al are:

- first clean and intense ^{23}Al source was produced
- the decay scheme was established
- established unambiguously the g.s. spin and parity $J^{\pi}=5/2^{+}$ (not $1/2^{+}$) [2], which allowed to evaluate that the reaction ^{22}Mg(p,g)^{23}Al is not important for depletion of ^{22}Na in novae
- absolute branching ratios were measured (not easy or common!)
- measured $T_{1/2}$=446(4) ms (<1% accuracy) - w. gamma-ray multiscaling
- and from these absolute log *ft* were determined
- the IAS was identified at E*=7802.9(5) keV by its *logft=3.31(2) – measured, not assumed!*

- used IMME to get new ^{23}Al mass [2] ⇒ $S_p(^{23}Al)$=143(3) keV
- after ^{23}Al mass measured in Jyvaskyla, the A=23 isobar multiplet became best IMME check
- observed for the first time in the same experiment the IAS and a state at E*=7787 keV with $J^{\pi}=(7/2)^+$ (only 16 keV apart)
- measured proton transitions with energies as low as Ep=207 keV from the E*=7787 keV state (see figure 1). This state is an important resonance in $^{22}Na(p,\gamma)^{23}Mg$ reaction. From the gamma and proton branchings measured here, and the lifetime from Ref. [5] we could determine its resonance strength: ωγ=1.4(5) meV [4].
- Determined the total proton branching =1.26(5)%.

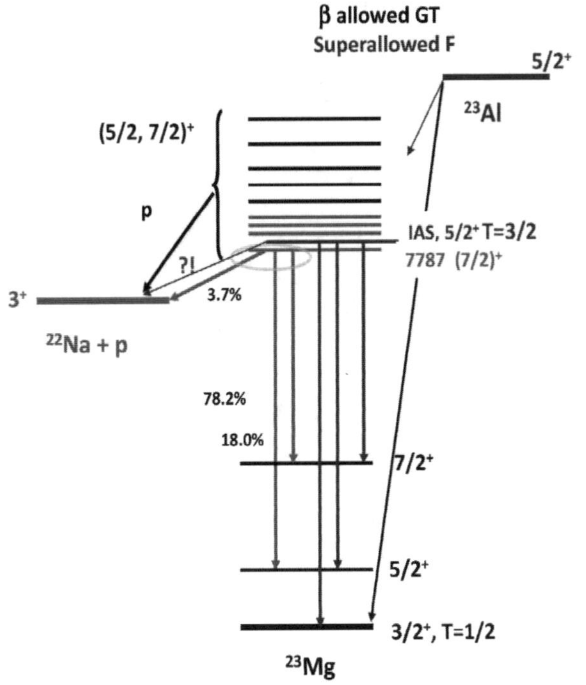

Figure 1. Partial decay scheme of ^{23}Al. Only the transitions to the first 3 levels of daughter nucleus ^{23}Mg are shown in the lower part to justify the spin and parity assignment of its ground state. The decay of the doublet around E*=7.8 MeV (the IAS and the E=7787 keV state), including a 3.7% proton branch is also shown. It is very rare that the decay of a state by both gamma-ray and proton emission is observed. More recently a weak proton decay branch was observed from the IAS (the ?! on figure; in first order this transition is isospin forbidden).

Very similar results were obtained from the study of ^{31}Cl, separated as in figure 2 below. The position and four gamma-ray decay branches of the IAS were found, a new mass excess could be determined using IMME, new resonances were found and resonance energies could be determined with +/-1.5 keV uncertainties (reducing the uncertainties of the astrophysical reaction rates). No proton-decay branch was observed from the IAS at this time.

While our technique of using very thin Si strip detectors allowed a very good sensitivity to protons, at energies below 400 keV there is a large background arising from the energy loss of the positrons in the Si detector, which peaks at ~12 – 20 keV but extends down in the region of interest. This is particularly important in the cases, as those described here, where the proton branchings are very small. We succeeded to reduce this background by selecting the events to occur in a single detector pixel, rather than in a full strip. Later, a considerable reduction was obtained when the volume of a pixel was reduced from 3 mm x 3 mm x 65 μm in the W1-65 detector to 1 mm x 1 mm x 45 μm in the BB2-45 detector. However, the proton spectra around 200 keV could only be obtained with a careful background subtraction; background measured with the decay of a beta-only emitter implanted in the same detector.

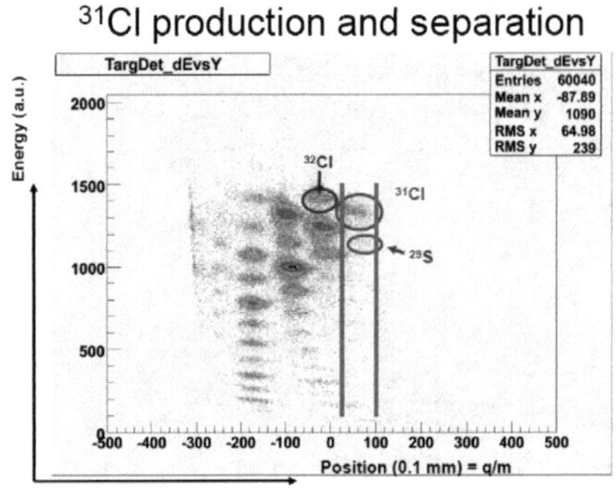

Figure 2. Representation of the isotope separation in the focal plane of MARS for the case of ^{31}Cl. After this position vs energy spectrum is taken, a couple of slits are moved to allow only ^{31}Cl through (red lines).

In a very recent experiment (only two weeks ago), using a new type of detector developed with E. Pollacco of CEA-Saclay, we could reduce drastically the beta background that has hindered the measurement of low energy proton branches while improving the resolution. The beta background was restricted to energies lower than about 80 keV. This will be reported elsewhere.

Acknowledgments

This work has been supported in part by the US DOE under Grant DE-FG02-93ER40773 and by the Academy of Finland under the Finnish Centre of Excellence Programme 2006-2011. PJW and TD would like to acknowledge support from the STFC.

References

[1] see Jose J., this *Proceedings*
[2] Iacob VE.. *et al.*, 2006 *Phys.Rev. C* **74**, 045810.
[3] Trache L. *et al.*, in *Proc. 10th Symposium on Nuclei in the Cosmos, July 27-Aug 1, 2008, Mackinac Island, MI,* **PoS (NIC X) 163.**
[4] Saastamoinen A. *et al.*, 2011 *Phys. Rev. C* **83**, 045808.
[5] Jenkins D.G. *et al.*, 2004 *Phys.Rev. Lett.* **92**, 031101.

Direct radiative proton capture ^{23}Al(p,γ)^{24}Si studied via one-proton nuclear breakup of ^{24}Si

A Banu[1], F Carstoiu[2], W N Catford[3], B Fernandez-Dominguez[4], M Horoi[5], N A Orr[6], B T Roeder[6], P Roussel-Chomaz[7], L Trache[1] and R E Tribble[1]

[1]Cyclotron Institute, Texas A&M University, College Station, TX 77843, USA
[2]National Institute of Physics and Nuclear Engineering "Horia Hulubei"(IFIN-HH), R-077125 Magurele-Bucharest, Romania
[3]Department of Physics, University of Surrey, Guildford GU2 5XH, UK
[4]Oliver Lodge Laboratory, University of Liverpool, Liverpool L69 7ZE, UK
[5]Department of Physics, Central Michigan University, Mount Pleasant, Michigan 48859, USA
[6]LPC-ENSICAEN, IN2P3-CNRS et Universite de Caen, 14050 Caen Cedex, France
[7]GANIL, BP 55027, 14076 Caen Cedex 5, France

E-mail: banula@jmu.edu

Abstract. We present the results of an experimental work that studied the direct component of the radiative proton capture reaction ^{23}Al(p,γ)^{24}Si for its relevance in X-ray burst nucleosynthesis. The experiment was performed at GANIL using one-proton nuclear breakup reaction at intermediate energies to determine the Asymptotic Normalization Coefficient (ANC) of the system ^{24}Si→^{23}Al + p. Using the ANC, we evaluated the corresponding astrophysical *S* factor and reaction rate.

1. Motivation

X-ray bursts are among the energetic events (~10^{40}-10^{42} ergs) observed in cosmos. It is thought that X-ray bursts are thermonuclear explosions occurring in binary systems consisting of a neutron star that accretes matter from a main sequence companion star. The extreme density found near the neutron star surface provides the environment for explosive Hydrogen and Helium burning allowing radioactive nuclei far from stability to be synthesized.

With a better understanding of the mechanism governing the explosive X-ray bursts, it is possible to determine the final abundance distribution that falls back onto the neutron star crust. To understand this explosive mechanism of X-ray burst production, it is necessary to determine reaction rates for the nuclear processes that occur during these events.

The nuclear astrophysics process considered as the onset for the explosive X-ray bursts is the rp-process (rapid proton capture). In the rp-process, proton capture is favored compared to β-decay until a nucleus with a small (p,γ) Q-value is reached. At that point, the thermo-equilibrium between (p,γ) radiative capture and its inverse reaction (γ,p) photodisintegration is established.

An example of such a radiative capture reaction with a small Q-value (only 142 keV) is the reaction ^{22}Mg(p,γ)^{23}Al which is thus in thermoequilibrium with its inverse reaction ^{23}Al(γ,p)^{22}Mg at X-ray burst temperature and density conditions. To allow to break out from the ^{22}Mg(p,γ)^{23}Al -

Published under licence by IOP Publishing Ltd

^{23}Al(γ,p)^{22}Mg equilibrium, it requires a fast ^{23}Al(p,γ)^{24}Si reaction rate. The influence of the fast sequential two-proton capture reactions on ^{22}Mg forming ^{24}Si becomes important during X-ray bursts [1]. Moreover, recent sensitivity studies on X-ray burst nucleosynthesis [2] have identified the radiative proton capture ^{23}Al(p,γ)^{24}Si among the influential reactions affecting the total energy output by more than 5%.

Following we describe a measurement performed recently at GANIL to determine with an indirect method the astrophysical *S* factor and the reaction rate for ^{23}Al(p,γ)^{24}Si.

2. Experiment and Method

It has been demonstrated that the momentum distributions of the core fragments measured in one-nucleon breakup reactions are powerful spectroscopic tools to determine single-particle structure of the nuclei far from stability [3]. The shapes (widths) of these momentum distributions provide information on the orbital angular momentum *l* of the removed nucleon [4], whereas the nuclear breakup cross section determines the Asymptotic Normalization Coefficient (ANC). This ANC is used to calculate the direct (non-resonant) component of the astrophysical *S* factor of radiative capture reactions [5].

To accomplish this for the radiative proton capture ^{23}Al(p,γ)^{24}Si, we have performed an experiment [6] at the GANIL coupled cyclotron facility that measured the longitudinal momentum distributions of the ^{23}Al fragments resulted from the nuclear breakup of a ^{24}Si secondary beam. As a matter of fact, a cocktail of secondary beams was produced via the fragmentation of an intense 95 MeV/nucleon ^{32}S primary beam on a thick carbon target. Fourteen ion species, among which was also ^{24}Si, with energies between 20 and 60 MeV/nucleon and intensities ranging from 30 and 7000 pps, were obtained. We had about 30 pps of ^{24}Si at 61 MeV/nucleon. A secondary reaction target of carbon, 175 mg/cm^2 thick, was used. To measure the breakup fragment momentum distributions, the SPEG energy-loss spectrometer was employed.

Ion identification at the focal plane of SPEG was achieved using the energy loss from a gas ionization chamber and the time-of-flight between a thick plastic stopping detector and the cyclotron radio frequency. Two large-area drift chambers straddling the focal plane of SPEG allowed the focal plane position spectra to be reconstructed. The longitudinal momentum of each particle was derived from the reconstructed focal plane position taking into account the beam central magnetic rigidity in SPEG. The momentum of the core fragments relative to the incident projectiles in the laboratory frame was transformed in the projectile rest frame using Lorentz transformation. To compare the measured momentum distributions with the theoretical ones, all broadening effects inherent in the measurement have been considered through Monte Carlo simulations. These effects include the energy spread in the beam, the differential energy losses of the projectiles and the fragments in the secondary target, the energy and angular straggling in the target, and the detector and spectrometer resolutions.

The reaction target was surrounded by eight EXOGAM Germanium clover detectors set up in a new configuration, for the first time in association with SPEG. This enabled us to measure also the coincidences with γ-rays from core fragments, left excited after the nuclear breakup occurred. However, this was not the case for the breakup of ^{24}Si because the core fragment, ^{23}Al, does not have bound excited states.

3. Results

Figure 1 shows the experimental momentum distribution of the ^{23}Al core fragments, in the center-of-mass reference frame, compared with extended Glauber-type theoretical calculations, which are explained in detail in Ref. [4]. For the calculations (full curve in figure 1), we used ^{24}Si mid-target energy of 53 MeV/nucleon. The single-particle wave functions are calculated in a Woods-Saxon proton binding potential with a set of radius and diffuseness parameters r_0 = 1.18 fm and a = 0.60 fm.

The comparison between the experimental momentum distribution and the calculated distribution enabled us to extract the ANC for the system $^{24}Si_{gs} \rightarrow ^{23}Al(5/2^+) + p$, and an experimental spectroscopic factor with a value of 2.5, which agrees reasonably [6] with a spectroscopic factor of 3.5 obtained from large-scale shell model calculations made with modern effective interactions. The squared value of the ANC for the ^{24}Si ground state configuration $|5/2^+_{gs} \otimes \pi 1d_{5/2}\rangle$ is found to be $C^2(^{24}Si_{gs}) = 62.2 \pm 7.1$ fm^{-1}.

Using this ANC, the direct component of the astrophysical *S* factor and the reaction rate for $^{23}Al(p,\gamma)^{24}Si$ can be evaluated. These results and their implication for X-ray burst nucleosynthesis are presented in a forthcoming paper. This is the first measurement of the direct (non-resonant) capture component of the radiative proton capture reaction $^{23}Al(p,\gamma)^{24}Si$.

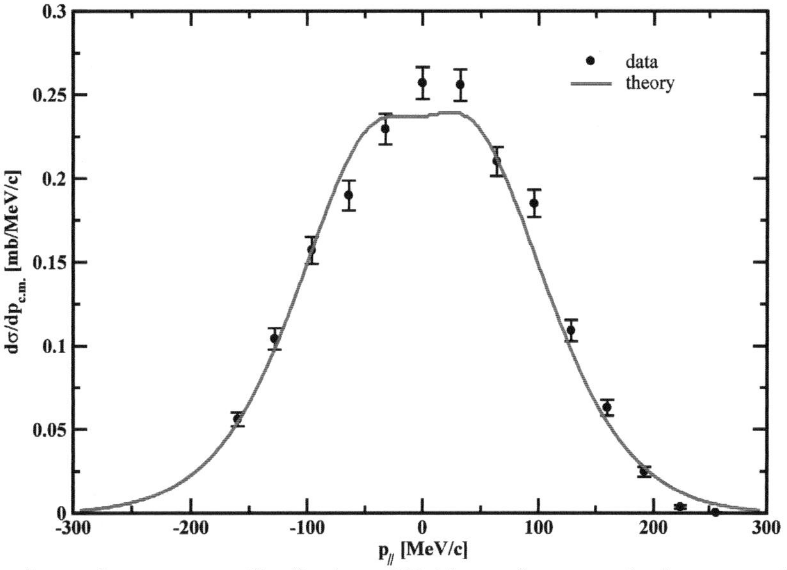

Figure 1. Experimental momentum distribution of 23Al core fragments in the center-of-mass frame, compared with a theoretical calculation (see text for details).

References

[1] Schatz H *et al.* 1997 *Phys. Rev. Lett.* **79** 3845
[2] Parikh A, Jose J, Moreno F and Iliadis C 2008 *ApJ Suppl. Series* **178** 110
[3] Hansen P G and Sherill B M 2001 Nucl. Phys. A **693** 133
[4] Sauvan E *et al.* 2004 Phys. Rev. C **69** 044603
[5] Trache L, Carstoiu F, Gagliardi C A and Trible R E 2001 Phys. Rev. Lett. **87** 271102
[6] Banu A *et al.* 2011 submitted to Phys. Rev. C

Study of the ^{15}N(p,γ)^{16}O at LUNA

Antonio Caciolli for the LUNA collaboration

Istituto Nazionale di Fisica Nucleare (INFN), Padova Section, Via Marzolo 8 - 35131 Padova, Italy.

Center for GeoTechnologies, Via Vetri Vecchi, 34 - 52027 San Giovanni Valdarno, Arezzo, Italy.

E-mail: caciolli@pd.infn.it

Abstract. The ^{15}N(p,γ)^{16}O reaction links the first CNO cycle with the second one allowing the production of the oxygen and fluorine isotopes during the hydrogen burning in novae and AGB and RGB stars. In recent years, this reaction has been studied deeply by LUNA (Laboratory for Underground Nuclear Astrophysics). As a matter of fact three different experiments have been run to measure the ^{15}N(p,γ)^{16}O S factor down to the energy of the novae and ABG hydrogen burning reaching for the first time the 72.8 keV in the center of mass. In this contribution the results of the first two experiments are summarized and the results of the recently performed experiment, which is dedicated to the study at very low energies, are presented.

1. Introduction

During hydrogen burning the p-p chain provides the energy supply in low mass stars whereas the CNO cycle is the principal nuclear process in the core of high Mass Sequence stars ($M \gtrsim 1.2\ M_\odot$). The ^{15}N(p,γ)^{16}O (Q-value 12.127 MeV) links the first CNO cycle to the second one providing the path to produce the ^{16}O in stellar hydrogen burning [1]. This reaction was studied in pioneering experiments by Hebbard [2] in 1960 by using a NaI detector and by Rolfs and Rodney [3] in 1974 using a Ge(Li) detector. The NACRE [4] and CF88 [5] database used only the most recent dataset since there is a discrepancy between the results of the two experiments. New results on the S factor have been recently published by Mukhamedzanov et al. [6]. They measured the Asymptotic Normalization Coefficients (ANC) and performed an R-Matrix fit that includes previous data from direct cross section measurements. This work reports an S factor lower by a factor of 2 than the one suggested by NACRE and in agreement with the Hebbards extrapolated S factor [2]. Another recent R-Matrix analysis, using again the available direct data but limited to the capture reaction to the ground state [7], also indicates a much lower S factor.

The LUNA collaboration decided to study the ^{15}N(p,γ)^{16}O reaction by performing three different experiments: one with a gas target setup and two others by using a solid target setup. The LUNA facility ([8] and reference therein), thanks to its underground position inside the Gran Sasso National Laboratory (LNGS), offers a favorable laboratory background for the γ-ray spectra with respect to analogue facilities on the Earths surface, allowing studies at energies close to the Gamow peak [9, 10, 11]. In this work the two setups where an high efficiency BGO detector was used are described in details. The other experiment was performed in a collaboration between the LUNA collaboration and the Notre Dame University and it was dedicated to explore a large energy range from 120 keV up to 1800 keV in the center of mass in order to perform an R-Matrix fit on a unique set of data. This experiment was described in a previous work and reference therein [12].

2. Gas Target Experiment

A reanalysis of γ-spectra acquired in a previous experiment was done to determine the cross section in a small energy range (from 230 keV down to 90 keV in the center of mass). The experiment was performed by using a window less gas target setup with the target chamber filled with nitrogen gas (natural isotopic aboundances) [13, 14].

A 4π-BGO detector array which surrounds the target chamber was used to detect the γ-rays emitted by the ^{15}N(p,γ)^{16}O reaction (the region of interest was from 10 up to 14 MeV). The detector efficiency was determined with a GEANT4 code to be 70% for 12 MeV γ-rays.

The beam induced background due to the ^{11}B(p,γ)^{12}C reaction, which produces a γ-rays ($E_\gamma \approx$ 11.5 MeV) in the same ROI of the ^{15}N(p,γ)^{16}O reaction , was the major source of systematic uncertainties since the signal was comparable with the one from the reaction studied (the ^{15}N in natural nitrogen is 0.4% [15]). The beam induced background was taken into account by acquiring γ-spectra with helium gas in the target chamber. As a matter of fact, with helium only the contribution due to the boron contamination is present in the ROI of the γ-spectra. The obtained data show a reduction of the cross section of a factor 2 with respect the Rolfs and Rodney results [17].

3. Solid target setup

The statistics contribution was improved by a factor 200 choosing solid TiN target enriched in ^{15}N. The targets were produced by the LNL of Padova and in HFZ of Karlsruhe using the reactive sputtering technique [16]. These targets were used in two different experiment: one performed with a HPGe detector in close geometry and dedicated to perform an R-Matrix fit with direct data acquired in a large energy range with the same setup [12] and another experiment performed with the same BGO detector already used with the gas target setup.

For the second experiment, since the target stability and the target properties are a fundamental parameter for the cross section analysis, the target irradiated at the LUNA accelerator were analyzed with SNMS (Secondary Neutral Mass Spectroscopy) [18] and high Z ERD (Elastic Recoil Detection) [19] in order to control the deterioration of the target and reduce the uncertainties on the isotopic abundances and the Ti/N stoichiometry ratio. Taking advantage of the high efficiency of the BGO detector for the first time, the cross section of the ^{15}N(p,γ)^{16}O reaction was measured down to 72.8 keV covering the total Gamow peak for nova explosion scenario. The S factor has been determined by integrating the target profiles with an a priori S factor and by comparing this integral with the measured yield. For the analysis, three different input S factors have been used: the one given in [12], a constant S-factor, and a value obtained from iterating the procedure. Identical results were obtained. This procedure is explained in details in [21]. Finally, the reduction by a factor 2 with respect the previous NACRE extrapolation observed at these energies provides a 30% reduction of the ^{16}O production by the hydrogen burning in nova [1].

4. Summary and Conclusion

The ^{15}N(p,γ)^{16}O reaction has been deeply studied by the LUNA collaboration performing three different experiments. We have measured the cross section of this reaction in a huge range of data and an R-Matrix has been produced to extrapolate the S factor at lower energies [12]. An other group has performed an R-Matrix fit on the same data obtaining an extrapolation in good agreement with the LUNA one [20]. Using a BGO detector the ^{15}N(p,γ)^{16}O S factor has been also obtained in the center of mass energy range from 70 keV to 370 keV. The S factor values are found to be lower by a factor of 2, compared to the previously accepted value [4]. The measured energy range covers, for the first time, the whole Gamow peak for novae explosions and the upper half of the Gamow peak of RGB and AGB stars are covered with precise experimental data.

Acknowledgments
Financial support by INFN and in part by the European Union (TARI RII3–CT–2004– 506222, AIM 025646 and SPIRIT 227012), the Hungarian Scientific Research Fund (K68801), and DFG (BE 4100/2–1) is gratefully acknowledged.

References

[1] C. Iliadis et al., Astrophys. J. Suppl. **142**(2002)105.
[2] D. F. Hebbard, Nucl. Phys. **15**(1960)289.
[3] C. Rolfs and W. Rodney, Nucl. Phys. **A235**(1974)450.
[4] C. Angulo et al., Nucl. Phys. **A656**(1999)3.
[5] G. R. Caughlan & W. A. Fowler, At. Data Nucl. Data Tables **40**(1998)283.
[6] A. M. Mukhamedzhanov et al., Phys. Rev. **C78**(2008)015804.
[7] F. C. Barker, Phys. Rev. **C78**(2008)044612.
[8] Broggini et al., Annual Review of Nuclear and Particle Science **60**(2010)53.
[9] D. Bemmerer et al., Eur. Phys. J **A24**(2005)313.
[10] A. Caciolli et al., Eur. Phys. J **A39**(2009)179.
[11] T. Szücs et al., Eur. Phys. J **A44**(2010)513.
[12] P. J. LeBlanc et al., Phys. Rev. **C82**(2010)055804.
[13] A. Lemut, Phys. Lett. **B634**(2006)483.
[14] D. Bemmerer et al., Nucl. Phys. **A779**(2006)297.
[15] T. B. Coplen et al., Pure Appl. Chem **74**(2002)1987.
[16] V. Rigato et al., Surface and Coating Technology **142-144**943.
[17] D. Bemmerer et al., J. Phys. **G36**(2009)045202.
[18] K. Vad et al., Spectroscopy Europe, **21**(2009)4.
[19] A. Bergmaier et al., Nucl. Inst. Meth. **B136-138**(1998)638.
[20] A. M. Mukhamedzhanov et al., Phys. Rev. **C83**(2011)044604.
[21] A. Caciolli et al., A&A **533**(2011)A66.

Astrophysical S factor for the ^{4}He(^{3}He,γ)^{7}Be reaction at medium energies

M Carmona-Gallardo[1], B S Nara Singh[2], O Tengblad[1], M Hass[3], M Alcorta[1], M J G Borge[1], J A Briz[1], C Cruz[1] M Cubero[1], R Domínguez-Reyes[1], B R Fulton[2], H Fynbo[5], N Gordillo[6], G Haquin[4], Y Nir-El[4], V Kumar[2], A Maira[6], J McGrath[2], A Muñoz-Martín[6], A Perea[1] and Z Yungreis[4]

[1]Instituto de Estructura de la Materia, CSIC, E-28006 Madrid, Spain
[2]University of York, York, UK
[3]Weizmann Institute of Science, Rehovot, Israel
[4]Soreq Research Center, Yavne, Israel
[5]Aarhus University, Aarhus, Denmark
[6]Centro de Microanálisis de Materiales- UAM, Madrid, Spain

E-mail: m.carmona.gallardo@csic.es

Abstract. The astrophysical **S** factor for the ^{4}He(^{3}He,γ)^{7}Be direct capture reaction plays a major role in the context of solar neutrino flux and primordial ^{7}Li abundances that demand accurate information on the reaction. We report here our recent cross section measurements using the activation method in the region of E_{CM}=900-2800 keV, that aim to shed light on the discrepancies in the existing data and lead to a more accurate extrapolation of the S factor.

1. Introduction

The ^{4}He(^{3}He,γ)^{7}Be reaction rate plays an important role in determining the primordial ^{7}Li abundance in the universe from the ^{7}Be disintegration. Furthermore, the high energy solar neutrino flux depends strongly on the ^{8}B abundance due to the ^{7}Be proton capture, and thus on the formation of ^{7}Be, which mainly occurs via the direct capture reaction ^{3}He(^{4}He,γ)^{7}Be. Currently, this reaction is one of the remaining major sources of uncertainty among the nuclear inputs required to calculate the high energy solar neutrino flux [1]. Present work is also timely in view of new results of the detection of solar anti-neurtrinos with the BOREXINO detector at Gran Sasso laboratory [2]. Precise measurements of the cross section σ are therefore necessary. The astrophysical S factor is extracted from σ via: $\sigma(E) = \frac{1}{E} \cdot S(E) \cdot e^{-2\cdot\pi\cdot\eta(E)}$, where E is the energy and η is the Sommerfeld parameter. The $S(E)$ gives the nuclear dependence of the astrophysical reaction rate and theoretical models use the extrapolation of $S(E)$ for the ^{4}He(^{3}He,γ)^{7}Be reaction at zero energy ($S_{34}(0)$) as an input parameter in determining the solar neutrino flux and ^{7}Li abundance. Presently, there are serious discrepancies between the existing measurements of S_{34} using different methods [3-7]. We focus mainly on the striking differences in the centre of mass (C.M.) energy range of 1.6 MeV to 3 MeV, where only data from Ref. [3] from 1963 and Ref. [4] from 2009 exist. Our measurements would help to constrain the experimental shape of $S_{34}(E)$ at medium energy and thus validate the theoretical calculations used for the extrapolation towards zero energy, thereby providing the accurate information needed for calculating solar neutrino and the ^{7}Li primordial abundances.

2. Experiment

In the ^{4}He(^{3}He,γ)^{7}Be capture reaction, prompt γ-rays from the capture state to ground state or to the first excited state (at 429 keV), as well as from the first excited state to the ground state are emitted. Subsequently, the ^{7}Be ions decay by electron capture to ^{7}Li with a half life of 53.35 days. Thus, three different alternatives appear for the determination of the amount of ^{7}Be created: direct detection and counting of the ^{7}Be ions, detection of prompt γ-rays, or detection of delayed 478 keV γ ray activity from the first excited state in ^{7}Li (Activation Method). The latter technique was used for the current work, utilizing the experimental setup from the Weizmann institute [8]. The experiment was carried out at the Nuclear Physics line of the 5 MV Tandem accelerator at the Centro de Microanálisis de Materiales (CMAM), in Madrid. Figure 1 shows a schematic of the setup. The ^{4}He gas target pressure in the chamber is isolated from the beam line high vacuum by using 1 μm thick Ni foil window. A ^{3}He beam with energies between 2.3 MeV and 5.3 MeV of 100 to 200 nA current passed through the Ni foil. The ^{7}Be nuclei, produced via the capture of the beam onto ^{4}He, recoiled through the chamber and were implanted into a Cu catcher placed at the end of the chamber. A silicon detector with a 0.27 mm radius collimator for monitoring the scattered beam was placed inside a chamber arm at 44.9°. A pressure gauge was used to maintain the pressure constant using a regulated gas flow system. The chamber was electrically isolated and a suppressor kept at -180 V upstream of the Ni-foil made possible to use the chamber as a Faraday cup and thus to directly measure the charge accumulated during the irradiations.

Figure 1. Sketch of the setup used in the experiment. See the text and Ref. [8] for more details.

3. Experimental analysis

Measurements, resulting in six Cu catchers with implanted ^{7}Be nuclei produced at different beam energies, were carried out in order to determine the reaction rate over a wide energy range. The amount of ^{7}Be recoiling ions, the number of target ions and the total incoming ion beam was measured precisely. The recoiling ^{7}Be atoms were collected onto the Cu catchers and the subsequent β delayed γ radiation was measured off-line using a low-background HPGe detection station at SOREQ laboratory, in Israel, similar to that of Ref. [8]. The measurement typically was carried out over a few days in order to obtain enough statistics in the 478 keV line in the γ spectrum. Due to the low pressure an ideal gas behaviours is assumed for the gas target and the number of target atoms can be expressed as: $N°(^{4}He) = 9.66 \cdot 10^{18} \cdot l \cdot \frac{P}{T_0+T_c}$, where P, l and T_0 are pressure in torr, room temperature in K and length between the Ni foil and the Cu catcher in cm, respectively. T_c is the correction to the temperature due to the beam heating the ^{4}He gas during the measurement. Pressure was kept constant over the different measurements ($\approx$50 Torr). The temperature correction (T_c) was interpolated using the data in reference [8]. In order to determine the number of incident ^{3}He ions two complementary techniques were being used: integration of the charge induced on the chamber and integration (IC) of the elastically scattered particles in the detector (RS).

4. Preliminary results and ongoing work

The results obtained after data analysis are shown in Table 1. The fourth and fifth columns show the S_{34} factors by determining the beam particles by charge integration (IC) and Rutherford scattering (RS), respectively. An agreement between the two methods is observed. Our results show a good agreement with those from ERNA data [4] at the highest energies. Around 2 MeV C.M. energy, the S_{34} factor fall between the results of [1] and [3]. New measurements with the same setup are ongoing at CMAM in order to reduce the statistical error and to check the result around 2 MeV C.M. energy, afterwards extrapolations will be carried out using different theoretical models. For a brief discussion on this data in relation with our TRIUMF work and the recent calculations by T. Neff [9] we refer to Ref. [10].

Table 1. First column gives the C.M. energy $E_{CM} = \frac{4}{7}(E_b - \Delta E_{Ni} - \frac{\Delta E_{He}}{2})$ where E_b is the incoming beam energy, ΔE_{Ni} is the energy loss in the Ni foil, and ΔE_{He} the energy loss in the gas target between the Ni foil and the Cu catcher. As we assume that the reaction takes place in the middle of the gas target, the last term in the equation is divided by two. The fourth and fifth columns show S-factor when the number of incoming particles are obtained from charge integration ($S_{34}(E)^{IC}$) and from Rutherford scattered particles ($S_{34}(E)^{RS}$), respectively. The statical errors associated to the delayed gamma radiation detection are shown. The systematic errors are negligible.

E_{CM} (keV)	P (Torr)	N°^{7}Be (10^6)	$S_{34}(E)^{IC}$ (keV·b)	$S_{34}(E)^{RS}$ (keV·b)
903.6	54.68	1.30	0.40±0.08	0.41±0.08
1487.5	63.77	4.02	0.31±0.02	0.31±0.02
2010.6	50.64	5.02	0.30±0.02	0.31±0.02
2256.0	50.66	4.71	0.37±0.04	0.37±0.04
2499.0	50.83	3.17	0.37±0.03	0.38±0.04
2791.1	56.69	6.01	0.41±0.02	0.41±0.02

References

[1] Adelberger E G et al. 2011 *Rev. Mod. Phys.* **83** 195
[2] Bellini G et al. 2011 *Physics Letter B***696** 191
[3] Parker P D and Kavanagh R W 1992 *Phys. Rev.* **131** 2758
[4] Di Leva A et al. *Phys. Rev. Lett* **102** 232502
[5] Krawinkel H et al. 1982 *Z. Phys.* A **304** 307
[6] Hilgemeier M et al. 1988 *Z. Phys.* A **329** 243
[7] Osborne J L et al. 1984 *Nucl. Phys.* A **419** 115
[8] Nara Singh B S et al. 2004 *Phys. Rev. Lett.* **93** 262503
[9] Neff T. 2011 *Phys. Rev. Lett.* **106** 042502
[10] Nara Singh B S et al. this conference

Journal of Physics: Conference Series **337** (2012) 012062 doi:10.1088/1742-6596/337/1/012062

Electron screening effects in nuclear reactions: still an unsolved problem

J Cruz[1,2], H Luís[1,2], M Fonseca[1,2] and A P Jesus[1,2]

[1] Departamento de Física da Faculdade de Ciências e Tecnologia da Universidade Nova de Lisboa, Portugal

[2] Centro de Física Nuclear da Universidade de Lisboa, Portugal

E-mail: jdc@fct.unl.pt

Abstract. Electron screening for nuclear reactions in metals plays an unexpected and important role in enhancing reaction cross sections in the ultra-low energy region. Even though there are still some discrepancies between experimental data from different authors, the enhanced screening effect in metallic environments is well established, and attributed to the quasi-free valence electrons in the metals. However, there is still no solid theory which can describe quantitatively the observed enhancements. In the present work, experimental and theoretical results obtained so far will be overviewed, and a proposal to improve our knowledge on this subject.

1. Introduction

Electron screening effects in charged particle induced nuclear reactions have been investigated for many years. Usually, the observed effect is very small, below 1%, due to the huge difference on the interaction energies; i.e., eV in an atomic system, versus MeV in a nuclear system. However, a surprisingly large effect has been measured when studying the D(d,p)T reaction in metallic environments at ultra-low energies, which stirred up attention more strongly on the influence of the environment where nuclear reactions take place.

2. Overview of D+D reaction in metallic environment

With the cold fusion claim in 1989, the enhancement of the reaction rates of the D+D reaction in metal host has been intentionally searched for by several authors. Yuki *et al* [1] measured the reaction rates at energies down to 2.5 keV, obtaining the first clear evidence that the fusion reaction rate can be enhanced in the metal environment during keV deuteron bombardment.

Such a large screening energy of the D+D reaction was found also in other materials by other groups. Czersky *et al* [2] has reported enhanced screening energy of the D+D reactions in Ta; it was about 300 eV, and was confirmed by Raiola *et al* [3]. Also Raiola *et al* [4] have carried out a systematic measurement of the screening potentials of the D+D reaction for a wide range of host materials including metals, semiconductors and insulators, in a total of 54 elements, clearly showing that the large screening energies ($U_e > 150$ eV) were obtained only for metals, while insulators and semiconductors show screening energies similar to the one obtained for a D_2 gas target ($U_e = 25 \pm 5$ eV [5]), being this last value in agreement with atomic physics models.

Kasagi [6] and Huke *et al* [7], reported an enhanced screening energy of the D+D reactions in different metals, showing discrepancies with Raiola *et al* [4] values. Huke *et al* claim that the differencies between their results and Raiola *et al* ones would arise from major experimental pitfalls in [4], namelly oxygen and/or carbon build-up on the target surface and unstable deuteron density profiles. For the carbon contamination, it is known that carbon can achive high deuteron densities but it does not show the enhanced electron screening effect [4]. So, a thin deuterated carbon layer could, in fact, simulate a screening enhancement due to its high capacity of retaining deuterium. However, taking into account that for each deuterium implanted target, data taking lasts several days, continuous carbon build-up would completely supress any enhancement after a few hours.

With respect to oxygen contamination, metal oxides show low deuterium solubilities, which according to [7] would promote an unstable deuterium depth profile that could simulate the exponential like screening enhancement. Once again, for long data taking, oxygen build-up would result on a thick oxide layer at the target surface. The presence of oxygen was investigated by Raiola *et al* [4] by Rutherford Backscattering Spectroscopy (RBS) of deuterated metal targets, and was below the detection limit which is quite low as can be inferred from the simulation of a RBS spectrum of ^{4}He particles with E_{lab} = 3038 keV [$^{16}O(\alpha,\alpha)^{16}O$ resonance - figure 1] hitting a 15 nm thick ($\approx$ 30 monolayers) TiO_2 layer on top of bulk Ti target. The simulation of the ^{4}He backscattered at 150° was done using the NDF code [8], and clearly shows the oxygen peak on top of the Ti barrier (figure 2). These results show that the discrepancies between these two experimental groups should not come from target contamination by either carbon or oxygen.

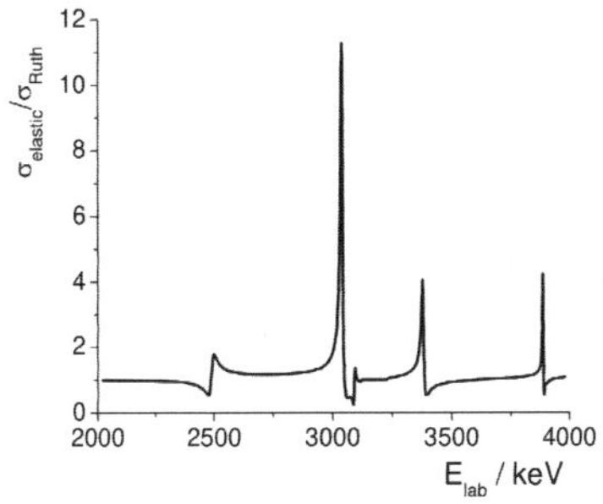

Figure 1. Non-Rutherford Elastic Scattering Cross Section of ^{4}He by ^{16}O - θ_{lab} = 150°.

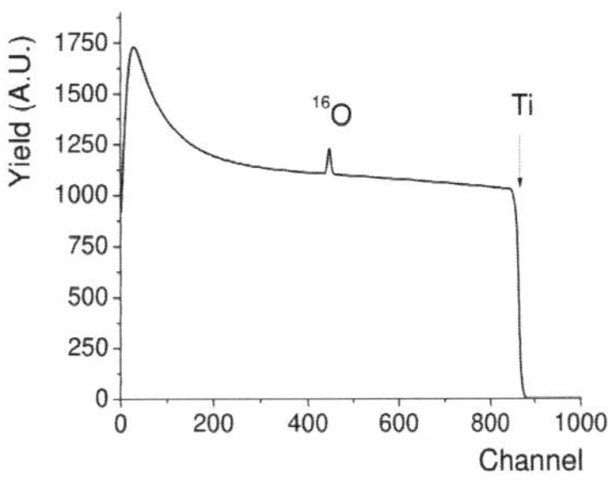

Figure 2. The 3038 keV ^{4}He backscattered at θ_{lab} = 150° simulated spectrum of a TiO_2 + Ti target.

Huke *et al* [7] claim that, as for the oxide, low deuteron density in the host metal also creates an unstable deuteron density profile that simulates a large screening, and, therefore, the correct screening is obtained only when the metal host is saturated with deuterium in a proportion close to 1:1. On the other hand, Kasagi [6] showed that there may be a relevant correlation: a large screening goes with a small deuteron density, suggesting the possibility of a dynamic screening mechanism during the deuteron bombardment and penetration into the host. However, for the ^{6}Li(p,α)^{3}He and ^{7}Li(p,α)^{4}He nuclear reactions, density dynamic effects can not explain the large screening energies obtained by Cruz *et al* [9], consistent with the D(d,p)T measurements, since the lithium mobility is orders of magnitude lower than for deuterium.

The screening energy for the deuteron in metals and non-metals was also theoretically studied. The simplest model is a static model since it is assumed an a priori electron distribution (at the Bohr radius) which does not change during the collision process. Here, the predicted electron screening energy is 27 eV. A refinement is obtained by introducing dynamic models as the one proposed by Bracci *et al* [10] which considered changes in the electron cloud configuration due to the incoming projectile. Here the maximum screening energy is 36.6 eV. Both these models predict a constant U_e value independent of the host material, so they fail in explaining the large screening energies measured in metals. A more revolutionary approach came from Raiola *et al* [4], which applied the classical Debye screening theory in metals, in which the quasi-free valence electrons form a Debye sphere around the deuterons yielding large U_e values compatible with experimental data. This model is, however, very controversial since is not applicable for low temperatures and dense plasmas (solid states). Czersky *et al* [11] calculated the screening energy for the D+D reactions in metal by applying a dielectric functional method which allows to treat the electron screening as a static polarization of the metallic medium induced by the positively charged deuteron. Their results including the cohesion effects give clearly larger values for the screening energy as compared with [10], but still tipically less than half of the observed values.

3. PIXE, XPS and plasmons

Such a striking effect with origin on the metal electronic cloud should somehow be noticeable in other analytical techniques which are also sensitive to the electronic configuration of the material being analysed. In fact, Proton Induced X-ray Emission (PIXE) studies using very high resolution setups have measured that the K_α diagram and satellite line energies depend crucially on dynamic factors [beam energy and type of beam particle (proton or α particle)] and static factors (the chemical state of the compound) [12]. Interpretation of these results is still an open question, but it has been proposed very recently by Reis *et al* [13] that plasmons can help giving an interpretation of these results. On the other hand, plasmons in metals play a well known effect in X-ray Photoelectron Spectroscopy (XPS), as the outgoing photoelectron may suffer quantized energy losses or energy gains before escaping out of the material. The effects that plasmons may have on the screening enhancement are still under study and we hope to present new results soon.

4. Conclusions

Low energy D+D and ^{6,7}Li + p reactions are greatly enhanced when the reactions occurs in a metal environment. The mechanism of the enhancement is not fully understood, even though it is well established that it is due to the quasi-free valence electrons of the metal. These arguments motivate further studies. Screening due to plasmons is one of the studies which is being carried out.

References

[1] Yuki H *et al* 1997 *J. Phys.* G **23** 1459
[2] Czersky K *et al* 2001 *Europhys. Lett* **54** 449
[3] Raiola F *et al* 2002 *Eur. Phys. J.* A **13** 377
[4] Raiola F *et al* 2005 *J. Phys.* G **31** 1141
[5] Greife U *et al* 1995 *Z. Phys.* A **351** 107
[6] Kasagi J 2007 *Surf. Coat. Tech.* **201** 8574-78
[7] Huke A, Czersky K, Heide P, Ruprecht G, Targosz N and Zebrowski W 2008 *Phys. Rev.* C **78** 015803
[8] Barradas N P , Jeynes C, Homewood K P , Sealy B J and Milosavljevic M 1998 *Nucl. Instr. Meth.* B **139** 235
[9] Cruz J *et al* 2005 *Phys. Lett* B **624** 181-85
[10] Bracci L, Fiorentini G, Melezhik V S, Mezzorani G and Quarati P 1990 *Nuc. Phys.* A **513** 316
[11] Czersky K *et al* 2006 *Eur. Phys. J.* A **27** 83
[12] Kavcic M, Karydas A G and Zarkadas Ch 2004 *Nucl. Instr. Meth.* B **222** 601-8
[13] Reis M A, Chaves P C and Taborda A 2011 *X-ray Spectrom.* **40** 141-6

Towards in-beam (α, γ) cross section measurements for the astrophysical γ-process

Z Halász, J Farkas, Zs Fülöp, Gy Gyürky, E Somorjai, T Szücs, M P Takács

Institute of Nuclear Research (ATOMKI), H-4001 Debrecen, POB.51., Hungary

E-mail: gyurky@atomki.hu (Gy. Gyürky)

Abstract. The activation technique has proven to be a very successful tool in measuring charged particle induced reaction cross sections relevant for the astrophysical γ-process. This method is, however, only applicable when the product nucleus of the reaction is radioactive and its decay can well be measured. In other cases the technically much more challenging in-beam measurements are inevitable. Here we present the first steps of a new research program aiming at the in-beam measurements of (α, γ) cross sections. As a first test case, the study of the $^{116}Sn(\alpha, \gamma)^{120}Te$ reaction has been started.

1. Introduction

The astrophysical γ-process is probably the most important sub-process of the more general p-process, which is the production mechanism of the so-called p-isotopes, those heavy, proton rich isotopes which are not produced by the s- and r-processes [1]. The p-process is one of the least understood processes of nucleosynthesis, the models are not able to reproduce well the p-isotope abundances observed in nature. The experimental study of nuclear reactions involved in a γ-process network is therefore very important for eliminating a possible source of error in the model calculations.

Detailed calculations show that γ-process models are very sensitive to the cross sections of α-induced reactions. On the other hand, very limited experimental information is available for these reactions in the relevant mass and energy range. Therefore, (α, γ) cross section measurements are needed to provide data for the models. In recent years, an increased effort has been made to carry out such measurement [2, and references therein]. These measurements were almost exclusively carried out with the activation technique. In this method the cross section is determined by measuring the radioactive decay of an isotope produced by α-irradiation. This means that the method is limited to those cases where the reaction product is unstable and its decay can be measured.

In a γ-process network, reactions leading to stable isotopes are of course also involved and the results of the calculations can be sensitive to these reactions, too [3]. Therefore, the measurement of these reactions is also important even though they can only be studied with in-beam technique. The in-beam technique is technically much more challenging than the activation, and results obtained with this method for the γ-process are practically limited to proton-induced reactions (see e.g. [4]). Realizing the importance of (α, γ) reactions, the method is to be extended to these reactions, and such a program is being carried out at different laboratories using various approaches like the application of a 4π summing crystal or a HPGe detector array [5, 6]. Here we present some details of our research initiatives towards this goal.

2. The ^{116}Sn(α,γ)^{120}Te reaction

For the first reaction to be studied with in-beam cross section measurement we have chosen ^{116}Sn(α,γ)^{120}Te based on the following reasons. In ref. [3] this reaction is suggested for measurements with high priority and since ^{120}Te is stable, it cannot be measured with activation. The target of this reaction is an even-even nucleus, thus a simpler γ-decay scheme of the ^{120}Te compound nucleus is expected. The isotope is available in high (close to 100 %) enrichment, so the disturbing effect of other Sn isotopes can be avoided. Target preparation tests showed that high purity self supporting Sn targets can be prepared with vacuum evaporation reducing this way the beam-induced background on target or backing impurities.

3. Gamma-detector calibration

The γ-radiation from the target has been measured with a 100 % relative efficiency HPGe detector and a composite Clover detector. The Q value of the ^{116}Sn(α,γ)^{120}Te reaction is 0.3 MeV and the alpha energy range to be studied is around about 10 MeV. Therefore, the prompt γ-radiation from the reactions includes high energy γ-lines up to around 10 MeV. For an absolute cross section measurements the absolute detector efficiency at these energies must be known. At low energies the 100 % detector has been calibrated using standard radioactive sources while at high energies some well known resonances of different proton capture reactions emitting cascade γ-transitions have been used. Figure. 1 shows the measured efficiency curve and lists the used sources. The resonant reactions were the following: ^{27}Al(p,γ)^{28}Si, ^{23}Na(p,γ)^{24}Mg, ^{39}K(p,γ)^{40}Ca. With these measurements the detector efficiency has been measured up to 11.6 MeV and a fourth degree polynomial has been fitted to the high energy data.

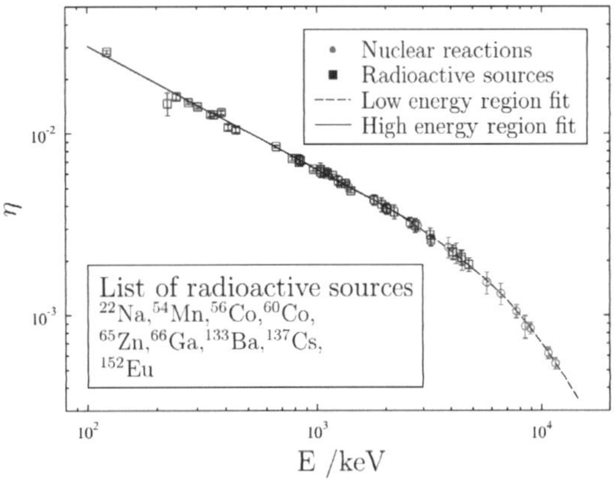

Figure 1. Efficiency curve of the 100 % HPGe detector measured with calibrated sources and resonant reactions.

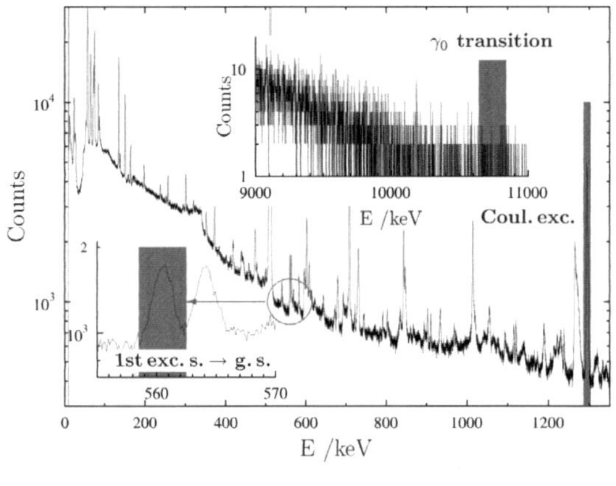

Figure 2. A γ-spectrum acquired at 10.7 MeV α-energy. The regions indicated by the red rectangles are discussed in the text.

4. The first in-beam measurements, conclusion and outlook

For the first test measurements the cyclotron accelerator of ATOMKI provided a 10.7 MeV α-beam with a typical intensity of 200 nA. The beam bombarded an enriched self-supporting ^{116}Sn target. The thickness of the target has been measured with α energy loss and Rutherford

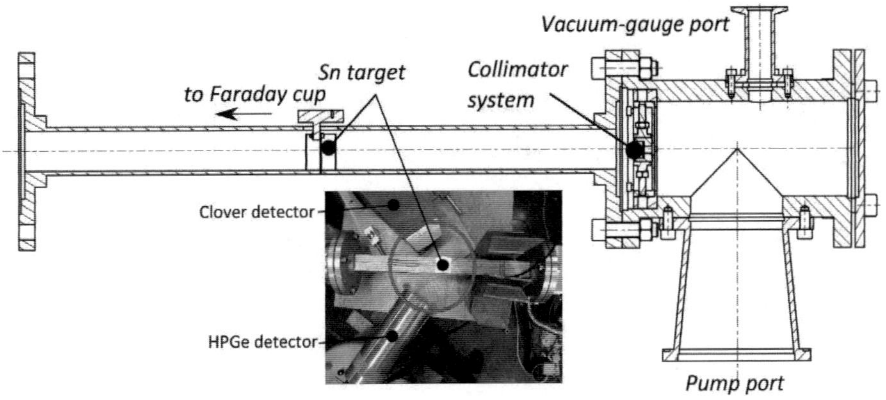

Figure 3. Schematic view and a photo of the setup.

Backscattering. The target has been put into a target chamber developed for in-beam γ-spectroscopy measurements. The inner surface of the chamber is covered with Ta in order to reduce the background caused by scattered beam. The γ-detectors have been placed next to the chamber at 55° with respect to the beam direction. The schematic view of the target chamber and a photo of the setup can be seen in Fig. 3.

Figure 2 shows a γ-spectrum which we acquired with the 100% detector. At 560 keV the transition between the first excited state and ground state in ^{120}Te is well visible and indicated by an arrow. The Coulomb-excitation peak of ^{116}Sn (the decay of the 1294 keV first excited state) is also shown in the figure. This peak can be used to monitor the target stability. Both at low and high energies the spectrum is, however, dominated by beam-induced background. In the inset of the figure the region where the γ_0-transition (the direct decay of the compound state to the ground state) is indicated and due to the background no peak can be identified. The identification of the sources of the beam induced background and its reduction is in progress.

Acknowledgments

This work was supported by the European Research Council StG. 203175 and OTKA grants K68801 and NN83261(EuroGENESIS).

References

[1] Arnould M and Goriely S 2003 *Phys. Rep.* **384** 1
[2] Gyürky Gy *et al.* 2010 *J. Phys. G* **37** 115201
[3] Rauscher T 2006 Phys. Rev. C **73** 015804
[4] Galanopoulos S *et al.* 2003 *Phys. Rev. C* **67** 015801
[5] Spyrou A *et al.* 2008 *Phys. Rev. C* **77** 065801
[6] Hasper J *et al.* 2010 *J. Phys. Conf. Ser.* **202** 012005

The hunt for permanent electric dipole moments

W Korsch

Department of Physics and Astronomy, University of Kentucky, 177 Chemistry-Physics Building, Lexington, KY 40506, USA

E-mail: korsch@pa.uky.edu

Abstract. The search for permanent electric dipole moments in non-degenerate systems has become a very active field of research in recent years. The experimental sensitivity has reached limits to probe physics beyond the Standard Model with high accuracy. This talk will review and summarize some of the ongoing efforts.

1. Introduction

The possible existence of permanent electric dipole moments (EDM) in nucleons, atoms, molecules, or other fundamental particles has motivated experimental physicists to push the limits of low energy precision studies for more than five decades. A non-zero EDM simply implies that electric charge is not uniformly distributed "inside" such a particle, giving rise to the violation of time-reversal(T) symmetry. Already in 1950 E. M. Purcell and N. F. Ramsey argued that a non-zero EDM could arise due to parity violation [1]. It is easy to comprehend this argument by studying the interaction Hamiltonian between external magnetic and electric fields and a particle with spin $\mathbf{s}$:

$$H = -(\boldsymbol{\mu} \cdot \mathbf{B} + \mathbf{d} \cdot \mathbf{E}) = -(d_M \frac{\mathbf{s}}{|\mathbf{s}|} \cdot \mathbf{B} + d_E \frac{\mathbf{s}}{|\mathbf{s}|} \cdot \mathbf{E}) \ . \quad (1)$$

Here $\boldsymbol{\mu}$ and $\mathbf{d}$ denote the magnetic and electric dipole moments of the particle. It is worth emphasizing that the direction of $\boldsymbol{\mu}$ as well as $\mathbf{d}$ has to be along the particle's spin axis. While the magnetic field is even under parity transformation, the electric field is not, so a finite value of $\mathbf{d}$ (or d_E) implies that H does not conserve parity. In more recent years the concept of parity violation was expanded to the "CPT Theorem". It is believed, even beyond the Standard Model of particle physics, that nature is invariant under a combined symmetry operation consisting of charge conjugation (C) and parity transformation (P) and time reversal (T). Indeed, the Hamiltonian in Eq. 1 does violate time reversal, and hence CP symmetry, if $|\mathbf{d}| \neq 0$. In our discussion, we assume that no degenerate states exist in the system of interest.

The discovery of a finite EDM at present experimental sensitivities would unambiguously yield physics beyond the Standard Model (SM) of particle physics. Some recent experimental results, together with (CKM) Standard Model predictions are listed in table 1. Unless new physics increases the values of EDMs by orders of magnitude, present and near future experimental techniques will not suffice to test the SM predictions.

Table 1. Recent experimental limits on permanent EDMs (no claim of completeness is made).

Particle	Limit [e·cm] ($\geq 90\%$ C.L.)	Experiment	SM Prediction [e·cm]
e^-	10.5×10^{-28}	Hinds *et al.* [2]	$< 10^{-38}$ [3]
μ	1.8×10^{-19}	Bennett *et al.* [4]	$< 10^{-36}$ [5]
n	2.9×10^{-26}	Baker *et al.* [6]	$< 10^{-32}$ [7]
^{199}Hg	3.1×10^{-29}	Griffith *et al.* [8]	$\sim 10^{-33}$ [9]
^{129}Xe	3.4×10^{-27}	Rosenberry *et al.* [10]	$\sim 10^{-34}$ [11]
^{205}Tl	9.4×10^{-25}	Regan *et al.* [12]	

The existence of a permanent EDM has numerous and far reaching implications, e.g. it could, at least partially, explain the baryon – anti-baryon asymmetry universe and it would reveal new sources of CP violation.

Section 2 describes some ongoing experimental efforts on neutron EDM searches with a focus on an experiment that will take place at the newly build Spallation Neutron Source (SNS, Oak Ridge National Laboratory (ORNL)). Section 3 summarizes briefly efforts on atomic and molecular systems.

2. Neutron EDM searches

The first neutron EDM measurement dates back to 1957 when J. H. Smith, E. M. Purcell, and N. F. Ramsey published their result of an experiment that was performed at ORNL using a magnetic resonance technique [13]. An upper bound of $(0.1 \pm 2.4) \times 10^{-20}$ e·cm was reported for the magnitude of the EDM. This was just the beginning of still ongoing efforts to find violation of fundamental time-reversal symmetry. During the past fifty years the limit was improved continuously at a rate of roughly one order of magnitude per decade. Basically all neutron EDM searches utilized magnetic resonance techniques and, since about 1980, the use of ultra-cold neutrons (UCNs) has become favorable over thermal or cold neutron beams. The kinetic energy of UCNs is so low that they can be easily trapped in containers (assuming coatings with large repulsive Fermi-potentials are applied), strong magnetic fields, or even gravitationally. The best limit on a neutron EDM was recently achieved at the ILL in Grenoble using polarized UCNs stored in a cylindrical trap. [6]. Figure 1 shows the basic components of the ILL experimental setup. The experimental procedure can be summarized as follows: First, longitudinally polarized UCNs were injected for about 20 s into a 21 l volume (trap). Inside the trap the neutrons were exposed to a combination of a weak uniform static magnetic field (≈ 10 mGauss) and an electric field with a magnitude of 10 kV/cm. The electric and magnetic field were aligned either parallel or anti-parallel to each other. After filling of the trap, an oscillating magnetic field acted on the neutrons for a short time interval ($\Delta T \approx 2$ s) rotating the spins perpendicular to the static fields. The frequency of these fields was close to the neutron Larmor frequency. A free spin precession followed for about 130 s. Then a second pulse was applied to rotate the neutron spins back to their original directions. Finally, the neutron polarization was measured to determine whether the electric field influenced the precession frequency. This technique is known as the Ramsey separated-oscillatory-field magnetic resonance method. Figure 2 shows a typical Ramsey fringe pattern. The final result reported by the collaboration constrains d_n to less than 2.9×10^{-26} e·cm [6].

The goal of upcoming neutron EDM experiments is to improve the present limit by a factor of about 100. Although the Standard Model of particle physics predicts only a value of $10^{-32}-10^{-31}$ e·cm for d_n, an improvement of the present bound by two orders of magnitude could have interesting consequences. For example, the masses of new elementary (SUSY) particles

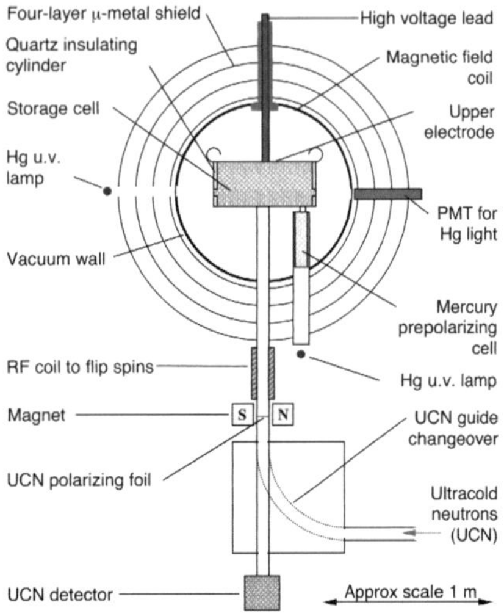

Figure 1. The Sussex-RAL-ILL EDM appartus.

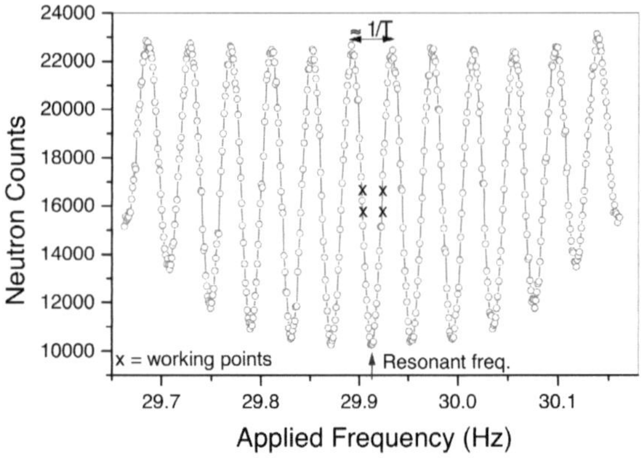

Figure 2. A typical Ramsey resonance curve for spin-up neutrons.

necessary to generate such a sizable neutron EDM are comparable to energies accessible at the LHC assuming that the CP violating phases are of $\mathcal{O}(1)$ [14]. Therefore, the discovery potential of SUSY particles at the LHC could be closely related to the discovery of a permanent EDM. Three parallel efforts have emerged. One experiment is presently constructed at the PSI in Switzerland [15], a second one will take place at the ILL in France [16], and finally the SNS-EDM experiment in the US [17]. All three experiments aim for comparable sensitivity. The following section describes the basic concept of the SNS-EDM experiment.

2.1. The SNS experiment

Inspection of the interaction Hamiltonian, Eq. 1, suggests that d_n can be extracted by monitoring the change in spin precession frequency of the neutron when the electric field, $\mathbf{E}$, is reversed relative to the magnetic field, $\mathbf{B}$. Solving Eq. 1 for d_n yields:

$$d_n = \frac{h\Delta\nu}{4E} , \tag{2}$$

where h is the Planck constant, $\Delta\nu$ is the change in precession frequency when $\mathbf{E}$ is reversed relative to $\mathbf{B}$, and E is the magnitude of the electric field. The statistical uncertainty on δd_n is bounded by the uncertainty principle:

$$\delta d_n = \frac{\hbar}{4ET_m\sqrt{mN}} , \tag{3}$$

where T_m is the time to perform a neutron spin precession measurement, and m denotes the number of separate complete measurements with N neutrons. It is obvious from this formula that long neutron spin precession times, many measurement cycles, and a large number of neutrons are needed to achieve a minimum statistical uncertainty.

The newly constructed SNS at the Oak Ridge National Laboratory commenced operation in 2006 and appears to be an ideal environment to conduct a high precision neutron EDM experiment. A 1.4 MW pulsed proton beam (E_p = 1.0 GeV) impinges on a high power mercury target and spallation neutrons are generated with unprecedented (pulsed) intensity. After moderation the neutrons have a typical temperature of $T_n \sim 20$ K. In 1975, R. Golub and R. Pendlebury pointed out that the local density of free neutrons can be increased significantly when they are converted to ultra-cold neutrons (UCNs) [18]. Specifically, neutrons with a wavelength of 8.9 Å can be down-scattered to milli-Kelvin temperatures via single-phonon excitations in superfluid helium-4 [19]. We plan to exploit this concept with the goal to capture and store such UCNs in a container.

The SNS cold neutron beam will be used in combination with a system of velocity filters (choppers) to select neutrons with a wavelength around 8.9 Å. These neutrons will then be transported and polarized using standard neutron super-mirror techniques, and finally they will be injected into two target cells containing superfluid helium-4. Any neutron that experiences a conversion to a UCN will have a chance to be trapped. As in other EDM searches, the effect of a (strong) electric field (up to 50 kV/cm) on the precession frequency of the neutrons will serve as an indication of a non-zero EDM. In addition to the electric field a very uniform small magnetic field ($\mathcal{O}$(10 mG)) will serve as a quantization axis. The main challenge of the experiment is to measure the precession frequency of the neutrons very precisely. To accomplish this goal we plan to take advantage of the strong spin dependence in the reaction: $\overrightarrow{n} + {}^3\overrightarrow{He}^{++} \rightarrow t + p + 764$ keV. For that purpose, a small amount of polarized ^{3}He will be injected into the helium-4 cell as well. Both the neutrons and ^{3}He nuclei will be prepared in such a way that all spins are aligned along the same direction (perpendicular to the magnetic field) at the beginning of a measurement cycle. Due to the difference in the magnetic moments of the two species, their spins will precess at different rates and therefore the rate at which the above reaction occurs is modulated accordingly. The protons released in this process carry enough energy to generate scintillation light in the superfluid medium. We plan to detect this scintillation light and use it as a measure of the relative spin precession frequency. A non-zero neutron EDM will manifest itself in a correlation between the light modulation and the direction of the electric field. Note that the effect of an external electric field on the precession of the helium-3 spins is vastly reduced due to the shielding effect of the atomic electrons. We also intend to monitor the precession frequency of the ^{3}He spins independently (using SQUIDs). This frequency information will help to keep track of the stability of the magnetic field.

The experimental techniques described in this paper appear to be promising to overcome statistical limitations experienced in previous EDM searches. Given a realistic funding scenario data taking is expected to start around 2017. A more complete description of the whole experiment can be found in Ref. [20].

3. EDM searches in atomic and molecular systems

The apparent sensitivity to physics beyond the Standard Model has revived a large number of EDM searches on atomic systems in recent years. Current searches of time-reversal violation in atomic systems can be split into two categories: highly polar diatomic (paramagnetic) molecular systems and diamagnetic atoms with, ideally, highly deformed nuclei. Polar molecules with unpaired electrons exhibit an increased sensitivity to the electron EDM since the bound electron is preferentially located along the internuclear axis and internal electric fields can reach tens of gigavolts per centimeter. In addition, relatively low external electric fields are needed (of order 10 kV/cm or less) to align such molecules. However, the main challenge with polar molecules is finding an efficient way to generate large quantities with low rotational states since only molecules with sufficiently low values of angular momentum have a detectable sensitivity to EDMs ($\mathbf{E} \cdot \mathbf{d_{mol}} > E_{rot}$). Systems like HfH$^+$, PbF, PbO, ThO, WC, and YbF are studied as

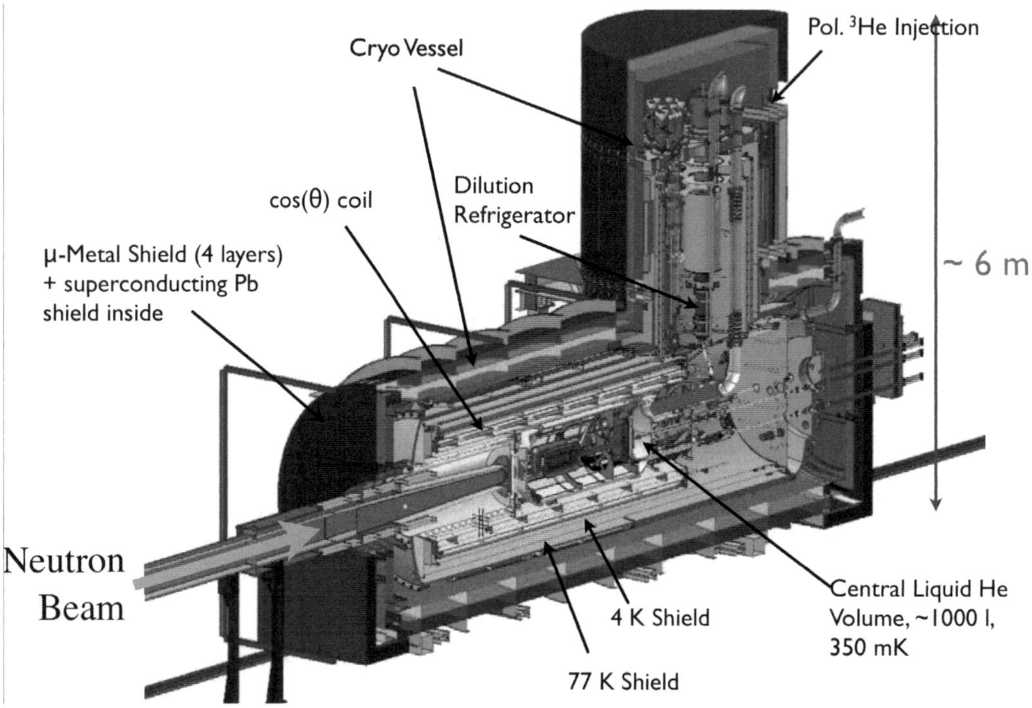

Figure 3. Conceptual design of the SNS-EDM apparatus.

possible candidates. Hinds and collaborators published recently a new result on d_e using YbF. Their measurement yielded a value consistent with zero and an upper limit of $10.5 \times 10^{-28} e\cdot$cm was reported [2].

Diamagnetic atoms, on the other hand, reveal a higher sensitivity to nuclear EDMs. Especially atoms that can be easily spin-polarized are excellent candidates for atomic EDM searches. The most precise limit on a diatomic atom thus far has been reached at the University of Washington (Seattle). Using optically pumped ^{199}Hg in a multi-cell configuration it was found that $d_{^{199}Hg} < 3.1 \times 10^{-29} e\cdot$cm [8]. This result by itself puts severe constraints on many supersymmetry inspired models. It turns out that ^{225}Ra has emerged as a very promising candidate for EDM searches. This radium isotope possesses a large Schiff moment which increases the sensitivity to T-violating effects significantly. Due to its large octupole deformation, the Schiff moment predicted for ^{225}Ra [21] is expected to yield an enhancement factor somewhere between 100 and 1000 over ^{199}Hg. Two groups at Argonne National Lab (ANL) [22] and at KVI (Groningen) [23] are in the process of developing an EDM experiment using the $I = 1/2$ radium isotope. Although ^{225}Ra is radioactive, its half life of 14.9 days is sufficiently long to be useful for EDM studies. The ANL group has already demonstrated the trapping of laser cooled Ra atoms in a magneto-optical trap (MOT) as well as successful transfer to a far off resonant optical dipole trap (ODT) - two essential steps in performing the measurement. The plan is to transport about 10^4 atoms into a region of uniform electric and magnetic fields by means of a mobile ODT. Within a small volume of about 4×10^{-3} mm^3 the atoms will be polarized by optical pumping. As described above the existence of a non-zero EDM will be probed by monitoring the spin precession as a function of the electric field direction. Compared to other EDM searches this experiment has several systematic advantages; e.g. the small size of the sample volume simplifies the generation of uniform electric and magnetic fields; therefore, effects due to motional electric fields ($\mathbf{E} = \mathbf{v} \times \mathbf{B}$) will be minimal. Further, electric fields of order 100 kV/cm can be produced relatively easily since the electrode separation is only about 2 mm. All components of the experiment appear to be working and first results can be expected within the next two or three years.

4. Summary

In recent years the hunt for permanent electric dipole moments has emerged as a very active field of research. Many groups around the world are in the process of developing new devices and setups in order to improve EDM limits on the electron, atoms, neutron, and other fundamental particles. Even new electrostatic particle storage rings are being considered to measure the EDM of the deuteron. Within the next decade several new results should emerge putting severe constraints on many new physics models.

Acknowledgments

This work is supported by the U.S. Department of Energy (DOE), Office of Nuclear Physics, under contracts DE-FG02-99ER41101 and DE-AC02-06CH11357.

References

[1] Purcell E M and Ramsey N F 1950 *Phys. Rev.* **78** 807
[2] Hudson J J et al. 2011 *Nature* **473** 493
[3] Khriplovich I B and Pospelov M E 1991 *Sov. J. Nucl. Phys.* **53** 638
[4] Bennett G W et al. 2009 *Phys. Rev. D* **80** 052008
[5] Farley F J M et al. 2004 *Phys. Rev. Lett.* **93** 052001
[6] Baker C A et al. 2006 *Phys. Rev. Lett.* **97** 131801
[7] Khriplovich I B and Zhitnitsky A R 1982 *Phys. Lett. B* **109** 490
[8] Griffith W C et al. 2009 *Phys. Rev. Lett.* **102** 101601
[9] Pospelov M and Ritz A 2005 *Annals Phys.* **318** 119
[10] Rosenberry M A and Chupp T E 2001 *Phys. Rev. Lett.* **86** 22
[11] Ramsey-Musolf M J and Su S 2006 Low energy precision tests of supersymmetry *Preprint* hep-ph/0612057v1
[12] Regan B C et al. 2002 *Phys. Rev. Lett.* **88** 071805
[13] Smith J H, Purcell E M , and Ramsey NF 1957 *Phys. Rev.* **108** 120
[14] Hamzaoui C, Pospelov M, and Roiban R 1997 *Phys. Rev. D* **56** 4295
[15] see http://nedm.web.psi.ch/
[16] see https://www.neutronedm.org/index.html
[17] see http://www.phy.ornl.gov/nedm/
[18] Golub R and Pendlebury J 1975 *Phys. Rev.* **53A** 133
[19] Golub and Pendlebury J 1977 *Phys. Rev.* **62A** 337
[20] Golub R and Lamoreaux S K 1994 *Phys. Rep.* **237** 62
[21] Auerbach N, Flambaum V V and Spevak V 1996 *Phys. Rev. Lett.* **76** 4316
[22] see http://www.phy.anl.gov/mep/atta/research/radiumedm.html
[23] see http://www.rug.nl/kvi/Research/kviTrimp/researchDescription/edm/radiumEDM

Nuclear Physics in Astrophysics V IOP Publishing
Journal of Physics: Conference Series **337** (2012) 012065 doi:10.1088/1742-6596/337/1/012065

Double Beta Decay, Nuclear Structure and Physics beyond the Standard Model.

Amand Faessler
Institute für Theoretische Physik der Universität Tübingen,
D-72076 Tübingen, Germany

E-mail: amand.faessler@uni-tuebingen.de

Abstract. The Neutrinoless Double Beta Decay ($0\nu\beta\beta$) allows to determine the absolute scale of the neutrino masses. For this one must assume, that the light Majorana neutrino exchange is the leading mechanism for $0\nu\beta\beta$ and that the matrix element of this transition can ba calculated reliably. The different methods: Quasi-particle Random Phase Approximation (QRPA), Shell Model (SM), Projected Hartree-Fock-Bogoliubov (PHFB) and Interacting Boson Model (IBM2) to calculate these matrix elements are reviewed. The question, how one can determine the leading mechanism or mechanisms from the data of the $0\nu\beta\beta$ decay in different nuclei is studied in the later part of this article. Possible interference terms allow to test CP (Charge and Parity conjugation) violation.

1. Introduction

The neutrinoless double beta decay ($0\nu\beta\beta$) provides a method to determine the absolute scale of of the neutrino masses. The determination of the masses is possible, if one assumes, that the light left handed Majorana neutrino exchange is the leading mechanism for the neutrinoless double beta decay and one is also able to calculate reliably the transition matrix element. To determine the absolute masses the matrix element $\mathcal{M}_{\nu}^{0\nu}$ is as important as the data for the $0\nu\beta\beta$ transition. The different methods used to calculate these matrix elements are presented and compared with their advantages and their drawbacks [1, 2, 3, 4, 5, 6], [7], [8, 9, 10], [11].

If one gives up the assumption, that the light left handed Majorana neutrino exchange is the leading mechanism [12, 13] one can have cases, where two or even more equally strong mechanisms interfere. There the relative phases can test the CP (combined Charge conjugation and Parity) conservation or violation.

In addition to the light left handed Majorana neutrino exchange, one has other possible mechanisms as cause for the neutrinoless double beta decay: Grand Unification (GUT), Supersymmerty (SUSY) and extensions to extra dimensions. We shall discuss here extension to GUT's and SUSY.

$$\mathcal{T}^{0\nu} = \mathcal{M}_{\nu}^{0\nu} \cdot < m_{\nu} > + \mathcal{M}_{\vartheta} < \tan\vartheta > + \mathcal{M}_{W_R} < (\frac{M_1}{M_2})^2 > + \\ \mathcal{M}_{SUSY} \cdot \lambda_{111}^{\prime 2} + \mathcal{M}_{NR} < \frac{m_p}{M_{MR}} > + ... \quad (1)$$

$$< m_\nu >= \sum_{k=1,2,3} (U_{ek})^2 \cdot m_{k\nu} = \sum_{k=1,2,3} e^{2i\alpha_k} \cdot |U_{ek}|^2 \cdot m_{k\nu} \quad (2)$$

The first term on the right hand side of eqn. (1) with the matrix element $\mathcal{M}_\nu^{0\nu}$ and the effective Majorana mass $< m_\nu >$ of eqn. (2) is often called the gold plated term.

2. The different Many Body Approaches for the $0\nu\beta\beta$ Matrix Elements.

The groups in Tübingen, Bratislava and Jyväskylä [1, 2, 3, 4, 5, 6] use the Quasi-particle Random Phase Approach (QRPA), the group in Strabourg-Madrid the Shell Model (SM) [7], Tuebingen, Rath et al. and Martinez-Pinedo et al. HFB and methods built on it [8, 9, 10] and Iachllo [11] the Interacting Boson Model.

The expressions for the matrix elements $\mathcal{M}_\nu^{0\nu}$ and the corresponding $0\nu\beta\beta$ transition operators are given, e.g., in Ref. [2]:

$$\mathcal{M}_\nu^{(0\nu)} = \mathcal{M}_{GT}^{0\nu} - (\frac{g_V}{g_A})^2 \mathcal{M}_F^{0\nu} - \mathcal{M}_T^{0\nu} \quad (3)$$

The SM approach has been applied by the Strasbourg-Madrid group [7]. It is compared with QRPA in fig.1 and in fig. 2.

The projected HFB approach for a real transformation, with axial symmetry and no parity mixing is restricted [16] to contributions of neutron pairs with angular momenta $0^+, 2^+, 4^+, \ldots$. In addition the contributions of transition of higher angular momentum neutron to proton pairs $2^+, 4^+, \ldots$ are drastically reduced compared to the QRPA and the SM. The reason for this is obvious: in a spherical nucleus the HFB solution contains only seniority zero and no stronger higher angular momentum pairs.

The IBM (Interacting Boson Model) [11] can only change 0^+ (S) and 2^+ (D) fermionic pairs from two neutrons into two protons.

3. Including the Nuclear Deformation in QRPA.

Table 1. The values of the deformation parameter of the Woods-Saxon mean field β_2 for initial (final) nuclei fitted to reproduce the experimental quadrupole moment (labeled as "1"). The spherical limit is labeled as "0". The particle-particle strength parameters g_{pp} are listed. They multiply the Brückcner particle-particle nucleon-nucleon G matrix elements of the BONN CD force. They are fitted to the $2\nu\beta\beta$ decay half lives. The axial charge is assumed to be the vacuum value $g_A = 1.25$. The particle-hole strength parameter, with which the nucleon-nucleon BONN CD Brückner particle-hole matrix elements are multiplied, is adjusted to the excitation energy of the Gamow-Teller resonance in the intermediate nucleus $g_{ph} = 0.90$. The BCS overlap factor $\langle BCS_f|BCS_i\rangle$ between the initial and final BCS vacua is given in the last column.

initial (final) nucleus	β_2		g_{pp}	$\langle BCS_i\|BSC_f\rangle$
^{76}Ge (^{76}Se)	0.10 (0.16)	"1"	0.71	0.74
	0.0 (0.0)	"0"	0.68	0.81
^{150}Nd (^{150}Sm)	0.240 (0.153)	"1"	1.05	0.52
	0.0 (0.0)	"0"	1.01	0.85
^{160}Gd (^{160}Dy)	0.303 (0.292)	"1"	1.00	0.74

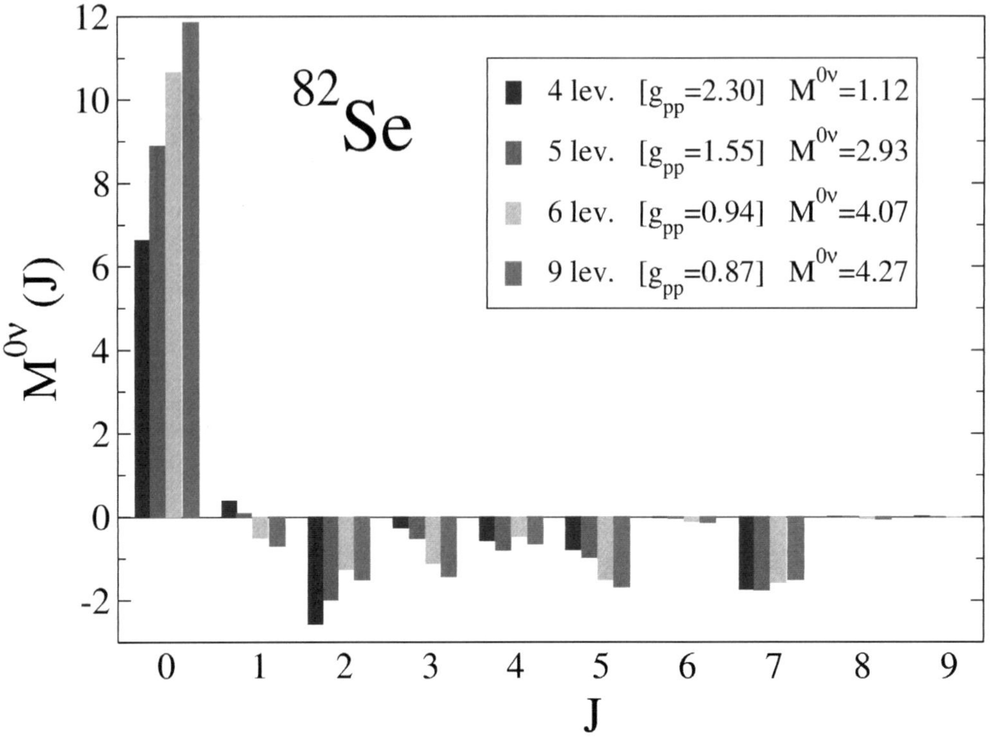

Figure 1. (Color online) Contributions of the transforming neutron pairs with different angular momenta J^π to the total $M_\nu^{0\nu}$ calculated within the QRPA and different basis sizes for the $0\nu\beta\beta$ decay ^{82}Se$\rightarrow^{82}$Kr. The left bar is calculated with the same basis of four levels, $1p_{3/2}, 0f_{5/2}, 1p_{1/2}$ and $0g_{9/2}$, used in the shell model calculations [7]. The Ikeda Sum Rule (ISR) [14] is exhausted only by 50%. The second bar from the left includes in addition the $1f_{7/2}$ level, one of the two missing spin-orbit partners given for the ^{82}Se nucleus in ref. [7] for the shell model. The ISR is exhausted by 66%. The third bar from the left includes both missing spin-orbit partners $0f_{7/2}$ and $0g_{7/2}$ amounting in total to 6 single-particle levels. The ISR is fulfilled by 100%. This leads to the increase in the neutrinoless matrix element from 1.12 to 4.07. The right bar represents the QRPA result with 9 single-particle levels ($1f_{7/2}, 2p_{3/2}, 1f_{5/2}, 2p_{1/2}, 1g_{9/2}, 2d_{5/2}, 3s_{1/2}, 2d_{3/2}, 1g_{7/2}$.). The matrix element gets only slightly increased from 4.07 to 4.27. The spin-orbit partners are essential to fulfill the Ikeda Sum Rule (ISR). In all four QRPA calculations the QRPA "renormalization" factor g_{pp} (given in the figure) of the particle-particle strength of the Bonn CD nucleon-nucleon interaction is adjusted to reproduce the experimental $2\nu\beta\beta$ decay rates.

We have also calculated [5, 6] the transition matrix elements in deformed nuclei for the light left handed Majorana neutrino exchange $\mathcal{M}_\nu^{0\nu}$. Different deformations for the inital and the final nuclei are allowed. The BCS overlaps are listed in table 1. The quadrupole deformations

Table 2. The total calculated nuclear matrix elements (NME) $\mathcal{M}_\nu^{0\nu}$ for $0\nu\beta\beta$ decays ^{76}Ge$\rightarrow^{76}$Se, ^{150}Nd$\rightarrow^{150}$Sm, ^{160}Gd$\rightarrow^{160}$ Dy including deformations. The BCS overlaps from table 1 are taken into account. In the last two columns the $0\nu\beta\beta$ matrix element $\mathcal{M}_\nu^{0\nu}$ and the half-lives for assumed $< m_\nu >= 50$ meV (see eqn. 2) are shown.

A	Def.	g_A	$M_\nu^{0\nu}$	$T_{1/2}^{0\nu} \cdot [10^{26}y]$ $< m_\nu >$=50 meV
76	"1"	1.25	4.69	7.15
	"0"	1.25	5.30	5.60
150	"1"	1.25	3.34	0.41
	"0"	1.25	6.12	0.12
160	"1"	1.25	3.76	2.26

are taken from the reorientation Coulomb excitation of the 2^+ states.

4. How to find the Leading Mechanisms for the Neutrinoless Double Beta Decay?

Normally one assumes, that the first term of eq. (1) is the leading one and with the experimental data and the matrix element for the light left handed Majorana neutrino exchange $\mathcal{M}_\nu^{0\nu}$ one can determine the effective Majorana neutrino mass $< m_\nu >$ (2). But in Grand Unification (GUT) and Supersymmetry (SUSY) additional mechanisms for the neutrinoless Double Beta Decay ($0\nu\beta\beta$) are possible [12, 13].

The inverse half life is given by:

$$\frac{1}{T_{1/2}^{0\nu}} = \frac{w^{0\nu}}{\ln 2} \approx G^{0\nu}(E_0, Z) \cdot |[\eta_\nu \mathcal{M}_\nu^{0\nu} + \eta_{NL}\mathcal{M}_{NL}^{0\nu} + \eta_{\lambda'}\mathcal{M}_{\lambda'}^{0\nu}]^2 + |\eta_{NR}|^2|\mathcal{M}_{NR}^{0\nu}|^2| \quad (4)$$

with η_l given in ref. [13].

To test, if the light left handed Majorana neutrino exchange is the leading mechnism, one meeds at least experimental data of the neutrinoless double beta decay and reliable transition matrix elements in two systems. To determine the absolute values of the two strength parameters for two non-interfering mechanisms e.g. η_ν and η_{NR}, one needs at least two decay systems. To verify, that these are indeed the leading mechanisms one needs at least a third measurement. But if one forms ratios of half lives using the Tuebingen matrix elements for the light Majorana neutrino exchange and the heavy right handed neutrino exchange one obtaines a very restricted allowed interval for these ratios. These ratios are quite insensitive to the different parameters of the model.

$$0.15 \leq \frac{T_{1/2}^{0\nu}(^{100}Mo)}{T_{1/2}^{0\nu}(^{76}Ge)} \leq 0.18; \quad 0.17 \leq \frac{T_{1/2}^{0\nu}(^{130}Te)}{T_{1/2}^{0\nu}(^{76}Ge)} \leq 0.22; \quad 1.14 \leq \frac{T_{1/2}^{0\nu}(^{130}Te)}{T_{1/2}^{0\nu}(^{100}Mo)} \leq 1.24; \quad (5)$$

If the two leading mechanisms like the light Majorana neutrino exchange and the SUSY mechanism with gluino or neutralino exchange can interfer, the situation is a bit more complicated: One needs three decay systems to determine the absolute values of the three parameters $\eta_\nu, \eta_{\lambda'}$ and the relative phase angle $\vartheta_{\nu,\lambda'}$. At least one additional system is needed to verify, that indeed these two mechanisms are the leading ones. Again the ratios of the half lives are allowed to lie only in narrow regions [13]. If this is not the case, the chosen mechanisms are

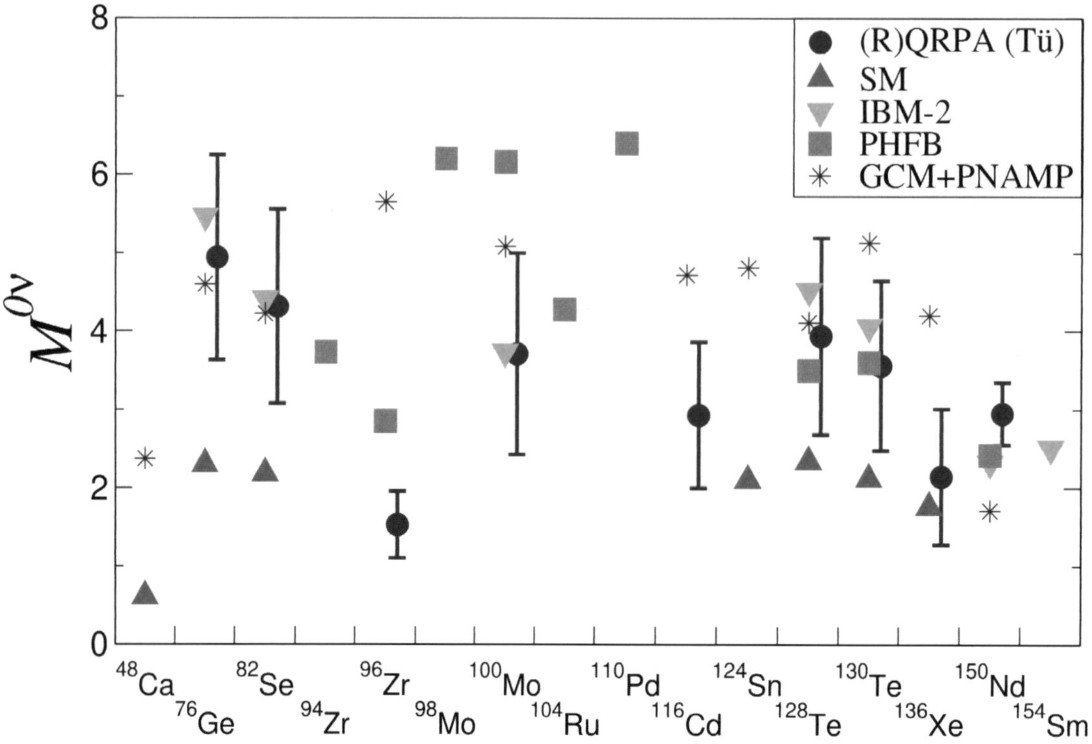

Figure 2. (Color online) Neutrinoless double beta decay transition matrix elements for the different approaches: QRPA [2, 3], the SM [7], the projected HFB method [9], the projected HFB with the Generator Coordinate Method (GCM) over deformations [10] (GCM-PNAMP) and the IBM2 [11]. The error bars of the filled circles for the QRPA are calculated as the highest and the lowest values for three different single-particle basis sets, two forces (Bonn CD and Argonne V18), two different axial charges $g_A = 1.25$ and the quenched value $g_A = 1.00$ and two different treatments of short range correlations (Jastrow-like [18] and the Unitary Correlator Operator Method (UCOM) [19]). The radius parameter is as in this whole work $r_0 = 1.2$ fm. The triangle with the tip up are the SM results [7]. The triangle with the tip down represent the transition matrix element of the Interacting Boson Model 2 (IBM2) [11]. The squares have been calculated by Pradfulla Rath and coworkers [9]. The star (GCM-PNAMP) is a projected HFB calculation with the Gogny force by Rodriguez and Martinez-Pinedo [10] allowing for supeposition of different deformations with the Generator Coordinate Method (GCM).

not the leading ones [13]. With CP conservation the strength parameters η must be real and thus the relative phase angle is zero or 180 degrees. So the determination of $\vartheta_{\nu,\lambda'}$ allows to test CP conservation or violation.

5. The effective Majorana Neutrino Mass.

Before we summarize the results let us assume Klapdor-Kleingrothaus et al. [17] have indeed measured the neutrinoless double beta decay in ^{76}Ge, although the general belief is, that this still needs confirmation. From the half life given by Klapdor et al. [17] one can derive with our matrix elements the effective Majorana neutrino mass under the assumption, that the light Majorana exchange is the leading mechanism.

$$< m_\nu >= 0.24[eV](exp \pm 0.02; theor. \pm 0.01)[eV] \tag{6}$$

The uncertainty (error) from experiment is 0.02 [eV], while the theoretical error originates from the uncertainties of the QRPA matrix elements as indicated in figure 2. The theoretical error is 0.01 [eV].

6. Conclusions

The Shell Model (SM) [7] has a severe handicap due to the restricted single-particle basis. The matrix elements in the ^{76}Ge region are by a factor 2 smaller than the results of the Quasiparticle Random Phase Approximation (QRPA) [1, 2, 3, 4], the projected Hartree Fock Bogoliubov approach [9, 10] and the Interacting Boson Model (IBM2) [11]. With the same restricted basis as used by the SM the QRPA obtains roughly the same results as the SM, but the Ikeda sum rule [14] is strongly violated due to the missing spin-orbit partners in the SM single-particle basis.

The angular momentum projected Hartee-Fock-Bogoliubov (HFB) method [9] is restricted in its scope. With a real Bogoliubov transformation without parity mixing and with axial symmetry, one can only describe neutron pairs with angular momenta and parity $0^+, 2^+, 4^+, 6^+, \ldots$ [16] changing into two protons for ground state-to-ground state transitions. The restriction for the Interacting Boson Model (IBM) [11] is even more severe: one is restricted to 0^+ and 2^+ neutron pairs changing into two protons.

Acknowledgments

This contribution is based on work with J. Engel, A. Escuderos, D.-L. Fang, G. Fogli, E. Lisi, A. Meroni, S. T. Petcov, V. Rodin, E. Rottuno, F. Simkovic, J. Vergados and P. Vogel [1, 2, 3, 5, 6, 12, 13]. I want to thank for their collaboration. I also acknowledge support of the Deutsche Forschungsgemeinschaft within both the SFB TR27 "Neutrinos and Beyond" and the Graduiertenkolleg GRK683.

References

[1] Escuderos A, Faessler A, Rodin V, Simkovic F 2010 *J. Phys. G.* **37** 125108 and **arXiv:1001.3519** [nucl-th]
[2] Rodin V A, Faessler A, Šimkovic F, and Vogel P 2006 *Nucl. Phys.* A **766** 107; *Erratum-ibid.* 2007 **793** 213; Šimkovic F, Faessler A, Müther H, Rodin V A and Stauf M 2009 *Phys. Rev.* C **79** 055501
[3] Šimkovic F, Faessler A, Rodin V, Vogel P and Engel J 2008 *Phys. Rev.* C **77** 045503
[4] Kortelainen M and Suhonen J, 2007 *Phys. Rev.* C **75** 051303(R); *ibidem* 2007 **76** 024315
[5] Fang D-L, Faessler A, Rodin V, Simkovic F 2010 *Phys. Rev.* **C82** 051301
[6] Fang D-L, Faessler A, Rodin V, Simkovic F 2010 *Phys. Rev.* **C83** 034320
[7] Caurier E, Nowacki F, Menendez J and Poves A 2008 *Phys. Rev. Lett.* **100** 052503
[8] Tomoda T, Faessler A, Schmid K W and Gruemmer F 1986 *Nucl. Phys.* **A452** 591
[9] Rath P K, Chandra R, Chaturvedi K, Raina P K and Hirsch J G 2010 *Phys. Rev.* C **82** 064310
[10] Rodriguez T R, Martinez-Pinedo G 2010 *Phys. Rev. Lett.* **105** 253503
[11] Barea J and Iachello F 2009 *Phys. Rev.* C **79** 044301
[12] Simkovic F, Vergados J, Faessler A 2010 *Phys. Rev.* D **82** 113015
[13] Faessler A, Meroni A, Petcov S T, Simkovic F, Vergados J 2011 *arXiv:* **1103.2434** [hep-ph]
[14] Ikeda K 1964 *Prog. Theor. Phys.* **31** 434
[15] Baranger M and Kumar K 1968 *Nucl. Phys.* **A110** 490
[16] Schmid K W 2004 *Prog. Part. Nucl. Phys.* **52** 565
[17] Klapdor-Kleingrothaus H V, Krivosheina I V 2006 *Mod. Phys. Lett* . A **21** 1547
[18] Miller G A and Spencer J E 1976 *Ann. Phys. (NY)* **100** 562
[19] Feldmeier H, Neff T, Roth T and Schnack J 1998 *Nucl. Phys.* **A632** 61

The status of the CUORE experiment

Andrea Giachero[1] on behalf of the CUORE collaboration

[1] Sezione INFN and University of Milano Bicocca, Piazza della Scienza 3, I-20126, Milan, Italy

E-mail: `Andrea.Giachero@mib.infn.it`

Abstract. The Cryogenic Underground Observatory for Rare Events (CUORE) is an experiment to search for neutrinoless double beta decay ($0\nu\beta\beta$) in ^{130}Te and other rare processes. The observation of $0\nu\beta\beta$ would indicate that neutrinos are Majorana particles and would provide information about the absolute neutrino mass scale. CUORE is a bolometric detector composed of 988 TeO_2 crystals, with the total mass of about 750 kg of Tellurium. We will discuss the status of the CUORE experiment, including recent R&D efforts, anticipated sensitivity, and present the most recent results from CUORICINO, the predecessor experiment operated in Gran Sasso National Laboratories in Italy.

1. Introduction

The nature of neutrino mass is one of the frontier problems of fundamental physics. Neutrinoless Double Beta Decay ($0\nu\beta\beta$) is a powerful tool to investigate the mass hierarchy and possible extensions of the Standard Model. The $0\nu\beta\beta$ is a rare spontaneous nuclear transition, never observed, involving a change of the nuclear charge Z by two units $(A, Z \pm 2)$ with the emission of two electrons and no neutrinos, resulting in a peak at the summed energy spectrum of the electrons. The decay is possible only if the neutrino is a Majorana massive particle. Its transition width is proportional to the square of the effective Majorana mass, $|\langle m_{ee}\rangle|$ [1]. In case of neutrino mixing the Majorana mass experiments measure a specific mixture of neutrino mass eigenvalues, $|\langle m_{ee}\rangle|^2 = |\sum_i U_{ei}^2 m_i|^2$, i summed over all mass eigenstates. From the $0\nu\beta\beta$ half-life it is therefore possible to infer important information concerning the mass hierarchy and the absolute mass scale of neutrinos.

In the $0\nu\beta\beta$ all the energy is shared between the two electrons and the nucleus is so heavy that the recoil is negligible. For these reasons the experimental signature is a monochromatic line at the Q-value of the decay. The experimental Sensitivity ($S^{0\nu}$), defined as the decay time corresponding to the minimum number of detectable events above background (B), in given by

$$S^{0\nu} = \ln 2 \cdot \epsilon \cdot \frac{\text{i.a.}}{A} \cdot \sqrt{\frac{M \cdot T}{B \cdot \Gamma}} \tag{1}$$

where i.a. is the isotopic abundance of the a chosen $\beta\beta$-emitter isotope, M its mass, ϵ the efficiency of the detector, Γ the energy resolution (around the Q-value), B is the background, A the molecular mass, and finally, T the measurement time. According to eq. 1, it is straightforward that to obtain the best sensitivity a double beta decay experiment must have a very large mass, high efficiency, a good resolution, a long measurement time, a very low background and the chosen $\beta\beta$-emitter isotope should have a high natural isotopic abundance (if enrichment is not necessary the result is a significant cost savings).

The bolometric technique, proposed for the first time in the 1986 [2], is a powerful tool to study neutrinoless double beta decay. A bolometer is composed by two main parts: the *Energy Asorber* and the *Phonon Sensor*. An impinging particle, which releases an amount of energy ΔE in the *Absorber*, is detected through a small temperature rise ΔT (eq. 2) that is converted by the *Sensor* to a voltage pulse signal ΔV (eq. 2).

$$\Delta T = \frac{\Delta E}{C} e^{-\frac{t}{\tau}} \quad , \quad \frac{\Delta V}{\Delta E} \simeq 0.1 \frac{\text{mV}}{\text{MeV}} \tag{2}$$

This temperature rising is directly proportional to the released energy ΔE and inversely proportional to the absorber thermal capacity C. To maximize the temperature rise a very small thermal capacity is needed. This is possible using dielectric and diamagnetic crystals, in which the thermal capacity, at very low temperature, is proportional to the cube of the temperature, following the Debye Law (eq. 3).

$$C(T) \propto \left(\frac{T}{T_D}\right)^3 \tag{3}$$

The main features of the bolometric technique are the very high efficiency ($\simeq$ 100%, a true calorimeter), a very large detection volume, and finally, a high intrinsic energy resolution (about 7 keV FWHM @ 2615 KeV).

2. The CUORE Experiment

CUORE [3, 4] (Cryogenic Underground Observatory for Rare Events) aims at searching for $0\nu\beta\beta$ of ^{130}Te using the bolometric technique. The CUORE detector is an array of 988 cryogenic bolometers arranged in 19 towers of 52 crystals. The bolometer is a TeO_2 cubic crystal (*absorber*), $5 \times 5 \times 5$ cm^3 in size, with an NTD-Ge thermistor (*sensor*) glued onto the crystal surface. Tellurium is an advantage in this instance because of the relatively high natural abundance (34.2% [5]) of the $0\nu\beta\beta$ candidate isotope, which means that enrichment is not necessary to achieve a reasonably large active mass. Also, the Q-value (2528 keV [6, 7]) of the decay falls between the peak and the Compton edge of the 2615 keV gamma line of ^{208}Tl, the highest-energy gamma from the natural decay chains; this leaves a relatively clean window in which to look for the signal. Moreover TeO_2 crystals are easy to be grown big with low radioactive contaminations [8], they have good mechanical properties and low heat capacity at low temperature (dielectric and diamagnetic). The entire detector will be maintained at the temperature of about 10 mK through a Pulse Tube Assisted Cryostat.

The CUORE program, started in the 2003 with the pilot experiment CUORICINO, includes an intesive R&D and three experimental main steps: CUORICINO [9–11], CUORE-0 (also known as CUORE One Tower), that will take data for three years starting at the end of 2011, and, the final step, CUORE. The R&D program is aimed at selecting and validating all the needed materials and procedures from a radioactive point of view in order to obtain the lowest possible background.

3. The first step: CUORICINO

CUORICINO [9–11] was a tower array of 62 TeO_2 crystals, the largest bolometric experiment ever realized. It took data form April 2003 to June 2008. The array was operated underground, in a dilution refrigerator located at the Laboratori Nazionali del Gran Sasso (LNGS, Italy), which provides an average coverage of 1400 m of rock (3650 m.w.e. [12–14]). CUORICINO was composed by 11 modules of 4 TeO_2 crystals $5 \times 5 \times 5$ cm^3 (790 g) and by 2 modules of 9 TeO_2 crystals $3 \times 3 \times 6$ cm^3 (330 g). These 2 modules, in turn, were composed by 14 crystals of natural

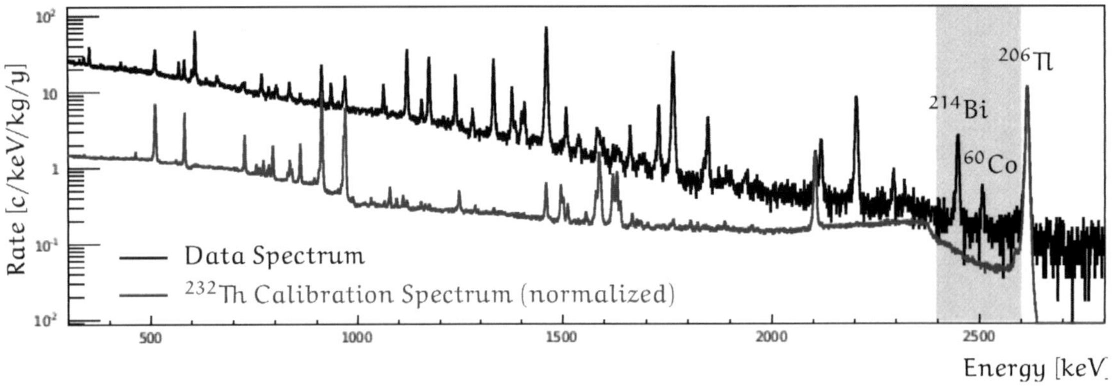

Figure 1. Anticoincidence total energy spectrum of all CUORICINO detectors (black) and the total energy spectrum of all CUORICINO detectors during calibration measurements (red). For convenience, it is normalized to have the same intensity of the 2615 keV line of ^{208} Tl as measured in the non-calibration spectrum.

^{130}Te (34.2% [5]), by 2 crystals enriched in ^{130}Te (75% [15]) and by 2 crystals enriched in ^{128}Te (82.3%). The total mass of TeO_2 was 40.7 kg for a total mass of ^{130}Te equal to 11.3 kg.

After a total exposure of 19.75 kg·y of ^{130}Te the CUORICINO final results, for the half life and for the majorana mass, are summarized with the following limits [16]:

$$\begin{array}{rl} \text{Lower Limit, Half-life:} & \tau^{0\nu}_{1/2}(^{130}\text{Te}) > 2.8 \cdot 10^{24}\ \text{y (90\% C. L.)} \\ \text{Upper Limit, Majorana Mass:} & m_{\nu_e} < (300 \div 710)\ \text{meV} \end{array} \tag{4}$$

These limits were obtained using the Nuclear Matrix Elements cumputed in [17–20]. The background in the Region Of Interest (ROI: 2474÷2580 keV) resulted [16]:

$$B = (0.153 \pm 0.006)\ \text{c/keV/kg/y} \tag{5}$$

The main sources for this background were the natural uranium (^{235}U) and thorium (^{232}Th) decay chains from contamination of the detector materials. There were two main components to this background: surface contamination of the detector components, and bulk contamination of the cryostat materials. The surface contamination (70%) produced a flat α background; the main contributors were the surfaces of the copper support structures facing the bolometers and of the crystals themselves. The principal background contribution due to bulk contamination was the tail of the 2614.5 keV gamma produced by the decay of ^{232}Th in the cryostat materials (30%).

These values, verified by extensive Monte Carlo studies, were obtained with a anti-coincidence analysis. In fact, the granularity of the CUORICINO detector allows to perform a coincidence analysis which lets to discriminate the location of the contamination, identifying single hits events (only one detector at time) and multi hits events (two or more events at the same time). The anti-coincidence technique leads a background reduction of about 15% in the ROI. In the case of CUORE, in order to reduce this background, more stringent material selection, production, cleaning, handling, and storage procedures have been established for all detector components to be used.

4. The near Future: CUORE-0

CUORE-0 will be a single CUORE-like tower realized with the same procedures defined for CUORE and, in particular, crystals from the same production line, the same copper and the same teflon, the same copper surface cleaning technique and the same assembly line. The tower will be installed in the Hall A cryostat at LNGS, the same as CUORICINO. CUORE-Zero will be not only an engineering run, useful to verify the CUORE assembly procedure, but also a new double beta decay experiment with a total mass of tellurium ^{130}Te of about 11 kg and a resolution around $\Delta E \simeq (5 \div 6)$ keV. Considering an aimed background of $B \simeq 0.12$ c/keV/kg/y, in one year of live time CUORE-0 will double the CUORICINO limit in sensitivity. The array will be assembled starting on May 2011 at the LNGS and will start to take data before the end of 2011.

5. Conclusion

Bolometers are a powerful tool for the search of Double Beta Decay and CUORICINO has demonstrated the feasibility of CUORE. Main concepts are to have about 20 times the active mass of CUORICINO, stringent controls on radioactivity of materials and assembly and to have a high efficency in background rejection [21]. The goal is to reach a background of 0.01 count/keV/kg/year that, for a live time of five years, means a lower limit for the half life $\tau^{0\nu}_{1/2} = 1.5 \cdot 10^{26}$ y, which gives an upper limit for the effective majorana mass in the range $41 \div 82$ meV. In this scenario CUORE will have the capability to explore the inverse hierarchy mass region. The experiment is under construction at the LNGS and the data taking is foreseen in 2015.

References

[1] Bilenky S M *et al.* 1999 *Physics Letters B* **465** 193 – 202
[2] Fiorini E and Niinikoski T O 1984 *Nucl. Instrum. Meth. A* **224** 83
[3] CUORE Collaboration 2005 (*Preprint* hep-ex/0501010v1)
[4] Arnaboldi C *et al.* 2004 *Nucl. Instrum. Meth. A* **518** 775–798 (*Preprint* hep-ex/0212053v1)
[5] Fehr M A *et al.* 2004 *International Journal of Mass Spectrometry* **232** 83 – 94
[6] Redshaw M *et al.* 2009 *Physical Review Letters* **102** 212502
[7] Scielzo N D *et al.* 2009 *Physical Review C* **80** 025501
[8] Arnaboldi C *et al.* 2010 *Journal of Crystal Growth* **312** 2999–3008
[9] Arnaboldi C *et al.* 2004 *Physics Letters B* **584** 260–268
[10] Arnaboldi C *et al.* 2005 *Physical Review Letters* **95** 142501 (*Preprint* hep-ex/0501034v1)
[11] Arnaboldi C *et al.* 2008 *Physical Review C* **78** 035502
[12] Ambrosio M *et al.* 1995 **52** 3793–3802
[13] Ambrosio M *et al.* 2003 *Astroparticle Physics* **19** 313 – 328
[14] Mei D and Hime A 2006 *Physical Review D* **73** 053004 (*Preprint* astro-ph/0512125)
[15] Alessandrello A *et al.* 1998 *Physics Letters B* **433** 156 – 162
[16] Andreotti E *et al.* 2011 *Astroparticle Physics* **34** 822–831 (*Preprint* 1012.3266)
[17] Šimkovic F, Faessler A, Rodin V, Vogel P and Engel J 2008 *Physical Review C* **77** 045503
[18] Civitarese O and Suhonen J 2009 *Journal of Physics: Conference Series* **173** 012012
[19] Menendez J, Poves A, Caurier E and Nowacki F 2009 *Nuclear Physics A* **818** 139–151 ISSN 0375-9474
[20] Barea J and Iachello F 2009 *Physical Review C* **79** 044301
[21] Bellini F *et al.* 2010 *Astroparticle Physics* **33** 169 – 174

Direct Dark Matter detection with CUORE

F. Orio, on behalf of the CUORE collaboration
Sapienza Università di Roma and INFN Sezione di Roma, Roma, I-00185, Italy
E-mail: filippo.orio@roma1.infn.it

Abstract. The Cryogenic Underground Observatory for Rare Events (CUORE) is an experiment to search for neutrinoless double beta decay in ^{130}Te and other rare processes. CUORE is a bolometric detector composed of 988 TeO_2 crystals, with the total mass of about 1 tonne. The large detector mass, low backgrounds, and the low energy threshold of a few keV make the experiment well suited for direct detection of galactic dark matter particles and solar axions. We discuss the development of a novel low-energy trigger that enables such searches, and present the preliminary results from a test run with four CUORE-like crystals at Gran Sasso National Laboratory in Italy.

1. The CUORE experiment

The CUORE experiment is designed to search for the neutrinoless double beta decay (0νDBD) of ^{130}Te [1]. CUORE will be made of 988 TeO_2 bolometers of 750 g each. Bolometers are detectors in which the energy from particle interactions is converted into heat and measured via the resulting rise in temperature. Operated at a temperature of about 10 mK, these detectors maintain an energy resolution of a few keV over their energy range, extending from a few keV up to several MeV. CUORE will be installed in the Gran Sasso National Laboratory (LNGS) and will start the data taking in about 3 years. The first CUORE tower, CUORE-0, composed by 52 bolometers, is under preparation and will start the data taking at the end of the year.

CUORE could also search for dark matter (DM) interactions, provided that the energy threshold is sufficiently low. DM candidates such as weak interacting massive particles (WIMPs) [2] are expected to produce signals at energies below $\sim$ 30 keV, and the interaction rate increases as the energy decreases. The energy threshold of CUORICINO bolometers, achieved using a trigger algorithm applied to the raw data samples, was of the order of tens of keV. If the threshold were of few keV, the CUORE experiment could be sensitive to DM interactions and thus play an important role in this growing research area. In the following we present a method we developed to lower the energy threshold, and the projected CUORE sensitivity to WIMP spin independent interactions.

2. Experimental setup

A CUORE bolometer is composed of two main parts, a TeO_2 crystal and a neutron transmutation doped Germanium (NTD-Ge) thermistor [3]. The crystal is cube-shaped (5x5x5 cm^3) and held by Teflon supports in copper frames. The frames are connected to the mixing chamber of a dilution refrigerator, which keeps the system at the temperature of $\sim$ 10 mK. The thermistor is glued on the crystal and acts as thermometer. When energy is released in the crystal, its temperature increases and changes the thermistor's resistance. The thermistor is biased with a constant current, and the voltage across it constitutes the signal [4]. The data presented in this paper comes from a four-bolometers array operated by the CUORE collaboration at LNGS in 2009. The main purpose of the detector was to check one of the first production batches of CUORE crystals [5]. Two Joule heaters were glued on two of the crystals and used to inject controlled amounts of energy, in order to emulate signals produced by particles [6, 7].

3. Lowering the energy threshold

At low energies the detection capability is limited by the detector noise. Electronics spikes, mechanical vibrations, and temperature fluctuations can produce pulses that, if not properly identified, generate nonphysical background. The lower the energy released in the bolometer, the more difficult it is to discriminate between physical and nonphysical pulses.

We developed a trigger and a pulse shape identification algorithm [8] that operate on data samples processed using the matched filter technique [9]. This filter is designed to estimate the amplitude of a signal, maximizing the signal to noise ratio. The transfer function is matched to the signal shape, and pulses with different shape are suppressed. The noise power spectrum of the detector, $N(\omega_k)$, and the signal shape, s_i, are needed to build the transfer function:

$$H(\omega_k) = h \frac{s^*(\omega_k)}{N(\omega_k)} e^{-\jmath \omega_k i_M}, \tag{1}$$

where $s(\omega_k)$ is the Discrete Fourier Transform (DFT) of s_i, i_M is the maximum position of s_i in the acquisition window, and h is a normalization constant that leaves unmodified the amplitude of the signal at the filter output. The estimation of $N(\omega_k)$ is made by averaging the power spectrum of a large set of data windows acquired with a random trigger, while the signal shape s_i is obtained by averaging a set of high-energy particle signals in the 1-3 MeV range. Pulses are triggered when the data samples exceed a positive threshold, and a new trigger is possible when the data samples return below threshold, or after that a local maximum is found. The detection efficiency is estimated on pulses generated by the heater, performing an energy scan from about 1 keV up to 100 keV. In the case of bolometers without the heater, the estimation is made on simulated data. Three of the bolometers have a 3 keV energy threshold and a detection efficiency in excess of 80%, while the fourth has a threshold of $\sim$ 10 keV. To ease the analysis job we ignored this bolometer. To remove nonphysical pulses we fit the filtered pulses with the expected shape of the filtered signal. The χ^2/ndf of the fit is used as shape parameter. We apply a cut on this variable to select signal events in the energy region below 30 keV. The probability of loosing a signal event after this cut has been estimated to be less than 3%.

4. Sensitivity to dark matter interactions

The four bolometers array took 19 days of data. The low energy spectrum obtained combining the three bolometers having 3 keV threshold, after cutting on the shape parameter and correcting for the detection efficiencies, is shown in Fig. 1. The data in the range 3-5 keV are not displayed, since they are still being studied; in place of the data we show their approximate behavior, that is an exponential decay. The counting rate ranges from $\sim$ 20 cpd/keV/kg at 3 keV to $\sim$ 2 cpd/keV/kg at 25 keV, and is well described by a double exponential decay. We estimate the CUORE sensitivity to WIMPs assuming that the background differential rate will be equal to the measured value in the four bolometers array, and that all bolometers will have a 3 keV threshold. It has to be stressed that CUORE bolometers, unlike many other detectors, cannot distinguish nuclear recoils from β/γ interactions. Nevertheless CUORE, by means of its mass and to the low background expected, could detect dark matter searching for an annual modulation of the counting rate at low energies.

We performed toy Monte Carlo simulations generating background events from the fit of the measured distribution in Fig. 1, and WIMP events from the theoretical distribution described in

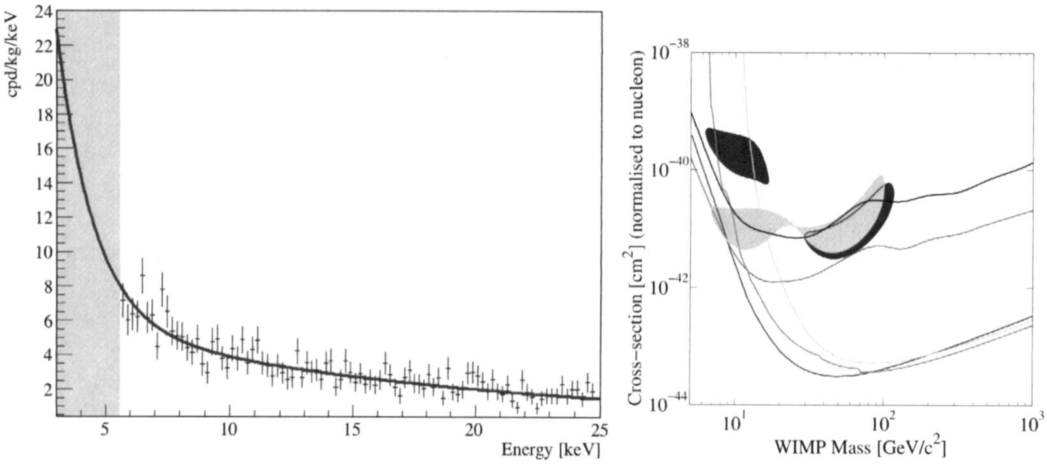

Figure 1. On the left, differential rate at low energies obtained combining data from the three bolometers having the energy threshold at 3 keV. The region below 5 keV is still under study. The data are fitted with a double exponential decay, and the fit function is used as model of the CUORE background. On the right, CUORE(5y) (in red) and CUORE-0(3y) (in black) expected sensitivities compared to the actual limits on WIMP-nucleon elastic scattering cross section: DAMA/LIBRA 3σ evidence with ion channeling in dark red and without ion channeling in pink [12], Edelweiss II in green [13], XENON100 in blue [14], CDMS in magenta [15].

Ref. [10], using a quenching factor for nuclear recoils in TeO_2 equal to 1 [11]. We included the dependence of the WIMP interaction rate on the time in the year, and estimated the background+signal asymmetry subtracting the 3-months integrated spectrum across 2 December from the 3-months integrated spectrum across 2 June. The 90% C.L. sensitivity to the cross-section for spin independent interactions normalized to nucleon, as a function of the WIMP mass, for CUORE-0 in 3 years and CUORE in 5 years of data taking, is shown in Fig. 1 together with the present results of the leading experiments of the field. As it can be seen from the figure, CUORE-0 and CUORE will not be as sensitive as experiments able to discriminate nuclear recoils. CUORE, however, could investigate the same parameter space of the DAMA/LIBRA experiment, and could be the only experiment other than DAMA/LIBRA looking for an annual modulation of dark matter interactions.

References

[1] Ardito R *et al.* 2005 (*Preprint* `hep-ex/0501010`)
[2] Bertone G, Hooper D and Silk J 2005 *Physics Reports* **405** 279–390
[3] Itoh K M *et al.* 1996 *Phys. Rev. Lett.* **77** 4058–4061
[4] Arnaboldi C *et al.* 2002 *IEEE Trans. Nucl. Sci.* **49** 2440–2447
[5] Arnaboldi C *et al.* 2010 *J. Cryst. Growth* **312** 2999
[6] Alessandrello A *et al.* 1998 *Nucl. Instrum. Meth. A* 454–464
[7] Arnaboldi C, Pessina G and Previtali E 2003 *IEEE Trans. Nucl. Sci.* **50** 979–986
[8] Di Domizio S, Orio F and Vignati M 2011 *JINST* **6** P02007
[9] Gatti E and Manfredi P F 1986 *Rivista del Nuovo Cimento* **9** 1–146
[10] Lewin J D and Smith P F 1996 *Astroparticle Physics* **6** 87 – 112
[11] Alessandrello A *et al.* 1998 *Nucl. Instrum. Meth. A* **409** 451–453
[12] Savage C, Gelmini G, Gondolo P and Freese K 2009 *JCAP* **0904** 010
[13] Armengaud E *et al.* (EDELWEISS-II) 2010 (*Preprint* `1011.2319`)
[14] Aprile E *et al.* (XENON100) 2010 (*Preprint* `1005.0380`)
[15] Ahmed Z *et al.* (The CDMS-II) 2009 (*Preprint* `0912.3592`)

Journal of Physics: Conference Series **337** (2012) 012068 doi:10.1088/1742-6596/337/1/012068

Latest results from the Pierre Auger Observatory

Hans P Dembinski[1], for the Pierre Auger Collaboration[2]

[1]Karlsruhe Institute of Technology (KIT), IEKP, Karlsruhe, Germany
[2]Observatorio Pierre Auger, Av. San Martin Norte 304, 5613 Malargüe, Argentina
(Full author list: http://www.auger.org/archive/authors_2011_05.html)

E-mail: hans.dembinski@kit.edu

Abstract. The Pierre Auger Observatory, located in the Province of Mendoza, Argentina, is the World's largest detector for cosmic rays at ultra-high energies. In its seven years of operation it has collected an exposure of more than 20000 km^2 sr yr, larger than all previous experiments combined. Its original design, optimized for the energy range 10^{18} eV to 10^{20} eV, is currently enhanced to cover energies down to almost 10^{17} eV. We give an overview of the latest results with a focus on the prospect to study nuclear interactions with cosmic rays and conclude with a brief outlook on developments and extensions of the observatory.

1. Introduction

Cosmic rays are light to heavy nuclei (up to iron). The Pierre Auger Observatory uses two techniques to study them, exploiting the induced extensive air showers in Earth's atmosphere. Charged particles and photons of the shower which arrive at ground are measured with more than 1600 water-Cherenkov detectors, distributed on a 1.5 km triangular-grid over 3000 km^2 (Surface Detector Array, SD [1]). In addition, charged particles in the air generate ultra-violet light by excitation/de-excitation of nitrogen, which is observed by 27 fluorescence telescopes (Fluorescence Detector, FD [2]) under suitable atmospheric conditions.

The FD follows the longitudinal development of the shower in the air. It allows one to reconstruct the shower energy in an almost model-independent way and the slant depth X_{max} of the shower maximum [3]. The latter is sensitive to the mass of the cosmic ray and the properties of its hadronic interactions in the atmosphere. The FD can only operate in dark, moonless nights with a field of view free of clouds, which limits its duty cycle to about 1/10 of the total time. The SD has a duty cycle of almost 100 % and collects the main bulk of events. Its observes the lateral profile of particle density around the shower axis at ground level. Its energy scale is derived from coincident measurements with the FD [4].

2. Cosmic ray flux and sources

Two of the main objectives of the observatory are to measure the cosmic ray flux between 10^{18} eV and 10^{20} eV and to discover its extra-galactic sources. In conventional scenarios, cosmic rays well above 10^{18} eV are accelerated in distant extragalactic objects and travel through the intergalactic medium until they eventually reach Earth. The rather fundamental prediction of the Greisen-Zatsepin-Kuz'min (GZK) cut-off [5] limits the maximum distance which cosmic rays of 6×10^{19} eV and higher can travel to about 50-100 Mpc. Interactions with the cosmic microwave background become frequent above this energy that strip energy from the cosmic ray.

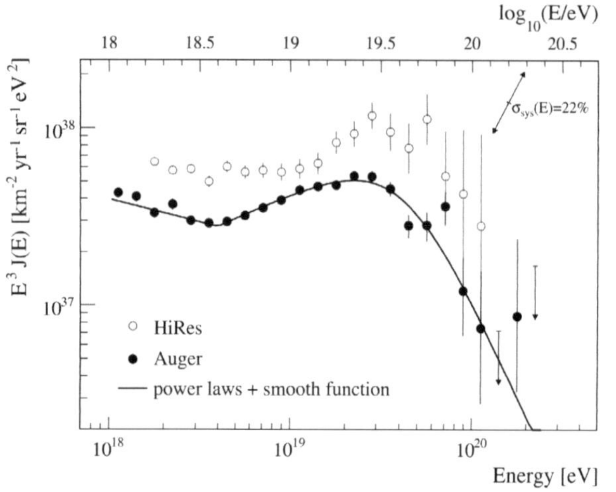

Figure 1. Cosmic ray flux derived from data of the Pierre Auger Observatory [4]. Empirically, the cosmic ray flux is well described by a broken power law model with a break at 4×10^{18} eV and a smooth Fermi-type suppression at higher energies. The HiRes result [6] is compared. Both results are compatible if shifted within the systematic uncertainties of their respective energy scales (22 % for Auger [7]).

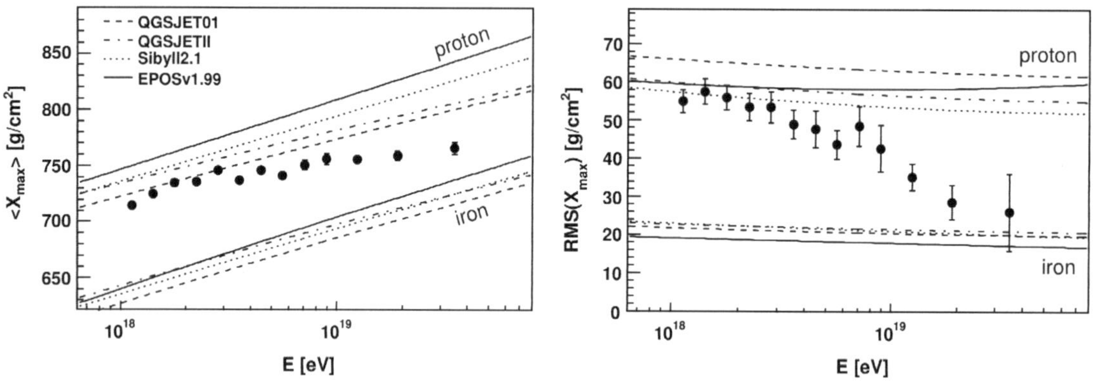

Figure 2. Measurements of the average depth $\langle X_{\max} \rangle$ of the shower maximum and the fluctuations RMS($X_{\max}$) as a function of the cosmic ray energy E [11]. Compared are simulations with a selection of hadronic interaction models [12, 13, 14, 15] for proton (top) and iron showers (bottom).

The GZK effect severely reduces the horizont for sources that could reach Earth at the highest energies and therefore it should lead to a flux suppression.

The cosmic ray flux measured by the Pierre Auger Observatory is shown in figure 1. It depicts a break around 4×10^{18} eV and a suppression at the highest energies compatible with the GZK effect. The features are compatible with earlier results from the HiRes experiment [6].

An anisotropy was found in the arrival directions of cosmic rays with the highest energies [8, 9]. Flux suppression and anisotropy make conventional acceleration likely, but some tension remains as the anisotropy signal has not become stronger since its original publication [10].

3. Mass composition and hadronic interactions

The slant depth $X_{\max}$ of the shower maximum is measured by the FD. It depends on the mass of the cosmic ray and scales with the logarithm of the energy E as more energetic showers penetrate deeper into the atmosphere. The $X_{\max}$ of a particular shower is the sum of the depth λ of its first interaction (anti-proportional to the cross-section) and the slant depth ΔX that the shower needs to develop to the maximum.

The mean value $\langle X_{\max} \rangle$ and the spread RMS($X_{\max}$) of $X_{\max}$ are depicted in figure 2. Cuts on the data set pick events of sufficiently quality without biasing the sample in favor of deeply penetrating (proton-like) or shallow (iron-like) showers [11]. Soft hadronic interactions drive

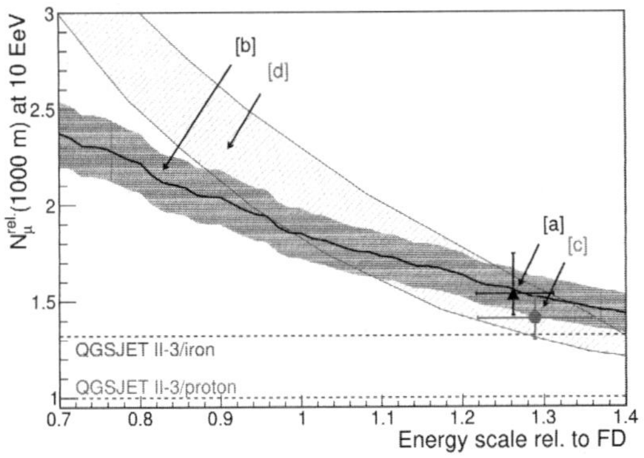

Figure 3. Number of muons in showers of 10^{19} eV at a radius of 1000 m relative to QGSJET-II/proton prediction [13] vs. energy scale from [a] the universality method (triangle); [b] the jump method (filled area); [c] the smoothing method (circle); [d] the hybrid analysis (dashed area). According to the tested model, iron primaries have 1.32 as many muons as protons (horizontal lines in the figure). Details are given in [17].

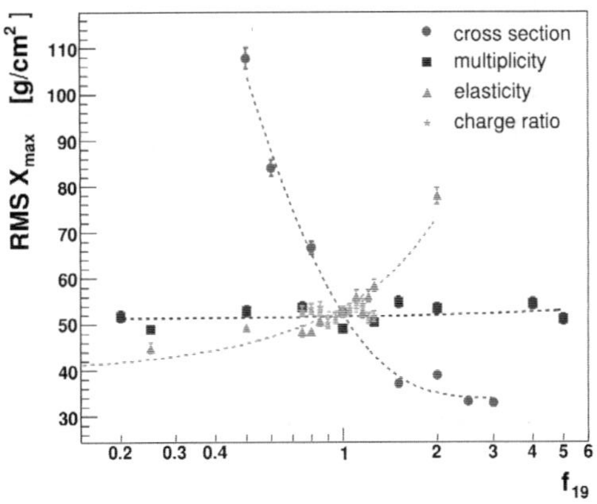

Figure 4. Change in RMS($X_{\max}$) if the parameters listed in the legend of the hadronic interaction model SIBYLL [14] are rescaled by an energy-dependent factor f_{19}. The points are individual simulations, the lines are empirical fits to guide the eye. The plot indicates that RMS($X_{\max}$) is very sensitive to the cross-section, but rather insensitive to other details of the hadronic interaction.

the development of the shower which cannot be computed from first principles. Thus, λ cannot be extracted directly from $X_{\max}$, because theoretical predictions for ΔX are rather uncertain. Also uncertain is the functional relation between λ and the energy E and mass of the cosmic ray. For these reasons, the $X_{\max}$ predictions from different hadronic interaction models show a considerable spread in figure 2. Nevertheless, we observe a trend compatible with a gradual increase of the average mass of cosmic rays with energy up to 5.9×10^{19} eV.

SD observables that depend on the mass composition are also under study [16]. SD observables benefit from the higher duty cycle of the SD which improves the statistics, but typically they are less sensitive to the mass composition.

The most striking difference between observation and predictions is an apparent muon deficit in simulations. The SD does not contain dedicated muon detectors so far, but analyses of the time traces of the signal in individual detectors, the use of shower universality properties and hybrid events (simultaneous observation of the shower in SD and FD) allow one to extract the muon component [17]. The apparent muon deficit is depicted in figure 3 in form of a rescaling factor required to adjust the number of muons in simulations to the data as a function of a shifted FD energy scale. The analyses prefer a FD energy scale increased by about 25 % (still compatible with its systematic uncertainty of 22 % [7]) and about 50 % more muons.

An interesting idea [18] is to study the the impact of changes to the hadronic interaction models on shower observables. An example is shown in figure 4. If inverted, such a relation could be used to fit a parameter like multiplicity or cross-section directly to air shower data. This would allow to study these parameters in energy ranges inaccessible to collider experiments. However, if the cosmic rays are a mixture of elements, such an analysis will become rather complex.

4. Limits on photons and neutrino fluxes

The Observatory also is a powerful photon and a competitive neutrino detector in the ultra-high energy range. Photons of high energy induce air showers like normal cosmic rays, but with a much deeper $X_{\max}$ and few muons since the first interaction is not strong but electromagnetic. Photon-induced showers are easier to predict, since electromagnetic interactions can be calculated from first principles. The main obstacle in photon searches is the large background of normal cosmic rays. The sensitivity mainly depends on the efficiency of the background rejection. So far, no photons have been found and limits are placed on their fractional flux [19].

Neutrinos barely interact with atmospheric nuclei at all. Almost horizontal showers yield the best separation between the rare neutrino events and the background of normal cosmic rays. Even slightly up-going showers are considered, caused by tau-neutrinos that skim Earth's crust where they are converted into τ's that leave Earth to decay over the observatory. The sensitivity of the observatory is significantly increased by including this detection channel. No neutrinos have been found and limits are placed on their flux [20, 21].

5. Summary and outlook

The cosmic ray flux and the $X_{\max}$ distribution have been determined with the Pierre Auger Observatory to unprecedented precision. An anisotropy in the arrival directions of the cosmic rays with the highest energies has been observed. Competitive limits have been placed on the fluxes of photons and neutrinos above 10^{18} eV. The analyses are continuously refined with the help of the large statistics collected. Systematic uncertainties decrease and new ways of extracting the mass composition and parameters of the hadronic interactions are developed.

The observatory recently has been enhanced to study cosmic rays down to about 10^{17} eV. The SD now includes an infill array with a 750 m grid, which will feature dedicated muon detectors [22]. Three High Elevation Auger Telescopes (HEAT) have been added to the FD, which can be tilted to look at higher elevations and observe air showers with shallower $X_{\max}$ [23].

Finally, there is an ongoing effort to study the radio emission of air showers with the goal to use this signal for cosmic ray detection [24].

References

[1] Allekotte I *et al.* 2008 *Nucl. Instr. Meth.* A **586** 409
[2] Pierre Auger Collaboration 2010 *Nucl. Instr. Meth.* A **620** 227
[3] Unger M *et al.* 2008 *Nucl. Instrum. Meth.* A **588** 433
[4] Pierre Auger Collaboration 2010 *Phys. Lett.* B **685** 239
[5] Greisen K 1966 *Phys. Rev. Lett.* **16** 748; Zatsepin GT and Kuzḿin VA 1966 *JETP Lett.* **4** 78
[6] Hires Collaboration 2009 *Astropart. Phys.* **32** 53
[7] Di Giulio C *et al.* (Pierre Auger Collaboration) 2009 *Proc. 31st ICRC (Łódź, Poland)* arXiv:0906.2189
[8] Pierre Auger Collaboration 2007 *Science* **318** 938
[9] Pierre Auger Collaboration 2008 *Astropart. Phys.* **29** 188
[10] Pierre Auger Collaboration 2010 *Astropart. Phys.* **34** 314
[11] Pierre Auger Collaboration 2010 *Phys. Rev. Lett.* **104** 091101
[12] Kalmykov NN and Ostapchenko SS 1993 *Phys. At. Nucl.* **56** 346
[13] Ostapchenko SS 2006 *Nucl. Phys.* B *Proc. Suppl.* **151** 143
[14] Ahn EJ, Engel R, Gaisser TK, Lipari P and Stanev T 2009 *Phys. Rev.* D **80** 094003 arXiv:0906.4113
[15] Pierog T and Werner K 2008 *Phys. Rev. Lett.* **101** 171101
[16] Wahlberg H *et al.* (Pierre Auger Collaboration) 2009 *Proc. 31st ICRC (Łódź, Poland)* arXiv:0906.2319
[17] Castellina A *et al.* (Pierre Auger Collaboration) 2009 *Proc. 31st ICRC (Łódź, Poland)* arXiv:0906.2319
[18] Ulrich R, Engel R and Unger M 2011 *Phys. Rev.* D **83** 054026 arXiv:1010.4310
[19] Pierre Auger Collaboration 2009 *Astropart. Phys.* **31** 399
[20] Pierre Auger Collaboration 2008 *Phys. Rev. Lett.* **100** 211101
[21] Pierre Auger Collaboration 2009 *Phys. Rev.* D **79** 102001
[22] Platino M *et al.* (Pierre Auger Collaboration) 2009 *Proc. 31st ICRC (Łódź, Poland)* arXiv:0906.2354
[23] Kleifges M *et al.* (Pierre Auger Collaboration) 2009 *Proc. 31st ICRC (Łódź, Poland)* arXiv:0906.2354
[24] van den Berg AM *et al.* (Pierre Auger Collaboration) 2009 *Proc. 31st ICRC (Łódź, Poland)* arXiv:0906.2354

Journal of Physics: Conference Series **337** (2012) 012069 doi:10.1088/1742-6596/337/1/012069

Highlights from Gran Sasso Laboratory

Lucia Votano

LNGS-INFN, s.s. 17 bis Km 18,910 - 67100 Assergi (AQ) Italy

lucia.votano@lngs.infn.it

Abstract. The Gran Sasso National Laboratory of INFN, between the cities of L'Aquila and Teramo, Italy, is a research infrastructure mainly dedicated to astroparticle and neutrino physics. It offers the most advanced underground Laboratory in terms of dimensions, complexity and completeness of its infrastructures. A review of the main experiments carried out at LNGS - devoted to neutrino and to nuclear astrophysics - will be given, together with the most recent and relevant scientific results achieved.

1 LNGS

Underground Laboratories are the main infrastructures for astroparticle and neutrino physics to explore the highest energy scales not accessible with accelerators, by searching for extremely rare phenomena. INFN Gran Sasso National Laboratory (LNGS) is the largest underground laboratory in the world devoted to astroparticle physics. It is one of the four INFN National Laboratories and it is a worldwide facility for scientists working in one of the twenty experiments currently set there. Located between L'Aquila and Teramo - at about 120 km far from Rome - the underground structures are on one side of the highway tunnel (10 km long) which crosses the Gran Sasso massif (A24 Teramo-Rome Highway) and consist of three huge experimental halls (each one 100 m long, 20 m large and 18 m high) linked by service tunnels, for a total volume of about 180.000 m^3 and a surface of 17.800 m^2.
Access to experimental halls is horizontal and it is made easier by the highway tunnel. Halls are equipped with all technical and safety plants required for the experimental activities and to ensure proper working conditions. For instance ventilation plants allow, in normal conditions, a change of air in less than 3 hours, and air radon content is kept in the 20-80 Bq/m^3 range.
The 1400 metre-rock thickness above provides a cosmic ray flux reduction by one million times; moreover, the flux of neutrons in the underground halls is about thousand times less than on the surface, due to the very small amount of uranium and thorium of the Dolomite calcareous rock of the mountain. Outside, next to the highway tollgate of Assergi, an area of more than 23 acres hosts the external laboratories, the Computing Centre, the Directorate and various Offices. Presently LNGS staff consists of 90 people; besides, more than 950 scientists from 29 different Countries take part in its experimental activities.

2 Physics at LNGS

At LNGS, many of the experiments currently underway or to be started soon are looking at fundamental and fascinating questions which are still unanswered. What is the Universe made up of and, in particular, Dark Matter? In this area, are we any nearer to open a door to a physics that goes

Journal of Physics: Conference Series **337** (2012) 012069 doi:10.1088/1742-6596/337/1/012069

beyond the Standard Model of particle physics? What is the intimate nature and what are the latest characteristics of the neutrino and what role have they played in the evolution of the Universe? What do we know of the interior of the Sun and the stars, of the mechanisms that produce energy and how they evolve and die? What do we know about the interior of our own Earth, can we find out more through the study of geoneutrinos? LNGS research activities thus range from neutrino physics to dark matter search, to nuclear astrophysics, but also to geophysics, biology and fundamental physics.

3 Neutrino Physics

The study of the intrinsic properties of neutrino is of prime interest in elementary particle physics and has provided by now the only experimental evidence of phenomena beyond in the Standard Model of elementary particles by measuring neutrino oscillations. In order to properly comprehend the mechanism of the oscillation; various neutrino sources, both natural (the Sun or other stars) and artificial (particle accelerators) can be used.

Besides, the study of a phenomenon as the neutrino-less double-beta decay could allow us to find out if the neutrino overlaps with its antiparticle, thus providing a very significant answer towards the comprehension of the evolution of the Universe. Finally, neutrinos from the cosmos are very important messengers which transport fundamental information to understand the functioning of the stars as energy sources, their evolution and what happens when they "turn off".

LNGS activities range among various aspects of the neutrino physics study.

3.1 BOREXINO and solar neutrinos

Borexino is a large unsegmented liquid scintillator detector built for real time measurement of low-energy solar neutrinos. The inner part of the detector is a Stainless Steel Sphere (SSS) 13.7 m in diameter that represents both the container of the scintillator and the mechanical support of the 2200 PMTs. Within the sphere, two nylon vessels separate the scintillator volume in three shells of increasing radiopurity. The SSS is enclosed within a large tank filled with ultrapure water that represents a powerful shielding against external background and is used as a Cherenkov muon counter and muon tracker. The experiment can then rely on an optimal energy and spatial resolution together with a very low energy threshold (40 KeV).

The main goal of the experiment is the detection of the monochromatic ^{7}Be neutrinos, even if the extreme radiopurity of the detector resulted in a broadening of the scientific goals. Borexino now aims at the spectral study of other solar neutrino components, such as the CNO, pep and, possibly, pp.

Low energy neutrinos of all flavours are detected by their elastic scattering of electrons while electron antineutrinos by means of their inverse beta decay on protons or carbon nuclei. As the shape of the energy spectrum of recoil electrons is the only signature available for neutrinos, γ or β background events due to natural radioactivity cannot be distinguished from the signal on an event by event analysis, but only through their spectral shape. This fact, together with the low event rate expected, demands an extreme radiopurity of the detector and actually the inner core of Borexino is 9-10 order of magnitude less radioactive than anything else on Earth.

After the first result published in 2007 with a 30% precision, and the second one in 2008 with 10% precision, the best value for the interaction rate of ^{7}Be neutrino published in 2011, corresponding to 750 days of data, is 46± 1.5stat ±1.3syst counts/(dayx100t), with 4.3% accuracy [2]. The expected signal for non oscillated solar $\nu_{\square}$ in the high metallicity model is 74 ±4 counts /(dayx100t), while in the MSW-LMA scenario of solar neutrino oscillation, it is 48+/-4 counts /(dayx100t), in very good agreement with the Borexino measurement. In this scenario, neutrino oscillations are dominated by matter effects above 3 MeV and by vacuum effects below 0.5 MeV. The ^{7}Be neutrinos lie in the lower edge of this transition region. Borexino has also been able, for the first time in a liquid scintillator detector, to detect ^{8}B neutrinos above 3 MeV [1]. The first simultaneous measurement of the neutrino survival probability Pee in the vacuum and in the matter enhanced oscillations regions is of great importance and could be improved in the near future.

Thanks to the extreme high radiopurity achieved, the high photon yield, and the large number of free target protons, Borexino is also a very sensitive detector for antineutrinos in the MeV energy range where geo-neutrinos lie. Geo neutrinos are electron antineutrinos produced in β decays of ^{40}K and of several nuclides in the chains of long-lived ^{238}U and ^{232}Th present in the Earth. They are direct messengers of the abundances and distribution of radioactive elements within our planet. By measuring their flux and spectrum it is possible to reveal the distribution of the natural radioactive isotopes and to assess the radiogenic contribution to the total heat balance of the Earth. Nowadays the existence of large mass scintillation detectors made their detection feasible. KamLAND has opened the way to geo-neutrino detection in 2005 while in 2010 the geo-neutrino observation has definitely been performed by Borexino [3]. The best estimate of the geo-antineutrinos has been determined with an unbinned maximum likelihood analysis of the twenty-one $\bar{\nu}$ candidates that passed all selections cuts, including the reactor $\bar{\nu}$ that represent an irreducible source of background. The KamLAND 2008 and Borexino 2010 combined analysis gives an evidence at more than 5σ for the detection of geoneutrinos.

3.2 CNGS and OPERA

Started in 2006, the project CNGS consists of an artificial neutrino beam, produced by the protons accelerator SPS of CERN and directed towards Gran Sasso. The main experiment of Gran Sasso National Laboratory devoted to CNGS neutrino detection is OPERA. The goal of the experiment is the detection of neutrino oscillations in direct appearance mode through the study of $\nu_\mu \rightarrow \nu_\tau$ channel.
Neutrino oscillations at the atmospheric scale have been studied up to now through atmospheric and accelerator neutrinos in disappearance mode, while the direct observation of flavour transition through the detection of the corresponding lepton has never been observed before OPERA first evidence. The challenge of the OPERA experiment is then the detection of the short lived τ lepton produced in the charge current interaction of ν_τ. Detection of tau neutrinos in the beam CNGS, originally constituted by muon neutrinos only, provides the first *direct* evidence of the so called 'oscillation' mechanism of these particles. The ν_μ beam from CERN is optimized for the observation of ν_τ CC interactions, the average energy being 17 GeV. With a total CNGS beam intensity of 22.5×10^{19} protons on target, about 24.300 neutrino events would be collected. The experiment should observe ≈10 ν_τCC events for the present Δm^2_{23} allowed region with a background of less than one event.
The huge apparatus mainly consists of 150.000 'bricks' made up of lead layers interleaved with nuclear emulsions, historically called Emulsion Cloud Chamber (ECC). OPERA is an hybrid detector made of a veto plane followed by two identical super modules (SM) each consisting of a target section of about 625 tons made of 75.000 bricks, and of a scintillator Target Tracker Detector (TT) to trigger the read-out and to localize neutrino interactions within the target, followed by a muon spectrometer. The observation of a first ν_τ candidate event in the experiment has been reported in June 2010 based on the analysis of part of the data taken during 2008 and 2009 runs.
The appearance of the τ lepton is identified in OPERA by the detection of its characteristic decay topologies, either in one prong (electron, muon or hadron) or in three prongs. The detection of decay topologies is triggered by the observation of a track with a large impact parameter with respect to the primary vertex. In the event samples analysed so far, one candidate has been identified with measured characteristics fulfilling the selection criteria a priori defined for the ν_τ interaction search. The primary neutrino interaction consists of seven tracks, one of which showing a visible kink. Two electromagnetic showers caused by γ-rays, associated with the event, have been located. The event passes all cuts with the presence of at least one gamma pointing to the secondary vertex, and it is therefore a candidate to the τ □□prong hadron decay mode. The invariant mass of the two detected gammas is consistent with the π^0 mass value. The invariant mass of the (π−γγ) system has a value compatible with that of the ρ(770). The two main sources of background to this channel are: the decay of charmed particles produced in ν_μ interactions, estimated for the analyzed sample to 0.007

±0.004syst and the one prong inelastic interactions of primary hadrons produced in ν_μ interactions, estimated to 0.011 [4].
By considering the 1-prong hadron channel only, the probability to observe 1 event due to a background fluctuation is 1.8%, for a statistical significance of 2.36 σ. If all τ decays modes which were included in the search are taken into account, the probability to observe one event due to the background fluctuation is 4.5%; this corresponds to a significance of 2.01 σ. OPERA data taking will continue at least up to 2012.

3.3 CNGS and ICARUS

Another experiment able to detect CNGS beam is ICARUS an innovative apparatus consisting of a big mass (about 600 tons) of liquid Argon, at a temperature of –186 °C. In particular conditions and by means of proper devices this liquefied gas is able to act as an extraordinary particle detector, allowing a 3D reproduction of any interactions of charged particles inside its volume. The commissioning of ICARUS was successfully completed in 2010 and in May the first CNGS neutrino events were recorded [5]. Such a massive liquid argon experiment running in an underground laboratory is, so far, the most important milestone for the LAr-TPC technology towards the design of a much more massive multikiloton LAr detector with unique imaging capability, and spatial/calorimetric resolutions.

3.4 The research of neutrinoless double beta decay

Neutrino oscillation experiments have clearly shown that neutrinos do oscillate and that they are massive, though this is not enough to determine the nature of such mass (Dirac or Majorana) and to determine the absolute mass scale. In fact neutrino oscillation experiments can only measure the absolute value of the difference of the square of the neutrino masses while two different hierarchical mass arrangements of neutrino masses (Direct and Inverted), besides the obvious quasi-degenerate option, are possible. Many theories beyond the standard Model suggest a mass generation mechanism that implies a Majorana character of neutrinos.
Neutrinoless double-beta decay is a process by which two neutrons in a nucleus undergo beta decay by exchanging a virtual Majorana neutrino, emitting an electron each. This would violate lepton number conservation ($\Delta L = 2$) and would necessarily require neutrinos to be Majorana particles; therefore this represents a unique tool to test this hypothesis and nowadays, thanks to the discovery of neutrino oscillations, this makes it the object of a renewed interest.
From an experimental point of view, the only available information is carried by the daughter nucleus and the two possible emitted electrons and then the possible measurable quantities are: sum of the electron energies, single electron energy and angular distributions, identification and/or counting of the daughter nuclei.
In most cases the experimental signature rely on the electron sum energies and on the discrimination of the different distribution expected for: a continuous bell distribution for $\beta\beta 2\nu$, and a sharp line at the transition energy for $\beta\beta 0\nu$). Energy resolution and level of measured background are the critical issues.
In 2001 evidence for a $\beta\beta 0\nu$) signal has been claimed by a small subset (KHDK) of the HDM ("Heidelberg-Moscow") collaboration at Gran Sasso National Laboratory. The result is based on a re-analysis of the HDM data. Such a claim has raised some criticisms that cannot be dismissed though as none of the existing experiments is able to rule it out. Therefore there is great interest on next generation and more sensitive experiments.

3.4.1 GERDA

GERDA [6] is designed to search for $\beta\beta 0\nu$-decay of ^{76}Ge using high purity germanium detectors (HPGe), enriched (~ 85%) in ^{76}Ge, directly immersed in LAr which acts both as shield and as cooling medium. The cryostat is located in a stainless steel water tank providing an additional shield against external background. GERDA experiment is scheduled to proceed in two phases. For Phase I, eight

reprocessed enriched HPGe detectors from the past HdM experiments (~ 18 kg) and six reprocessed natural HPGe detectors (~ 15 kg) from the Genius Test-Facility will be deployed in strings.
At the expected background rate of 10^{-2} cts/(kg keV y) at the Q-value of the ^{76}Ge decay (2.039 keV) the resulting sensitivity after one-year exposure for the half-life of the neutrinoless double beta decay is 2 x 10^{25} y. This is sufficient to confirm or refute the existing claim from KHDK. In Phase II new diodes, which are able to discriminate between single- and multi-site events, will be added to increase the active mass up to 37.5 kg. The expected background reduction down to 10^{-3} cts/(kg keV y) will allow to increase the sensitivity to the half-life of the process by one order of magnitude for the total exposure of 100 kg y. In this case it would be also possible to probe the effective neutrino masses (m_{ee}) at the level of 150 meV.
The installation of the experiment in Gran Sasso Laboratory has been accomplished and now the experiment is in the commissioning phase. The first non-enriched Ge detectors have been deployed for a technical run aimed to perform background evaluation.

3.4.2 CUORE
The CUORE experiment (Cryogenic Underground Detector for Rare Events) [7] is the most recent and ambitious development of the 'TeO_2 bolometers' technique in which INFN has over than 20 years experience. In the bolometer detectors, the energy from particle interactions is converted into heat and measured via the resulting rise in temperature.
CUORE will consist of a rather compact cylindrical structure of 988 cubic natural TeO2 crystals of 5 cm side (750 g), arranged into 19 separated towers (13 planes of 4 crystals each) and operated at a temperature of 10 mK. At such temperature, these detectors have an energy resolution of a few keV over their energy range, extending from a few keV up to several MeV. The measured resolution in the region of interest (2527 keV) is about 5 keV FWHM; a background level of the order of 0.01 c/keV/kg/y is expected by extrapolating the CUORICINO background results and the dedicated CUORE R&D measurements. Considering the high mass of the experiment, the CUORE predicted limit, after 5 years of running sensitivity, for the half-life is 2.1 x 10^{26} y (90% C.L.) corresponding to $m_\nu \leq$(24-83) MeV. CUORE will therefore allow a close look at the IH region of neutrino masses.
As well as being protected by the Gran Sasso rock, the apparatus is protected by a series of shields, including ancient Roman lead for the utmost reduction of residual radiation in the Laboratory and to allow the detection of the very rare decay. Very recently the Laboratory has received additional 120 lead bricks (4 tons) from an ancient Roman ship that sunk off of the coast of Sardinia 2.000 years ago. CUORE will be installed in the Gran Sasso National Laboratory and will start the data taking in 2014. The first CUORE tower, CUORE-0, composed by 52 bolometers, is under preparation and will start the data taking in 2011.

4 Nuclear astrophysiscs and LUNA
Nuclear astrophysics aims at the knowledge of thermonuclear reactions responsible for the stellar luminosity and for the synthesis of the chemical elements. In the energy region of interest (generally below 100 keV, lower than the Coulomb energy) the cross section undergoes an exponential decrease as the energy itself decreases. Its extremely small value has always prevented from direct measuring in a surface laboratory where background events, produced by the interactions of cosmic rays, are by far dominant.
In order to start measuring in this unexplored nuclear astrophysics region, LUNAcollaboration has installed two accelerators - 50kV and 400kV - in the Gran Sasso underground Laboratory. The qualifying characteristics of both the accelerators are the very high beam current and the very small energy spread. High pure solid and gaseous targets have been developed, while silicon, ultrapure germanium and a segmented BGO detector have been used. The cross sections of the key reactions of the proton–proton chain and of the Carbon–Nitrogen–Oxygen (CNO) cycle have been measured by

LUNA right down to the energies of astrophysical interest [8]. Currently only the 400 KV accelerator is being used whereas the 50KV one has been dismissed in 2005.
The measurement of key reactions of the He burning cycle is relevant at higher temperatures than reactions belonging to the hydrogen-burning studied by LUNA so far, and it requires high energy accelerators. The collaboration has recently submitted to LNGS Scientific Committee an update of a previous letter of intent for a research program devoted to the He burning reactions using a 3.5MV accelerator; in this very LoI the experimental conditions and neutron production rates have been re-evaluated and the machine characteristics have been better specified. A possible location at the LNGS interferometric zone has been agreed, a detailed technical study of the "B node" preparation in LNGS was completed, GEANT4 simulations for neutron fluxes calculations were accomplished and the layout and shielding of the accelerator are under study.
Moreover a successful round table was organized at LNGS in order to gather new interested collaborators.

5 Conclusions

Neutrino and astroparticle physics would not been able to make the massive progress seen in the last thirty years without the necessary great infrastructures like underground laboratories. The present scientific program of the Gran Sasso Laboratory is at the top of its life thanks to a very broad spectrum of competitive experiments. Last years have been very successful: new experiments have come into operation and relevant scientific results have been published.

References

[1] G. Bellini et al., Phys. Rev. D 82(2010), 033006
[2] G. Bellini et al., arXiv:1104.1816v1 [hep-ex]
[3] G. Bellini et al., Physics Letters B 687(2010), 299-304
[4] N. Agafonova et al., Physics Letters B 691(2010), 138-145
[5] C. Rubbia et al., arXiv:1106.0975 [hep-ex]
[6] S. Schönert et al., Nucl. Phys. B 145(2005), 242
[7] C. Arnaboldi et al., Phys. Rev. Lett 95(2005), 14501
[8] C. Broggini et al. Ann.Rev.Nucl.Part.Sci. 60 (2010) 53-73

Journal of Physics: Conference Series **337** (2012) 012070 doi:10.1088/1742-6596/337/1/012070

Approaching r-process nuclei at N=126

J. Benlliure[1], H. Alvarez-Pol[1], T. Kurtukian-Nieto[1], A.I. Morales[1], L. Audouin[2], F. Becker[3], B. Blank[4], I. Borzov[3], E. Casarejos[1], D. Cortina-Gil[1], B. Fernandez[5], J. Giovinazzo[4], D. Henzlova[3], B. Jurado[4], L. Langanke[3], G. Martinez-Pinedo[3], P. Napolitani[2], J. Pereira[1], F. Rejmund[5], K.-H. Schmidt[3], O. Yordanov[3] and the RISING collaboration

[1] Universidade de Santiago de Compostela, 15782 Santiago de Compostela, Spain
[2] IPN, IN2P3/CNRS, Université Paris-Sud 11, UMR 8608, 91406 Orsay, France
[3] GSI Helmholtzzentrum für Schwerionenforschung GmbH, Planckstr. 1, 64291 Darmstadt, Germany
[4] Université Bordeaux I, IN2P3-CNRS, CENBG, 33175 Gradignan, France
[5] GANIL, CEA/DSM-CNRS/IN2P3, BP 55027, 14076 Caen, France

E-mail: j.benlliure@usc.es

Abstract. The production of heavy neutron-rich nuclei approaching the r-process waiting point at N≈126 has been investigated in fragmentation reactions of relativistic ^{238}U and ^{208}Pb projectiles. Using this technique we were able to identify 73 new heavy neutron-rich nuclei expanding considerably the north-west frontier of the chart of nuclide. Moreover, we were able to determine the half lives of 13 of those nuclides. The measured values are significantly shorter than the predictions used for r-process model calculations. These shorter half lives are understood as due to the role of first-forbidden transitions in the decays of these nuclei. The confirmation of these results for r-process nuclei at N≈126 would indicate that the r-process at this point is faster than expected, leading to a larger production of the heaviest nuclei.

1. Introduction

One of the most challenging contributions of nuclear physics to the understanding of the stellar nucleosynthesis r-process is probably the production and determination of ground properties of r-process nuclei at the N=126 shell closure. During the last years, existing radioactive beam facilities contributed significantly to the production and characterization of medium-mass r-process nuclei [1, 2, 3]. However, the region of heavy neutron-rich nuclei remained almost unexplored despite its interest. This region plays an important role in fundamental nuclear structure investigations, because of the presence of the neutron shell N=126, but it is also extremely relevant for the characterization of the r-process at the waiting point around A≈195. Indeed, the r-process matter flow towards the heaviest elements is to a large extend determined by the half lives of the r-process nuclei at the N=126 waiting point. Unfortunately, present model descriptions of the beta half lives of heavy neutron-rich nuclei are rather discordant [5, 6]. It was then the purpose of this work to investigate the production of heavy neutron-rich nuclei but also to measure the β half lives of some of them in order to benchmark the existing theoretical predictions.

Journal of Physics: Conference Series **337** (2012) 012070 doi:10.1088/1742-6596/337/1/012070

2. Production of heavy neutron-rich nuclei

Cold fragmentation reactions of relativistic projectiles was used to produce heavy neutron-rich nuclei [10]. In particular we used beams of ^{238}U and ^{208}Pb projectiles accelerated at 1 A GeV at the SIS18 synchrotron in GSI. Pulsed beams with a typical bunch length of 3 seconds, a repetition rate of 10 seconds and a typical intensity of $\approx 10^8$ ions/s were extracted from the accelerator and directed to a 2 g/cm^2 thickness beryllium target. Projectile fragments, flying forward because of the kinematic conditions, were analyzed with the magnetic spectrometer FRagment Separator (FRS) [8]. This is a zero-degree magnetic spectrometer with two symmetric sections in order to preserve the achromatism of the system. A profiled energy degrader placed at the intermediate image plane was fundamental for the separation of the transmitted nuclei [9] and the identification of atomic charge states [10, 11]. The nuclei traversing the spectrometer were identified by determining their magnetic rigidity, velocity and atomic number. A description of the identification method for these nuclei can be found in [12, 13].

In this experiments we were able to produce 73 new neutron-rich isotopes of elements between Ytterbium and Francium. Some of these results were previously reported in [14, 15]. The present results constitute a major step towards the production of nuclei at the A$\approx$195 r-process waiting point. Indeed, the present experiment allowed us to produce in few days of beam time as many new nuclei in this region of the chart of nuclide as in the previous fifty years.

In this work we could also determine the production cross sections of these nuclei by normalizing the production yields to the integrated beam current and the number of target nuclei. The beam current was continuously measured during the experiment using a secondary-electron monitor placed before the reaction target. This monitor was carefully calibrated during the experiment using a plastic scintillator [12]. The measured yields were also corrected by the losses inherent to the experimental technique used in this work. The main corrections were due to the limited momentum acceptance of the spectrometer, in particular, for the nuclei with magnetic rigidities far from the central value defined by the FRS tune, the fraction of non fully stripped nuclei, and secondary reactions in all layers of matter placed along the spectrometer. Smaller corrections were due to the acquisition dead time and detectors efficiency. These cross sections were used to benchmark model calculations predicting the production of heavy neutron-rich nuclei in future radioactive beam facilities.

3. Half-lives of heavy neutron-rich nuclei

Some of the nuclide produced in this experiment were slowed down and implanted into an array of three double-sided silicon strip detectors with a thickness of 1 mm each [16]. This array provided the position and arriving time of the nuclide and the position and emission time of the subsequent β particles. From the position and time correlations we were able to determine the β half lives of the nuclide of interest. Because of the pulsed nature of the beam and the relatively long half lives of the nuclide we were investigating, we could not use standard techniques for the determination of the half lives. In this case, we proposed to use backward-time correlations for evaluating our background and numerical fitting functions based on Monte Carlo simulations. A detailed description of this method can be found in Ref. [17].

In figure 1 we reported the measured half lives normalized to the predictions obtained with three model calculations, the gross theory (green symbols)[19] , calculations based on the finite-range drop model by Moller et al.[5] (blue symbols) and calculations based in the energy-density functional by Borzov [6] (red symbols). As can be seen, the Gross theory and Borzov's calculations provide a rather good description of the data. However, the calculations based on the finite-range droplet model of Moller et al. overestimate the half lives by almost two orders of magnitude for most of the nuclide investigated in this work [18].

The main reason for the different half lives predicted by the models of Moller and Borzov seems to be the role of first-forbidden transitions. According to Borzov, in this region of the

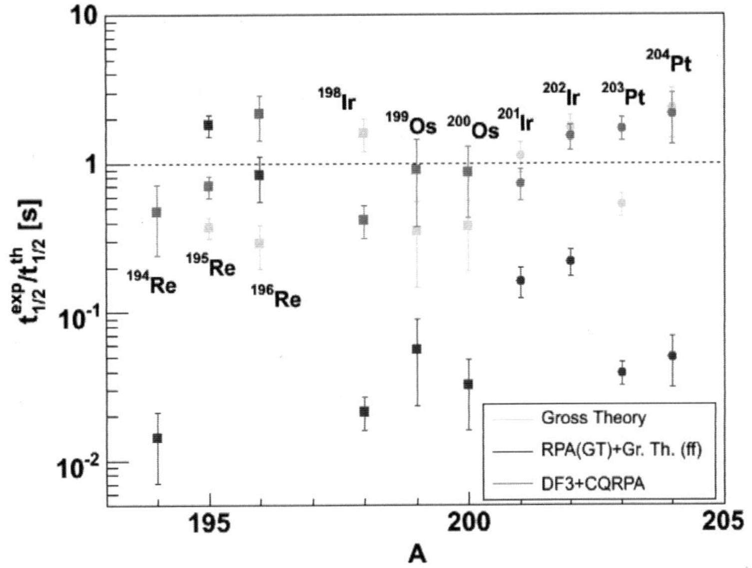

Figure 1. Half lives of heavy neutron-rich nuclei measured in this work normalized to different theoretical predictions, gross theory (green symbols), finite-range droplet model [5] (blue symbols) and energy-density functional (red symbols) [6]. The measured half-lives for ^{198}Ir(8±2 s), ^{199}Ir(6±4 s) and ^{204}Au(37±1 s) are not shown for the clarity of the figure.

chart of nuclides, decays to the lower lying states in the daughter nuclei seems to be governed by forbidden decays while Gamow-Teller transitions populate states at higher excitation energies. Under these conditions these forbidden decays compete with the Gamow-Teller.

We investigated this issue taking advantage of the RISING high-resolution gamma-ray spectrometer to establish the level scheme of the low-lying excited states of the daughter nuclei populated in these decays. As shown in figure 2 we represent the beta-delayed gamma spectrum populated in the decay of ^{204}Pt to ^{204}Au and propose a level scheme which is compared to shell model calculations [20]. In these calculations single-particle energies and the effective residual interaction were taken from Ref. [21] whith a ^{208}Pb core being the valence space for neutrons ($1h_{9/2}$, $2f_{7/2}$, $2f_{5/2}$, $3p_{3/2}$, $3p_{1/2}$, $1i_{13/2}$) and for protons ($1g_{7/2}$, $2d_{5/2}$, $2d_{3/2}$, $3s_{1/2}$, $1h_{11/2}$). The results of the calculations were used to tentatively assign spins and parities to the different levels. Finally, the spin and parities together with the apparent feeding of the states determined in the experiment, confirmed that transitions to the lower lying states involved changes in parity with a non negligible strength compatible with a forbidden character.

Although we did not measure nuclides directly at the r-process, calculations for r-process Osmium isotopes indicate a similar difference between the model predictions [18]. The confirmation of the tendency we observed in this work as a general trend for r-process nuclei, would indicate that finite-range droplet model calculations would underestimate the speed of the r-process at the waiting point A≈195. Since these calculations are widely used in r-process simulations, these results would indicate that the r-process matter flow towards the heaviest elements would be faster than presently thought.

4. Summary and conclusions

In this work we have demonstrated that it is possible to produce heavy neutron-rich nuclei approching the r-process waiting point A≈195 using fragmentation reactions of heavy stable

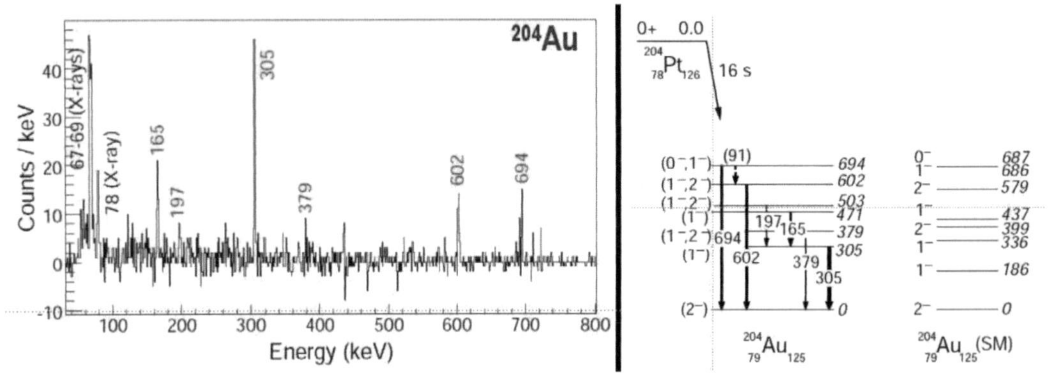

Figure 2. Beta-delayed gamma-ray spectrum obtained in the decay of ^{204}Pt to ^{204}Au and the proposed level scheme compared to shell model calculations

projectiles such us ^{238}U and ^{208}Pb at relativistic energies. Indeed, in and experiment performed at GSI with modest primary beam intensities we were able to produce 73 new isotopes of elements between Ytterbium and Francium. Moreover, we could determine the β half lives of thirteen of these nuclei. The measured half lives are rather well described with calculations based on the energy-density functional while calculations based on the finite-range droplet model largely overestimate the measured values. The confirmation of this overestimation of the β half lives for nuclei at the A≈195 r-process waiting point would indicate that the matter flow at this waiting point is faster than predicted by present model calculations based on the finite-range droplet model. Therefore, the production of the heaviest r-process nuclei would also be larger than presently expected.

References

[1] J. Benlliure et al., Phys. Rev. C 78, 054605 (2008)
[2] K. Helariutta et al., Eur. Phys. J. A 17,181 (2003)
[3] T. Ohnishi et al., J. Phys. Soc. Jpn. 79, 073201 (2010)
[4] A. Jungclaus et al., Phys. Rev. Lett. 99, 132501 (2007)
[5] P. Moller et al., Phys. Rev. C 67, 055802 (2003)
[6] I. Borzov, Phys. Rev. C 67, 025802 (2003)
[7] J. Benlliure et al., Nucl. Phys. A 660, 87 (1999)
[8] H. Geissel et al., Nucl. Instr. and Methods B 70, 286 (1992)
[9] K.-H. Schmidt et al., Nucl. Instr. and Methods A 260, 287 (1987)
[10] J. Benlliure et al., Nucl. Phys. A 660 (1999) 87
[11] A.I. Morales et al., Phys. Rev. C 84, 011601 (2011)
[12] E. Casarejos et al., Phys. Rev. C 74, 044612 (2006)
[13] J. Taïeb et al., Nucl. Phys. A 724, 413 (2003)
[14] H. Alvarez-Pol et al., Eur. Phys. J. A 42, 485 (2009)
[15] H. Alvarez-Pol et al., Phys. Rev. C 82, 041602 (2010)
[16] R. Kumar et al., Nucl. Instrum. and Methods A 598, 754 (2009)
[17] T. Kurtukian-Nieto et al., Nucl. Instrum. and Meth. A 589, 472 (2008)
[18] T. Kurtukian-Nieto et al., Phys. Lett. B submitted (arXiv:0711.0101v1)
[19] H. Nakata et al. Nucl. Phys. A 625 (1997) 521
[20] M. Gorska, private communication
[21] L. Rydström et al., Nucl. Phys. A 512 (1990) 217

Novae -The study of the reactive flow

S A Glasner [1] and J W Truran [2,3]
[1]Racah Institute of Physics, The Hebrew University, Jerusalem, ISRAEL
[2]Department of Astronomy and Astrophysics, Enrico Fermi Institute, University of Chicago, Chicago, IL 60637, USA
[3]Physics Division, Argonne National Laboratory, Argonne, IL 60439, USA

E-mail: glasner@phys.huji.ac.il

Abstract. There is a wide consensus in the astrophysics community that the mechanism underlying the observed Classical Nova eruptions is a surface thermonuclear runaway. We start this short review with the main observational facts that lead to the theoretical model of a thermonuclear runaway that takes place in an accreted hydrogen rich envelope placed on top of a cool degenerate core of a white dwarf. According to the theory, the accreted envelope becomes unstable to convection days to weeks prior to the runaway. During the extreme stages of the runaway itself, when the burning is most efficient, the envelope is fully convective. Therefore, the elements processed under such extreme conditions are lifted to the outermost regions of the star. A significant fraction of the envelope is ejected during the outburst. The complicated combination of hydrodynamic instabilities and explosive hydrogen burning, close to the surface of the star, gives us a unique opportunity to study this complex reactive flow. The range of core masses, core temperatures and accretion rates introduce a whole range of burning temperatures and densities. Following the description of the "standard" cases, we then focus on rare, but still possible, portions of the relevant parameter space, in which "breakout" of the traditional CNO cycle can occur and lead to heavy element enrichment patterns caused only by breakout burning. We conclude our review with the main challenges that nova theorists face today, with special emphasis on problems related to the nucleosynthesis issues.

1. Introduction-Classical novae the observational constraints

Classical Novae are extremely bright outbursts. A rise to peak brightness is followed by steady decline. The time taken for a nova to decay by 2 magnitudes from maximum optical brightness is used to classify a nova by its speed class. A fast nova will typically take less than 25 days to decay by 2 magnitudes and a slow nova can take over 80 days. The bolometric luminosity stays almost constant, at the level of a few 10^4 solar luminosities ($L_\odot$) for a period that extends from a few months to a few years [1]. In spite of their violence, the amount of material ejected in a nova outburst is usually only about 10^{-6}-10^{-4} of a solar mass ($M_\odot$). Matter is ejected at velocities as high as several thousand kilometers per second—higher for fast novae than for slow novae. Spectroscopic observations of nova ejecta show that they can be enriched relative to solar abundances in helium, carbon, nitrogen, oxygen, neon, magnesium and even heavier elements. Classical novae eruptions occur on Cataclysmic Variable stars (CV) that consist of binary star systems in which the primary, a white dwarf, accretes hydrogen rich matter from a secondary, which is a low mass main sequence star. The entire binary system is usually the size of the Earth-Moon system – with an orbital period of 1 to 10 hours. A much broader review with further details can be found in [2] and references therein.

2. The "Standard Model" for Classical Novae

The standard model for classical novae consists of a thermonuclear runaway (TNR) occurring in a degenerate hydrogen rich envelope accreted on a carbon-oxygen (CO) or oxygen-neon-magnesium (ONeMg) white dwarf in a close binary system. As long as the accretion process continues, the degenerate matter that accumulates at the base of the hydrogen rich envelope is compressed and heated. Under the prevailing degenerate conditions, the heated matter is burning without hydrodynamic regulation by expansion. Therefore, an explosive runaway occurs only once the relevant timescale for heat release by hydrogen burning in the degenerate envelope becomes shorter than the typical cooling time by any of the available cooling mechanisms. The runaway phase ends when the energy produced by the burning lifts degeneracy and the pressure decreases. The runaway time is much shorter than the burning time, and therefore only a few percent of the fuel is consumed at this stage. At later stages, the burning sets into a quasi equilibrium. In the quasi equilibrium phase, the burning rate is approximately equal to the outward luminosity. Many 1D hydrodynamic models for hydrogen TNRs on white dwarfs give very good agreement between theory and the main features of the observed phenomenon ([3]-[14]).

3. The reactive flow -mixing

As stated above, observations show that for nova outbursts the accreted mass is of hydrogen rich whereas the ejected mass is enriched by CNO and heavier elements. Since the timescales and the thermodynamic conditions (density, temperature) during the runaway do not predict any significant production of CNO elements, we conclude that there must be a mechanism that mixes the accreted matter with the cold white dwarf core prior to, or during, the runaway. A few mechanisms for such mixing were proposed up to now. All of them claim to predict the right observational amounts of mixing:

- The diffusion layer mechanism. The base of this layer is at the deepest point where hydrogen sinks into the core [15]-[19].
- The shear instability induced by differential rotation during the accretion phase [20]-[25].
- Shear gravity wave breaking on the white dwarf surface [26],[27].
- Undershoot of the convective flow during the runaway [28].

In all the published models, once the temperature at the base of the layer is about 20 million degrees, the CNO burning can't be controlled any more by radiative losses, and the whole envelope becomes convective. In the first three cases mentioned above, at this stage, much before the runaway takes place, the abundance of the whole envelope is totally mixed. In the fourth model, most of the mixing occurs during the runaway itself.

In a series of papers we [29]-[31] studied the convective undershoot mechanism (figure 1). The most significant result of this study is the observed universal behavior of all the models as it is presented in [31]. Multi-D effects give rise to mixing that shortens the rise-time to TNR, and the numerical tests support the conclusion that the overall mixing is at the level of 35-50 % . Mixing in the early stages is small and sums up to about 10%. The fact that at the late stages of the runaway all the models converge to an almost universal model, can be explained by the great sensitivity of the burning rate enhancement by CNO mixing to the temperature. Observations of nova ejecta abundances provide evidence for dredge-up that enriches the envelope with heavy elements up to 30%-40% relative to solar ([32] and references therein). Recently [33],[34] made 2D and 3D detailed surveys that confirm those results.

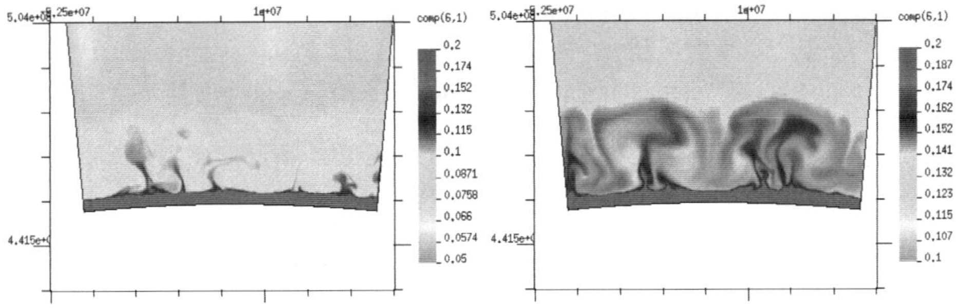

Figure 1. Color map of the ^{12}C abundance at various stages of the runaway. Left Log(Q)=42 erg/s. Right: Log(Q)= 43 erg/s.

4. The reactive flow - unique issues related to the nuclear burning

4.1. β^+decay time vs. convective turnover time

CNO isotopes unstable to β^+ decay play a major role in the energetics and nucleosynthesis of the outburst [35]. The four nuclei (^{13}N , ^{14}O , ^{15}O and ^{17}F) have lifetimes of the order of 100-1000 seconds. Once the temperature at the base of the envelope increase above 10^8 K, the lifetime against proton capture becomes shorter than the β^+ decay time. Therefore, the abundances of these nuclei increase rapidly. At this stage of the runaway, most of the envelope is convective and the nuclei circulate within the convective cells. As long as no fresh proton capturing CNO elements (^{12}C , ^{14}N , ^{16}O) appear in the entire convective zone, the energy generation rate is determined by the lifetime of the four β^+ unstable nuclei and therefore is not temperature dependent. Since the temperature at the upper regions of the convective zone is much lower than 10^8 K, the only nuclear reactions that occur there are the delayed decays of the β^+ unstable nuclei. Under these conditions, it is clear that the convective turnover time and the size of the convective cells are the most important parameters impacting the energy generation rate and the nucleosynthesis. After one or two turnover times, the outer part of the envelope expands with very high velocities (about 10^8 cm/sec). The entropy of the outer parts increases due to the deposition of energy released by the decay of unstable nuclei. Under such conditions the outer zones are detached from the main convective cells underneath.

4.2. Limits on energy production rate and undershoot mixing

For any mechanism other than the convective undershoot mechanism, the mixing with fresh ^{12}C takes place much before the runaway. Therefore, as was stated above, during the runaway itself, when the convective time-scales are much shorter than the β^+ decay time virtually all abundant stable CNO isotopes are transformed to β^+ unstable proton rich isotopes and the burning rate is limited to a constant value related to the total amount of CNO nuclei:

$$q_{\max} = 5.8\times10^{13}\times(Z_{cno}/0.01) \quad \text{(erg/gr/sec)}$$

If, on the other hand, there is dredge-up of CNO elements during the runaway itself, there is no such limit on the burning rates. In this case, rates could be much higher and temperature dependent once the rate of ingestion of fresh ^{12}C into the convective envelope is high enough. The enhanced burning rate increases the burning temperature and by that can have an effect on the abundances of the processed material. The processed matter is advected outwards by the convective flux and can later be observed in the ejecta. A demonstration of this effect is given in figure 2 and figure 3. The immediate capture of protons and the short period of time for lifting these nuclei to the outermost zones, above the main convective cells, (10-15 seconds) have the following consequences:

- Most of the dredged up ^{12}C will end up as ^{12}C , ^{13}C or ^{14}N.
- Most of the β^+ products (^{13}C and ^{15}N) are born in the upper part of the envelope and will not have the chance to capture another proton. Therefore, the $^{13}C/^{12}C$ and the $^{15}N/^{14}N$ ratios are relatively high.
- Novae may contribute significantly to the abundance of the ^{13}C , ^{15}O or ^{17}O.

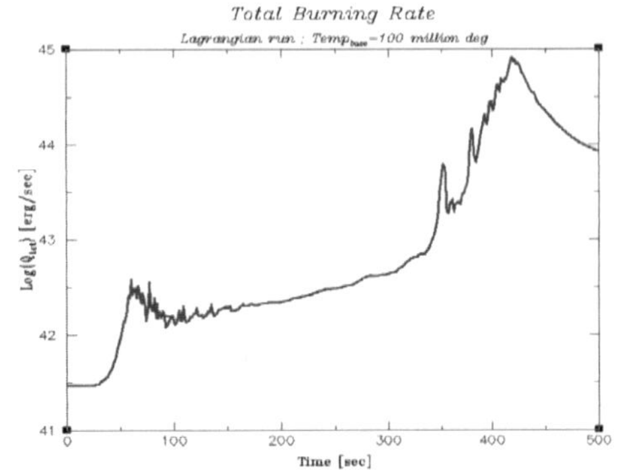

Figure 2. The logarithm of the total energy generation rate [ers/sec] vs. time.

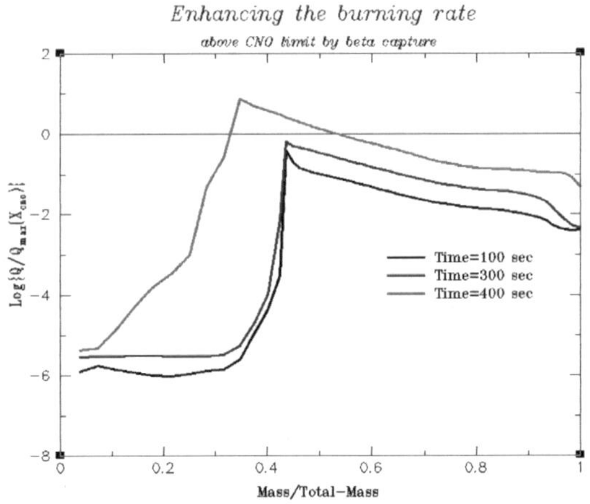

Figure 3. The ratio of the specific burning rate [erg/gr/sec] to the limiting rate vs. mass at various stages of the runaway (see left figure).

5. Burning under extreme conditions – the possibility of breakout from the CNO cycle

5.1. Breakout – the relevant parameters

For very slow white dwarf accretors in cataclysmic variables, Townsley&Bildsten [37] found a relation between the accretion rate $\dot{M}$ and the central temperature Tc of the white dwarf. According to this relation, for $\dot{M}$ less than 10^{-10} $M_{\odot}/yr$, T_c is much lower than 10^7K. Motivated by this study, we followed the thermonuclear runaway on massive white dwarfs (M_{WD} = 1.25–1.40 $M_{\odot}$) with T_c lower than 10^7K, accreting hydrogen rich matter. In [38] we demonstrated that in this range of the relevant parameter space, the accreted envelope is extremely massive, and the slope of the relation between the peak temperatures achieved during the runaway and T_c becomes much steeper than its value for T_c above 10^7K. The peak temperatures we derived were above $5 \cdot 10^8$K and they stay above a critical value for breakout from the conventional "hot carbon–nitrogen–oxygen" cycle (T_{crit}~$4 \cdot 10^8$K) for a few hours. When breakout conditions are achieved, the heavy-element abundances can exhibit a much wider variety of enrichments than what is possible with the common enrichment mechanisms. If indeed observed, such rare novae events will offer a unique opportunity to examine CNO breakout abundances in environments other than X-Ray burst on neutron stars.

In order to ensure that we indeed achieved "breakout" from the conventional "hot CNO" cycle, we conducted a definitive test. We examined a model similar to the nominal model (M = 1.35 $M_{\odot}$), for which the accreted matter was assumed to include solar concentrations of nuclei only up to fluorine. The energetics of this model were similar to those of the original model. With regard to the nucleosynthesis, the final concentration of iron group elements was determined to be comparable to that of the original model. There is only a small concentration of intermediate-mass elements in the final stages, which confirms that indeed CNO nuclei were burned all the way to iron (figures 4-5).

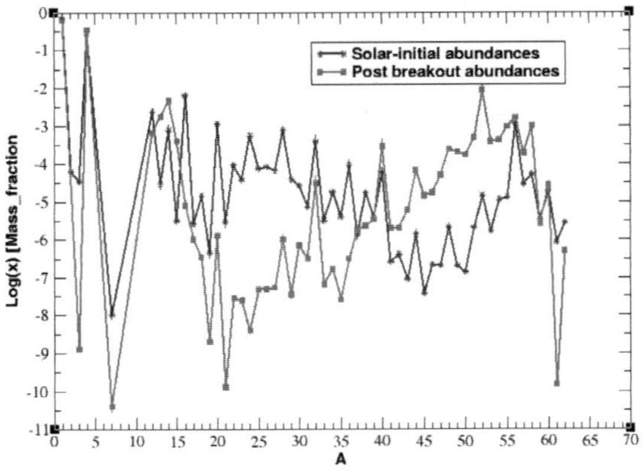

Figure 4. Abundances of the elements in the reaction network as a function of the mass number A. Presented are both the initial (solar) abundances (blue) and the "final" abundances (red) when the peak temperature has fallen below T_{crit} (M=1.35 $M_{\odot}$).

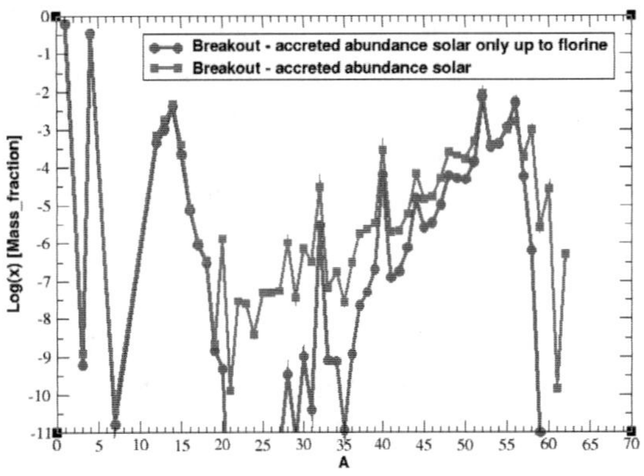

Figure 5. Abundances of the elements in the reaction network as a function of the mass number A. Presented are the "final" abundances for the nominal model (red here and in the left figure) and a model with accreted solar abundances, only up to fluorine (green).

5.2. Breakout - sensitivity

For our attempts to demonstrate the sensitivity to the exact value of specific cross sections of the burning reactions, we define two sensitivity levels. The first and major level is a sensitivity that alters the overall energetic of the runaway i.e., the temperature history of the burning, the total burning rate and light curve. We define those reactions as reactions with global sensitivity. In the second case, the sensitivity can show itself only by altering the abundances of specific isotopes without having any substantial effect on the energetics of the runaway. We define those as reactions with local sensitivity. The only global sensitivity we found up to now, in a full runaway model, is for the reaction $\alpha + {}^{15}O \rightarrow {}^{19}Ne + \gamma$. The effect of multiplication of the cross section by a factor of 10 on the overall energy production rate was profound (figure 6).

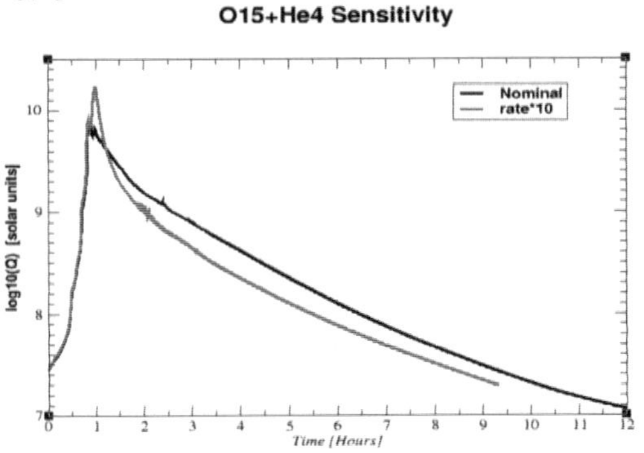

Figure 6. The effect of multiplication of the cross section for the reaction $\alpha + {}^{15}\mathrm{O} \rightarrow {}^{19}\mathrm{Ne} + \gamma$ on the energy production rate.

6. Conclusions - Challenges to novae theorists

The main problems facing novae theoretical investigators are the following (the problems related to the nuclear physics are marked in bold letters):

- Modeling of the early luminosity histories of novae in outburst.
- **Identification of the mechanism by which novae envelopes are enriched in heavy elements.**

- Clarifying the consequences of the phase of common envelope evolution that characterizes all novae at maximum light.
- **Systematic interpretation of the observed compositions of novae ejecta.**
- **Understanding of the timescale for post-outburst nova systems to return to minimum.**
- **Gathering accurate data base of abundance observations that is good enough for sensitivity examination of nuclear cross sections.**

7. References

[1] Gallagher, J.S, & Starrfield, S.: 1976, MNRAS 176, 53
[2] Bode M. F., Evans A., 2008, Classical Novae, Cambridge Astrophysics Series, No. 43. Cambridge Univ. Press, Cambridge
[3] Starrfield,S. Sparks,W.M. & Truran,J.W. 1974, Apjs, 28,247.
[4] Prialnik, D., Shara, M. M. & Shaviv, G. 1978, A&A, 62, 339
[5] Shara, M.M., Prialnik, D. & Shaviv, G., 1980, ApJ, 239, 586
[6] MacDonald, J. 1980, MNRAS, 191,
[7] Nariari,K., Nomoto,K.& Sugimoto,D. 1980, Pub. Ast. Soc. Japan 32,473
[8] Iben, I.Jr., 1982 Apj, 259,244
[9] Truran,J.W. 1982, Nuclear Theory of Nova- Essays in Nuclear Astrophysics ;ed. C.A.Barnes, D.D.Clayton & D.N. Schramm (Cambridge University Press).
[10] MacDonald, J. 1983, Apj, 267, 732
[11] Starrfield,S. Sparks,W.M. & Truran,J.W. 1985, Apj, 291,136
[12] Prialnik, D. 1986 Apj, 310,222
[13] Jos´e, J., & Hernanz, M. 1998, ApJ, 494, 680
[14] Yaron, O., Prialnik, D., Shara, M. M., & Kovetz, A. 2005, ApJ, 623, 398
[15] Prialnik, D., & Kovetz, A. 1984, ApJ, 281, 367
[16] Kovetz, A., & Prialnik, D. 1985, ApJ, 291, 812
[17] Iben, I., Jr., Fujimoto, M.-Y., & MacDonald, J. 1991, ApJ, 375, L27
[18] Iben, I., Jr., Fujimoto, M.-Y., & MacDonald, J. 1992, ApJ, 388, 521
[19] Fujimoto, M.-Y., & Iben, I., Jr. 1992, ApJ, 399, 646
[20] Durisen, R. H. 1977, ApJ, 213, 145
[21] Kippenhahn, R., & Thomas, H.-C. 1978, A&A, 63, 265
[22] MacDonald, J. 1983, ApJ, 273, 289
[23] Livio, M., & Truran, J. W. 1987, ApJ, 318, 316
[24] Kutter, G. S., & Sparks, W. M. 1987, ApJ, 321, 386
[25] Sparks, W. M., & Kutter, G. S. 1987, ApJ, 321, 394
[26] Rosner, R., Alexakis, A., Young, Y., Truran, J. W., & Hillebrandt, W. 2001, ApJ,562, L177
[27] Alexakis, A., Calder, A. C., Heger, A., et al. 2004, ApJ, 602, 931
[28] Woosley, S. E. 1986, in Nucleosynthesis and Chemical Evolution, ed. B. Hauck, A. Maeder, & G. Meynet (Sauverny: Geneva Observatory), 1
[29] Glasner,S.A., Livne,E., & Truran,J.W. 1997,ApJ, 475,754
[30] Glasner,S.A., Livne,E., & Truran,J.W. 2005,ApJ, 625,347
[31] Glasner,S.A., Livne,E., & Truran,J.W. 2007,ApJ, 665,1321
[32] Truran,J.W. & Livio,M. 1986, ApJ, 308, 721.
[33] Casanova, J., José, J., García-Berro, E., Calder, A., & Shore, S. N. 2010,A&A, 513, L5
[34] Casanova, J., José, J., García-Berro, E., Calder, A., & Shore, S. N. 2011,A&A, 527, A5
[35] Starrfield,S., Truran,J,W., Sparks,W.M.,Kutter,G.S., 1972, ApJ, 176,169
[36] Clayton,D.D., 1968, Principles of Stellar Evolution and Nucleosynthesis, (New-York; McGraw Hill)
[37] Townsley, D. M., & Bildsten, L. 2004, ApJ, 600, 390
[38] Glasner,S.A., Truran,J.W. 2009,ApJL, 692,L58

Search for Dark Matter in the sky in the Fermi era

Aldo Morselli

Via della Ricerca Scientifica 1

E-mail: aldo.morselli@roma2.infn.it

Abstract. To uncover the dark matter, to connect what is astrophysically observed to what will be seen as new particles produced in the LHC, we need new measurements. The Fermi Large Area Telescope is providing the measurement of the high energy (7 GeV to 1 TeV) cosmic ray electrons and positrons spectrum with unprecedented accuracy. This measurement represents a unique probe for studying the origin and diffusive propagation of cosmic rays as well as for looking for possible evidences of Dark Matter. In this talk we focus mainly on astrophysical sources of cosmic ray electrons and positrons which include the standard primary and secondary diffuse galactic contribution, as well as nearby point-sources which are expected to contribute more significantly to higher energies. In this framework, we discuss possible interpretations of Fermi results in relation with other recent experimental data on energetic electrons and positrons and with recent results in the searches of gamma-ray fluxes coming from WIMP pair annihilations in the sky.

1. The Cosmic Ray Electron spectrum

Recently the experimental information available on the Cosmic Ray Electron (CRE) spectrum has been dramatically expanded as the *Fermi* LAT Collaboration [1] has reported a high precision measurement of the electron spectrum from 7 GeV to 1 TeV performed with its Large Area Telescope (LAT) [2], [3]. The spectrum shows no prominent spectral features and it is significantly harder than that inferred from several previous experiments. These data together with the PAMELA data on the rise above 10 GeV of the positron fraction [4] are quite difficult to explain with just secondary production [5],[6], [7], [8].

The temptation to claim the discovery of dark matter from detection of electrons from annihilation of dark matter particles is strong but there are competing astrophysical sources, such as pulsars, that can give a strong flux of primary positrons and electrons (see [9], [10], [11] and references therein). At energies between 100 GeV and 1 TeV the electron flux reaching the Earth may be the sum of an almost homogeneous and isotropic component produced by Galactic supernova remnants and the local contribution of a few pulsars with the latter expected to contribute more and more significantly as the energy increases.

Two pulsars, Monogem, at a distance of 290 pc and Geminga, at a distance of 160 pc, can give a significant contribution to the high energy electron and positron flux reaching the Earth and with a set of reasonable parameters of the model of electron production the *Fermi* LAT data and the PAMELA positron fraction can be well fit fraction [4] (see figure 1). However we have a lot of freedom in the choice of these parameters because we still do not know much about these processes, so further study on high energy emission from pulsars is needed in order to confirm or reject the pulsar hypothesis.

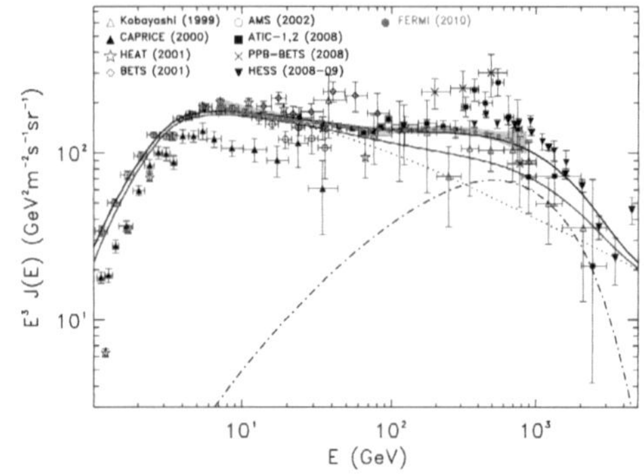

Figure 1. Cosmic ray electron + positron spectrum as measured by *Fermi* Large Area Telescope for one year of observations (filled circles), along with other recent high energy results. The gray band represents systematic errors on the *Fermi* LAT data [2], [3]. The solid line is the computed conventional GALPROP model but with an injection index Γ= 1.6/2.7 below/above 4 GeV (dotted line). An additional component with an injection index Γ = 1.5 and exponential cut-off is shown by the dashed line. Blue line shows e^- spectrum only.

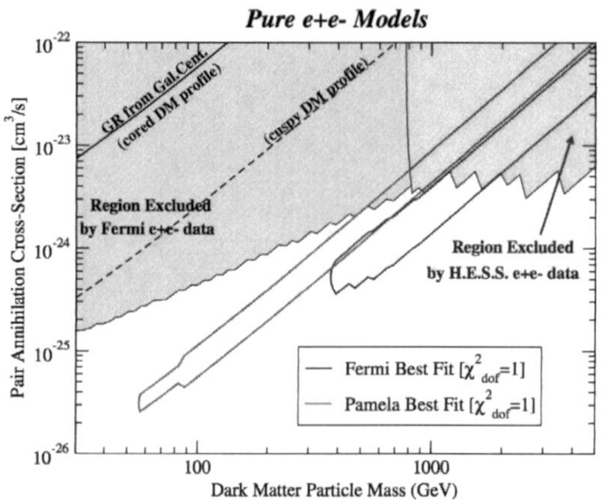

Figure 2. The parameter space of dark matter particle mass versus pair-annihilation rate, for models where dark matter annihilates into monochromatic $e^{\pm}$. Models inside the regions shaded in gray and cyan over-produce $e^{\pm}$ from dark matter annihilation with respect to the *Fermi* LAT and H.E.S.S. measurements, at 2-σ level. The red and blue contours outline the regions where the χ^2 per degree of freedom for fits to the PAMELA and *Fermi* LAT data is less than 1.

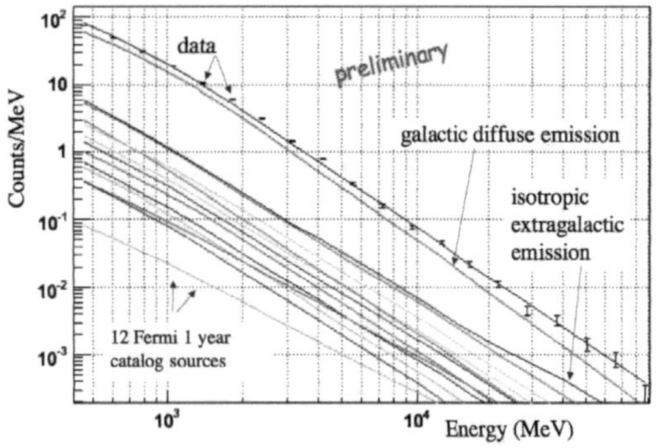

Figure 3. Counts spectra from the likelihood analysis of the *Fermi* LAT data (number of counts vs reconstructed energy) in a 7°×7° region around the Galactic Center (number of counts vs reconstructed energy

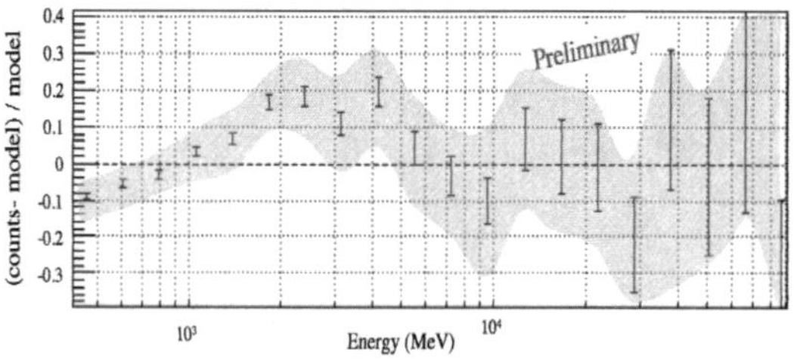

Figure 4. Residuals ((exp.data - model)/model) of the above likelihood analysis. The blue area shows the systematic errors on the effective area.

Nevertheless a dark matter interpretation of the *Fermi* LAT and of the PAMELA data is still an open possibility. Figure 2 shows the parameter space of dark matter particle mass versus pair-annihilation rate, for models where dark matter annihilates into monochromatic $e^{\pm}$ [10]. The preferred range for the dark matter mass lies between 400 GeV and 1-2 TeV, with larger masses increasingly constrained by the H.E.S.S. results [12]. The required annihilation rates, when employing a particular dark matter density profile imply typical boost factors ranging between 20 and 100, when compared to the value $\langle \sigma v \rangle \sim 3 \times 10^{-26}$ cm^3/sec expected for a thermally produced dark matter particle relic.

How can one distinguish between the contributions of pulsars and dark matter annihilations? Most likely, a confirmation of the dark matter signal will require a consistency between different experiments and new measurements of the reported excesses with large statistics.

Another possibility is to look for anisotropies in the arrival directions of the electrons. The *Fermi* LAT) detected more than 1.6 million cosmic-ray electrons/positrons with energies above 60 GeV during its first year of operation. The arrival directions of these events were searched for anisotropies of angular scale extending from $\sim 10°$ up to $90°$, and of minimum energy extending from 60 GeV up to 480 GeV. An upper limit for the dipole anisotropy has been set to 0.5 - 10% depending on the energy [14]. The levels of anisotropy expected for Vela-like and Monogem-like sources (i.e. sources with similar distances and ages) seem to be greater than the scale of anisotropies excluded by the results. However, it is worth to point out that the model results are affected by large uncertainties related to the choice of the free parameters.

2. The gamma-ray signals

A strong leptonic signal should be accompanied by a boost in the γ-ray yield providing a distinct spectral signature detectable by *Fermi* LAT. The Galactic center (GC) is expected to be the strongest source of γ-rays from DM annihilation, due to its coincidence with the cusped part of the DM halo density profile [15], [16]. A preliminary analysis of the data, taken during the first 11 months of the Fermi satellite operations, is shown in figures 3 and 4. The diffuse gamma-ray backgrounds and discrete sources, as we know them today, can account for the large majority of the detected gamma-ray emission from the Galactic Center. Nevertheless a residual emission is left, not accounted for by the above models [17], [18]. Improved modeling of the Galactic diffuse model as well as the potential contribution from other astrophysical sources (for instance unresolved point sources) could provide a better description of the data. Analyses are underway to investigate these possibilities. An excess in gamma-ray from dark matter annihilation also should be seen in the Galactic diffuse spectrum. Overall, the agreement between the LAT-measured spectrum and the model shows that the fundamental processes are consistent with our data, thus providing a solid basis for future work in understanding the Diffuse Galactic emission (DGE) [19], [20] .

3. Dwarf spheroidal galaxies and Clusters of galaxies

Local Group dwarf spheroidal galaxies, the largest galactic substructures predicted by the cold dark matter scenario, are attractive targets for dark matter indirect searches because they are nearby and among the most extreme dark matter dominated environments. With the data taken during the first 11 months no significant γ-ray emission was detected above 100 MeV from any dwarf galaxies. So we can determine upper limits to the γ-ray flux assuming both power-law spectra and representative spectra from WIMP annihilation. The resulting integral flux above 100 MeV is constrained to be at a level below around 10^{-9} photons cm^{-2}s^{-1} [21]. Using recent stellar kinematic data, the γ-ray flux limits can be combined with improved determinations of the dark matter density profiles in 8 of the 14 candidate dwarfs to place limits on the pair annihilation cross-section of WIMPs in several widely studied extensions of the standard model, including its supersymmetric extension and other models that received recent attention. With the present data we are able to rule out large parts of the parameter space where the thermal relic density is below the observed cosmological dark matter density and WIMPs (neutralinos here) are dominantly produced non-thermally. These γ-ray limits also constrain some WIMP models proposed to explain the *Fermi* LAT and PAMELA e^+e^- data, including low-mass wino-like neutralinos and models with TeV masses pair-annihilating into muon-antimuon pairs. The same kind of analysis can be made for the clusters of galaxies [22].

Finally a line at the WIMP mass, due to the 2γ production channel, could be observed as a feature in the astrophysical source spectrum [13]. Such an observation would be a "smoking gun" for WIMP DM as it is difficult to explain by a process other than WIMP annihilation or decay and the presence of a feature due to annihilation into γZ in addition would be even more convincing. Up to now however no lines have been observed and we obtain γ-ray line flux upper limits in the range $0.6 - 4.5 \times 10^{-9}\text{cm}^{-2}\text{s}^{-1}$ and corresponding DM annihilation cross-section and decay lifetime limits [23] .

4. Conclusion

Fermi Gamma-ray Space Telescope has opened a new era in DM searches and a large variety of analyses have been developed for clusters of galaxies, DM satellites, DM subhalos, cosmological DM and spectral lines. No significant detections have been made, but constraints that start to probe the available phase space have been put on the annihilation cross-section and decay lifetimes. The CRE spectrum measured by *Fermi* LAT is significantly harder than what was expected on the basis of previous data. Adopting the presence of an extra $e^{\pm}$ primary component with ~ 2.4 spectral index and $E_{cut} \sim 1$ TeV allows a consistent interpretation of the *Fermi* LAT CRE data, HESS and PAMELA. Such an extra-component can be produced by nearby pulsars for a reasonable choice of relevant parameters or by annihilating dark matter for models with $M_{DM} \sim 1$ TeV. Improved analysis and complementary observations (CRE anisotropy, spectrum and angular distribution of diffuse γ, DM sources search in γ) are required to possibly discriminate the right scenario. The dark matter origin of any exotic signal has to be confirmed by complementary findings in γ-rays by *Fermi* LAT and atmospheric Cherenkov telescopes, and by LHC in the debris of high-energy proton destructions. On the other hand, if the signal is due to a conventional astrophysical source of cosmic rays, it will mean a direct detection of particles accelerated at an astrophysical source, again a major breakthrough. However, independent of the origin of these excesses, exotic or conventional, we can expect a very exciting several years ahead of us.

References

[1] W.B.Atwood et al. [Fermi Coll.] ApJ 697 No 2 (2009 June 1) 1071-1102 [arXiv:0902.1089]
[2] A.A.Abdo et al. [Fermi Coll.], PRL 102, 181101 (2009) [arXiv:0905.0025]
[3] M.Ackermann, et al. [Fermi Coll.], PRD 82, 092004 (2010) [arXiv:1008.3999]
[4] O. Adriani et al. [PAMELA Coll.] Nature 458, 607, 2009 [arxiv:0810.4995]
[5] A. W. Strong and I. V. Moskalenko, ApJ 509 (1998) 212, ApJ 493 (1998) 694
[6] A. Lionetto, A. Morselli, and V. Zdravkovic, JCAP09 (2005) 010 [astro-ph/0502406]
[7] A. Morselli, I.Moskalenko, PoS(idm2008)025 [arXiv:0811.3526]
[8] A.Morselli, Progress in Particle and Nuclear Physics 66 (2011) 208215
[9] F. A. Aharonian, A. M. Atoyan and H. J. Völk, Astron. Astrophys. 294 (1995) L41
[10] D.Grasso et al. Astropart. Phys. 32 (2009), pp. 140-151 [arXiv:0905.0636]
[11] D.Grasso et al., Nuclear Instruments and Methods in Physics Research A 630 (2011) 4851
[12] F. Aharonian et al. [H.E.S.S. Coll.], Phys. Rev. Lett. 101 (2008) 261104,
[13] E. Baltz et al. , JCAP07 (2008) 013 [arXiv:0806.2911]
[14] M.Ackermann et.al. [Fermi Coll.], PRD 82, 092003 [arXiv:1008.5119]
[15] A. Morselli at al., Nucl.Phys. 113B (2002) 213
[16] A.Cesarini, F.Fucito, A.Lionetto, A.Morselli, P. Ullio, Astropart. Phys. 21 (2004) 267 [astro-ph/0305075]
[17] V.Vitale and A. Morselli for the Fermi/LAT Collaboration, 2009 Fermi Symposium [arXiv:0912.3828]
[18] V.Vitale, A.Morselli, Nuclear Instruments and Methods in Physics Research A 630 (2011) 147150
[19] M.Ackermann, et al. [Fermi Coll.], PRL 103, 251101 (2009) [arXiv:0912.0973]
[20] A.Abdo et al. [Fermi Coll.], ApJ, 726, (2011) 81 [arXiv:1011.0816]
[21] A.A.Abdo et al., [Fermi Coll.], ApJ 712 (2010) 147-158 [arXiv:1001.4531]
[22] M.Ackermann et al. [Fermi Coll.], JCAP 05, 025 (2010) [arXiv:1002.2239]
[23] A.A.Abdo et al., [Fermi Coll.], PRL 104, 091302-08 (2010) [arXiv:1001.4836]

Nuclear Physics in Astrophysics V IOP Publishing
Journal of Physics: Conference Series **337** (2012) 012073 doi:10.1088/1742-6596/337/1/012073

Constraints on the Equation-of-State of neutron stars from nearby neutron star observations

R Neuhäuser, V V Hambaryan, M M Hohle, T Eisenbeiss

Astrophysikalisches Institut, University Jena, Schillergäßchen 2-3, D-07745 Jena, Germany

E-mail: `rne@astro.uni-jena.de`

Abstract. We try to constrain the Equation-of-State (EoS) of supra-nuclear-density matter in neutron stars (NSs) by observations of nearby NSs. There are seven thermally emitting NSs known from X-ray and optical observations, the so-called Magnificent Seven (M7), which are young (up to few Myrs), nearby (within a few hundred pc), and radio-quiet with blackbody-like X-ray spectra, so that we can observe their surfaces. As bright X-ray sources, we can determine their rotational (pulse) period and their period derivative from X-ray timing. From XMM and/or Chandra X-ray spectra, we can determine their temperature. With precise astrometric observations using the Hubble Space Telescope, we can determine their parallax (i.e. distance) and optical flux. From flux, distance, and temperature, one can derive the emitting area - with assumptions about the atmosphere and/or temperature distribution on the surface. This was recently done by us for the two brightest M7 NSs RXJ1856 and RXJ0720. Then, from identifying absorption lines in X-ray spectra, one can also try to determine gravitational redshift. Also, from rotational phase-resolved spectroscopy, we have for the first time determined the compactness (mass/radius) of the M7 NS RBS1223. If also applied to RXJ1856, radius (from luminosity and temperature) and compactness (from X-ray data) will yield the mass and radius - for the first time for an isolated single neutron star. We will present our observations and recent results.

1. Introduction: The Magnificent Seven Neutron Stars

With deep optical follow-up observations of bright X-ray sources from the ROSAT All-Sky Survey, seven neutron stars (NSs) were identified, which share their main properties (called *Magnificent Seven* M7), but are different from radio pulsars: Bright X-ray sources, faint optical sources, thermal X-ray spectra, X-ray pulsation, isolated without companions nor nearby supernova remnants, no radio emission, strong magnetic fields, and intermediate spin-down ages. See Haberl (2007) for a recent reviews. The two brightest M7 NSs, RXJ1856 and RXJ0720, were identified as such by Walter et al. (1996) and Haberl et al. (1997).

Mass determinations for M7 NSs would be of special interest, because these M7 NSs may well be different from radio pulsars in binaries, for which masses could be determined before from orbital solutions, e.g. they do not show radio emission and rotate slower (few seconds) than normal radio pulsars, but still have strong magnetic fields.

2. Basic X-ray and optical observations to obtain the radius

Since the X-ray spectra of the M7 NSs are consistent with blackbodies (some also show broad absorption features suggested to be proton cyclotron absorption), we observe their cooling surfaces. With X-ray spectroscopy, we can determine the temperature of the the X-ray emitting surface. E.g., for RXJ1856 and RXJ0720, the blackbody temperatures are 63 ± 1 eV for RXJ1856 (Burwitz et al. 2001, 2003) and 90 ± 1 eV for RXJ0720 (Hohle et al. 2009).

The distance towards these two brightest M7 NSs have been observed directly as trigonometric parallaxe with the Hubble Space Telescope (HST), even though they are faint (V=25.7 mag for RXJ1856 and V=26.8 mag for RXJ0720). The results have been a matter of debate for the last years: For RXJ1856, Kaplan et al. (2002) found 140 ± 40 pc with three HST PC data points; Walter et al. (2002) got 117 ± 12 pc with four HST PC data points; then after having obtained eight HST ACS data points, Kaplan et al. (2007) obtained 167 ± 17 pc, while Kaplan et al. (2007) gave 161 ± 16 pc as result; the latter two papers, however, did not give any details as to how the results were obtained; then, Walter et al. (2010) re-reduced the eight HST ACS data points with various techniques, all yielding 123 ± 13 pc. The latter distance is also consistent with 135 ± 25 pc obtained by Posselt et al. (2007) from a 3D N_H extinction model, so that RXJ1856 is just foreground to the CrA dark cloud located at around ~ 130 to ~ 140 pc.

For RXJ0720, Kaplan et al. (2007) published ~ 360 pc (250 - 530 pc) based on eight HST ACS data points. Posselt et al. (2007), using their 3D N_H extinction model, determined ~ 250 pc for the distance towards RXJ0720. Recently, Eisenbeiss (2011) re-determined the parallaxe of the NS RXJ0720 using the archival eight HST data points yielding $\pi = 3.6 \pm 1.6$ mas (Eisenbeiss 2011), i.e. ~ 278 pc (192-500 pc); this is consistent with an earlier determination by Kaplan et al. (2007), (270-530 pc) within large error bars.

Once flux in different wavelength, extinction, and distance are known, one can calculate the lumonisity. Luminosity and temperature can immediately yield the emitting area if blackbody. This would be true and correct, if both the X-ray emission (used to determine the temperature) and the optical emission (used to determine the distance) come from the same (emitting) area. However, both RXJ1856 and RXJ0720, like all M7 NSs, show a so-called optical excess, i.e. they emit more optical emission (by a factor of 7 for RXJ1856, Burwitz et al. 2003, and 10 for RXJ0720, Eisenbeiss et al. 2010) than expected from extrapolating their X-ray blackbody spectra. Hence, it might well be that the X-ray emission comes from one or two small emitting areas, while the optical emission comes from the whole warm surface.

Since the temperature distribution across the surface is not known, e.g. the spot(s) could have some non-uniform temperature distribution with or without a thin atmosphere (e.g. H), there is so far no unique way to derive the radius. It was determined first by Walter et al. (1999) obtaining ~ 14 km, and most thoroughly by Trümper (2003), who got ~ 16.5 km (for radius at infinity, see Fig. 1) from a two-blackbody optical and X-ray sepctral energy distribition at 117 pc, and Ho et al. (2007), who found ~ 17 km ($\pm 30\%$) for a magnetic H atmosphere (at 140 pc). For any simple distribution, one would obtain 16.8 ± 1.25 km radius for RXJ1856 (Walter et al. 2010), using the most recent distance of 123 ± 13 pc, where the error bar includes all errors from brighness, temperature (both around 1 %) and distance (~ 10 %).

For RXJ0720, one obtains a radius (at infinity) of 17 ± 3 km (280 pc distance), where the error budget does not yet include the large distance error.

Those two radius determinations (both radii as seen at infinity, see Fig. 1) are so far the only two direct radius measurements for NSs. Given that the peaks in the optical emission are not seen in both NSs, those estimates might be considered as lower radius limit.

3. X-ray spectroscopy to obtain the redshift

Both RXJ1856 and RXJ0720, as well as other M7 NSs, are observed often with both the X-ray satellites XMM and Chandra, partly for (cross-)calibration purposes. Hence, one can combine all spectroscopic observations of either XMM or Chandra to obtain a spectrum with high resolution and high S/N. A measurement of the gravitational redshift of spectral features in the spectrum of thermal emitting NSs would yield the so-called compactness, i.e. mass-to-radius ratio.

Hambaryan et al. (2009) detected a narrow absorption feature at 0.57 keV in the co-added

Journal of Physics: Conference Series **337** (2012) 012073 doi:10.1088/1742-6596/337/1/012073

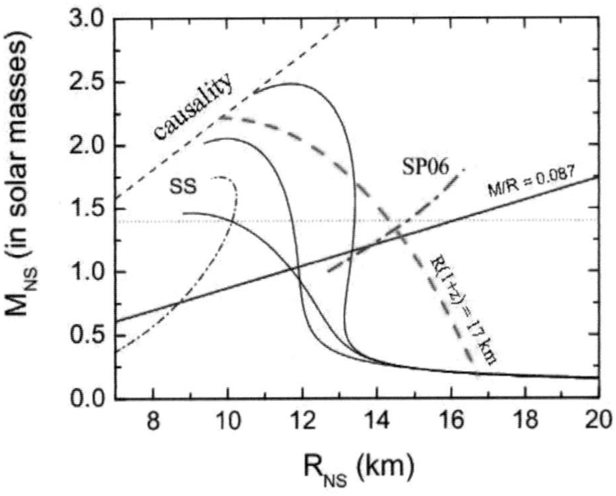

Figure 1. Mass-radius relations for NSs and several EoS (Haensel et al. 2007) as thin solid curves, plus a strange star EoS as thin dash-dotted. The thick dashed curve is for a constant radius at infinity of $R^{\infty} = R(1+z) = R/\sqrt{1-2GM/Rc^2} = 17 \pm 3$ km as found for both RXJ1856 and RXJ0720, see Sect. 2. The thick dash-dotted line gives a model-dependent estimation of the M-R relation from boundary layer spectra modeling in LMXBs (Suleimanov & Poutanen 2006). The straight solid line gives M [$M_{\odot}$] / R [km]=0.087 from rotational phase-resolved X-ray spectroscopy of the M7 NS RBS 1223.

XMM/RGS spectrum of RXJ0720. The feature was identified with an absorption line of highly ionized oxygen O VII resonance Kα line, possibly originating in the interstellar medium (in the foreground of RXJ0720) or even more likely in the nearby ambient circumstellar medium around RXJ0720. If the line would have originated in the thin NS atmosphere, it would be very difficult to identify it, because the gravitational redshift is still poorly known for very large magnetic fields like in M7 NSs. A redshift of ~ 0.16 (Hambaryan et al. 2009) would correspond to a compactness of mass (in solar masses) devided by radius (in km) of ~ 0.087 (e.g., for the canonical NS mass of 1.4 solar masses, this would correspond to 16 km radius). In the latter case, one should also detect other lines redshifted by the same amount, but they are not detected.

4. X-ray rotational phase resolved spectroscopy to measure the compactness

The light curve of the M7 NS RBS1223 shows not only the highest pulse fraction with 18 % (Schwope et al. 2007), but also different double-humped light curves at different energy intervalls. This behaviour may indicate the existence of two emitting areas, which are not exactly identical in size, temperarure, and location with respect to the rotation axes, e.g. they could be non-antipodal with a small shift angle. Hence, it should be possible to perform rotational phase resolved spectroscopy, i.e. to investigate (and fit) light curves at different energy intervalls and spectra at different spectral ranges simultaneously. Given that RBS1223 also shows the highest pulse fraction, it is best suited for such an attempt.

Hambaryan et al. (2011) could determine from the best fit several parameters of the NS: Not only the sizes, temperatures, magnetic fields, and locations of the two emitting areas, together with the energy and strength of the broad proton cyclotron absorption feature and the absorbing foreground column density, but also the gravitational redshift or compactness of the NS.

The result here is gravitational redshift $z = 0.16 \pm 0.02$ (or 0.15 ± 0.02 for a condensed iron surface with partially ionized H atmosphere as described in Suleimanov et al. (2010), or 0.17 ± 0.03 for a pure blackbody with electron scattering) corresponding to a compactness (mass/radius) of $\sim 0.087 \pm 0.011$, e.g. a radius of 16 ± 1 km for 1.4 solar masses.

5. Summary and Outlook

The first two radius determinations for NSs, namely for RXJ1856 and RXJ0720, provide strong constraints for the EoS, as they exclude quark stars, but are consistent with a very stiff EoS. Even if these radii are interpreted as lower limits, due to the uncertainties in the model assumptions, they still exclude quark or strange stars with Pion and Kaon condensates.

The attempts to identify atomic absorption lines in M7 X-ray spectra to determine the gravitational redshift, have not yielded unique results so far, because the detected faint lines could also have originated in the interstellar or circumstellar material.

Rotational phase resolved spectra yield the gravitational redshift or compactness of M7 NSs. Combined with a radius (or mass) determination, such as the ones above for RXJ1856 or RXJ0720, they would provide the possibility to determine for the very same NS(s) at the same time both radius and mass. This would be a stringent constraint.

5.1. Acknowledgments

We would like to thank DFG in SFB TR 7 for financial support.

6. References

[1] Burwitz V, Zavlin V E, Neuhäuser R, et al. 2001 *A&A* **379** L35
[2] Burwitz V, Haberl F, Neuhäuser R, et al. 2003 *A&A* **399** 1109
[3] Eisenbeiss E 2011 PhD thesis U Jena
[4] Eisenbeiss T, Ginski C, Hohle M M, Hambaryan V V, Neuhäuser R, Schmidt T O B 2010 AN *331* 243
[5] Haberl F 2007 *ApSS* **308** 181
[6] Haberl F, Motch C, Buckley D A H, Zickgraf F J, Pietsch W. 1997 *A&A* **326** 662
[7] Haensel P, Potekhin A Y, Yakovlev D G 2007 *ASSL* **326**
[8] Hambaryan V V, Neuhäuser R, Haberl F, Hohle M M, Schwope A 2009 *A&A* **497** L9
[9] Hambaryan V V, Suleimanov V, Schwope A P, Neuhäuser R et al. 2011 *A&A* **534**, 74
[10] Ho W C G, Kaplan D L, Chang P, van Adelsberg M, Potekhin A Y 2007 *MNRAS* **375** 821
[11] Hohle M M, Haberl F, Turolla R, Hambaryan V V, Zane Sm de Vries C P, Mendez M 2009 *A&A* **498** 811
[12] Kaplan D L, van Kerkwijk M H, Anderson J 2002 *ApJ* **571** 447
[13] Kaplan D L, van Kerkwijk M H, Anderson J 2007 *ApJ* **660** 1428
[14] Posselt B, Popov S B, Haberl F, et al. 2007 *ApSS* **308** 171
[15] Schope A, Hambaryan V V, Haberl F, Motch C 2007 *ApSS* **308** 619
[16] Suleimanov V & Poutanen 2006 *MNRAS* **369** 2036
[17] Suleimanov V, Hambaryan V V, Potekhin A Y, van Adelsberg M, Neuhäuser R, Werner K 2010 *A&A* **522** 111
[18] Trümper J, Burwitz V, Haberl F, et al. 2003 *Nucl. Phys. B Proc. Suppl.* **132** 560
[19] Walter F W, Wolk S J, Neuhäuser R 1996 *Nature* **379** 233
[20] Walter F W & Matthews L D 1999 *Nature* **389** 358
[21] Walter F W & Lattimer J M 2002 *ApJ* **576** L145
[22] Walter F W, Eisenbeiss T, Lattimer J M, et al. 2010 *ApJ* **724** 669

Cosmic chemical evolution with intermediate mass pop III stars

Elisabeth Vangioni
Institut d'Astrophysique de Paris, UMR 7095 CNRS, University Pierre et Marie Curie, 98bis Boulevard Arago, Paris 75014, France

E-mail: vangioni@iap.fr

Abstract. We explore the consequences of an early population of intermediate mass stars (IMS) in the 2 – 8 $M_{\odot}$ range on cosmic chemical evolution. We discuss the implications of this population as it pertains to several cosmological and astrophysical observables. Some very metal-poor galactic stars show large enhancements of carbon, typical of the C-rich ejecta of IMS ; moreover, halo star carbon and oxygen abundances show a scatter, which imply a wide range of star-formation and nucleosynthetic histories contributed to the first generations of stars. Also, recent analyses of the He abundance in metal-poor extragalactic H II regions suggest an elevated primitive abundance of Helium, Yp ≃ 0.256 by mass, higher than the predicted result from big bang nucleosynthesis assuming the baryon density determined by WMAP, Yp = 0.249. This offset suggests a prompt initial enrichment of He in early metal-poor structures, and IMS Pop III stars are again good candidates. We also discuss the effect of these Pop III stars on global cosmic evolution for example the reionization of the Universe. We conclude that if IMS are to be associated with some Population III stars, their relevance is limited to low mass structures involving a little fraction of the total baryon content of the Universe typical at redshift 10 [1].

1. Introduction

Recent hydrodynamical simulations and analytical calculations have led to a general scenario [4] for the first star formation. We implement this typical scheme into the cosmic evolution code developed in [5, 9] to calculate (Press and Schechter scenario) the cosmic star formation rate (SFR) (Fig1) and the optical depth (Fig 2) as a function of the redshift (z), considering Pop II/I (0.1-100 $M_{\odot}$, blue solid line) and Pop III (2-8 $M_{\odot}$, IMS, black dotted line or 40-100 $M_{\odot}$, MS, red dashed line) modes. In figure 2, we plot the integrated optical depth from z = 0 to z. As one might expect, the normal mode without a massive component (the solid blue curve) is not able to produce enough ionizing UV photons to obtain a sufficiently large optical depth. The massive mode (the red dashed curve) does manage to reionize the intergalactic medim (IGM), but only barely. Of course more reionization is possible by increasing the SFR for the massive mode, but this would lead to an overproduction of metals. The IM mode (the black dotted curve) is able to reionize the Universe compared to other models. The escape fraction of UV photons produced by stars going to the IGM is about 0.1–0.2. Ionizing photons from a 6 $M_{\odot}$ star vs that of a 20 $M_{\odot}$ star is small but the SFR in the IM scenario (fig. 1) is high enough to produce the required ionizing photons.

2. Cosmic evolution of ^{4}He and CNO elements

In this context, we consider the evolution of the abundance of He, CNO as a function of the redshift, z, and/or metallicity. In Fig 3 plotted is the ^{4}He mass fraction, Y, vs. [O/H]. The solid blue curve corresponds to the model with only a Pop II/I contribution. This result is indistinguishable from that produced by models including the massive Pop III contribution. This can be understood as the bulk of the ^{4}He and oxygen are derived from the same stars in the two models. In both cases, the helium abundance begins at the BBN primordial value and remains rather flat until the oxygen abundance is roughy 1/10th of solar. In contrast, as shown by the black dotted curve, in the IM Pop III model, ^{4}He is produced early in IM stars with little or no accompanied oxygen. As a result, the ^{4}He mass fraction begins to grow at very low [O/H] and for the model parameters chosen, plateaus at value close to Y ≈ 0.256, in good agreement with recent determinations ; this model, therefore produces an effective prompt initial enrichment of ^{4}He in low metallicity galaxies. Moreover, concerning CNO evolution, the evolutionary behaviour of the three models as a function of the redshift is well understood. Normal mode stars produce a moderate abundance of heavy elements at high redshift. The same stars, nevertheless are effective at low redshift and produce the bulk of metals observed in the Milky Way. The Massive Pop III component produces C, O (and Mg) at high redshift, though these abundances are diluted at low redshift by infall as structures continue to grow. The IM stellar component produces a specific nucleosynthetic signature: essentially C and N. For details see the paper [1].

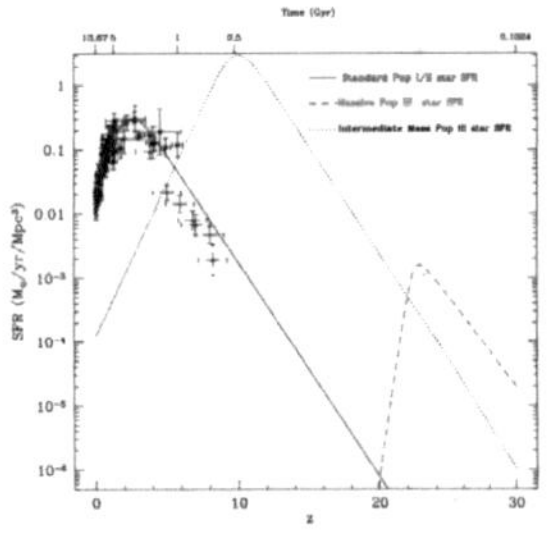

Fig 1. Cosmic SFR as a function of redshift for each mode of star formation. Observations come from [6] (black points) and [7] (black dashed point).

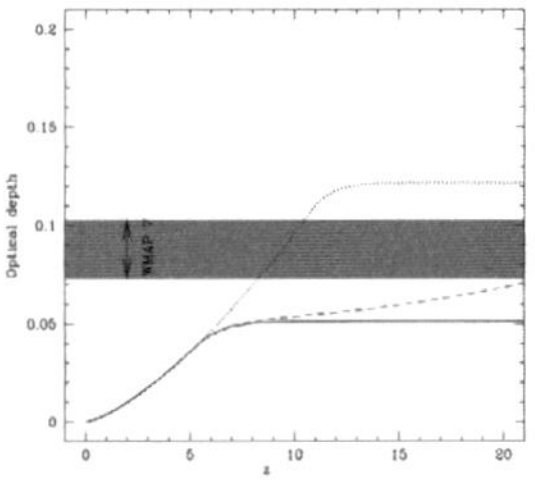

Fig 2. The optical depth as a function of the redshift. The red range corresponds to the observed results from WMAP7 [2]. The red dashed line corresponds to model with an addition of 40 – 100 M⊙ stars, the blue solid line to model with a normal mode (0.1-100 M⊙) and the black dotted line to model with an addition of 2- 8 M⊙ stars. This last scenario is able to reproduce observations.

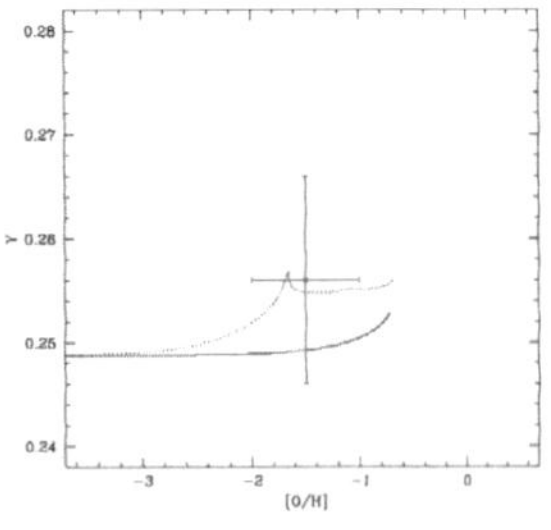

Fig 3. Cosmic evolution of ^{4}He as a function of [O/H] for the different stellar modes.

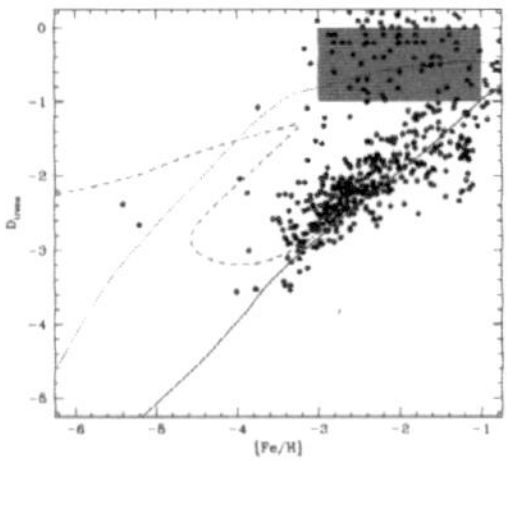

Fig 4. Evolution of Dtrans as a function of the metal enrichment [Fe/H] for the three models. The black circles come from the compilation of [3]. The red box corresponds to the region associated to C rich stars.

3. The transition discriminant Dtrans and the Carbon-enhanced Metal-Poor stars (CEMPS)

The transition from the massive star mode to the normal one is described in terms of a critical metallicity. Ref [8] have shown the importance of C and O in this transition; it is argued that low mass star formation requires Dtrans (= log(10**[C/H] + 0.3X10**[O/H]) > -3.5 ± 0.2.
This representation (Fig 4) allows one to distinguish the effects of the different stellar mass ranges: the normal mode (blue solid line) fits the bulk of standard stellar data. The massive mode (red dashed line) can explain the few very low mass stars [9] and the IM stellar mode (black dotted line) can explain the bulk of stars very enriched in C between -3 < Fe/H < -1 (red box) as shown in [3] ; about 10 – 20% fall in the C-rich red box fit by IM stars. Since our IM model has 10% of baryons processed through this mode, the C-rich stars would in turn imply that ~ 1–2% of baryons participate in IM star formation.

4. Conclusion

By definition, as the first stars formed in a metal-free environment, Pop III stars existed and played a role in our past history. However, the identity and mass distribution of these stars are largely unknown. They played an important role in re-ionizing the Universe at high redshift, as well as laid the chemical seeds for future generations of stars (population II/I). A Salpeter IMF, typical of population II/I with a mass range of 0.1 - 100 $M_{\odot}$ is not capable of reionizing the Universe neither of explaining specific abundance patterns in EMPs. We have shown that an early generation of IM stars is indeed capable of providing an effective prompt initial enrichment of He, a sufficient number of ionizing photons to the early IGM and large abundances of C, N relative to Fe as observed in EMP stars. Regarding the Dtrans parameter which represents a measure of carbon and oxygen produced in prior generations of stars, the data present two distinct populations: the bulk of the data show increasing Dtrans with [Fe/H] (explained by the standard mode) and a second population with very large Dtrans between −3 < [Fe/H] < −1. The latter is well explained by the Pop III IM stellar mode. We roughly estimate that a Pop III IM mode operates at ~ 10 – 20% of our global model rate. Since our IM mode processes about 10% of total baryons, this in turn implies that ~ 1 – 2% of baryons participate in IM star formation. On the other hand, a massive mode (MS) is capable of reionizing the Universe and producing sufficiently high ratios of [C, O, Mg/Fe] to explain the observations of extremely iron-poor stars, the high ratios occur almost instantly after the first stars are born, and thus require these stars to be born at that time at very high redshift.

References

[1] Vangioni E. et al 2011 arXiv10105726, MNRAS, 351, in press
[2] Komatsu, E., et al. 2010, arXiv:1001.4538
[3] Frebel et al. 2007, MNRAS, 380, L40
[4] Greif et al 2010 ApJ, 716, 510 and references therein
[5] Daigne et al., 2006, ApJ 647, 773
[6] Hopkins and Beacom, 2006, ApJ 651, 142
[7] Bouwens et al, 2007, ApJ 670, 928, 2010 ApJL, 709, L33
Gonzales et al , 2010, ArXiv 1008.3901
[8] Bromm and Loeb 2003, Nat, 425, 812
[9] Rollinde, E., Vangioni, E., Maurin, D., et al 2009, MNRAS, 398, 1782

AUTHOR INDEX

AUTHOR INDEX

AUTHOR INDEX

AUTHOR INDEX

Institute of Physics
Dirac House, Temple Back
Bristol BS1 6BE UK

CURRAN ASSOCIATES INC
proceedings.com